Methods in Molecular Biology

Series Editor
John M. Walker
School of Life and Medical Sciences
University of Hertfordshire
Hatfield, Hertfordshire, UK

For further volumes:
http://www.springer.com/series/7651

For over 35 years, biological scientists have come to rely on the research protocols and methodologies in the critically acclaimed *Methods in Molecular Biology* series. The series was the first to introduce the step-by-step protocols approach that has become the standard in all biomedical protocol publishing. Each protocol is provided in readily-reproducible step-by-step fashion, opening with an introductory overview, a list of the materials and reagents needed to complete the experiment, and followed by a detailed procedure that is supported with a helpful notes section offering tips and tricks of the trade as well as troubleshooting advice. These hallmark features were introduced by series editor Dr. John Walker and constitute the key ingredient in each and every volume of the *Methods in Molecular Biology* series. Tested and trusted, comprehensive and reliable, all protocols from the series are indexed in PubMed.

NMR-Based Metabolomics

Methods and Protocols

Edited by

G. A. Nagana Gowda

Northwest Metabolomics Research Center, Mitochondria and Metabolism Center, Department of Anesthesiology and Pain Medicine, University of Washington, Seattle, WA, USA

Daniel Raftery

Northwest Metabolomics Research Center, Mitochondria and Metabolism Center, Department of Anesthesiology and Pain Medicine, University of Washington, Fred Hutch Cancer Research Center, Seattle, WA, USA

Humana Press

Editors
G. A. Nagana Gowda
Northwest Metabolomics
Research Center, Mitochondria
and Metabolism Center, Department
of Anesthesiology and Pain Medicine
University of Washington
Seattle, WA, USA

Daniel Raftery
Northwest Metabolomics
Research Center, Mitochondria
and Metabolism Center, Department
of Anesthesiology and Pain Medicine
University of Washington, Fred Hutch
Cancer Research Center
Seattle, WA, USA

ISSN 1064-3745 ISSN 1940-6029 (electronic)
Methods in Molecular Biology
ISBN 978-1-4939-9689-6 ISBN 978-1-4939-9690-2 (eBook)
https://doi.org/10.1007/978-1-4939-9690-2

This Humana imprint is published by the registered company Springer Science+Business Media, LLC, part of Springer Nature.
The registered company address is: 233 Spring Street, New York, NY 10013, U.S.A.

Preface

The field of metabolomics focuses on the measurement of large numbers of metabolites in biological systems, from cells and tissues to animal models and human subjects. Metabolites encompass a family of small molecules, such as carboxylic acids, amines, alcohols, amino acids, carbohydrates, and lipids, that are typically <1000 Da in molecular weight. Metabolites represent the downstream products of genes, transcripts, and protein function and provide an instantaneous snapshot of biological status. Hence, their measurement provides critical insights for understanding biological phenotypes, deciphering mechanisms, and identifying disease biomarkers or drug targets. Over the past decade, the metabolomics field has witnessed significant technological advances for measuring and interpreting metabolic changes in a wide variety of biological systems. As a result, metabolomics applications have been growing rapidly and simultaneously in a number of areas, including, but not limited to, early disease diagnosis, drug target identification, toxicology, pharmacology, environmental studies, agriculture, and food and nutrition.

The analysis of many metabolites in parallel is pivotal to the area of metabolomics. The widely used methods for such analysis are nuclear magnetic resonance (NMR) spectroscopy and mass spectrometry, and the two methods are largely complementary. Numerous characteristics, including NMR's excellent reproducibility and quantitative accuracy, and its ability to identify structures of unknown metabolites offer unique strengths for metabolomics applications. NMR's ability to distinguish between geometrical isomers, positional isomers, and positional isotopomers is invaluable in the field.

NMR-Based Metabolomics: Methods and Protocols is intended to provide a broad coverage of NMR-based methods and applications along with detailed step-by-step procedures to aid beginners as well as advanced practitioners in the field. Each chapter provides a brief introduction to the methodology, followed by a detailed protocol including sample preparation, NMR experiments, data analysis, and statistical analysis to enable the readers to easily learn, reproduce, and use the methods in their research. Our goal is to cover the breadth of the field; hence, in-depth discussions of NMR theory or biological interpretation are out of our scope. Interested readers are, however, advised to refer to many excellent books and reviews for in-depth discussions.

The topics chosen are representative of the many and growing number of methods and applications in the field. The biological samples used in the studies range from biofluids, cells, animal models, human, and plants to food and nutrients. And the applications range from mechanistic understanding, biomarker discovery, environmental studies, and drug discovery to food and nutrition. NMR methods include global, targeted, and isotope tracer-based techniques.

We thank our colleagues for succinctly describing the advances along with the detailed methodologies within their respective areas to help others learn from their expertise and experience. We hope that the compilation of the diverse areas of NMR-based metabolomics covered in these chapters will serve as a wealth of information for the readers and also stepping stones for further advances in the field of metabolomics.

Seattle, WA, USA

G. A. Nagana Gowda
Daniel Raftery

Contents

Contributors

MARYAM TABATABAEI ANARAKI • *Department of Physical and Environmental Sciences, University of Toronto Scarborough, Toronto, ON, Canada*

TRYGVE ANDREASSEN • *Department of Circulation and Medical Imaging, Norwegian University of Science and Technology (NTNU), Trondheim, Norway*

JAN H. ARDENKJÆR-LARSEN • *Department of Health Technology, Technical University of Denmark, Kgs. Lyngby, Denmark*

HANUDATTA S. ATREYA • *NMR Research Centre, Indian Institute of Science, Bangalore, India*

RUBÉN BARRILERO • *Metabolomics Platform, IISPV, DEEEA, Universitat Rovira i Virgili, Tarragona, Spain*

MONICA BASTAWROUS • *Department of Physical and Environmental Sciences, University of Toronto Scarborough, Toronto, ON, Canada*

TONE F. BATHEN • *Department of Circulation and Medical Imaging, Norwegian University of Science and Technology (NTNU), Trondheim, Norway*

MARK BERJANSKII • *Department of Biological Sciences, University of Alberta, Edmonton, AB, Canada*

HANNE CHRISTINE BERTRAM • *Department of Food Science, Aarhus University, Årslev, Denmark*

TEDROS BEZABEH • *College of Natural and Applied Sciences, University of Guam, Mangilao, GU, USA*

ANIL BHATIA • *Department of Biochemistry, University of Missouri at Columbia, Columbia, MO, USA; MU Metabolomics Center, University of Missouri at Columbia, Columbia, MO, USA; Bond Life Sciences Center, University of Missouri at Columbia, Columbia, MO, USA*

FATEMA BHINDERWALA • *Department of Chemistry, University of Nebraska-Lincoln, Lincoln, NE, USA; Nebraska Center for Integrated Biomolecular Communication, University of Nebraska-Lincoln, Lincoln, NE, USA*

ANA CAPATI • *College of Natural and Applied Sciences, University of Guam, Mangilao, GU, USA; Morsani College of Medicine, University of South Florida, Tampa, FL, USA*

CECILIA CASTRO • *Department of Biochemistry and Cambridge Systems Biology Centre, University of Cambridge, Cambridge, UK*

LEO L. CHENG • *Department of Radiology, Massachusetts General Hospital, Harvard Medical School, Boston, MA, USA; Department of Pathology, Massachusetts General Hospital, Harvard Medical School, Boston, MA, USA*

XAVIER CORREIG • *Metabolomics Platform, IISPV, DEEEA, Universitat Rovira i Virgili, Tarragona, Spain; Spanish Biomedical Research Centre in Diabetes and Associated Metabolic Disorders, CIBERDEM, Madrid, Spain*

HESAM DASHTI • *Department of Medicine, Brigham and Women's Hospital, Harvard Medical School, Boston, MA, USA*

FRANK DELAGLIO • *Institute for Bioscience and Biotechnology Research, National Institute of Standards and Technology, University of Maryland, Rockville, MD, USA*

ABHINAV DUBEY • *NMR Research Centre, Indian Institute of Science, Bangalore, India; Department of Biochemistry and Molecular Pharmacology, Dana-Farber Cancer Institute, Harvard Medical School, Boston, MA, USA*

Timothy M. D. Ebbels • *Computational and Systems Medicine, Department of Surgery and Cancer, Imperial College, London, UK*
Arthur S. Edison • *Department of Biochemistry, University of Georgia, Athens, GA, USA; Department of Genetics, University of Georgia, Athens, GA, USA; Complex Carbohydrate Research Center, University of Georgia, Athens, GA, USA*
Nina Eggers • *Department of Food Science, Aarhus University, Årslev, Denmark*
Hamid R. Eghbalnia • *Department of Biochemistry, University of Wisconsin Madison, Madison, WI, USA*
Teresa W.-M. Fan • *Center for Environmental and Systems Biochemistry, University of Kentucky, Lexington, KY, USA; Department of Toxicology and Cancer Biology, University of Kentucky, Lexington, KY, USA*
Miriam Gil • *Biosfer Teslab SL, Plaça Prim, Reus, Spain*
Patrick Giraudeau • *CEISAM, Université de Nantes, CNRS, UMR 6230, Nantes, France; Institut Universitaire de France, Paris, France*
Guro F. Giskeødegård • *Department of Circulation and Medical Imaging, Norwegian University of Science and Technology (NTNU), Trondheim, Norway*
G. A. Nagana Gowda • *Northwest Metabolomics Research Center, University of Washington, Seattle, WA, USA; Department of Anesthesiology and Pain Medicine, University of Washington, Seattle, WA, USA; Mitochondria and Metabolism Center, University of Washington, Seattle, WA, USA*
Gonçalo Graça • *Computational and Systems Medicine, Department of Surgery and Cancer, Imperial College, London, UK*
Julian L. Griffin • *Department of Biochemistry and Cambridge Systems Biology Centre, University of Cambridge, Cambridge, UK*
Maria T. Grinde • *Department of Circulation and Medical Imaging, Norwegian University of Science and Technology (NTNU), Trondheim, Norway*
Piet Habbel • *Department of Hematology and Oncology, Charité – Universitätsmedizin Berlin, Berlin, Germany*
Omkar B. Ijare • *Department of Chemistry, University of Winnipeg, Winnipeg, MB, Canada; Houston Methodist Research Institute, Houston, TX, USA*
Naranamangalam R. Jagannathan • *Department of NMR and MRI Facility, All India Institute of Medical Sciences, New Delhi, India*
Amy Jenne • *Department of Chemistry, University of Toronto Scarborough, Toronto, ON, Canada*
Pernille R. Jensen • *Department of Health Technology, Technical University of Denmark, Kgs. Lyngby, Denmark*
Ibrahim Karaman • *Department of Epidemiology and Biostatistics, School of Public Health, Imperial College London, London, UK; UK Dementia Research Institute, Imperial College London, London, UK*
Magnus Karlsson • *Department of Health Technology, Technical University of Denmark, Kgs. Lyngby, Denmark*
Jun Kikuchi • *RIKEN Center for Sustainable Resource Science, Yokohama, Kanagawa, Japan; Graduate School of Bioagricultural Sciences, Nagoya, Aichi, Japan; Graduate School of Medical Life Science, Yokohama City University, Yokohama, Kanagawa, Japan*
Ēriks Kupče • *Bruker UK Ltd., Coventry, UK*
Andrew N. Lane • *Center for Environmental and Systems Biochemistry, University of Kentucky, Lexington, KY, USA; Department of Toxicology and Cancer Biology, University of Kentucky, Lexington, KY, USA*

Daniel Lane • *Department of Chemistry, University of Toronto Scarborough, Toronto, ON, Canada*
Adrien Le Guennec • *Complex Carbohydrate Research Center, University of Georgia, Athens, GA, USA; NMR Facility, Guy's Campus, King's College London, London, UK*
Zhentian Lei • *Department of Biochemistry, University of Missouri at Columbia, Columbia, MO, USA; MU Metabolomics Center, University of Missouri at Columbia, Columbia, MO, USA; Bond Life Sciences Center, University of Missouri at Columbia, Columbia, MO, USA*
Mathilde H. Lerche • *Department of Health Technology, Technical University of Denmark, Kgs. Lyngby, Denmark*
Penghui Lin • *Center for Environmental and Systems Biochemistry, University of Kentucky, Lexington, KY, USA*
Matthias Lipfert • *Department of Biological Sciences, University of Alberta, Edmonton, AB, Canada*
John L. Markley • *Department of Biochemistry, University of Wisconsin Madison, Madison, WI, USA*
Estelle Martineau • *CEISAM, Université de Nantes, CNRS, UMR 6230, Nantes, France; SpectroMaitrise, CAPACITES SAS, Nantes, France*
Siver A. Moestue • *Department of Circulation and Medical Imaging, Norwegian University of Science and Technology (NTNU), Trondheim, Norway*
Fausto Carnevale Neto • *Department of Anesthesiology and Pain Medicine, Northwest Metabolomics Research Center, University of Washington, Seattle, WA, USA*
Kumar Pichumani • *Department of Neurosurgery, Kenneth R. Peak Brain and Pituitary Tumor Treatment Center, Houston Methodist Hospital and Research Institute, Houston, TX, USA; Weill Cornell Medical College, New York, NY, USA*
Alan Cesar Pilon • *Núcleo de Pesquisa em Produtos Naturais e Sintéticos (NPPNS), Departamento de Física e Química, Faculdade de Ciências Farmacêuticas de Ribeirão Preto, Universidade de São Paulo, Ribeirão Preto, São Paulo, Brazil*
Robert Powers • *Department of Chemistry, University of Nebraska-Lincoln, Lincoln, NE, USA; Nebraska Center for Integrated Biomolecular Communication, University of Nebraska-Lincoln, Lincoln, NE, USA*
Shivanand M. Pudakalakatti • *NMR Research Centre, Indian Institute of Science, Bangalore, India; Department of Cancer Systems Imaging, University of Texas MD Anderson Cancer Center, Houston, TX, USA*
Daniel Raftery • *Northwest Metabolomics Research Center, University of Washington, Seattle, WA, USA; Department of Anesthesiology and Pain Medicine, University of Washington, Seattle, WA, USA; Mitochondria and Metabolism Center, University of Washington, Seattle, WA, USA; Fred Hutchinson Cancer Research Center, Seattle, WA, USA*
Renuka Ranjan • *Centre of Biomedical Research, Lucknow, India; School of Biotechnology, Institute of Science, Banaras Hindu University, Varanasi, India*
Manoj Kumar Rout • *Department of Biological Sciences, University of Alberta, Edmonton, AB, Canada*
Sara Samino • *Biosfer Teslab SL, Plaça Prim, Reus, Spain*
Saurav J. Sarma • *Department of Biochemistry, University of Missouri at Columbia, Columbia, MO, USA; MU Metabolomics Center, University of Missouri at Columbia,*

Columbia, MO, USA; Bond Life Sciences Center, University of Missouri at Columbia, Columbia, MO, USA

DENISE MEDEIROS SELEGATO • *Núcleo de Bioensaios, Biossíntese e Ecofisiologia de Produtos Naturais (NuBBE), Departamento de Química Orgânica, Instituto de Química, Universidade Estadual Paulista (UNESP), Araraquara, São Paulo, Brazil*

UMA SHARMA • *Department of NMR and MRI Facility, All India Institute of Medical Sciences, New Delhi, India*

ANDRÉ J. SIMPSON • *Department of Physical and Environmental Sciences, University of Toronto Scarborough, Toronto, ON, Canada; Department of Chemistry, University of Toronto Scarborough, Toronto, ON, Canada*

NEERAJ SINHA • *Centre of Biomedical Research, Lucknow, India*

LLOYD W. SUMNER • *Department of Biochemistry, University of Missouri at Columbia, Columbia, MO, USA; MU Metabolomics Center, University of Missouri at Columbia, Columbia, MO, USA; Bond Life Sciences Center, University of Missouri at Columbia, Columbia, MO, USA*

ULRIK K. SUNDEKILDE • *Department of Food Science, Aarhus University, Årslev, Denmark*

MAY-BRITT TESSEM • *Department of Circulation and Medical Imaging, Norwegian University of Science and Technology (NTNU), Trondheim, Norway*

MARLON TILGNER • *Department of Radiology, Massachusetts General Hospital, Harvard Medical School, Boston, MA, USA; Department of Pathology, Massachusetts General Hospital, Harvard Medical School, Boston, MA, USA; Department of Hematology and Oncology, Charité – Universitätsmedizin Berlin, Berlin, Germany*

TIM S. VATER • *Department of Radiology, Massachusetts General Hospital, Harvard Medical School, Boston, MA, USA; Department of Pathology, Massachusetts General Hospital, Harvard Medical School, Boston, MA, USA; Department of Hematology and Oncology, Charité – Universitätsmedizin Berlin, Berlin, Germany*

JONATHAN R. WEDELL • *Department of Biochemistry, University of Wisconsin Madison, Madison, WI, USA*

WILLIAM M. WESTLER • *Department of Biochemistry, University of Wisconsin Madison, Madison, WI, USA*

DAVID S. WISHART • *Department of Biological Sciences, University of Alberta, Edmonton, AB, Canada; Department of Computing Science, University of Alberta, Edmonton, AB, Canada*

DABAO ZHANG • *Department of Statistics, Purdue University, West Lafayette, IN, USA*

MIN ZHANG • *Department of Statistics, Purdue University, West Lafayette, IN, USA*

Part I

Historical and Methodological Overview

Chapter 1

Overview of NMR Spectroscopy-Based Metabolomics: Opportunities and Challenges

G. A. Nagana Gowda and Daniel Raftery

Abstract

The fast-growing field of metabolomics is impacting numerous areas of basic and life sciences. In metabolomics, analytical methods play a pivotal role, and nuclear magnetic resonance (NMR) and mass spectrometry (MS) have proven to be the most suitable and powerful methods. Although NMR exhibits lower sensitivity and resolution compared to MS, NMR's numerous important characteristics far outweigh its limitations. Some of its characteristics include excellent reproducibility and quantitative accuracy, the capability to analyze intact biospecimens, an unparalleled ability to identify unknown metabolites, the ability to trace in-cell and in-organelle metabolism in real time, and the capacity to trace metabolic pathways atom by atom using ^{2}H, ^{13}C, or ^{15}N isotopes. Each of these characteristics has been exploited extensively in numerous studies. In parallel, the field has witnessed significant progress in instrumentation, methods development, databases, and automation that are focused on higher throughput and alleviating the limitations of NMR, in particular, resolution and sensitivity. Despite the advances, however, the high complexity of biological mixtures combined with the limitations in sensitivity and resolution continues to pose major challenges. These challenges need to be dealt with effectively to better realize the potential of metabolomics, in general. As a result, multifaceted efforts continue to focus on addressing the challenges as well as reaping the benefits of NMR-based metabolomics. This chapter highlights the current status with emphasis on the opportunities and challenges in NMR-based metabolomics.

Key words NMR, Metabolomics, Biomarkers, Pathways, Historical perspective

1 Metabolomics

Metabolomics involves analysis of small molecules, commonly known as metabolites, in biological mixtures such as blood, tissue, urine, cells, plants, and food. Also known widely as metabolite profiling and sometimes termed metabonomics, metabolomics is a relatively new member of systems biology and the omics fields. Metabolites are the downstream products of the functions of genes, transcripts, and enzymes and their levels represent a sensitive measure of the effects of disease, drugs, toxicity, and the environment. Measurement of metabolite levels in humans and animal and cell models of human diseases offers avenues for understanding,

G. A. Nagana Gowda and Daniel Raftery (eds.), *NMR-Based Metabolomics: Methods and Protocols*, Methods in Molecular Biology, vol. 2037, https://doi.org/10.1007/978-1-4939-9690-2_1, © Springer Science+Business Media, LLC, part of Springer Nature 2019

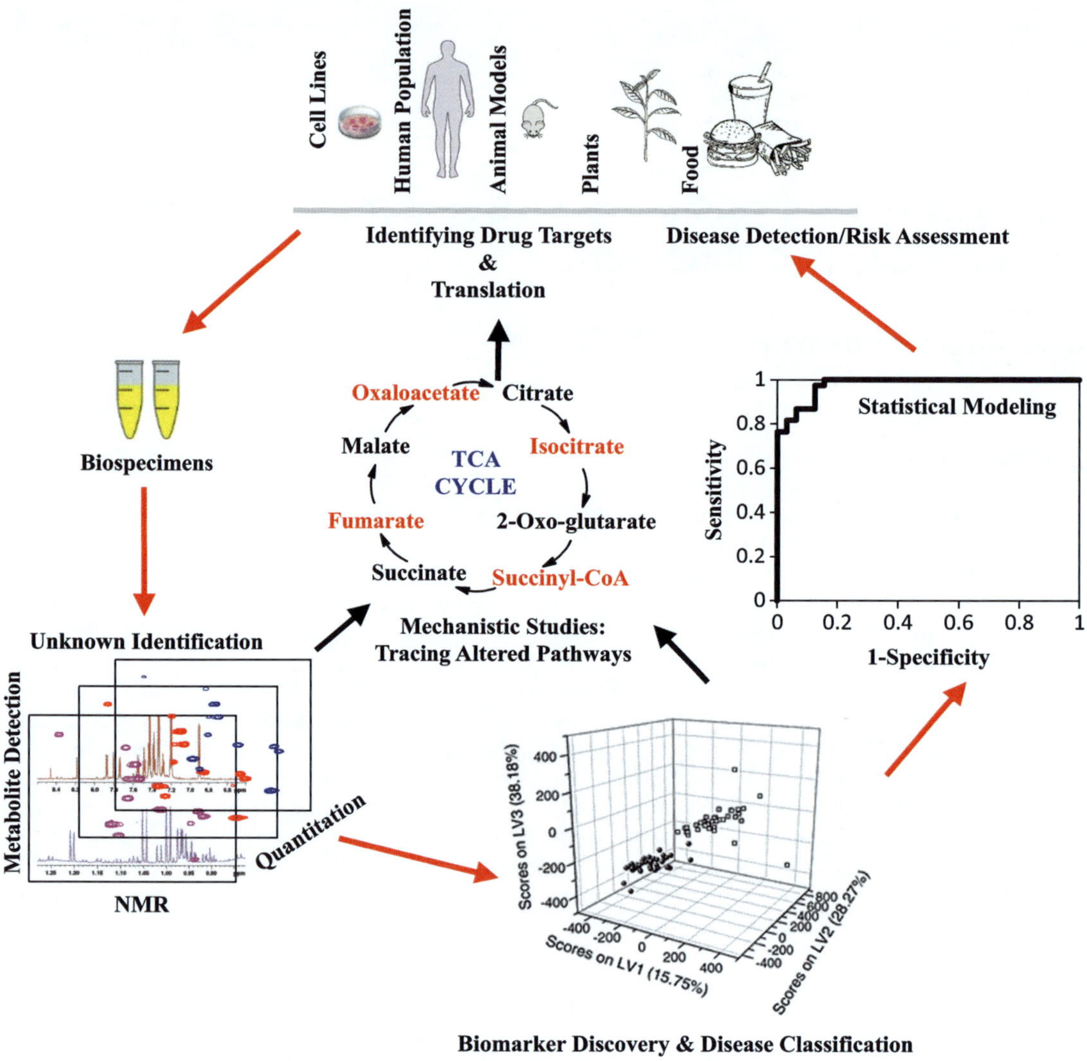

Fig. 1 Schematic diagram depicting the application of NMR-based metabolomics for early disease diagnosis, systems biology research, and drug target discovery

diagnosing, and managing human diseases (*see* Fig. 1). Further, metabolite levels enable objective assessment of disease risk factors associated with demography, drugs, toxins, and the environment and help guide the development of personalized medicine. To date, numerous disciplines including the life, food, and plant sciences, toxicology, environmental science, and drug development have been the beneficiaries of the field. However, in view of the important need to prevent human diseases or detect them early so therapy can be more effective, life sciences continues to be the primary focus of metabolomics field.

2 Historical Perspectives of Metabolomics

Conceptually, the phenomenon of metabolomics dates back 6000 years, when the fascinating history of urine analysis reportedly began [1]. Authentic documentation, however, exists only for the last 2500 years, when the ancient Indian medical science known as "Ayurveda" (meaning "life science") is described in a series of treatises on the diagnosis and treatment of diseases. Based on "Ayurveda," many diseases were diagnosed using characteristics of urine such as color, taste, texture, temperature, and precipitation. For example, one such disease, "madhumeha" (meaning "honey urine") now known as "diabetes," was diagnosed based on the sweet urine that attracted insects and flies [1–5]. For many centuries since then, the characteristics of urine were exploited by numerous ancient human civilizations for the diagnosis of diseases [1]. Metabolomics studies somewhat closer to what is practiced today dates back to the early 1950s, when Roger Williams and his coworkers showed that metabolic patterns in biological samples could be fingerprinted [6]. The authors utilized more than 200,000 paper chromatograms to demonstrate the variation of metabolic patterns of taste thresholds and excretion in individuals. However, further studies of multicomponent analysis had to wait for technological advances in analytical methods. In particular, advances in gas chromatography (GC), liquid chromatography (LC), and mass spectrometry (MS) in the 1970s provided the spark for the rapid progress in multicomponent analysis and quantitative metabolite profiling studies [7, 8]. Using gas-liquid partition chromatography Pauling et al. performed quantitative analysis of >250 substances in breath and urine vapor [9]. Over the past 20 years, in particular, the metabolomics field has been growing exponentially with advances in NMR and MS instrumentation, databases, automation, and software tools for data analysis.

3 Historical Development of NMR in Metabolomics

NMR in bulk samples was discovered in the mid-1940s [10, 11]. However, it was only after the introduction of superconducting magnets and the use of the Fourier transform in the mid-1960s did NMR gain the much needed impetus for application to numerous disciplines of basic and applied sciences. Analysis of metabolites in biological samples using in vivo NMR began in the mid-1970s. Radda from Oxford and Shulman from Yale were the first to investigate metabolism in humans using NMR, noninvasively [12]. Identification of separate ^{31}P NMR signals was first made in 1973 by Moon and Richards. Subsequently, in 1974, metabolic changes in muscle tissue was investigated by Hoult et al. [13]. Initially, studies were focused mainly on ^{31}P NMR and nearly a dozen phosphorus-containing metabolites were detected in numerous live biospecimens including tumor cells, red blood cells, tissue, yeast, and bacteria using ^{31}P NMR [13–19]. Subsequently, the use of ^{13}C-labeled substrates dramatically improved the detection of metabolites using ^{13}C NMR. Detection of metabolites using ^{1}H NMR was still challenging at that time due to the interference from the huge water signal. The early 1980s witnessed advances in NMR instrumentation and high-resolution techniques, which enabled analysis of intact biofluids using ^{1}H NMR by suppressing the unwanted signal from water [20]. NMR-based metabolomics has been growing exponentially ever since, with significant advances occurring regularly in both methods development and multifaceted applications.

4 NMR Characteristics

NMR exhibits numerous beneficial characteristics for the analysis of metabolites in complex biological mixtures. Important strengths of NMR include:

- Does not need sample separation.
- Detects up to hundreds of metabolites in a single measurement.
- Highly quantitative.
- Highly reproducible.
- Nondestructive to samples.
- Enables unknown metabolite identification.
- Enables analysis of intact biofluids, tissue, cells, and subcellular organelles.
- Enables real-time metabolism studies of live animals, cells, and subcellular organelles.
- Single internal reference is sufficient for absolute quantitation of all detected metabolites.
- High throughput: Enables measurement of hundreds of samples per day.
- Using isotope tracer studies, NMR enables tracking isotope distribution within a formed metabolite.
- Numerous nuclei such as ^{1}H, ^{2}H, ^{13}C, ^{31}P, and ^{15}N can be independently used for analysis of the same sample.
- Well suited for analysis of metabolites that are unstable and sensitive to harsh treatments, such as applying high voltage in MS analysis.

- Metabolite profiles derived are independent of the operator and instrument used.

NMR also exhibits a number of challenges that are continually being addressed to alleviate them. Major drawbacks of NMR include:

- Relatively poor sensitivity compared to MS (limit of detection ~100 nM).
- Limited spectral resolution since chromatography is not used.
- Difficulty in measuring many lipids individually.
- High field instrument cost.

5 NMR-Based Metabolomics

Methods development in NMR-based metabolomics has focused on improved metabolite detection, better unknown identification, and more accurate quantitation of larger numbers of metabolites. Advances in NMR instrumentation offer improvements in resolution and sensitivity and enable analysis of complex biological mixtures with increasing depth and detail. Advances have also led to the realization that sample matrices such as proteins deleteriously affect both quantitation accuracy and the number of metabolites detected in samples such as blood serum or plasma. Protein removal prior to analysis not only enabled expanding the pool of quantified metabolites to an unprecedented level, but also provided common protocols for analysis using both NMR and MS methods. In terms of sensitivity, NMR instrumentation now enables analysis of metabolites down to 100 nM without using micro-coil probe or hyperpolarization methods [21]. Such advances now enable the analysis of many coenzymes that are fundamental to virtually all biochemical reactions using simple NMR methods [22, 23].

In terms of applications, numerous areas of life and basic sciences have been greatly impacted by the advances in metabolomics. Applications of NMR-based metabolomics to life sciences have gained major interest due to the continued and urgent need for better understanding of human diseases. In particular, biomarker discovery and mechanistic understanding of pathogenesis of diseases have been the major focus, driven by the promise of earlier disease detection and concomitantly better treatment outcomes, as well as new targets for drug development. Widespread diseases including cancer, diabetes, cardiovascular disease, infections, and obesity have thus been the subjects of numerous investigations [24]. Metabolomics has also impacted significantly the drug discovery process. In particular, the ability to follow dynamic changes in the metabolome from drug treatment and toxicity offers immense value for the development of new therapeutic approaches

[25]. Investigations of the food metabolome that make up our diets and the food production process that also affects the metabolome have gained major attention, as diet significantly impacts human health and well-being [26]. The plant metabolome is extremely complex, with an estimated number of metabolites exceeding well over 200,000. Research interest in this area is focused on correlating metabolites with genotype or phenotype in part to improve plant production, or natural-product-based drug discovery [27, 28].

6 Advanced NMR Approaches in Metabolomics

A number of advanced methods are being applied to NMR-based metabolomics and are yielding new and more in-depth information. Several of these methods are discussed in the following section.

6.1 Isotope-Enhanced NMR to Track Metabolism

Isotope tracers provide tremendous opportunities to track metabolic pathways and measure fluxes connected to specific substrates and thereby promise important clues to disease mechanisms. NMR spectroscopy exhibits a unique capability in this approach, as it allows the determination of atom-specific positional isotopomer distributions derived from precursors enriched with stable isotopes. Substrates with numerous stable isotopes such as ^{13}C, ^{15}N, and ^{2}H have been extensively investigated. For example, analysis of cells supplied with ^{13}C-labeled glucose is extensively used to measure and quantify the downstream metabolic products of numerous pathways including the glycolysis, TCA cycle, and pentose phosphate pathways. Not only do these studies track the active metabolic pathways, but they provide rates at which the isotope-labeled substrates are consumed and their products formed [29, 30]. Studies using cells provide the means for understanding metabolic pathways under controlled conditions, whereas studies using animal models or even human subjects enable understanding of pathogenesis directly in the relevant organs [31]. Using multiple substrates, simultaneously, provides the advantage of multiplexing and enables simultaneous quantification of utilization of individual substrates under different pathological conditions [32]. Such investigations continue to attract major attention because of their ability to answer a number of key mechanistic questions in systems biology and biochemistry. Recent advances in this area have provided a more sensitive approach for real-time monitoring of cancer cell metabolism [33]. Dynamic changes in cellular metabolism can be measured quantitatively in cancer and control cells in intervals of 5 min or less using NMR. More recently, this fascinating approach has been extended to real-time monitoring of metabolism of live mammalian cellular organelles such as mitochondria

[34]. Investigation of cancer metabolism is a major topic in metabolomics, and NMR has become a core tool to study altered cellular metabolism specifically using stable isotope tracers. The growing number of applications in this area indicates the importance of this approach for mechanistic understanding of cancer metabolism.

6.2 Micro-Coil NMR

It is well recognized that increasing the sensitivity is a critical requirement for the realization of the full benefits of NMR-based metabolomics. Use of micro-coil NMR is one of the important routes for enhancing the sensitivity and it is particularly promising for mass-limited samples [35–37]. Micro-coil NMR is often preceded by liquid chromatography for separating metabolites from complex mixtures followed by direct online detection [38, 39], after online pre-concentration [40, 41], or offline pre-concentration [37]. The throughput of this approach, however, is not high, and it is more suited for analysis of metabolites in mass-limited samples. Currently, commercially available micro-coil flow probes enable high-throughput analysis of metabolites [42]. Samples may be pre-concentrated prior to analysis to improve sensitivity. Pre-concentration, however, may have deleterious consequences due to the varying solubility of metabolites [36] and resulting high concentrations of salt that may need to be removed. Owing to the significant interest and growth in the demand for miniaturization of NMR, nowadays, major instrument vendors offer micro-coil probes with multinuclear capability.

6.3 Fast NMR Methods

High-throughput analysis is another critical requirement in metabolomics. A number of developments have led to fast data acquisition in 2D NMR. HSQC and HMQC are the most often used 2D NMR experiments in metabolomics and their combination with nonuniform sampling and forward maximum entropy reconstruction can reduce the data acquisition time by nearly an order of magnitude [43, 44]. Further, combining this approach with the J-compensated experiment reduces the data acquisition time by a factor of >20 [45]. The SOFAST NMR approach enhances the steady-state magnetization by combining an accelerated T_1 relaxation time and optimized flip angle. The SOFAST-HMQC 2D experiment can be performed within 10–15 s and such fast data acqusition enables monitoring of metabolism virtually in real time in live cells [46]. An approach that achieves significantly enhanced speed of data acquisition for 2D experiments is covariance NMR spectroscopy, which is based on minimal data points in the T_1 dimension [47]. The fastest in terms of data aqusition, however, is the single-scan acquisition method and several single-scan 2D experiments have demonstrated their potential applications in metabolomics [48, 49].

6.4 Hyperpolariztion Methods

Boosting sensivitiy based on the hyperpolarization of nuclear spins continues to draw major interest for studying metabolsim in real time in vivo. Of several such methods, parahydrogen-induced polarization (PHIP) and dynamic nuclear polarization (DNP) have been found to be the most useful for metabolism studies as they can boost the polarization of nuclei such as ^{13}C or ^{15}N in a number of metabolic substrates. The hyperpolarized substrates are then infused into biologocal systems and the formed downstream metabolites that still maintain the highly polarized nuclear spin state can be detected with high sensitivity. PASADENA (parahydrogen and synthesis allow dramatically enhanced nuclear alignment) is a more often used PHIP method since it generally involves a fast preparation step of about 1 min [50]. The method, however, works with only unsaturated substrates, and hence important substrates from the TCA cycle, such as pyruvate, are not amenable for hyperpolarization by this method. There are, however, methods to circumvent this challenge through metabolites that are closely linked to pyruvate metabolism [51]. The DNP method, on the other hand, is devoid of the challenges met by the PHIP method and can be used to hyperpolarize many different metabolites. In particular, the dissolution DNP approach, wherein a hyperpolarized solid consisting of the substrate of interest, a glassing agent, and a polarizing agent is rapidly melted and injected into cells, tissue, or organs, is more useful for real-time metabolism studies [52]. Owing to the advantages of DNP over PHIP, more studies have been focused on advancing and using the DNP-based methods in metabolomics. Overall, these technologies are still at a relatively early stage, though their potential promise to the field is immense.

7 Data Analysis

Multivariate statistical methods continue to be important for unraveling the highly complex data in metabolomics. Principal component analysis (PCA) and partial least squares discriminant analysis (PLS-DA) are the most routinely used methods. PCA is an unsupervised method that is most useful for assessing data clusters and outliers. Logistic regression is widely used by epidemiologists and is heavily used in diagnostic applications and for risk prediction. PLS-DA is also a supervised method and allows the construction of predictive models based on the regression of the data matrix (X) against sample class matrix (Y). PLS-DA works well with correlated metabolite variables, unlike logistic regression. The models derived from supervised methods are typically tested using cross-validation, such as the popular approach known as Monte Carlo cross-validation (MCCV) [53]. For more stringent validation, however, the use of independent sample sets when available is highly advised. Metabolomics data are often met with challenges from the effects of confounding factors such as those arising from sample collection and preparation as well as the effects of demographic and clinical variables on metabolite levels. These factors have affected the validation of putative biomarkers or classification models deleteriously. New approaches that take into account the effects of numerous demographic covariates such as gender, BMI, alcohol, and smoking status promise new avenues to alleviate the challenges arising from confounding effects in the metabolomics field [54, 55].

8 Databases, Software, and Automation Tools

A growing number of databases, software, and automation tools have enabled spectral assignments of unknown metabolites, metabolite quantitation, data analysis, and interpretation. Such tools are becoming critically important considering that advances in analytical methods generate increasingly large and complex datasets in metabolomics. Importantly, many of these tools are publicly available and facilitate data analysis and interpretation even by nonexpert users. Many platforms enable automated analysis of NMR spectra with virtually no need for any detailed knowledge of NMR [24]. The latest version of the MetaboAnalyst platform enables data analysis including statistical analysis, biomarker discovery, validation, and pathway analysis all in a single platform, which has been very useful for a large percentage of the investigators in the field. Similarly, many other resources such as NMRBox [56], BMRM, MMCD, COLMAR, and PRIMEe [57] are also freely available and provide a variety of options for data analysis and interpretation. It is also increasingly becoming common to deposit raw data at major international public repositories such as the Metabolomics Workbench [58] and MetaboLights [59]. Such data sharing resources promise a great impact on the metabolomics field going forward.

9 Opportunities, Challenges, and Future Directions

NMR spectroscopy offers a plethora of opportunities to understand systems biology, predict disease risk, and discover early disease biomarkers and targets for therapy. While human health and diseases have been the major subjects of NMR-based metabolomics, numerous other areas including agriculture, food and nutrition, environmental science, drug discovery, pharmacology, and toxicology continue to benefit from the advancing field. NMR continues to play a major role in unknown metabolite identification, which is a critical branch of the metabolomics field. The growing number of applications demonstrates the ability of NMR to follow metabolism in real time using specimens such as live cells and cellular organelles and promise new avenues for mechanistic insights into cellular and subcellular functions. Many challenges associated with NMR, however, stem from its long known limited sensitivity and resolution. Despite significant efforts to alleviate these challenges, the high complexity of biological mixtures continues to demand greater improvements for the detection, identification, and quantitation of a large pool of metabolites. In view of its numerous opportunities as well as the challenges, NMR-based metabolomics will continue to witness advances in techniques and methods development as well as their applications. Boosting sensitivity and resolution and establishing the identity for unknown metabolites will continue to be at the forefront in methodological advances. Such advances are anticipated to greatly impact various disciplines of the NMR-based metabolomics field including the understanding of systems biology and managing human health and diseases.

References

1. Antic T, DeMay RM (2014) The fascinating history of urine examination. J Am Soc Cytopathol 3:103e107
2. Bhishagratna KKL (1911) An English translation of "The Sushruta Samhita", vol II. The Bharat Mihir Press, Calcutta
3. Jacob R (2015) Sickening sweet. Distillations 1 (4):12–15
4. Das S (2001) Susruta, the pioneer urologist of antiquity. J Urol 165(5):1405–1408
5. Karamanou M, Protogerou A, Tsoucalas G, Androutsos G, Poulakou-Rebelakou E (2016) Milestones in the history of diabetes mellitus: the main contributors. World J Diabetes 7 (1):1–7
6. Williams RJ et al (1951) Individual metabolic patterns and human disease: an exploratory study utilizing predominantly paper chromatographic methods. Biochemical Institute Studies IV, University of Texas Publication No. 5109, University of Texas, Austin, 204 pp
7. Gates SC, Sweeley CC (1978) Quantitative metabolic profiling based on gas chromatography. Clin Chem 24(10):1663–1673
8. Nagana Gowda GA, Djukovic D (2014) Overview of mass spectrometry-based metabolomics: opportunities and challenges. Methods Mol Biol 1198:3–12
9. Pauling L, Robinson AB, Teranishi R, Cary P (1971) Quantitative analysis of urine vapor and breath by gas-liquid partition chromatography. Proc Natl Acad Sci U S A 68(10):2374–2376
10. Bloch F, Hansen WW, Packard M (1946) Nuclear induction. Phys Rev 69:127
11. Purcell EM, Torrey HC, Pound RV (1946) Resonance absorption by nuclear magnetic moments in a solid. Phys Rev 69:37–38
12. Becker ED (1993) A brief history of nuclear magnetic resonance. Anal Chem 65 (6):295A–302A
13. Hoult DI, Busby SJ, Gadian DG, Radda GK, Richards RE, Seeley PJ (1974) Observation of tissue metabolites using 31P nuclear magnetic resonance. Nature 252(5481):285–287
14. Navon G, Ogawa S, Shulman RG, Yamane T (1977) High-resolution 31P nuclear magnetic resonance studies of metabolism in aerobic Escherichia coli cells. Proc Natl Acad Sci U S A 74(3):888–891
15. Navon G, Ogawa S, Shulman RG, Yamane T (1977) 31P nuclear magnetic resonance studies of Ehrlich ascites tumor cells. Proc Natl Acad Sci U S A 74(1):87–91
16. Burt CT, Glonek T, Barany M (1976) Analysis of phosphate metabolites, the intracellular pH, and the state of adenosine triphosphate in intact muscle by phosphorus nuclear magnetic resonance. J Biol Chem 251:2584–2591
17. Henderson TO, Costello AJR, Omachi A (1974) Phosphate metabolism in intact human erythrocytes: determination by phosphorus-31 nuclear magnetic resonance spectroscopy. Proc Natl Acad Sci U S A 71:2487–2490

18. Moon RB, Richards JH (1973) Determination of intracellular pH by 31P magnetic resonance. J Biol Chem 248:7276–7278
19. Salhany JM, Yamane T, Shulman RG, Ogawa S (1975) High resolution 31P nuclear magnetic resonance studies of intact yeast cells. Proc Natl Acad Sci U S A 72:4966–4970
20. Nicholson JK, Wilson ID (1989) High resolution proton magnetic resonance spectroscopy of biological fluids. Prog Nucl Magn Reson Spectrosc 21(4–5):449–501
21. Nagana Gowda GA, Gowda YN, Raftery D (2015) Expanding the limits of human blood metabolite quantitation using NMR spectroscopy. Anal Chem 87(1):706–715
22. Nagana Gowda GA, Abell L, Lee CF, Tian R (2016) Raftery D (2016) simultaneous analysis of major coenzymes of cellular redox reactions and energy using ex vivo (1)H NMR spectroscopy. Anal Chem 88(9):4817–4824
23. Nagana Gowda GA, Abell L, Lee CF, Tian R (2019) Extending the scope of 1H NMR spectroscopy for the analysis of cellular coenzyme A and acetyl coenzyme A. Anal Chem 91 (3):2464–2471
24. Nagana Gowda GA, Raftery D (2017) Recent advances in NMR-based metabolomics. Anal Chem 89(1):490–510
25. Leenders J, Frédérich M, de Tullio P (2015) Nuclear magnetic resonance: a key metabolomics platform in the drug discovery process. Drug Discov Today Technol 13:39–46
26. Johanningsmeier SD, Harris GK, Klevorn CM (2016) Metabolomic technologies for improving the quality of food: practice and promise. Annu Rev Food Sci Technol 7:413–438
27. Sumner LW, Lei Z, Nikolau BJ, Saito K (2015) Modern plant metabolomics: advanced natural product gene discoveries, improved technologies, and future prospects. Nat Prod Rep 32:212–229
28. Mahrous EA, Farag MA (2015) Two dimensional NMR spectroscopic approaches for exploring plant metabolome: a review. J Adv Res 6(1):3–15
29. Lloyd SG, Zeng H, Wang P, Chatham JC (2004) Lactate isotopomer analysis by 1H NMR spectroscopy: consideration of long-range nuclear spin-spin interactions. Magn Reson Med 51:1279–1282
30. Lane AN, Fan TWM (2007) Quantification and identification of isotopomer distributions of metabolites in crude cell extracts using 1H TOCSY. Metabolomics 3:79–86
31. Lane AN, Fan TW, Bousamra M 2nd, Higashi RM, Yan J, Miller DM (2011) Stable isotope-resolved metabolomics (SIRM) in cancer research with clinical application to nonsmall cell lung cancer. OMICS 15(3):173–182
32. Mashimo T, Pichumani K, Vemireddy V, Hatanpaa KJ, Singh DK, Sirasanagandla S et al (2014) Acetate is a bioenergetic substrate for human glioblastoma and brain metastases. Cell 159(7):1603–1614
33. Wen H, An YJ, Xu WJ, Kang KW, Park S (2015) Real-time monitoring of cancer cell metabolism and effects of an anticancer agent using 2D in-cell NMR spectroscopy. Angew Chem Int Ed Engl 54(18):5374–5377
34. Xu WJ, Wen H, Kim HS, Ko YJ, Dong SM, Park IS, Yook JI, Park S (2018) Observation of acetyl phosphate formation in mammalian mitochondria using real-time in-organelle NMR metabolomics. Proc Natl Acad Sci U S A 115(16):4152–4157
35. Kc R, Henry ID, Park GH, Aghdasi A, Raftery D (2010) New solenoidal microcoil NMR probe using zero-susceptibility wire. Concepts Magn Reson Part B Magn Reson Eng 37B (1):13–19
36. Grimes JH, O'Connell TM (2011) The application of micro-coil NMR probe technology to metabolomics of urine and serum. J Biomol NMR 49:297–305
37. Bird SS, Sheldon DP, Gathungu RM, Vouros P, Kautz R, Matson WR et al (2012) Structural characterization of plasma metabolites detected via LC-electrochemical coulometric array using LC-UV fractionation, MS, and NMR. Anal Chem 84:9889–9898
38. Cloarec O, Campbell A, Tseng LH, Braumann U, Spraul M, Scarfe G et al (2007) Virtual chromatographic resolution enhancement in cryoflow LC-NMR experiments via statistical total correlation spectroscopy. Anal Chem 79:3304–3311
39. Lenz EM, Wilson ID (2007) Analytical strategies in metabonomics. J Proteome Res 6:443–458
40. Djukovic D, Liu S, Henry I, Tobias B, Raftery D (2006) Signal enhancement in HPLC/microcoil NMR using automated column trapping. Anal Chem 78:7154–7160
41. Djukovic D, Appiah-Amponsah E, Shanaiah N, Nagana Gowda GA, Henry I, Everly M et al (2008) Ibuprofen metabolite profiling using a combination of SPE/column-trapping and HPLC-micro-coil NMR. J Pharm Biomed Anal 47:328–334
42. Garcia E, Andrews C, Hua J, Kim HL, Sukumaran DK, Szyperski T et al (2011) Diagnosis of early stage ovarian cancer by 1H NMR metabonomics of serum explored by use of a microflow NMR probe. J Proteome Res 10:1765–1771

43. Hyberts SG, Heffron GJ, Tarragona NG, Solanky K, Edmonds KA, Luithardt H et al (2007) Ultrahigh-resolution (1)H-(13)C HSQC spectra of metabolite mixtures using nonlinear sampling and forward maximum entropy reconstruction. J Am Chem Soc 129 (16):5108–5116
44. Hyberts SG, Arthanari H, Wagner G (2012) Applications of non-uniform sampling and processing. Top Curr Chem 316:125–148
45. Rai RK, Sinha N (2012) Fast and accurate quantitative metabolic profiling of body fluids by nonlinear sampling of 1H–13C two-dimensional nuclear magnetic resonance spectroscopy. Anal Chem 84 (22):10005–10011
46. Motta A, Paris D, Melck D (2010) Monitoring real-time metabolism of living cells by fast two-dimensional NMR spectroscopy. Anal Chem 82(6):2405–2411
47. Bruschweiler R, Zhang F (2004) Covariance nuclear magnetic resonance spectroscopy. J Chem Phys 120:5253–5260
48. Frydman L, Scherf T, Lupulescu A (2002) The acquisition of multidimensional NMR spectra within a single scan. Proc Natl Acad Sci U S A 99:15858–15862
49. Giraudeau P, Remaud GS, Akoka S (2009) Evaluation of ultrafast 2D NMR for quantitative analysis. Anal Chem 81(1):479–484
50. Bhattacharya P, Chekmenev EY, Perman WH, Harris KC, Lin AP, Norton VA et al (2007) Towards hyperpolarized (13)C-succinate imaging of brain cancer. J Magn Reson 186:150–155
51. Shchepin RV, Coffey AM, Waddell KW, Chekmenev EY (2012) PASADENA hyperpolarized 13C phospholactate. J Am Chem Soc 134 (9):3957–3960
52. Ardenkjaer-Larsen JH, Fridlund B, Gram A, Hansson G, Hansson L, Lerche MH et al (2003) Increase in signal-to-noise ratio of > 10,000 times in liquid-state NMR. Proc Natl Acad Sci U S A 100(18):10158–10163
53. Girard DA (1989) A fast 'Monte-Carlo cross-validation' procedure for large least squares problems with noisy data. Numer Math 56 (1):1–23
54. Chen C, Deng L, Wei S, Nagana Gowda GA, Gu H, Chiorean EG et al (2015) Exploring metabolic profile differences between colorectal polyp patients and controls using seemingly unrelated regression. J Proteome Res 14 (6):2492–2499
55. Chen C, Nagana Gowda GA, Zhu J, Deng L, Gu H, Chiorean EG et al (2017) Altered metabolite levels and correlations in patients with colorectal cancer and polyps detected using seemingly unrelated regression analysis. Metabolomics 13:125. https://doi.org/10.1007/s11306-017-1265-0
56. Maciejewski MW, Schuyler AD, Gryk MR, Moraru II, Romero PR, Ulrich EL et al (2017) NMRbox: a resource for biomolecular NMR computation. Biophys J 112 (8):1529–1534
57. Nagana Gowda GA, Raftery D (2015) Can NMR solve some significant challenges in metabolomics? J Magn Reson 260:144–160
58. Sud M, Fahy E, Cotter D, Azam K, Vadivelu I, Burant C et al (2016) Metabolomics workbench: an international repository for metabolomics data and metadata, metabolite standards, protocols, tutorials and training, and analysis tools. Nucleic Acids Res 44(D1): D463–D470
59. Kale NS, Haug K, Conesa P, Jayseelan K, Moreno P, Rocca-Serra P et al (2016) MetaboLights: an open-access database repository for metabolomics data. Curr Protoc Bioinformatics 53:14.13.1–14.13.18

Part II

Metabolite Analysis Methods

Chapter 2

Analysis of Plasma, Serum, and Whole Blood Metabolites Using ^{1}H NMR Spectroscopy

G. A. Nagana Gowda and Daniel Raftery

Abstract

Blood is the most widely used biological specimen in the metabolomics field. With its unique characteristics of high reproducibility and excellent quantitation, NMR spectroscopy offers immense benefits for the analysis of blood metabolites. In the metabolomics field, intact blood serum and plasma have been widely used for many years. However, such analysis has met with challenges arising from the deleterious effects of the abundant proteins in serum and plasma. Recent advances have led to the development of improved NMR methods that involve removal of protein before analysis. In particular, protein removal by precipitation using methanol alone or using a mixture of methanol and chloroform was shown to be an optimal method for metabolite recovery and for producing highly resolved NMR spectra. This has led to the absolute quantitation of nearly 70 metabolites in serum and plasma and nearly 80 in whole blood. In this chapter, we describe protocols for the analysis of blood serum, blood plasma, and whole blood metabolites using 1D ^{1}H NMR spectroscopy methods.

Key words Quantitative NMR, Serum, Plasma, Whole blood, Metabolomics, NMR spectroscopy, Protein precipitation, Absolute quantitation

1 Introduction

Analysis of human blood is widely used in the metabolomics field owing to its clinical relevance that arises from its association with essentially every living cell in the human body [1–3]. Nuclear magnetic resonance (NMR) spectroscopy and mass spectrometry (MS) are the two most widely used analytical techniques to analyze blood metabolites. MS is a highly sensitive method and detects several hundreds of metabolites (or more) in a single measurement. However, absolute quantitation of metabolites in MS is challenging owing to the fact that each metabolite requires an isotope-labeled analogue as an internal standard. Hence, in the MS-based metabolomics studies of blood, only relative peak areas are typically employed. The drawbacks of such analysis are that peak areas for the same metabolite vary between instruments and, owing to

G. A. Nagana Gowda and Daniel Raftery (eds.), *NMR-Based Metabolomics: Methods and Protocols*, Methods in Molecular Biology, vol. 2037, https://doi.org/10.1007/978-1-4939-9690-2_2, © Springer Science+Business Media, LLC, part of Springer Nature 2019

differences in the ionization efficiency between metabolites, intensity ratios between different metabolites do not represent their concentration ratios. It can thus be difficult to combine data from different studies or compare them to known clinical values. NMR spectroscopy on the other hand provides absolute concentrations for metabolites. A single internal standard is sufficient to determine concentrations of all detected metabolites.

Conventionally, blood metabolite analysis refers to the analysis of blood serum or plasma; for many years, serum and plasma were analyzed using intact samples. Serum and plasma samples contain a massive amount of protein (6–8 g/dL), and hence metabolite analysis using these samples often involved the suppression of NMR signals from proteins, generally, using the CPMG (Carr–-Purcell–Meiboom–Gill) NMR pulse sequence. Analysis of intact serum/plasma is generally attractive as it does not need any preprocessing; however, due to numerous challenges, this approach is increasingly recognized to be less informative for metabolite analysis. The challenges are that it detects far fewer metabolites, attenuates levels for many metabolites due to the presence of a massive amount of proteins, causes line broadening due to the shorter T_2 relaxation time that arises from the presence of proteins and exchange between protein-bound and free metabolites, and causes distorted baseline due to residual macromolecule signals. All of these factors deleteriously affect metabolite detection and quantitative analysis [4–7].

There have been numerous efforts focused on alleviating the challenges of analyzing intact serum and plasma by NMR spectroscopy. In particular, it has been recognized that physically removing proteins from serum or plasma results in better spectral quality and a significant improvement in the number of metabolites quantitated [7–13]. Broadly, two major approaches have been utilized for removal of proteins from serum/plasma: one is ultrafiltration using low molecular weight cutoff filters (ca. 3 kDa) and the other is protein precipitation using an organic solvent such as methanol, acetonitrile, acetone, perchloric acid, or trichloroacetic acid. Ultrafiltration using molecular weight cutoff filters provides high-quality spectra since proteins are removed very effectively from serum/plasma sample. However, further studies have shown that protein precipitation provides better recovery of metabolites from the protein matrix in serum/plasma (*see* Fig. 1) [7]. Comparison of protein precipitation using different organic solvents including those most often used for MS analysis (methanol and acetonitrile) has shown that methanol performs better [8]. In particular, methanol at 1:2 v/v ratio was found to be suitable for recovering most serum/plasma metabolites quantitatively, while acetonitrile performs poorly under similar conditions (*see* Fig. 2).

^{1}H NMR spectra of serum/plasma obtained after removing proteins by precipitation provided highly resolved spectra.

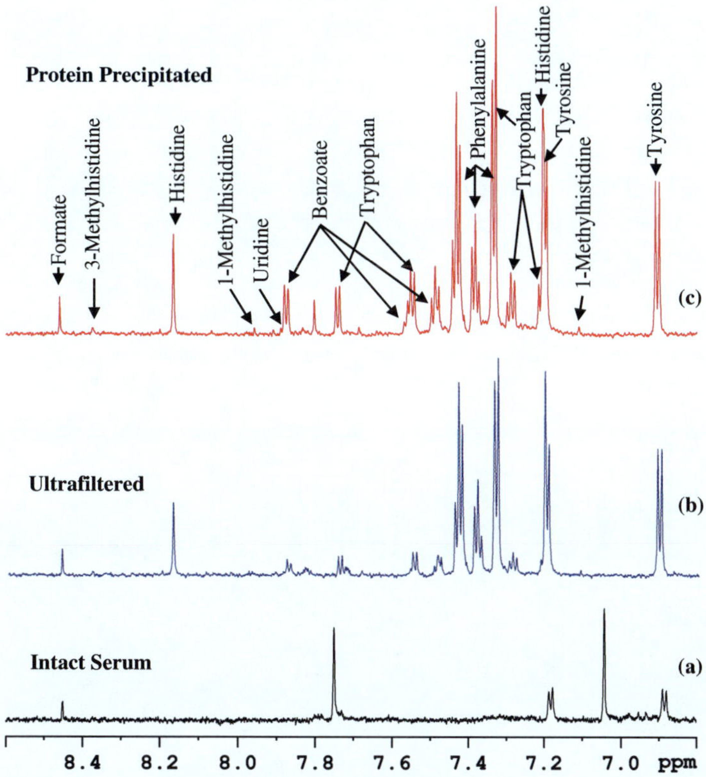

Fig. 1 Portions of 800-MHz ^{1}H NMR spectra of the same pooled human serum sample obtained by suppressing protein signals by (**a**) T_2 filtering using the CPMG pulse sequence, (**b**) ultrafiltration using a 3-kDa molecular weight cutoff filter, and (**c**) protein precipitation using methanol (1:2 v/v sample to solvent). Most of the metabolite signals in the displayed region are missing or significantly attenuated in the T_2-filtered spectrum and many including tryptophan, benzoate, and formate were significantly attenuated in ultrafiltered serum compared to the protein-precipitated serum (reproduced from ref. 7 with permission)

A comprehensive analysis of such spectra resulted in the identification of virtually all peaks in the spectra and identification of a large pool of quantifiable metabolites [8]. Nearly 70 metabolites were thus identified and characteristic peaks of all identified metabolites were annotated in the NMR spectra (*see* Fig. 3). Given that ^{1}H NMR spectra of all human serum and plasma samples are virtually identical (qualitatively) to that shown in Fig. 3, the annotated spectrum serves as a useful visual aid to identify and quantify blood metabolites. Chemical shifts for characteristic metabolite peaks are also listed in Table 1 as an additional aid for metabolite identification. It may however be noted that protein precipitation using organic solvents often involves sample drying to remove solvent before analysis and hence volatile endogenous serum/ plasma metabolites such as acetone, ethanol, dimethylamine, and

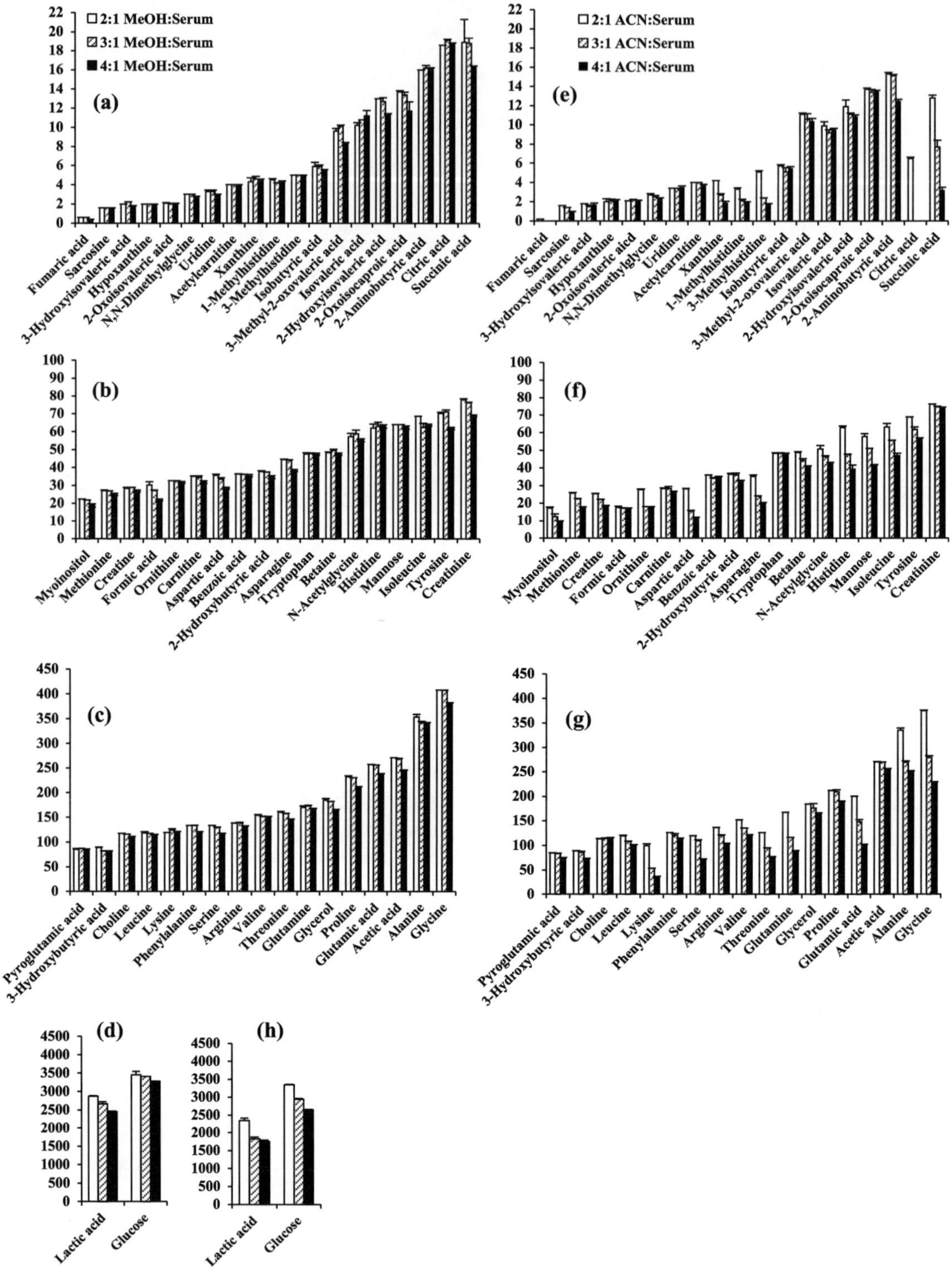

Fig. 2 Comparison of absolute concentrations (in μM) of metabolites detected in pooled human blood serum and quantitated using 800-MHz NMR spectroscopy after protein precipitation using methanol (MeOH) (**a**, **b**, **c**, and **d**) or acetonitrile (ACN) (**e**, **f**, **g**, and **h**) at a solvent to serum ratios of 2:1, 3:1, and 4:1. Methanol performs most optimally over a wide range, and the methanol to serum ratio of 2:1 provides the best performance (reproduced from ref. 8 with permission)

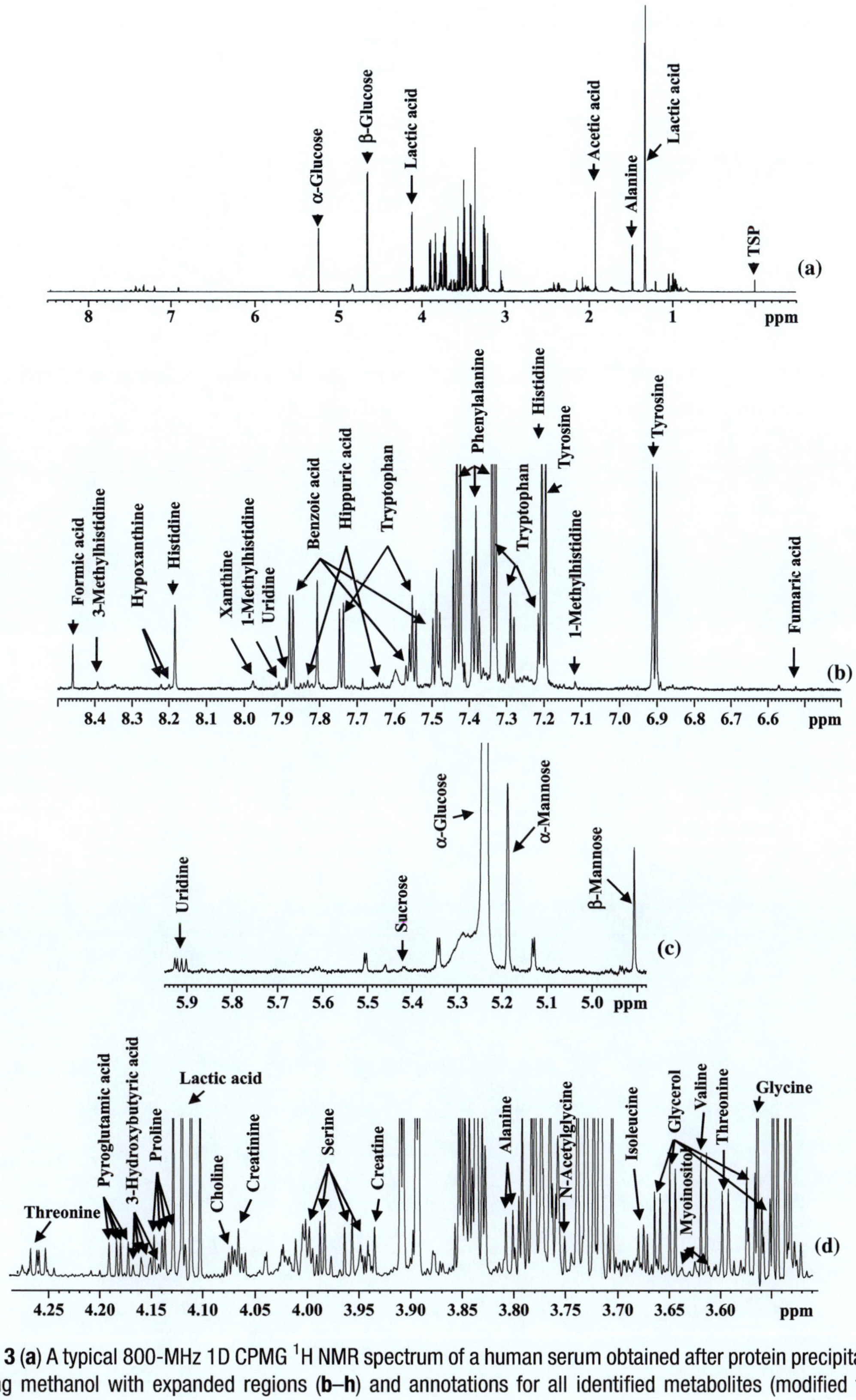

Fig. 3 (**a**) A typical 800-MHz 1D CPMG ^{1}H NMR spectrum of a human serum obtained after protein precipitation using methanol with expanded regions (**b–h**) and annotations for all identified metabolites (modified from ref. 8)

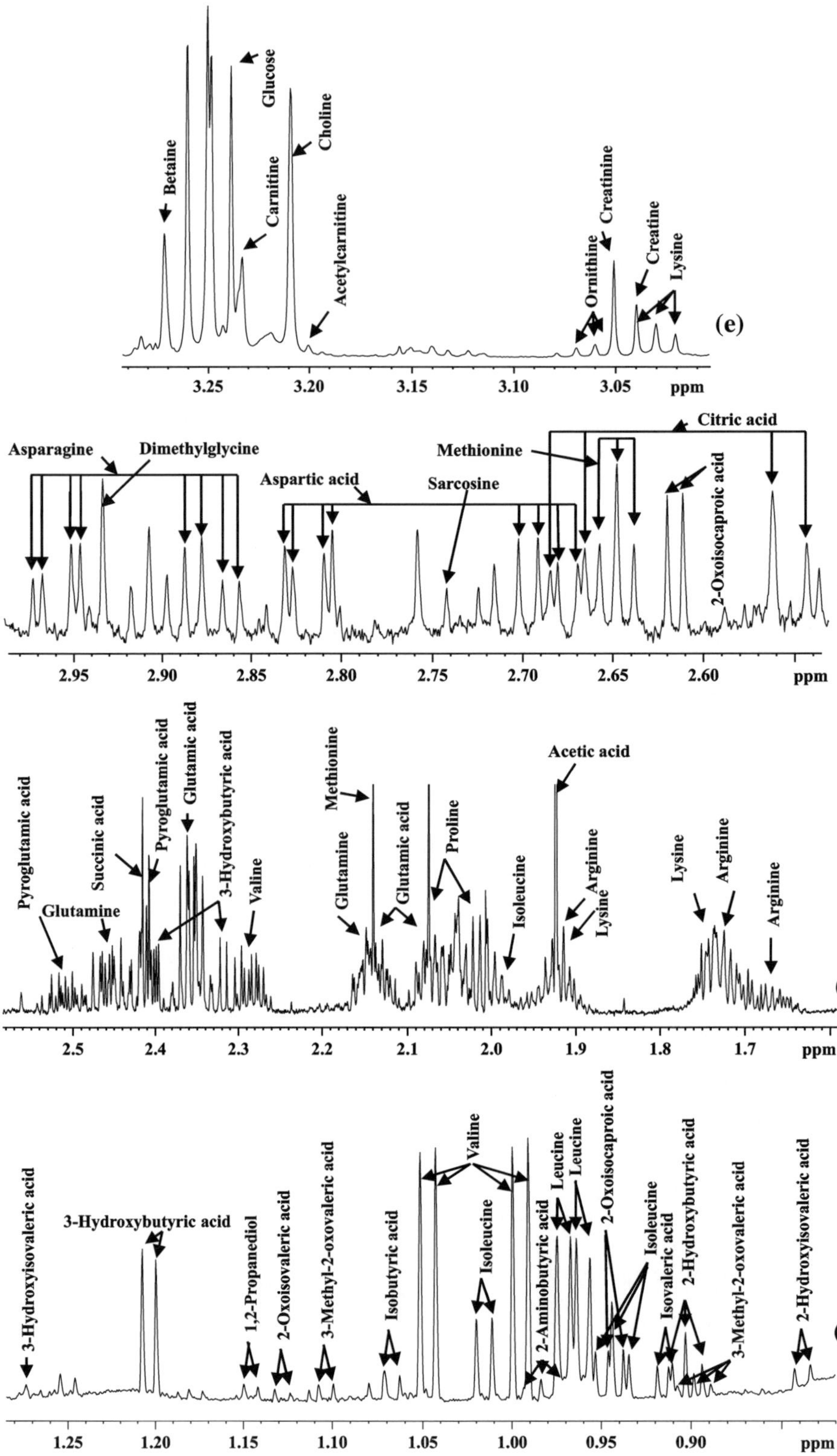

Fig. 3 (continued)

Table 1
^{1}H NMR chemical shifts (in ppm with reference to TSP), J couplings (in Hz) and peak multiplicities for blood plasma/serum metabolites. Chemical shifts for characteristic peaks of metabolites that provide unambiguous information for identification and quantitation using 1D ^{1}H NMR are shown in bold and number of protons (*n*) that each peak(s) represents is also shown in bold (modified from ref. 8)

Metabolite	Chemical shift/J-coupling	Metabolite	Chemical shift/J-coupling
1-Methylhistidine	**7.116** (s; ***n* = 1**), **7.908** (s; ***n* = 1**)	Glutamic acid	2.064 (m), 2.130 (m), **2.357** (m; ***n* = 2**), 3.766 (dd)
1,2-Propanediol	**1.146** (d; *J* = 6.461; ***n* = 3**), 3.457 (dd), 3.545(dd), 3.894(m)	Glutamine	2.145 (m), **2.459** (m; ***n* = 2**), 3.787 (t)
2-Hydroxybutyric acid	**0.903** (t; *J* = 7.459; ***n* = 3**), 4.001 (m)	Glycerol[a]	**3.555** (d; *J* = 6.636; ***n* = 0.42**), **3.570** (d; *J* = 6.636; ***n* = 0.58**), **3.647** (d; *J* = 4.341; ***n* = 0.58**), **3.662** (d; *J* = 4.341; ***n* = 0.42**)
2-Hydroxyisovaleric acid	**0.839** (d; *J* = 6.868)	Glycine	**3.564** (s; ***n* = 2**)
2-Aminobutyric acid	**0.984** (t; *J* = 7.606; ***n* = 3**)	Hippuric acid	**7.641** (t; ***n* = 1**), **7.837** (d; ***n* = 2**)
2-Propanol	**1.177** (d; *J* = 6.160; ***n* = 6**)	Histidine	**7.205** (s; ***n* = 1**), **8.183** (s; ***n* = 1**)
2-Oxoisocaproic acid	**0.941** (d; *J* = 6.633; ***n* = 6**), 2.097 (m), **2.616** (d; *J* = 7.021; ***n* = 2**)	Hypoxanthine	**8.199** (s; ***n* = 1**), **8.221** (s; ***n* = 1**)
2-Oxoisovaleric acid	**1.128** (d; *J* = 7.058; ***n* = 6**)	Isobutyric acid	**1.067** (d; *J* = 6.823; ***n* = 6**)
3-Hydroxybutyric acid	**1.204** (d; *J* = 6.233 Hz; ***n* = 3**), 2.311 (m), 2.414 (m), 4.156 (m)	Isoleucine	**0.944** (t; *J* = 7.485; ***n* = 3**), **1.015** (d; *J* = 6.994; ***n* = 3**), 3.678 (d; *J* = 4.173)
3-Hydroxyisovaleric acid	**1.274** (s; ***n* = 6**)	Isovaleric acid	**0.915** (d; *J* = 6.645; ***n* = 6**)
3-Methylhistidine	**8.391** (s; ***n* = 1**)	Lactic acid	**1.332** (d; *J* = 6.976; ***n* = 3**), **4.115** (q; *J* = 6.976; ***n* = 1**)
3-Methyl-2-oxovaleric acid	**0.899** (t; *J* = 7.518; ***n* = 3**), **1.104** (d; *J* = 6.736; ***n* = 3**)	Leucine	**0.961** (d; *J* = 6.227; ***n* = 3**), **0.972** (d; *J* = 6.050; ***n* = 3**), 1.718 (m), 3.740 (m)
Acetic acid	**1.924** (s; ***n* = 3**)	Lysine	1.452 (m), 1.512 (m), 1.733 (m), 1.913 (m), **3.030** (t; *J* = 7.627; ***n* = 2**)
Acetone	**2.236** (s; ***n* = 6**)	α-Mannose	**5.186** (d; *J* = 1.708; ***n* = 1**)

(continued)

Table 1
(continued)

Metabolite	Chemical shift/J-coupling	Metabolite	Chemical shift/J-coupling
Acetylcarnitine	**3.201** (s; n = **9**)	β-Mannose	**4.907** (d; J = 0.949; n = **1**)
N-Acetylglycine	**3.751** (s; n = **2**)	Methanol	**3.362** (s; n = **3**)
Alanine	**1.485** (d; J = 7.296; n = **3**), 3.805 (q)	Methionine	2.140 (s), **2.648** (t; n = **2**), 3.872 (m)
Arginine	**1.664** (m; n = **1**), **1.731** (m; n = **1**), **1.918** (m; n = **2**), 3.255 (t), 3.775 (t)	Myoinositol	**3.625** (t; n = **2**)
Asparagine[b]	**2.861** (d; J = 7.687; n = **0.38**), **2.883** (d; J = 7.687; n = **0.62**), **2.949** (d; J = 4.294; n = **0.62**), **2.970** (d; J = 4.294; n = **0.38**), 4.015 (dd)	Ornithine	**3.060** (t; J = 7.489; n = **2**)
Aspartic acid[b]	**2.675** (d; J = 8.823; n = **0.42**) **2.697** (d; J = 8.823; n = **0.58**), **2.807** (d; J = 3.713; n = **0.58**), **2.829** (d; J = 3.713; n = **0.42**), 3.912 (dd)	Phenylalanine	**7.333** (d; J = 7.466; n = **2**), **7.381** (t; J = 7.356; n = **1**), **7.432** (t; n = **2**)
Benzoic acid	**7.487** (t; n = **2**), **7.559** (t; n = **1**), **7.875** (d; J = 8.041; n = **2**)	Proline	**4.138** (dd; J = 6.372; J = 8.716; n = **1**)
Betaine	**3.272** (s; n = **9**)	Pyroglutamic acid	2.407 (m), 2.507 (m), **4.182** (dd; J = 5.904; J = 9.081; n = **1**)
Carnitine	**3.233** (s; n = **9**)	Sarcosine	**2.742** (s; n = **3**)
Choline	**3.209** (s; n = **9**), 3.526 (m), **4.070** (m; n = **2**)	Serine	**3.945** (d; J = 5.903; n = **0.33**), **3.960** (d; J = 5.903; n = **0.67**), **3.986** (d; J = 3.542; n = **0.67**), **4.000** (d; J = 3.542; n = **0.33**)
Citric acid	**2.553** (d; J = 15.124; n = **2**), **2.676** (d; J = 15.124; n = **2**)	Succinic acid	**2.415** (s; n = **4**)
Creatine	**3.040** (s; n = **3**), **3.935** (s; n = **2**)	Sucrose	**5.419** (d; n = **1**)

Creatinine	**3.051** (s; ***n*** = **3**), **4.066** (s; ***n*** = **2**)	Threonine	1.337 (d), 3.596 (d; *J* = 4.931), **4.261** (dq; *J* = 4.918; *J* = 6.536; ***n*** = **1**)
Dimethylamine	**2.726** (s; ***n*** = **6**)	Tryptophan	**7.208** (t; ***n*** = **1**), **7.289** (t; ***n*** = **1**), **7.329** (s; ***n*** = **1**), **7.546** (d; ***n*** = **1**), **7.741** (d; *J* = 8.017; ***n*** = **1**)
Dimethylglycine	**2.934** (s; ***n*** = **6**)	Tyrosine	**6.906** (d; *J* = 8.503; ***n*** = **2**), **7.199** (d; *J* = 8.503; ***n*** = **2**)
Ethanol	**1.188** (t; *J* = 7.106; ***n*** = **3**), 3.664 (q)	Urea	**5.787** (br.s; ***n*** = **4**)
Formic acid	**8.459** (s; ***n*** = **1**)	Uridine	4.250 (m), 4.366 (m), **5.906** (d; *J* = 8.010; ***n*** = **1**), **5.923** (d; *J* = 4.532; ***n*** = **1**), **7.886** (d; *J* = 8.010; ***n*** = **1**)
Fumaric acid	**6.524** (s; ***n*** = **2**)	Valine	**0.996** (d; *J* = 7.061; ***n*** = **3**), **1.047** (d; *J* = 7.061; ***n*** = **3**), 2.281 (m), 3.617 (d; *J* = 4.408)
α-Glucose	3.416 (m), 3.539 (m), 3.715 (m), 3.771 (m), 3.842 (m), **5.238** (d; *J* = 3.798; ***n*** = **1**)	Xanthine	**7.977** (s; ***n*** = **1**)
β-Glucose	3.251 (dd), 3.414 (m); 3.490 (m), 3.729 (m), 3.902 (m), **4.652** (d; *J* = 7.989; ***n*** = **1**),		

s singlet, *br.s* broad singlet, *d* doublet, *dd* doublet of doublets, *t* triplet, *dt* doublet of triplets, *q* quartet, *dq* doublet of quartets, *m* multiplet

Note: Metabolites, acetone, ethyl alcohol, and methanol are lost during sample drying after protein precipitation

[a]At 800 MHz: only peak splitting is shown; germinal *J* couplings are not shown

[b]Number of protons refers to 800-MHz NMR spectrum

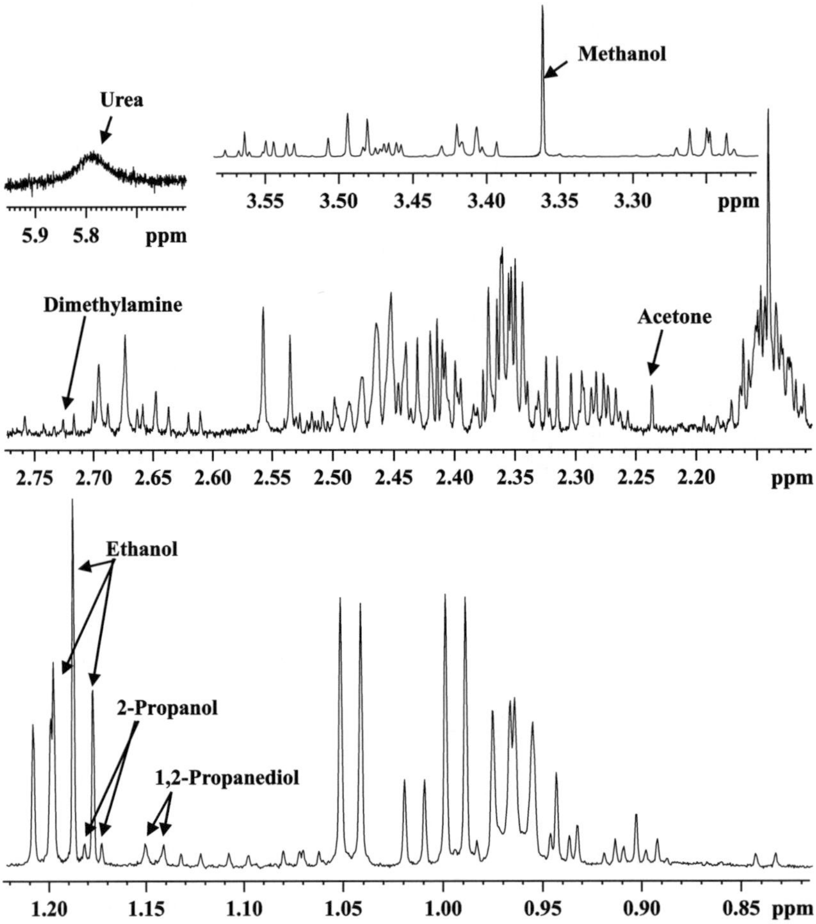

Fig. 4 Portions of the 700-MHz 1D ^{1}H NMR spectrum of a pooled human serum sample obtained after ultrafiltration using a 3-kDa filter with expanded regions highlighting volatile metabolites that were not detected in protein-precipitated serum because of sample drying (*see* Fig. 3) (modified from ref. 8)

methanol become undetectable. To detect such metabolites, the analysis needs to be performed either without drying samples or using ultrafiltration that does not require sample drying (*see* Fig. 4).

A new and alternative approach for analysis of blood metabolites is to use whole blood instead of serum or plasma. Analysis of whole blood metabolites is potentially interesting owing to the significant and alternative value it offers in the investigation of human health and diseases [14–17]. With virtually no additional effort, whole blood analysis provides opportunities to visualize metabolites that come from both blood serum or plasma and blood cells. In particular, blood cells consist of red blood cells (RBC), white blood cells, and platelets; RBC are the major constituents of blood cells and constitute more than 99% of all the blood cells and nearly 50% of the total blood volume. Interestingly, most of the metabolites from whole blood are common to serum and plasma,

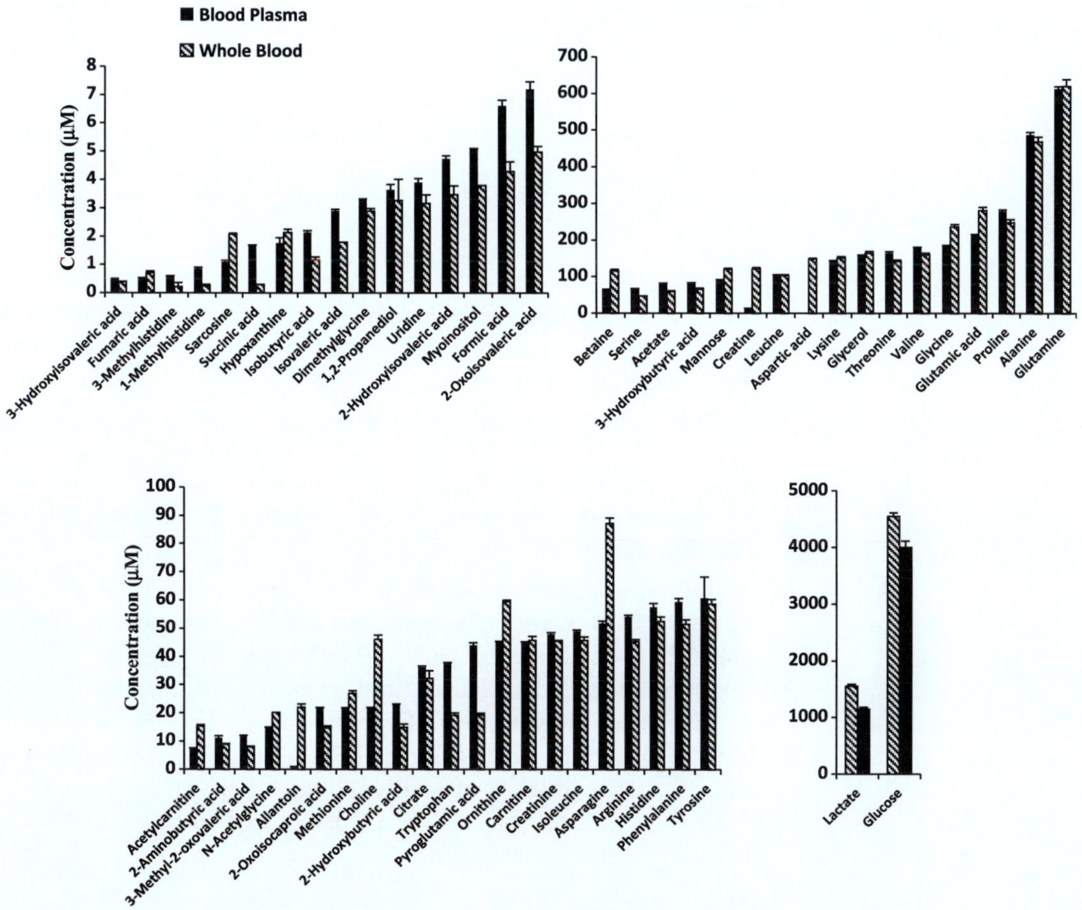

Fig. 5 Comparison of absolute concentrations of metabolites detected in a typical healthy human whole blood with those detected in plasma from the same blood (modified from ref. 18)

qualitatively (Fig. 5); however, whole blood analysis provides information on additional molecules that are important to cellular function [18]. An important characteristic of whole blood analysis is that it provides access to a metabolite pool that is complementary to serum and plasma, and it also avoids the confounding effect of hemolysis, which is often encountered by serum/plasma analysis. Unlike whole blood, obtaining serum or plasma involves processing of whole blood, which carries the risk of hemolysis and metabolites from blood cells seeping into the serum/plasma [15, 19].

This chapter describes protocols for the quantitative analysis of metabolites in serum, plasma, and whole blood using 1D ^{1}H NMR spectroscopy.

2 Materials

2.1 Serum, Plasma, and Whole Blood Samples

Biological safety guidelines must be followed strictly including undergoing blood-borne pathogen training and using appropriate personal protection equipment. Blood samples from overnight fasted individuals are collected following institutional ethical guidelines.

1. Blood serum: blood is collected in a tube designated for serum that has no additives, ideally a glass or plastic "red top" tube. A serum separator tube can also be used. The tube is kept aside for 30–60 min at room temperature for clotting (*see* **Note 1**). Serum is then separated by centrifuging the tube at 1500 × *g* for 15 min at 4 °C. Make serum aliquots of smaller volumes (≤1 mL) (*see* **Note 2**) and store at −80 °C until used for analysis or use directly for analysis when there is no need for storing.
2. Blood plasma: blood is collected in a designated tube with an anticoagulant such as heparin and gently mixed by inverting the tube 5–6 times. Heparin tubes should be used for NMR analysis since, unlike EDTA and citrate tubes, no peaks from heparin interfere with peaks from plasma metabolites. Plasma is then separated by centrifuging the tube at 1500 × *g* for 15 min at 4 °C. Make aliquots of smaller volumes (≤1 mL) and store at −80 °C until used for analysis or use directly for analysis when there is no need for storing.
3. Whole blood: blood is collected in a designated tube with heparin as an anticoagulant and mix gently by inverting the tube 5–6 times. Heparin tubes should be used for NMR analysis since, unlike EDTA and citrate tubes, no peaks from heparin interfere with peaks from blood metabolites. Make aliquots of smaller volumes (≤1 mL) and store at −80 °C until used for analysis or use directly for analysis when there is no need for storing.

2.2 Chemicals and Gases

1. Monosodium phosphate (NaH_2PO_4).
2. Disodium phosphate (Na_2HPO_4).
3. Sodium salt of 3-(trimethylsilyl)propionic acid-2,2,3,3-d_4 (TSP).
4. Sodium azide (NaN_3).
5. Helium gas (*see* **Note 3**).

2.3 Solvents

1. Methanol (analytical grade).
2. Chloroform (analytical grade).
3. Deionized (DI) water.
4. Deuterium oxide (D_2O).

2.4 Phosphate Buffer in D_2O

Buffer solution (100 mM; pH = 7.4) made by dissolving 1124.0 mg anhydrous Na_2HPO_4 and 249.9 mg anhydrous NaH_2PO_4 in 100 g D_2O. Add TSP and sodium azide to achieve a final concentration of 25 μM and 0.01%, respectively (*see* **Note 4**).

2.5 Lab Supplies

1. Pipette (1 mL, 200 μL).
2. Eppendorf tubes (1.5 or 2 mL).

2.6 Lab Equipment

1. Vortex mixer.
2. −20 °C refrigerator.
3. −80 °C freezer.
4. Vacuum concentrator/freeze drier.

2.7 NMR Instrument and Accessories

1. High-resolution NMR spectrometer (*see* **Note 5**).
2. NMR data processing software (*see* **Note 6**).
3. 5-mm or 3-mm NMR tubes (*see* **Note 7**).

3 Methods

3.1 Protocol to Remove Proteins from Blood Serum or Plasma (See Note 8)

1. Thaw blood serum or plasma samples at room temperature, if they were stored in the freezer prior to analysis.
2. Mix serum or plasma with methanol in the ratio 1:2 (v/v) (*see* **Note 9**), vortex for 30 s, and incubate at −20 °C for 20 min.
3. Centrifuge the mixture at ~13,000 × *g* for 30 min to pellet proteins (*see* **Note 10**).
4. Transfer the supernatant (solvent layer) to a fresh vial and dry using vacuum concentrator for 5 h to remove solvents (*see* **Note 11**).
5. Mix the dried samples with 200 or 600 μL D_2O buffer containing the internal reference (TSP).
6. Centrifuge the samples ~13,000 × *g* for 5 min to sediment any residue, and transfer the supernatant to NMR tube for analysis (*see* **Note 12**).

3.2 Protocol for Extraction of Metabolites from Whole Blood

1. Thaw whole blood samples at room temperature, if they were stored in the freezer prior to analysis.
2. Approximately 200–400 μL whole blood is mixed with cold methanol and chloroform in a 1:2:2 sample:methanol:chloroform (v/v/v) ratio (*see* **Note 13**), vortexed for 30 s, sonicated for 2 min at 4 °C, and incubated at −20 °C for 20 min. The mixture is centrifuged at ~13,000 × *g* for 30 min to pellet proteins and cell debris. The clear aqueous solution is transferred to a fresh vial (*see* **Note 14**) and dried using a vacuum concentrator for 5 h (*see* **Note 11**).

3. Dried samples are mixed with 200 or 600 μL phosphate buffer containing 25 μM TSP, spun to sediment any residue, and the supernatants transferred to 3- or 5-mm NMR tubes for analysis (*see* **Note 15**).

3.3 NMR Experiments: Shimming Using Standard Sample

1. Set NMR probe temperature to 298 K.
2. Insert a standard sample consisting of a mixture of H_2O and D_2O (90:10 v/v).
3. Tune the probe.
4. Read most recent shim file for the current probe.
5. Perform field-frequency locking using D_2O signal.
6. Perform 1D gradient shimming (*see* **Note 16**).
7. Perform automated 3D gradient shimming and then redo 1D gradient shimming.
8. Save shims as the most recent shim values.
9. Remove the standard sample from the magnet.

3.4 NMR Experiments: Extracts of Serum, Plasma, and Whole Blood

1. Insert sample into the magnet.
2. Create 1D NMR data parameter file for ^1H acquisition.
3. Perform probe tuning.
4. Read the most recent shim file.
5. Perform field-frequency locking using D_2O signal.
6. Perform 1D gradient shimming.
7. Calibrate 90° pulse.
8. Set pulse sequence: use the CPMG pulse sequence modified for residual water suppression using presaturation.
9. Set parameters: temperature, 298 K; spectral width, 10,000 Hz; recycle delay, 6 s; number of transients, 128 transients; time domain points (TD), 32 K (*see* **Note 17**).
10. Adjust offset frequency (O1) to the center of water signal.
11. Adjust receiver gain.
12. Acquire data.
13. Perform Fourier transformation after multiplying the FID (free induction decay) by exponential multiplication with a line broadening typically of 0.3–0.5 Hz.
14. Perform phase correction and baseline correction, and calibrate chemical shift scale by setting reference (TSP) peak to 0 ppm.

3.5 Metabolite Identification and Quantitation

1. Identify characteristic peaks for metabolites. For unambiguous identification, refer to the visual aids in Fig. 3 and the individual chemical shifts in Table 1.

2. Determine the area of characteristic peaks (Table 1) for metabolites relative to the area of the reference (TSP) peak.
3. Peak areas for metabolites from multiple NMR spectra can be obtained manually after converting processed NMR spectra into ASCII format or using automation software such as AMIX.
4. Determine absolute concentration of metabolite (C_m) using the following equation:

$$C_m = \frac{I_m n_r C_r}{I_r n_m}$$

where C_r is concentration of the reference compound (TSP) in the solution; I_m and I_r are the integrated peak areas of metabolite and reference (TSP), respectively; and n_m and n_r are number of protons that represent the metabolite and reference peaks, respectively (*see* **Note 18**).

5. Alternatively, use automation softwares such as Chenomx and MetaboAnalyst that enable obtaining metabolite concentrations for single or multiple spectra based on the concentration of the reference standard, TSP.

3.6 Data Analysis

Whole NMR spectra, relative peak integrals, or absolute concentrations of metabolites from multiple samples are subjected to statistical analyses such as univariate and multivariate analyses. Such analysis follows two major approaches: chemometric analysis and quantitative analysis [20].

Chemometric analysis: Chemometric analysis is a global method, in which the spectral data are used directly for multivariate statistical analysis without the knowledge of peak identification. Prior to the analysis, the data are generally subjected to preprocessing such as phase and baseline correction, peak alignment, and solvent peak removal. Distinguishing spectral features from the chemometric analysis are then identified based on established assignments or using additional NMR experiments and databases. A challenge for the chemometric approach is that imperfect peak alignments and spectral baselines can lead to results that often lack robustness.

Quantitative analysis: In this approach metabolites are first identified and the relative or absolute concentrations are then obtained before subjecting these data to multivariate statistical analysis or pathway analysis. A major benefit of this approach compared to the global chemometric analysis is that it can reduce potential errors arising from factors such as baseline distortions, strong solvent signals, and peak misalignments. The quantitative metabolomics approach is thus more often used for reliable insights into the mechanistic understanding of cellular function and associated diseases.

4 Notes

1. Samples kept aside for less than 30 min can have unwanted cellular components and those that are kept for longer than an hour can have contamination from cell lysis and glycolysis products.
2. Aliquots of smaller volume help reduce the number of freeze and thaw cycles for later use.
3. Nitrogen gas can be used instead of helium gas; however, oxygen contamination, if any, can deleteriously affect the stability of coenzymes.
4. The pH of the buffer solution should be 7.4. However, a variation of pH by about ~0.1 does not alter the chemical shifts appreciably.
5. A spectrometer frequency of 500 MHz or higher is preferred owing to the closely spaced chemical shifts for many metabolites.
6. NMR instrument vendors such as Bruker provide data processing software (Topspin NMR in this case) free of cost for academic use.
7. Either 5-mm or 3-mm NMR tubes can be used for metabolite analysis; however, use of 3-mm NMR tubes helps improve both resolution and sensitivity significantly. If the NMR instrument is not equipped with a 3-mm probe, experiments using 3-mm NMR tubes can still be performed using a 5-mm NMR probe with superior results.
8. This procedure is the same for blood serum and plasma.
9. The volume of serum or plasma used can vary depending on need and availability. Typically 200 μL is used for a 3-mm NMR tube and 400 μL for a 5-mm NMR tube. If the sample volume is limited (<200 μL), using a 3-mm NMR tube is recommended.
10. Both speed and time are not magic numbers. However, higher speed and longer centrifugation time help proteins pellet better.
11. Actual drying time depends on the condition and efficiency of the drier. Some driers dry samples faster than others. Make sure that samples are dry and no solvent is left in the sample tube. If not dried properly, you will see a strong peak for methanol in the NMR spectrum.
12. 200 μL is used for 3-mm NMR tubes and 600 μL is used for 5-mm NMR tubes.
13. Methanol alone is used for protein precipitation in serum or plasma. For whole blood, addition of chloroform helps prevent oxidation of coenzymes and helps the extraction process.

14. Cell debris and precipitated protein lie in between the top aqueous layer and bottom chloroform layer.
15. If the sample is dissolved in 200 μL buffer, use a 3-mm NMR tube. However, if the sample is dissolved in 600 μL buffer, use a 5-mm NMR tube to transfer the solution. Use of 3-mm NMR tubes is recommended since they provide significantly higher sensitivity as well as better resolution.
16. If the results of 1D gradient shimming is good (final standard deviation of B_o typically ~0.2 Hz), there is no need for 3D shimming. Skip Subheading 3.3, **step 7**.
17. These are typical parameters. The parameters can be altered depending on the sample amount and the requirement of specific resolution and quantitation accuracy.
18. For quantitation purposes the number of protons (n) for each of the characteristic metabolite peaks is given in Table 1. The number of protons for TSP is 9.

Acknowledgements

We acknowledge financial support from the NIH (National Institute of General Medical Sciences 2R01GM085291).

References

1. Beckonert O, Keun HC, Ebbels TM, Bundy J, Holmes E, Lindon JC et al (2007) Metabolic profiling, metabolomic and metabonomic procedures for NMR spectroscopy of urine, plasma, serum and tissue extracts. Nat Protoc 2(11):2692–2703
2. Psychogios N, Hau DD, Peng J, Guo AC, Mandal R, Bouatra S et al (2011) The human serum metabolome. PLoS One 6(2):e16957. https://doi.org/10.1371/journal.pone.0016957
3. Nagana Gowda GA, Raftery D (2013) Biomarker discovery and translation in metabolomics. Curr Metabolomics 1(3):227–240
4. Nicholson JK, Gartland KP (1989) 1H NMR studies on protein binding of histidine, tyrosine and phenylalanine in blood plasma. NMR Biomed 2(2):77–82
5. Chatham JC, Forder JR (1999) Lactic acid and protein interactions: implications for the NMR visibility of lactate in biological systems. Biochim Biophys Acta 1426(1):177–184
6. Bell JD, Brown JC, Kubal G, Sadler PJ (1988) NMR-invisible lactate in blood plasma. FEBS Lett 235:81–86
7. Nagana Gowda GA, Raftery D (2014) Quantitating metabolites in protein precipitated serum using NMR spectroscopy. Anal Chem 86(11):5433–5440
8. Nagana Gowda GA, Gowda YN, Raftery D (2015) Expanding the limits of blood metabolite quantitation using NMR spectroscopy. Anal Chem 87(1):706–715
9. Wevers RA, Engelke U, Heerschap A (1994) High-resolution 1H-NMR spectroscopy of blood plasma for metabolic studies. Clin Chem 40(7 Pt 1):1245–1250
10. Daykin CA, Foxall PJ, Connor SC, Lindon JC, Nicholson JK (2002) The comparison of plasma deproteinization methods for the detection of low-molecular-weight metabolites by (1)H nuclear magnetic resonance spectroscopy. Anal Biochem 304(2):220–230
11. Tiziani S, Emwas AH, Lodi A, Ludwig C, Bunce CM, Viant MR, Günther UL (2008) Optimized metabolite extraction from blood serum for 1H nuclear magnetic resonance spectroscopy. Anal Biochem 377(1):16–23
12. Fan TW (2012) In: Fan TW, Higashi RM, Lane AN (eds) The handbook of metabolomics,

Methods in pharmacology and toxicology. Springer, New York, pp 7–27

13. Bouatra S, Aziat F, Mandal R, Guo AC, Wilson MR, Knox C et al (2013) The human urine metabolome. PLoS One 8(9):e73076. https://doi.org/10.1371/journal.pone.0073076
14. Chaleckis R, Murakami I, Takada J, Kondoh H, Yanagida M (2016) Individual variability in human blood metabolites identifies age-related differences. Proc Natl Acad Sci U S A 113(16):4252–4259
15. Stringer KA, Younger JG, McHugh C, Yeomans L, Finkel MA, Puskarich MA et al (2015) Whole blood reveals more metabolic detail of the human metabolome than serum as measured by 1H-NMR spectroscopy: implications for sepsis metabolomics. Shock 44(3):200–208
16. Chaleckis R, Ebe M, Pluskal T, Murakami I, Kondoh H, Yanagida M (2014) Unexpected similarities between the Schizosaccharomyces and human blood metabolomes, and novel human metabolites. Mol BioSyst 10(10):2538–2551
17. Catalán Ú, Rodríguez MÁ, Ras MR, Maciá A, Mallol R, Vinaixa M et al (2013) Biomarkers of food intake and metabolite differences between plasma and red blood cell matrices; a human metabolomic profile approach.Solà, R. Mol BioSyst 9(6):1411–1422
18. Nagana Gowda GA, Raftery D (2017) Whole blood metabolomics by 1H NMR spectroscopy provides a new opportunity to evaluate coenzymes and antioxidants. Anal Chem 89(8):4620–4627
19. Gorman MW, Feigl EO, Buffington CW (2007) Human plasma ATP concentration. Clin Chem 53(2):318–325
20. Djukovic D, Nagana Gowda GA, Raftery D (2013) Mass spectrometry and NMR spectroscopy-based quantitative metabolomics. In: Issaq HJ, Veenstra TD (eds) Proteomic and metabolomic approaches to biomarker discovery. Elsevier, New York, pp 279–297

Chapter 3

Lipid Profiling Using ^{1}H NMR Spectroscopy

Miriam Gil, Sara Samino, Rubén Barrilero, and Xavier Correig

Abstract

Lipid profiling, which includes fatty acids, phospholipids, glycerides, and cholesterols is extremely important because of the essential role lipids play in the regulation of metabolism in animals. ^{1}H-NMR-based protocols for high-throughput lipid analysis in complex mixtures have been developed and applied to biological systems. Many classes of lipids can be quantitatively analyzed in many sample matrices including serum, cells, and tissues using a simple ^{1}H NMR experiment. In this chapter, we provide protocols for NMR-based lipid profiling including sample preparation, NMR experiments, and quantification using the LipSpin software tool.

Key words Lipid profiling, ^{1}H NMR spectroscopy, Lipid extracts

1 Introduction

Lipids are essential to such cellular functions as membrane composition and anchoring, signalling, cell metabolism, and energy storage [1]. The vast number of lipid species [2] includes a variety of fatty acids and their derivatives (including tri-, di-, and monoglycerides and phospholipids) and other sterol-containing lipids such as cholesterol. Lipids play an essential role in homeostatic processes and disease states. Lipidomics, a branch of metabolomics, deals with the large-scale analysis of many lipid classes in biological matrices [3]. Analysis of lipids is important for characterization of metabolic diseases such as diabetes and obesity, atherosclerosis and cardiovascular diseases, and cancer and inflammatory diseases [4]. Lipid levels can be used to improve risk prediction models, as a diagnostic, prognostic, or disease severity biomarker [5]. They can also be applied to assess the health benefits of diets and nutritional supplements [6] and to monitor the effects of physical exercise [7] or drug therapies in clinical trials [8].

In the area of biomedicine, currently, lipidomics is applied to the analysis of biofluids (such as blood serum/plasma), cells, and tissues. Mass spectrometry (MS) is the main analytical technique

G. A. Nagana Gowda and Daniel Raftery (eds.), *NMR-Based Metabolomics: Methods and Protocols*, Methods in Molecular Biology, vol. 2037, https://doi.org/10.1007/978-1-4939-9690-2_3, © Springer Science+Business Media, LLC, part of Springer Nature 2019

used in lipidomics because it is highly sensitive and hence it can detect a vast number of lipid species. The conventional lipid analysis method involves lengthy sample processing and separation using chromatography before analysis [9]. Recent advances in MS that use, for example, differential mobility spectrometry (DMS) enable analysis of several hundreds of lipids without the need for chromatography. This new method uses multiple internal standards for quantitation, and tedious sample processing prior to the analysis is still required in this approach. An alternative technique and an altogether different approach for lipid profiling that is well suited for high-throughput metabolomics applications is proton nuclear magnetic resonance (^{1}H NMR) spectroscopy [10]. This method is highly quantitative and reproducibfle. A single internal standard is sufficient for quantitative analysis of all lipids owing to the linear response of the NMR signals of lipids. It is also very robust and reliable, and hence it makes an ideal method for clinical applications. Owing to the intrinsic nature of the technique, NMR enables analysis of several major classes of lipids based on their characteristic functional groups.

The proprietary method of Ala-Korpela et al. can estimate lipids in fourteen lipoprotein subfractions [10] and fatty acids in serum/plasma without extraction. Some advanced lipoprotein tests on serum/plasma are being or have already been successfully commercialized [11–13]. However, this approach cannot quantify the total amount of lipid species in serum/plasma independently of the lipoproteins. To achieve this, an extraction step prior to ^{1}H NMR is a prerequisite for lipid profiling in complex mixtures. Using a biphasic system in which low molecular weight metabolites and proteins reside in the aqueous phase and lipids are dissolved in the organic phase, lipids can be extracted by simple phase separation [14]. The gold standard protocol for lipid extraction is the Folch method [15], initially developed to extract lipids from animal tissues.

In this chapter we describe a modified Folch method optimized to extract lipids from cells and tissue. In particular, as a result of the efforts made to develop efficient methods for extracting lipids from serum, Lögfren et al. [16] introduced the BUME method based on butanol and methanol for lipid analysis using LC-MS. This BUME method was improved so that it was suitable for NMR analysis. Specifically, heptane was replaced with diisopropyl ether, since the ^{1}H NMR spectrum of lipids is affected by solvents used in the extraction procedure and heptane gives rise to a signal in the methyl region that overlaps with the fatty acid signal and makes them difficult to quantify. In this chapter, we provide lipid extraction procedures for serum samples, cell cultures, and animal tissues.

An important aspect of lipid profiling is to make unambiguous peak identification and interpretation of the spectra. Analysis of lipid species in ^{1}H NMR spectra is very challenging because similarity in lipid structure, specifically of the fatty acid chain and unsaturation, causes severe signal overlap. To overcome this challenge and extract quantitative data from ^{1}H NMR spectra of lipids, we developed the LipSpin software. This software tool can automatically profile severely overlapped spectral regions and lipid signals with complex coupling patterns [17], as described later.

2 Materials

Prepare all solutions for lipid extraction and lipid reconstitution and store them at room temperature unless otherwise specified. The lipid extraction procedure does not require reagents of the highest purity. Diligently follow all waste disposal regulations.

2.1 Lipid Extraction

1. Lipid extraction of serum.
   (a) Solvent 1: butanol:methanol (3:1, v/v).
   (b) Solvent 2: diisopropyl ether:ethyl acetate (3:1, v/v). Store solvent 2 at 4 °C until needed.
   (c) Buffer: deionized water.
   (d) Glass vials with a capacity of 2 mL (*see* **Note 1**).
2. Lipid extraction of cell culture and tissue.
   (a) Solvents: dichloromethane, methanol, deionized water, phosphate buffer (PBS).
   (b) Trypsin.
   (c) Glass vials with a capacity of 2 mL (*see* **Note 1**).

2.2 Lipid Reconstitution

1. Deuterated solvent: chloroform-d:methanol-d_4:deuterium oxide (16:7:1, v/v/v) containing 2 mM tetramethylsilane (TMS). Use clean glass material to prepare the solution. Store the solution at 4 °C until needed.
2. 5-mm NMR tubes and caps without hole.

2.3 NMR Spectroscopy

1. Chloroform-d, methanol-d_4, deuterium oxide.
2. Tetramethylsilane (TMS).

2.4 Software

1. Matlab (ver. 7.5.0. The MathWorks, Inc., Natick, MA, USA).

3 Methods

3.1 Sample Preparation

Carry out all procedures at room temperature unless otherwise specified.

3.1.1 Lipid Extraction from Serum Samples

The following steps should be performed in two replicates for each serum sample (*see* **Note 2**):

1. Place 100 μL of serum in 2-mL glass vials.
2. Add 300 μL of solvent 1. Mix by vortexing. A white pellet is formed.
3. Add 300 μL of solvent 2 to the mixture. Mix by vortexing.
4. Add 300 μL of buffer to the mixture. Mix by vortexing.
5. Centrifuge the mixture at 1500 × *g* for 1 min at 15 °C (*see* **Note 3**).
6. Transfer 360 μL of the upper phase containing lipids into a new glass vial.

   After this step, perform two more rounds of extraction as described in the following and pool all lipid layers together for better recovery of lipids.
7. Add 300 μL of solvent 2 to the mixture (lower phase from **step 6**). Mix by vortexing.
8. Centrifuge the mixture at 22,600 × *g* for 1 min at 15 °C (*see* **Note 3**).
9. Transfer 320 μL of the upper phase into the previous vial (pool with upper phase from **step 6**).
10. Add 250 μL of solvent 2 to the mixture. Mix by vortexing.
11. Centrifuge the mixture at 1500 × *g* for 1 min at 15 °C (*see* **Note 3**).
12. Transfer 200 μL of the upper phase into the previous glass vial (pool with upper phase from **step 6**).
13. Dry the pooled lipid layer under a stream of N_2 gas to remove organic solvents (*see* **Note 4**).

3.1.2 Lipid Extraction from Cell Culture Samples

1. Dispense trypsin into culture vessel(s) and incubate the cells.
2. Transfer the whole cell suspension into a fresh tube.
3. Centrifuge them (*see* **Note 5**).
4. Remove the supernatant (trypsin).
5. Add PBS solution to wash cell pellet.
6. Centrifuge again (*see* **Note 5**).
7. Remove the supernatant.

8. Washed cell pellet is subjected to lipid extraction as described from **step 9** (*see* **Note 6**).
9. To the cell pellet, add 220 μL of methanol followed by 440 μL of dichloromethane.
10. Mix vigorously for 30 s.
11. Ultrasonicate the sample on ice for 1 min.
12. Add 140 μL of deionized water.
13. Mix the sample vigorously for 30 s.
14. Incubate samples on ice for 30 min (*see* **Note 7**).
15. Centrifuge the sample at 22,600 × *g* and 4 °C for 15 min (*see* **Note 8**).
16. Transfer 400 μL of lower phase containing lipids into a new glass vial.
17. Remove the organic solvent from the vial by drying under a stream of N_2 gas.

3.1.3 Lipid Extraction of Tissue Samples (See Note 9)

1. Place tissue sample in 2-mL Eppendorf tube and immediately freeze it.
2. Lyophilize samples overnight (*see* **Note 10**).
3. Place ~20 mg of dry tissue into a new glass vial.
4. Add 220 μL of methanol followed by 440 μL of dichloromethane.
5. Vortex samples vigorously for 30 s.
6. Place samples in an ultrasound ice bath for 5 min.
7. Add 140 μL of deionized water.
8. Mix samples vigorously for 30 s.
9. Centrifuge samples at 22,600 × *g* and 4 °C for 15 min.
10. Transfer 400 μL of lower phase containing lipids into a new glass vial.
11. Remove the organic solvent from the vial by drying under a stream of N_2 gas.

3.1.4 Reconstitution of Lipids

Deuterated solvent containing TMS reference is used for reconstituting dried lipid extracts from biological specimens. While the deuterated solvent provides lock and avoids strong solvent signal in the spectrum, peak from TMS serves as chemical shift reference (0 ppm). For lipid analysis the following steps are performed:

1. Add 700 μL of chloroform-d:methanol-d_4:deuterium oxide (16:7:1, v/v/v) containing 2 mM tetramethylsilane (TMS) to glass vials containing dried lipid extracts (*see* **Notes 11** and **12**).

2. Immediately transfer the solution to 5-mm NMR tubes and cover them with caps without holes (*see* **Note 13**).

3.2 ^{1}H NMR Data Acquisition

1. Acquire the ^{1}H NMR spectra using an Avance III-600 Bruker spectrometer or similar NMR spectrometer equipped with a 5-mm triple-resonance pulsed-field gradient cryoprobe (e.g., CPTCI Bruker probe).
2. Use a 90° pulse with water presaturation sequence (Bruker: zgpr).
3. Perform the measurements at 286 K to shift the residual water signal away from lipid peaks (to around 4.51 ppm).
4. Set the relaxation delay to 5 s. During the relaxation delay the water signal is irradiated with a low-power RF pulse.
5. Set the acquisition time at 3 s. The free induction decays (FIDs) consist of 64-k complex data points, leading to a spectral width of 18.6 ppm. For each spectrum, 128 scans are accumulated, which results in a total acquisition time of 17 min per sample.

3.3 Lipid Profiling and Quantification

LipSpin, an in-house developed software package based on Matlab, is a graphical user interface (GUI) software for quantitative profiling of lipids using the ^{1}H NMR spectra of lipid extracts in a user-friendly manner. LipSpin implements the spectral analysis workflow as depicted in Fig. 1. The program source code and user manual can be freely downloaded from https://github.com/rbarri/LipSpin.

3.3.1 Data Importation

1. Raw FIDs or 1D spectra from Topspin (or other third-party software exported in Bruker format) can be imported with LipSpin. When FIDs are used, the LipSpin importer requires the directory containing the NMR data for all samples starting from the experiment number (Bruker NMR file folder format).
2. If you use the FID for analysis, zero-fill to 128-k real data points and apodize using an exponential window function with 0.3-Hz line broadening prior to Fourier transformation. If you use processed 1D spectra as the input mode, the processing number must be provided (Bruker NMR file folder format).

3.3.2 Phase Correction

Proceed to correct the phase of the NMR spectra in order to set all spectral lines in pure absorptive mode. LipSpin provides two different options to correct the zero- and first-order phases. One option seeks to maximize the entropy of the spectrum [18]. The second option is based on a least-squares fitting, which minimizes residuals between a horizontal line (i.e., a flat baseline), and a set of user-defined spectral regions, in which no signals are expected. If the spectral baseline has drifts or rolls, the baseline correction tool

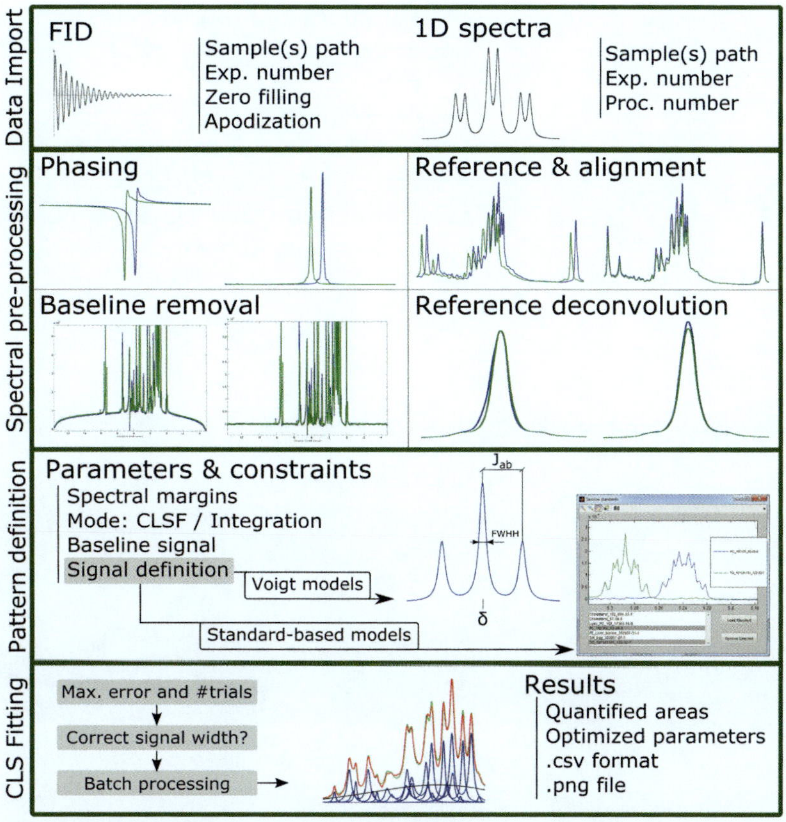

Fig. 1 Spectral analysis workflow in LipSpin (reproduced with permission from the American Chemical Society © 2018) [17]

removes them using interpolating functions (cubic spline, cubic Hermite, or polynomials) to a set of user- or automatically defined points. As an example, in the spectra of serum lipid extracts, spectra can be phase-corrected by flattening the void spectral regions: 9 to 7.8, 6.8 to 5.6, 0.5 to 0.2, and −0.2 to −1 ppm. Adjust the baseline using cubic Hermite interpolation with automatic detection of baseline points.

3.3.3 Spectra Reference

Set the chemical shift of the reference, TMS, peak to 0 ppm to calibrate the chemical shift scale.

3.3.4 Peak Area Determination: Pattern Definition

1. Deconvolution is a method that determines peak areas by fitting lines under spectral regions. Every spectral region fitting requires a set of parameters: spectral margins, baseline signal (to compensate for residual baseline distortions or spurious broad signals), and model definitions of underlying signals. Individual signals can be defined using either Voigt profiles (combination of Lorentzian and Gaussian functions) or spectral templates from the spectra of lipid standards (*see* **Note 14**).

Initial values and fitting constraints must be provided for each peak pattern (chemical shift, line width, Gaussian ratio, and J-coupling). Alternatively, bucket integration can also be used to determine signal area, in which case only spectral margins are considered and the sum of the intensity points within the region is computed instead. Bucket integration should be restricted to isolated signals without baseline artefacts to guarantee valid quantitative results.

2. If the line shape fitting algorithm is chosen, assign the lipid signals and choose constraints as in Fig. 2 and Table 1.

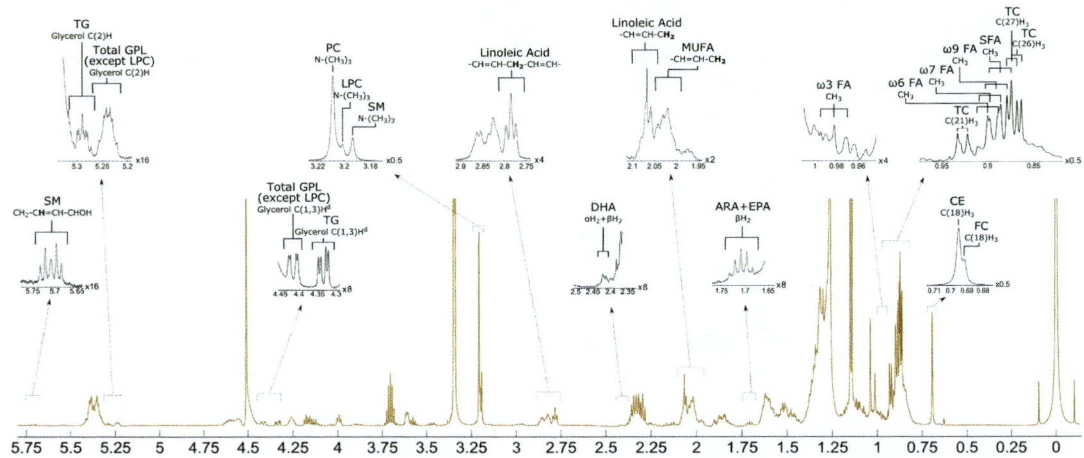

Fig. 2 ^{1}H NMR spectrum of lipophilic extract of plasma with labelling of signals of lipids used for quantification with LipSpin in a nutritional study. Other regions not included in the analysis but typically used in lipid analysis to estimate FA chain length and number of unsaturated bonds are [17] methylene protons at 1.27 ppm, diallylic protons at 2.84 ppm (PUFA other than linoleic acid), and olefinic protons at 5.4 ppm (double bonds) (reproduced with permission from the American Chemical Society © 2018) [17]

Table 1
Spectral regions and signals included in the signal patterns (reproduced with permission from the American Chemical Society © 2018)

Signal pattern (ppm)	Signal assignment[a]	Model	Protons	Chemical shift (ppm)	J-coupling (Hz)
C18 cholesterol (0.73–0.65)	**FC**	**Voigt**	**3**	**0.694**	s
	CE	**Voigt**	**3**	**0.691**	s
FA methyl[b] (0.95–0.85)	Chol C26	Voigt	3	0.868	d (6.6)
	Chol C27	Voigt	3	0.873	d (6.6)
	SFA	**Voigt**	**3**	**0.886**	**t (6.9)**
	ω-9	**Voigt**	**3**	**0.888**	**t (6.9)**
	ω-7	**Voigt**	**3**	**0.898**	**t (6.9)**
	ω-6	**Voigt**	**3**	**0.900**	**t (6.9)**
	Chol C21	Voigt	3	0.922	d (6.6)

(continued)

Table 1
(continued)

Signal pattern (ppm)	Signal assignment[a]	Model	Protons	Chemical shift (ppm)	J-coupling (Hz)
FA ω-3 methyl (1.01–0.95)	**ω-3**	**Voigt**	**3**	**0.982**	**t (7.2)**
	FC	Standard	–	0.982	m
	CE	Standard	–	0.982	m
FA ARA+EPA β-methylene (1.75–1.66)	**EPA**	**Standard**	**2**	**1.697**	**m**
	ARA	**Standard**	**2**	**1.699**	**m**
FA allylic[b] (2.10–1.93)	Chol	Standard	0.5	1.973	m
	Chol	Standard	0.5	2.001	m
	Chol	Standard	0.5	2.018	m
	MUFA	**Standard**	**4**	**2.024**	**m**
	Chol	Standard	0.5	2.040	m
	Linoleic	Voigt	4	2.063	qua (7.05)
FA DHA α-methylene (2.45–2.40)	**DHA**	**Standard**	**4**	**2.420**	**m**
FA diallylic (2.94–2.70)	**Linoleic**	**Voigt**	**2**	**2.783**	**t (7.0)**
	PUFA	Voigt	–	2.817	t (7.0)
	PUFA	Voigt	–	2.827	t (7.0)
	PUFA	Voigt	–	2.854	t (7.0)
	PUFA	Voigt	–	2.862	t (7.0)
PE alkyl (3.18–3.12)	**PE[c]**	**Standard**	**2**	**3.152**	**m**
PL cholines (3.25–3.18)	**SM**	**Voigt**	**9**	**3.192**	**s**
	LPC	**Voigt**	**9**	**3.201**	**s**
	PC	**Voigt**	**9**	**3.208**	**s**
Glycerol backbone sn-1,3[d] (4.45–4.31)	**TG**	**Standard**	**2**	**4.332**	**m**
	GPL (except LPC)	**Standard**	**2**	**4.412**	**m**
Glycerol backbone sn-2 (5.33–5.15)	**TG**	**Standard**	**1**	**5.237**	**m**
	GPL (except LPC)	**Standard**	**1**	**5.288**	**m**
PL olefinic (6–5.65)	**SM**	**Standard**	**1**	**5.708**	**m**
	PLA[c]	**Standard**	**1**	**5.922**	**m**

Key: *FC* free cholesterol, *EC* ester cholesterol, *FA* fatty acids, *SFA* saturated fatty acids, *EPA* eicosapentaenoic acid, *ARA* arachidonic acid, *DHA* docosahexaenoic acid, *MUFA* monounsaturated fatty acids, *PUFA* polyunsaturated fatty acids, *PE* phosphatidylethanolamine, *SM* sphingomyelin, *LPC* lysophosphatidylcholine, *PC* phosphatidylcholine, *TG* triglycerides, *GPL* glycerophospholipids, *PLA* plasmalogen, *s* singlet, *d* doublet, *t* triplet, *m* multiplet

[a]Signals used for lipid quantifications are given in bold

[b]BUME extraction could leave solvent residues: butanol triplet (7.35) at 0.929 ppm, diisopropyl ether doublet (6.7) at 0.8647 ppm, and ethyl acetate singlet at 2.066 ppm

[c]Rarely observed in ^{1}H NMR spectra of human serum and plasma samples

[d]The water suppression used in our NMR experiments affected signal intensities of glycerol sn-1,3 so signals from glycerol backbone sn-2 were used to quantify TG and GPL

3. If bucket integration is chosen, use the option "Correct signal width by" to compensate for line width variations between different spectra, due to differences in shimming and viscosity, by taking the line width from a previously quantified signal as a reference (normally a well-resolved singlet or solvent signal).
4. Provide a valid "Figures path" and a .png file will be created for every quantified region and sample, so that you can visually inspect the performance of the algorithm.
5. Finally, the result chart displays the quantified signal areas, which can be exported in .csv format for further analysis using the spreadsheet functions or other statistical tools.
6. After quantification, signal areas (in arbitrary units) can be converted into molar concentrations (mM), for example, by using an internal standard of known concentration [19, 20] or a calibrated synthetic signal introduced in the spectrum [20] or by normalization based on external calibration. The latter option is recommended as it corrects the variability of recovery volumes in manual extractions and compensates to avoid errors caused by the high volatility of internal standards, such as TMS [20].
7. Convert signal areas to mM before statistical analysis by using total cholesterol concentrations determined by other methods and the following equation [21]:

$$M_{\mathrm{x}} = A_{\mathrm{x}} \times \frac{M_{\mathrm{chol}}}{A_{\mathrm{chol}}}$$

where A_{chol} and A_{x} refer to the ^{1}H NMR areas of total cholesterol and signal x, respectively, and M_{chol} and M_{x} refer to the molar concentration of the externally measured total cholesterol and ^{1}H NMR signal x. LipSpin provides ^{1}H NMR areas normalized by the number of resonating protons.

4 Notes

1. It is important in NMR analysis to use glass material since organic solvents can degrade plastic material and generate contamination signals in the spectrum, which can compromise the analysis.
2. The original BUME method ensures high lipid recovery when extracting up to 100 μL of serum. For NMR analysis concentration is a critical point and at least 200 μL of serum is required for good sensitivity. Hence, performing the method twice in two replicates of 100 μL of serum and mixing the lipids before NMR measurement ensure a good signal-to-noise ratio in the spectra.

3. Centrifugation separates the mixture into two phases: the organic upper phase contains the lipids and the aqueous lower phase contains proteins and low molecular weight metabolites. If you observe a wide intermediate phase that sparkles, centrifuge for a longer period until the mixture separates into two phases.
4. At least 4 h of evaporation is recommended to prevent residual solvent signals from appearing in the NMR spectra of lipids.
5. The aim of this step is to separate the cell pellet from the supernatant. The conditions of the centrifugation will depend on the type of tube used.
6. At this point, you can store the cell pellet at −80 °C until ready for lipid extraction.
7. After 30 min, a biphasic layer can be observed.
8. After centrifugation, the lower organic phase contains the lipids. The interphase is well defined and a thinner layer can be observed.
9. This procedure has been tested for liver, muscle, white adipose tissue, and hypothalamus.
10. After overnight lyophilization, tissue can be pulverized mechanically.
11. The ^{1}H NMR spectrum is affected by whether the reconstitution solvent has been prepared with or without D_2O. In particular, D_2O affects the polar groups of phospholipids and maximizes spectral separation between choline phospholipids at 3.2 ppm.
12. For serum analysis in which the lipid extraction is performed in two replicates, the deuterated solvent solution should be transferred from one replicate to the other and then to the NMR tube.
13. Be sure tubes are tightly capped so that the deuterated solvent does not evaporate since chloroform is very volatile.
14. The spectra templates mentioned are a collection of signal patterns created for a 600-MHz spectrometer operating at 286 K in chloroform-d:methanol-d_4:deuterium oxide (16:7:1, v/v/v) solvents based on literature values and using spectra of standard lipids acquired in our lab or downloaded from public databases. LipSpin includes 30 spectra of lipid standards, which represent the most common lipids in serum, including palmitic acid, stearic acid, oleic acid, linoleic acid, eicosapentaenoic acid, arachidonic acid, docosahexaenoic acid, phosphatidylcholine, lysophosphatidylcholine, sphingomyelin, cholesterol, cholesteryl linoleate, and triglycerides, among others.

Acknowledgment

The authors wish to acknowledge the financial support of the Spanish Ministry of Economy and Competitiveness through project TEC2015-69076-P.

References

1. Watson AD (2006) Thematic review series: systems biology approaches to metabolic and cardiovascular disorders. Lipidomics: a global approach to lipid analysis in biological systems. J Lipid Res 47:2101–2111
2. Quehenberger O, Armando AM, Brown AH, Milne SB, Myers DS, Merrill AH et al (2010) Lipidomics reveals a remarkable diversity of lipids in human plasma. J Lipid Res 51:3299–3305
3. German JB, Gillies LA, Smilowitz JT, Zivkovic AM, Watkins M (2007) Lipidomics and lipid profiling in metabolomics. Curr Opin Lipidol 18:66–71
4. Murphy SA, Nicolaou A (2013) Lipidomics applications in health, disease and nutrition research. Mol Nutr Food Res 57:1336–1346
5. Soininen P, Kangas AJ, Würtz P, Suna T, Ala-Korpela M (2015) Quantitative serum nuclear magnetic resonance metabolomics in cardiovascular epidemiology and genetics. Circ Cardiovasc Genet 8:192–206
6. Miao H, Chen H, Pei S, Bai X, Vaziri ND, Zhao YY (2015) Plasma lipidomics reveal profound perturbation of glycerophospholipids, fatty acids, and sphingolipids in diet-induced hyperlipidemia. Chem Biol Interact 228:79–87
7. Goto-Inoue N, Yamada K, Inagaki A, Furuichi Y, Ogino S, Manabe Y et al (2013) Lipidomics analysis revealed the phospholipid compositional changes in muscle by chronic exercise and high-fat diet. Sci Rep 3:3267. https://doi.org/10.1038/srep03267
8. Meikle PJ, Wong G, Barlow CK, Kingwell BA (2014) Lipidomics: potential role in risk prediction and therapeutic monitoring for diabetes and cardiovascular disease. Pharmacol Ther 143:12–23
9. Sethi S, Brietzke E (2017) Recent advances in lipidomics: analytical and clinical perspectives. Prostaglandins Other Lipid Mediat 128–129:8–16
10. Soininen P, Kangas AJ, Würtz P, Tukiainen T, Tynkkynen T, Laatikainen R et al (2009) High-throughput serum NMR metabonomics for cost-effective holistic studies on systemic metabolism. Analyst 134:1781. https://doi.org/10.1039/b910205a
11. Mallol R, Amigó N, Rodríguez MA, Heras M, Vinaixa M, Plana N et al (2015) Liposcale: a novel advanced lipoprotein test based on 2D diffusion-ordered ^{1}H NMR spectroscopy. J Lipid Res 56:737–746
12. Silbernagel G, Pagel P, Pfahlert V, Genser B, Scharnagl H, Kleber ME et al (2017) High-density lipoprotein subclasses, coronary artery disease, and cardiovascular mortality. Clin Chem 63:1886–1896
13. Jeyarajah J, Cromwell EJ, Otvos WC (1981) Lipoprotein particle analysis by nuclear magnetic resonance spectroscopy. Clin Lab Med 26:847–855
14. Vinaixa M, Ángel Rodríguez M, Rull A, BeltrÁn R, Bladé C, Brezmes J et al (2010) Metabolomic assessment of the effect of dietary cholesterol in the progressive development of fatty liver disease. J Proteome Res 9:2527–2538
15. Folch J, Lees M, Stanley G (1957) A simple method for the isolation and purification of total lipides from animal tissues. J Biol Chem 226:497–509
16. Löfgren L, Ståhlman M, Forsberg GB, Saarinen S, Nilsson R, Hansson GI (2012) The BUME method: a novel automated chloroform-free 96-well total lipid extraction method for blood plasma. J Lipid Res 53:1690–1700
17. Barrilero R, Gil M, Amigó N, Dias CB, Wood LG, Garg ML et al (2018) LipSpin: a new bioinformatics tool for quantitative ^{1}H NMR lipid profiling. Anal Chem 90:2031–2040
18. Chen L, Weng Z, Goh L, Garland M (2002) An efficient algorithm for automatic phase correction of NMR spectra based on entropy minimization. J Magn Reson 158:164–168
19. Oostendorp M (2006) Diagnosing inborn errors of lipid metabolism with proton nuclear

magnetic resonance spectroscopy. Clin Chem 52:1395–1405

20. Bharti SK, Roy R (2012) Quantitative ^{1}H NMR spectroscopy. TrAC Trends Anal Chem 35:5–26
21. Serkova N, Florian Fuller T, Klawitter J, Freise CE, Niemann CU (2005) 1H-NMR–based metabolic signatures of mild and severe ischemia/reperfusion injury in rat kidney transplants. Kidney Int 67:1142–1151

Chapter 4

High-Resolution Magic Angle Spinning (HRMAS) NMR Methods in Metabolomics

Marlon Tilgner, Tim S. Vater, Piet Habbel, and Leo L. Cheng

Abstract

High-resolution magic angle spinning (HRMAS) NMR spectroscopy enables the evaluation of metabolite profiles of intact tissue with high spectral resolution. The ability to preserve the tissue after analysis permits subsequent histopathological examination and enables the analyses of correlations between tissue metabolites and pathologies, thus making HRMAS NMR spectroscopy a powerful tool in the metabolomics field. Improved methods for the elimination of spinning sidebands that appear at low spinning rates preserve the integrity of tissue structures better and allow measurement of delicate tissues, such as clinical biopsy core samples. In the metabolomics field, HRMAS NMR has been established as a valuable tool for both untargeted and targeted metabolite profiling. In this chapter, we present protocols to perform HRMAS NMR spectroscopy experiments, including sample preparation, acquisition procedures, measurement parameters, histopathological examination techniques, spectral processing, and metabolite quantification and statistical analyses.

Key words High-resolution magic angle spinning, NMR spectroscopy, Metabolomics, Metabolite quantification, Intact tissue

1 Introduction

High-resolution magic angle spinning (HRMAS) NMR spectroscopy, invented in 1996 for intact tissue analysis [1, 2], has been widely used in the metabolomics field. The advantage of this methodology in preserving the architectural structure of the tissue for pathology evaluation after the NMR measurements [3] has contributed greatly to the establishment of NMR-based metabolomics of tissue physiology and pathology [4, 5].

Applications of conventional one- or two-dimensional NMR spectroscopy for analysis of liquid samples of chemical and biological origin have increased with the advances in NMR and computer technologies. However, direct application of such NMR

Marlon Tilgner and Tim S. Vater contributed equally to this work.

G. A. Nagana Gowda and Daniel Raftery (eds.), *NMR-Based Metabolomics: Methods and Protocols*, Methods in Molecular Biology, vol. 2037, https://doi.org/10.1007/978-1-4939-9690-2_4, © Springer Science+Business Media, LLC, part of Springer Nature 2019

methods for analysis of nonliquid samples, such as biological tissue, is limited due to low spectral resolution caused by line broadening. Such line broadening is caused by "solid effects" which include residual chemical shift anisotropy and dipolar coupling, as well as magnetic susceptibility. Compared to solutions in which molecules undergo rapid molecular tumbling in a homogenous environment, molecules in tissue do not move rapidly enough to average the chemical shift anisotropy, dipolar couplings, and inhomogeneous environments that affect spectra due to magnetic susceptibility effects. The interactions of nuclei in the solid or semisolid samples are proportional to $(3\cos^2\theta - 1)$, where θ is the angle between the static magnetic field B_0 and the vector axis of an interaction, such as the dipole direction for a dipolar coupling. This factor approaches 0 at $\theta = 54.7°$ (known as the "magic angle") and under this condition the solid effects can be neglected to the first degree of approximation. Based on this principle, the magic angle spinning (MAS) approach for solid-state NMR was invented in 1959 [6, 7]. However, MAS alone is insufficient to obtain high-resolution spectra for nuclei such as ^1H in solids in the absence of other resonance line-narrowing approaches [8]. Nevertheless, since the residual solid effects for biological tissue are several orders of magnitude smaller than those in solids, application of MAS can drastically improve the resolution of tissue proton (^1H) NMR spectra. This led to the discovery of intact tissue HRMAS ^1H NMR; the design of HRMAS NMR probes emphasizes high spectral resolution rather than endurance of the applied high-power levels that characterize the classic MAS probes for solid-state NMR. A protocol for analyzing tissue with HRMAS NMR spectroscopy has been proposed with a focus on procedures for frozen tissues spun at high rates [9]. In this chapter, based on our experiences with HRMAS NMR since the inception of its concept, we present comprehensive protocols for preparation and measurement of various tissue types, spectral processing and quantification, histopathological evaluation, and metabolite data analyses.

2 Materials

2.1 Samples and Preparation Materials

When working with biological samples, particularly those of human origin, safety procedures including the use of gloves, goggles, mask, and disposable gown should be followed at all times.

1. Samples: fresh or frozen tissue, or inhomogeneous biofluids (*see* Subheading 3 for preparation).
2. Humidified chamber.
3. Sample preparation tubes and pipettes.
4. Ice, dry ice.

5. Deuterium oxide (D_2O, 99.9 atom % D, 0.05 wt. % TSP, sodium salt).
6. Tweezers, scalpel and cutting blades, metal cutting plate, surgical gauze sponges.
7. Analytical balance.

2.2 NMR Equipment, Software, and Accessories

1. NMR spectrometer with an HRMAS probe.
2. HRMAS rotors, Kel-F inserts, sealing screws and caps.
3. Rotor tools: insert screwdriver, forked screwdriver, and cap removal tool.
4. NMR software suite.
5. Potassium bromide (KBr).
6. Kimwipes™.

2.3 Histo-pathological Examination

1. Tissue processor, tissue embedder, microtome, tissue staining station, microscope.
2. Tissue cassettes and foam pads, microtome blades, charged glass slides, cover glasses.
3. Formalin, ethanol, xylene, paraffin, hematoxylin, eosin.

2.4 Data Processing and Analysis

1. Data processing software.
2. Statistical software packages.

3 Methods

3.1 Sample Preparation

Sample preparation encompasses various steps from tissue collection and storage to loading it into the HRMAS rotor. Intact tissue samples can generally be categorized as frozen or fresh, each involving a different procedure for preparation. Apart from tissue, HRMAS NMR spectroscopy can also be used to analyze biological fluid samples, such as blood, seminal fluid or sputum, with the advantage of requiring only minute amounts of samples and eliminating line-broadening effects caused by macromolecules in these fluids.

3.1.1 Frozen Tissue Samples

Frozen samples for intact tissue HRMAS NMR spectroscopy should be stored at −80 °C. Previous observations have shown that, in some cases, the process of freezing and thawing might alter both spectral resolution and metabolite intensities when compared with fresh tissues and reduce the quality of histopathological examinations [10]. However, since it is often impractical to measure fresh tissue, these procedural artifacts cannot be averted and must be accounted for. Therefore, prior to measurements of a new tissue type, we recommend scanning tissue in fresh (if possible) and frozen states of various storage durations to determine whether the potential artifacts can be neglected. It is important to keep the tissue samples frozen until analysis and prevent multiple freezing-thawing cycles.

1. Keep the sealed frozen tissue samples on dry ice.
2. Weigh an empty rotor and record the weight.
3. Lay a metal plate on dry ice and cover it with a surgical gauze sponge.
4. Place the tissue on top of the gauze with tweezers and wrap it into the gauze (*see* **Note 1**).
5. Use a scalpel to cut off a small piece of tissue up to 3 mm in diameter.
6. Place the frozen tissue piece at the bottom of the rotor's interior using tweezers.
7. Reweigh the rotor with the included tissue. Add or remove tissue to obtain the desired weight (~10 mg) of tissue.
8. Add 2.0 μL D_2O.
9. Place the insert into the rotor (*see* **Note 2**).
10. Place the sealing screw into the insert's threaded opening and tighten.
11. Remove overflowing liquid, if any, from the top of the insert with a gauze sponge.
12. Press the cap tightly onto the rotor.
13. Immediately insert the rotor into the probe which has been precooled to the appropriate temperature (*see* Subheading 3.2). Otherwise, keep the rotor on ice in a cryotube (*see* **Note 3**).

3.1.2 Fresh Tissue Samples

In a clinical environment, a fresh tissue specimen is typically collected either from a surgical procedure or a biopsy. Surgical samples are often large amount of tissue that can be snap-frozen immediately and stored for later analyses as described previously, while sensitive and structurally delicate biopsy samples need to be available for clinical reporting in a fresh and intact state after HRMAS NMR analysis. Thus, the procedure for analysis of fresh biopsy tissue needs to lay emphasis on analysis speed, avoidance of freeze-thawing, and protection of tissue integrity. To reduce potential tissue metabolite degradation, the sample has to be kept cold at all times until it is ready for analysis. Furthermore, since biopsy samples are very thin and susceptible to drying (particularly those from core needle biopsies), precautions against dehydration need to be taken. Therefore, the biopsy core should be transported in a humidified chamber as pictured in Subheading 2 (*see* Fig. 1).

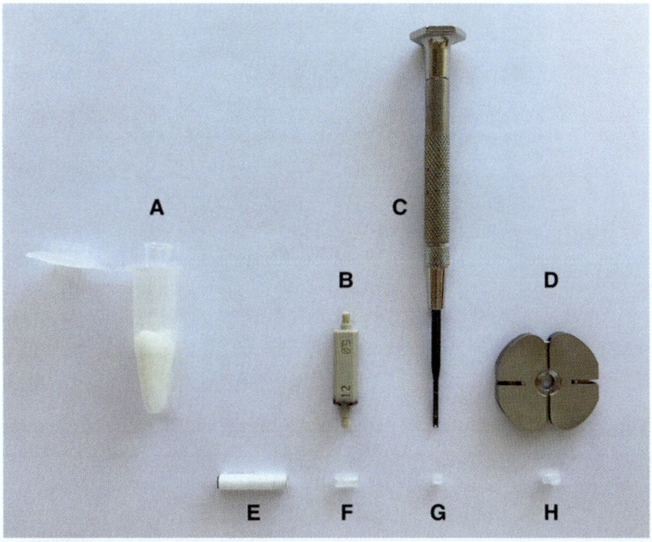

Fig. 1 Overview of rotor parts, their specific tools, and humidified chamber. (**a**) Humidified chamber (1.5-mL tube with Kimwipe™ soaked in D_2O and small tube for tissue inside), (**b**) insert screwdriver, (**c**) forked screwdriver, (**d**) cap removal tool, (**e**) rotor, (**f**) insert, (**g**) sealing screw, (**h**) rotor cap

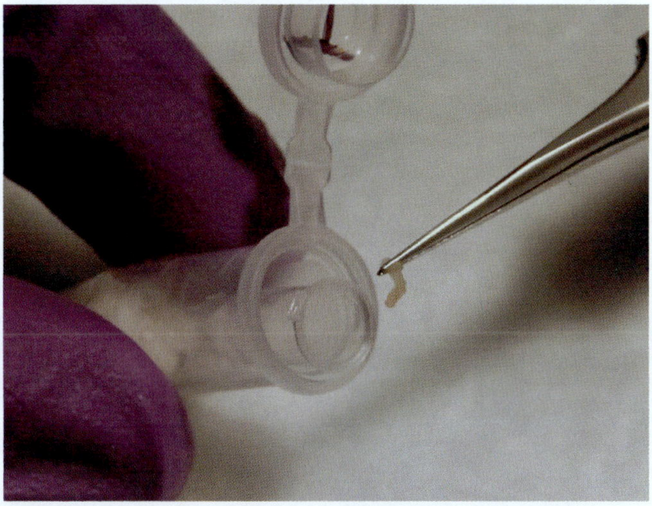

Fig. 2 Fresh needle biopsy tissue core and humidified chamber

1. Release the tissue core from the biopsy needle into a tube filled with 1.0 mL D_2O. Carefully swirl the tissue around with tweezers to wash away blood for no more than 3 s to prevent the loss of cellular metabolites into D_2O.
2. Pick up the tissue core carefully with tweezers and place it in the humidified chamber (*see* Fig. 2).
3. Transport the closed humidified chamber on ice.

4. Weigh an empty rotor and record the weight.
5. Place the frozen tissue piece at the bottom of the rotor's interior using tweezers.
6. Weigh the rotor including the tissue again (a needle biopsy tissue core generally weighs ~1–2 mg).
7. Add 10.0 μL D_2O.
8. Follow **steps 9–13** outlined in Subheading 3.1.1 for rotor assembly.

3.1.3 Fluid Samples

Almost all biological fluids, except urine, do not fall into the classical description of aqueous solutions due to the presence of cells, proteins, and macromolecules. Broadened NMR resonances in these fluids can be reduced by measuring them with the HRMAS method.

1. Load 10.0 μL of the fluid of interest into the rotor with a pipette.
2. Add 2.0 μL D_2O.
3. Follow **steps 9–13** outlined in Subheading 3.1.1 for rotor assembly.

3.2 HRMAS NMR Spectrometer Preparation

An NMR spectrometer equipped with an HRMAS probe requires adjustment of the magic angle after probe installation and a change in temperature prior to measurements. The magic angle can be calibrated by measuring the FID of potassium bromide (KBr) powder under spinning conditions for its visible signal intensity variations at varying angles. The magic angle adjustment should be carried out at the temperature required for the subsequent experiment (*see* **Note 4**). The temperature for tissue measurements is typically set at 4 °C to minimize degradation but prevent freezing of the tissue.

1. Fill a rotor with KBr powder.
2. Insert the rotor into the probe.
3. Set HRMAS spinning rate and temperature to the values required for the succeeding tissue measurement (e.g., 4 °C and 3600 Hz, respectively).
4. Wait until the temperature has reached the desired setting and ensure it is fully stabilized.
5. Use a standard 90° pulse program to collect a spectrum from the ^{13}C channel (*see* **Note 5**).
6. Run the continuous acquisition mode and observe the FID view (*see* **Note 6**).

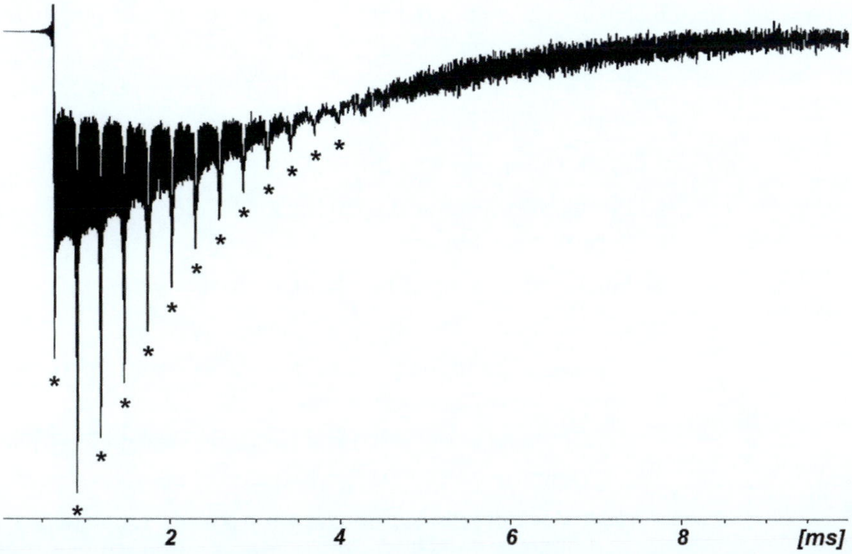

Fig. 3 FID view of KBr sample spinning in HRMAS probe. Maximum intensity and rightmost appearance of the spinning sidebands (∗) are reached at 54.7° inclination to the external magnetic field (magic angle)

7. Adjust the magic angle with the magic angle screw underneath the probe until the spinning sidebands, presenting as spikes in the FID, extend to the farthest distance in the FID (*see* Fig. 3).
8. End the continuous acquisition mode, but keep the KBr rotor spinning in the probe to maintain the desired temperature until the tissue rotor is ready for insertion (this applies to every change of sample).

3.3 HRMAS NMR Procedure

For tissue measurements, there are numerous methods that selectively emphasize metabolites of interest using different NMR pulse programs and parameters. Here, as an illustration of the application of HRMAS in intact tissue analysis, we focus on ^{1}H NMR and the basic parameters necessary for performing an HRMAS measurement.

The setup of a typical NMR spectroscopy experiment comprises the determination of a 90° pulse, application of water suppression, and selection of a pulse program and parameters. We will present a stepwise instruction for conducting a basic experiment using a standard one-dimensional ^{1}H 90° pulse sequence in combination with continuous wave (CW) water suppression. Considerations of additional pulse programs, experimental parameters, and the relevance of the spinning rate will be discussed in Subheading 3.4.

In this illustration, we will use a Bruker BioSpin NMR system as an example, but readers using different NMR systems can adopt these concepts accordingly.

3.3.1 One-Dimensional ^1H 90° Pulse Sequence

1. Eject the previous rotor (KBr or tissue sample) from the pre-cooled probe (4 °C).
2. Insert the rotor containing the tissue sample and start spinning at the desired rate.
3. Lock the field onto the D_2O signal.
4. Perform tuning and matching of the probe after sample temperature is stabilized.
5. Shim the magnet if needed (*see* **Note 7**).
6. Create a new dataset.
7. Choose a standard 90° pulse sequence pulse program (*zg*) and set the following parameters:
   (a) Pulse duration (P1), 3 μs; power level (PLdB1), −14 dB; recycle delay (D1), 5 s.
   (b) For determination of other parameters, such as the number of points in the time domain (TD = 4096), the spectral width (SW = 12 ppm), the number of scans (NS = 32), and the number of dummy scans (DS = 8), we refer to Subheading 3.4.

   Ensure the combination of pulse duration and power level results in a flip angle <90° (*see* **Note 8**).
8. Acquire a spectrum, perform Fourier transformation, correct the phase, and save.
9. Modify the pulse duration until a 90° pulse is reached (*see* **Note 9**).
10. Adjust the receiver gain.
11. Acquire a spectrum, correct the phase, and measure the full width at half maximum (FWHM) of the water peak (*see* **Note 10**).
12. If the FWHM is too broad or the water peak is asymmetric, re-shim the magnet (*see* **Note 7**).
13. Acquire a spectrum, perform Fourier transformation, and set the carrier frequency (O1) to the center of the water peak.
14. Acquire the spectrum.

3.3.2 Water Suppression

Since on average human tissues consist of more than 80% water, when measuring intact tissue samples, it is inevitable that the large water content often presents a strong signal that exceeds the resonance signals of metabolites by several orders of magnitude. Therefore, the water resonance should be suppressed to achieve a more accurate metabolite measurement. A common method is the adoption of a low-power CW pre-saturation pulse during the recycle delay.

1. Copy the dataset (including all adjusted parameters) from Subheading 3.3.1 to a new file.
2. Choose a pulse program that combines a 90° pulse sequence with a CW pre-saturation pulse (*zgpr*).
3. Determine the required power level (PLdB9) for CW applied at the carrier frequency O1 for appropriate water suppression (*see* **Note 11**).
4. Adjust the receiver gain.
5. Acquire the spectrum.

3.3.3 Rotor Disassembly and Cleaning

1. Stop the spinning sample and eject the rotor.
2. Remove the rotor cap with the cap removal tool (*see* **Note 12**).
3. Remove the sealing screw with the forked screwdriver.
4. Remove the insert with the insert screwdriver.
5. Carefully take out the tissue sample with the tweezers. Depending on the study design, place the tissue in formalin for fixation, keep it on ice, or discard it, with considerations of safety precautions.
6. Put rotor and cap into a beaker with water and detergent. Add insert and sealing screw, each in a small tube filled with water and detergent to avoid jamming. Make sure there are no air bubbles in any of the rotor parts.
7. Place the beaker in an ultrasound cleaner for 15 min.
8. Transfer all parts into distilled water following the same procedure as listed in **step 6**.
9. Place the beaker in an ultrasound cleaner for another 15 min.
10. Dry all parts thoroughly with Kimwipes™ and store them in a dry place.

3.4 Selection of Other Pulse Programs and Acquisition Parameters

3.4.1 Pulse Programs

In addition to the standard one-dimensional ^1H 90° pulse programs (*zg* and *zgpr*) discussed earlier, various other pulse programs can be used to acquire HRMAS NMR spectra depending on the need, such as NOESY (nuclear Overhauser effect spectroscopy), COSY (correlated spectroscopy), TOCSY (total correlation spectroscopy), DOSY (diffusion-ordered spectroscopy), and JRES (J-resolved spectroscopy).

Of particular interest for intact tissue analysis is the CPMG (Carr-Purcell-Meiboom-Gill) pulse sequence, which has been used to improve spectral resolution by filtering out the interference from macromolecules and probe background signals. Following a 90° pulse to bring the magnetization onto the x-y plane, a train of 180° echo pulses is applied as a T_2 filter [11]. For HRMAS NMR measurements, the applied 180° pulse train is rotor-synchronized to the HRMAS rate.

3.4.2 Relevance of the Spinning Rate

In general, when measuring solids, the higher the spinning rate, the better the spectral resolution. However, since most soft biological tissues are not classic solids and show only residual solid effects, high spectral resolution can be achieved with modest spinning rates. In fact, higher spinning rates are often unnecessary after the spectral resolution improvement reaches its plateau at a modest spinning rate. Instead, an appropriate spinning rate balances the prevention of spinning sidebands, particularly those from tissue water, in the spectral regions of metabolites, with the preservation of tissues from structural damages caused by high centrifugal forces. This consideration is particularly important when working with delicate tissues such as brain, liver, prostate, or kidney, where centrifugal damage can render histological evaluations after HRMAS NMR impossible.

Therefore, methods of acquiring spectra at reduced spinning rates without sideband interference have been developed for the purpose of preserving the delicate tissue structures intact. We recommend a method that we have used effectively over the past decade [12]. This method combines two spectra measured at two properly selected, different low spinning rates into one composite spectrum by using the lower value of the two spectra for each point in the frequency domain. Since these different spinning rates produce no sidebands at identical positions within the spectral region of interest, sidebands in that region will be eliminated. Only sidebands at identical positions (the common multiples of the two spinning rates) will remain, but are outside the spectral region of interest. For instance, on a 600-MHz NMR spectrometer, using spinning rates of 600 and 700 Hz will cause the common matching sidebands for 600 and 700 Hz to appear at ±4200 Hz (±7.0 ppm for ^{1}H). Assuming a water resonance of 4.7 ppm, the resulting 11.7 and − 2.3 ppm are way beyond the region of interest. Detailed instructions on how to combine the two spectra are given in Subheading 3.5.

If, however, no further histopathological examination is intended, high spectral resolution without spinning sideband interference can be achieved with high spinning rates (such as 3600 Hz for soft tissue, resulting in spinning sidebands at ±6.0 ppm on a 600-MHz NMR spectrometer). Tissues of higher solidity (e.g., skin or bone) can benefit from even higher spinning rates.

3.4.3 Basic Acquisition Parameters

1. Number of time domain points (TD): To prevent signal truncation, a full FID that returns to baseline should be acquired using 2^n TD points, in combination with a certain dwell time (DW) determined by the desired spectral width, $SW = 0.5 \times (1/DW)$, in Hz. Common values for TD are 4096 or 8192.

2. Spectral width (SW): The SW should include all measurable resonances without being too wide, as this reduces spectral resolution at a given number of points. In HRMAS, attention should be given to the sidebands, especially those from the water resonance, and to using a sufficient SW to prevent these sidebands from folding back into the measured spectral region. Common SW values for tissue ^1H NMR are between 15 and 20 ppm.
3. Number of scans (NS): An increase in the number of scans improves the signal-to-noise ratio (SNR) according to the relationship SNR ~ √(NS), i.e., quadrupling the NS will increase the SNR by twofold. The appropriate NS depends on the strength of the signal and the pulse program used. An acceptable SNR for an NMR signal should be greater than 4.
4. Power level (PL): The PL represents the energy level used to generate a radio-frequency pulse. It should be determined in combination with the pulse duration (P1) to result in the desired angle of rotation in a pulse program. In general, P1 for a ^1H 90° pulse should be around 7 μs.
5. Recycle delay (D1): D1 is the time delay between each measurement during signal averaging to allow magnetization to fully relax. Experimentally, D1 should be at least five times longer than the T1 of the molecules of interest. For intact tissue HRMAS ^1H NMR, a 5-s D1 is often sufficient.
6. Receiver gain (RG): RG should be adjusted for optimum SNR. If RG is too low, signals cannot be properly amplified; on the other hand, if it is too high, signals will be truncated.

3.5 Spectra Processing

The FID is a domain raw NMR data obtained as a function of time. These time domain signals need to be processed to generate the commonly recognizable NMR spectra of frequency domain, i.e., resonance intensities as a function of frequency ("chemical shift," unit = ppm). Many software packages can be used for processing the spectra data, with some including the option to batch-process multiple files automatically. Although this is a time-effective method, we caution users to monitor the spectra quality in detail. If in doubt, one should also process the FID files manually to ensure accuracy.

1. The FID data can be manipulated mathematically to increase either resolution or sensitivity (signal-to-noise ratio), each at the expense of the other. This process is called windowing (applying a window function) or apodization. Most often a better sensitivity with less noise is preferable (*see* **Note 13**). To achieve this goal, apply an exponential multiplication (EM) function with a line broadening (LB) value of 0.5 Hz (*see* **Note 14**).

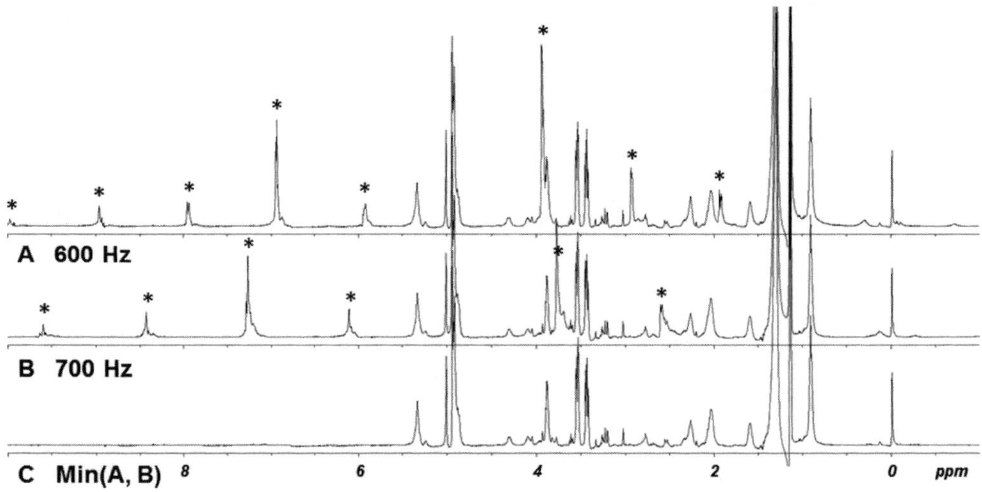

Fig. 4 Spectra of human prostate tissue acquired with CPMG pulse sequence with continuous wave (CW) presaturation of water resonance and spinning rates of 600 Hz (**a**) and 700 Hz (**b**). (**c**) Spectrum produced by Min (*A*, *B*) function without spinning sidebands. *Water spinning side bands

2. Zero-fill the data to digitally increase the resolution (*see* **Note 15**).
3. Perform Fourier transformation (FT). This essential step converts the information from the time domain to the frequency domain (*see* **Note 16**).
4. Adjust the phase and correct the baseline.

As introduced in Subheading 3.4.2, high spinning rates can damage tissue architecture and complicate the subsequent histopathological assessment. As mentioned previously, we have shown [12] a method for eliminating spinning sidebands, which inevitably occur at low spinning rates, by combining two spectra measured at different low spinning rates. The underlying method is a Minimum (A, B) function, where A, B are corresponding points in the two spectra spun at two different low spinning rates. Applying the calculation of $(A + B - |A - B|)/2 = \mathrm{Min(A, B)}$ for each point in the frequency domain and by selecting appropriate spinning rates (such as 600 and 700 Hz), a new spectrum without sidebands in the spectral region of interest can be calculated (*see* Fig. 4).

3.6 Data Quantification

Each resonance peak in a given NMR spectrum can be quantified based on peak integration. Peak area is proportional to the amount of the corresponding metabolite in a sample. Peak identification and integration can be used to estimate metabolite concentrations from spectra and compare among samples measured under the same condition. The following section provides a step-by-step approach for quantifying and preparing spectral data for statistical analyses.

1. Spectral alignment: Spectra need to be aligned against either an external reference standard or an internal known metabolite. If the spectrum contains only a few known metabolites or if only specific peaks are of interest, it may be sufficient to simply set a reference point (e.g., trimethylsilylpropanoic acid, TSP, at around 0 ppm) and shift the spectrum globally without using special spectral alignment methods.
2. Peak integration: A common way to select metabolite peaks for integration is through manual peak-picking, possibly with assistance from a reference database, such as the Human Metabolome Database (HMDB) [13]. Alternatively, automatic peak-identification algorithms apply thresholds of different types (e.g., peak height or area) to the spectrum. After selection, the area is calculated for each identified peak through integration. For overlapping peaks, curve-fitting methods with a combination of Lorentzian and Gaussian line shapes are preferable over integration because of their deconvoluting abilities. Further methods for peak-picking and integration are available (*see* **Note 17**).
3. Deconvolution: While curve fitting offers a method to deconvolute specific peak regions, certain software packages can deconvolute complete spectra by estimating the contribution of different metabolites to a certain peak, given that all metabolites contained in the sample are known [14].
4. Calculation of metabolite concentration: The concentration of a metabolite can be estimated by referencing the corresponding identified and quantified peak area to the area of an internal standard with known concentration or to the tissue weight.
5. Normalization: Normalization of peaks by total spectrum intensity or an internal reference standard provides for each sample the relative intensities needed for comparisons of different peaks within one spectrum or similar peaks from different spectra.

3.7 Histopathological Examination

A great advantage of HRMAS NMR spectroscopy is that it presents the possibility of conducting histopathological examination of intact tissues after the spectroscopic analysis to correlate microscopic observations (e.g., amount of cancerous formations in a tissue sample) with metabolite levels. Therefore, all tissue samples, particularly those known to contain heterogenous pathological components, should undergo histopathological evaluation after HRMAS NMR measurements (*see* **Note 18**).

1. Fixation: Store the tissue sample in formalin (5–10% formaldehyde solution) directly after the HRMAS NMR measurement for at least 4 h.

2. Processing: Enclose the tissue sample in a securely labeled cassette with foam pads. Use a tissue processor to automatically perform the dehydration (most commonly with alcohol solutions of increasing concentration), clearing (most commonly with xylene), and infiltration (with liquid paraffin) of the tissue sample (*see* **Note 19**). Immediately after processing, continue with embedding.
3. Embedding: Pour a thin layer of hot liquid paraffin in the mold of an embedding tray and let it harden a little. Transfer the tissue from the cassette onto the hardening paraffin layer and fill the mold to the top (*see* **Note 20**). Attach the cassette and fill it with more liquid paraffin. Let the paraffin cool down and harden, for example, on a cold plate, and then remove the tray. The tissue is now preserved in a paraffin block.
4. Sectioning: Place the cassette facedown on melting ice to moisten and cool the paraffin wax. Lock the cassette in a microtome and "face" the wax, i.e., remove the first layers of solid paraffin until the tissue is exposed (*see* **Note 21**). Then cut a ribbon of sections of 5-μm thickness at around 100-μm intervals (*see* **Note 22**). Use tweezers to place the ribbons in a 45 °C water bath. Pick each ribbon up with a labeled charged glass slide and dry the slides at 60 °C for 15–45 min. Normally, a tissue piece of 10–15-mg weight will generate 12–16 histology slides at 100-μm intervals.
5. Staining: Follow a standard hematoxylin and eosin protocol [15] to stain the slides. Depending on tissue type and study aim, other staining methods (such as immunohistochemical staining) can be used (*see* **Note 18**). After staining, cover the slides with cover glasses and let them dry for 24 h.
6. Microscopy: For each slide, determine the quantities of different pathology types (e.g., epithelium, gland cells, connective tissue, fat) as the percentage of the slide's total tissue area, by either visual estimations or computer measurements (*see* **Note 23**). Average the % values of all slides derived from one sample to estimate the volume percentages, i.e., the % of the specific pathology type in relation to the total tissue volume (*see* **Note 24**). Since NMR spectroscopy provides data on the entire sample, volume percentages are more appropriate for further data analyses and comparisons.

3.8 Data Analysis

High-throughput analysis of HRMAS NMR spectroscopy data for metabolomic assessment requires advanced statistical methods. The quantified NMR "features" (peaks, peak regions, or metabolite integrals/concentrations) need to be organized and combined with corresponding clinical and pathological data for different tests, chosen according to the respective study's objective.

Common aims include investigation of correlations between tissue metabolites and pathologies, differentiation of tissue or disease categories, and observation of changes in certain metabolic pathways.

1. Matrix structure and data merging: The structure of a data matrix should be developed pursuant to the study design, the data types, and the analytical software program used. Usually, rows are reserved for samples, such as cases, subjects, or patients, and columns represent all categories of data, such as quantified NMR features, histopathological volume percentage results, or clinical data.
2. Univariate analysis: Univariate methods are used to analyze each feature separately, possibly across several subgroups of the sample population (e.g., changes of one metabolite in healthy vs. diseased). Statistical tests for unpaired parametric data are the Student's *t*-test (for comparison of two groups) or the one-way ANOVA (for comparison of > two groups). For non-parametric data, the Mann-Whitney-Wilcoxon *U*-test (two groups) and the Kruskal-Wallis test (> two groups) are adequate (*see* **Note 25**). Different tests are needed for paired data (e.g., changes of one metabolite in benign vs. malign samples from one individual) [16]. Since NMR spectroscopy often results in a high number of features from limited sample numbers, coincidental false-positive errors can frequently occur (the "multiple testing" problem) and a correction method needs to be applied (e.g., Bonferroni or FDR correction) [14].
3. Multivariate analysis: Multivariate methods take multiple or all features into account at once and, unlike univariate analysis, can recognize correlations between individual features. Unsupervised and supervised methods are distinguished. Unsupervised multivariate methods cluster samples and reduce the dimensionality of the dataset without considering sample type or identity. A popular method is principal component analysis (PCA), which presents features according to maximum variance and enables an early distinction of groups. Other methods include independent component analysis (ICA) and hierarchical cluster analysis (HCA) [17]. Supervised methods, on the other hand, do consider sample identity and can be used for specific testing of known groups, but need to be modeled, evaluated, and validated. (Orthogonal) Partial least squares discriminate analysis ((O)-PLS-DA) and linear regression models belong to this category [14].
4. Data interpretation: Metabolomic interpretation of NMR spectroscopy data can be extended through analysis of the metabolic pathways that involve the observed metabolites, based on databases and further correlation analyses.

4 Notes

1. Covering frozen tissue with a surgical gauze sponge for cutting will prevent pieces of tissue from bouncing away.
2. Put the insert in with the thread facing up using the insert screwdriver, or it will be very difficult to remove it again from the rotor.
3. If the rotor is exposed to ice or water, dry it thoroughly with a Kimwipe™ before inserting into the HRMAS probe.
4. The mechanical parts supporting the magic angle and performing the adjustment can expand or contract at different temperatures, which could alter the magic angle. The change in spinning rate, however, does not affect the magic angle adjustment.
5. The resonance frequency of ^{79}Br (50.5% natural abundance) is only 0.36% lower than that of ^{13}C and can be observed in the ^{13}C channel. The KBr signal will reach its relative maximum at the angle of 54.7° without any additional tuning and matching of the channel.
6. If the carrier frequency is not at the center band of the ^{79}Br signal, the resulting FID may prevent correct observation of magic angle adjustment. If the FID appearance deviates from Fig. 3, set the carrier frequency (O1) to the middle of the center band.
7. Shimming is an important step that corrects inhomogeneities around the sample and minimizes NMR peak linewidths. Adjust the shims (e.g., Z, $Z^{2\text{-}5}$, X, Y) during acquisition for better line shapes.
8. A flip angle >90° will result in negative contributions to the resonance peaks.
9. Since a maximum peak intensity resulting from a 90° pulse is difficult to observe, we recommend identifying a pulse duration that produces a 180° pulse and divide the value by 2. In theory, no signal should be detected when a precise 180° pulse is applied. In practice, however, complete signal eradication can barely be realized. Hence, the corresponding value for a minimum intensity should be found where the integrals of positive and negative peaks are of equal size.
10. The width of the water peak measured at 50% of its maximum height reflects the linewidth and therefore the spectral resolution. The acceptable FWHM value for the tissue water peak depends on the tissue types analyzed.
11. The adequate value of pre-saturation power level can be adjusted until the water peak is of about the same intensity as

the metabolic peaks. If the water suppression is set too strong, interference with metabolic peaks may occur.

12. Use caution when removing the cap. The rotor is made up of soft material and may be damaged by the application of pressure, making it incapable of spinning.
13. Earlier parts of the FID file have a better signal-to-noise ratio than later parts. A window function "emphasizes" certain parts of the FID. An exponential multiplication (EM) function emphasizes earlier parts, thereby increasing sensitivity. On the other hand, window functions such as Gaussian multiplication and sine bell multiplication increase resolution [18].
14. A higher LB value forces a faster decay of the exponential function, which increases sensitivity at the cost of resolution [18]. A way to find a suitable LB value is to use the full width at half height (in Hz) of a narrow peak after performing a Fourier transformation (**step 3**). Another method is to calculate the spectrum's digital resolution in Hz/pt: (2 × spectrum width)/ number of points. However, using a LB value of 0.2–0.7 Hz should provide good results in most cases.
15. Correct zero-filling depends on a full decay to zero of the FID. That is why applying a window function beforehand is recommended.
16. If the spectrum appears wavy and without a straight baseline after performing a FT, check if the FID signal starts with a short delay. This occurs with digitally filtered FID data files from Bruker NMR spectrometers. Exclude the delay before performing a FT when processing these FID files.
17. Another possibility is to use a "binning" approach that divides the spectrum into equal-size snippets or "buckets." The information in one bucket (intensity values or area) is then summed. Thus, small spectral alignment shifts are compensated, with the disadvantage of reducing the spectral resolution. "Adaptive binning" methods are used to avoid the division of one peak into two snippets or the merging of several peaks into one snippet [19].
18. For immunohistochemical staining methods, frozen sectioning is more suitable than tissue fixation, processing, and embedding.
19. If a tissue processor is not available, all of these procedures can be performed manually inside a fume hood.
20. The first paraffin layer ensures that the tissue sample cannot sink to the bottom of the mold, where it would not be covered in wax entirely but exposed on one side. Pay attention to alignment and orientation of the tissue in the liquid paraffin, as this may be important for cutting. Use preheated tweezers to move the tissue.

21. Microtome blades need to be very sharp to ensure proper cutting. Use an old blade for facing the wax to prevent blunting a new blade before sectioning the actual tissue.
22. Sectioning is a rather delicate procedure. When embedding, create a few wax blocks without tissue and practice the cutting process with these. Use a slow, steady speed for cutting and carefully pull away the ribbon with a tweezer as it builds up. The length of the ribbons should be in accordance with the size of the glass slides. After facing and in between sectioning, rewetting and recooling the tissue on ice can be helpful.
23. Low slide quality can, amongst other things, result from dirty slides and dust, insufficient or excessive staining, and damage in the tissue coherence due to incorrect cutting.
24. For this calculation, it is important to derive slides from sections of equal intervals. Otherwise higher inaccuracy may occur when making assumptions on the entire tissue sample.
25. The Kolmogorov-Smirnov test can indicate whether a dataset is parametric (e.g., normally distributed) or nonparametric.

Acknowledgements

The authors gratefully acknowledge Lindsey A. Vandergrift for her editorial assistance. Research reported in this publication was supported by the National Cancer Institute of the National Institutes of Health under award number R01CA115746 (Cheng) and the Massachusetts General Hospital Athinoula A. Martinos Center for Biomedical Imaging. Marlon Tilgner and Tim S. Vater contributed equally to this work.

References

1. Cheng LL, Lean CL, Bogdanova A et al (1996) Enhanced resolution of proton NMR spectra of malignant lymph nodes using magic-angle spinning. Magn Reson Med 36:653–658
2. Cheng LL, Ma MJ, Becerra L et al (1997) Quantitative neuropathology by high resolution magic angle spinning proton magnetic resonance spectroscopy. Proc Natl Acad Sci 94:6408–6413
3. Cheng LL, Anthony DC, Comite AR et al (2000) Quantification of microheterogeneity in glioblastoma multiforme with ex vivo high-resolution magic-angle spinning (HRMAS) proton magnetic resonance spectroscopy. Neuro-Oncology 2:87–95
4. Cheng LL, Burns MA, Taylor JL et al (2005) Metabolic characterization of human prostate cancer with tissue magnetic resonance spectroscopy. Cancer Res 65:3030–3034
5. Vandergrift LA, Decelle EA, Kurth J et al (2018) Metabolomic prediction of human prostate cancer aggressiveness: magnetic resonance spectroscopy of histologically benign tissue. Sci Rep 8:4997
6. Andrew ER, Bradbury A, Eades RG (1959) Removal of dipolar broadening of nuclear magnetic resonance spectra of solids by specimen rotation. Nature 183:1802
7. Lowe IJ (1959) Free induction decays of rotating solids. Phys Rev Lett 2:285–287
8. Zheng L, Fishbein KW, Griffin RG et al (1993) Two-dimensional solid-state proton NMR and proton exchange. J Am Chem Soc 115:6254–6261

9. Giskeødegård GF, Cao MD, Bathen TF (2015) High-resolution magic-angle-spinning NMR spectroscopy of intact tissue. In: Bjerrum JT (ed) Metabonomics. Springer New York, New York, NY, pp 37–50
10. Wu C-L, Taylor JL, He W et al (2003) Proton high-resolution magic angle spinning NMR analysis of fresh and previously frozen tissue of human prostate. Magn Reson Med 50:1307–1311
11. Meiboom S, Gill D (1958) Modified spin-Echo method for measuring nuclear relaxation times. Rev Sci Instrum 29:688–691
12. Burns MA, Taylor JL, Wu C-L et al (2005) Reduction of spinning sidebands in proton NMR of human prostate tissue with slow high-resolution magic angle spinning. Magn Reson Med 54:34–42
13. Wishart DS, Tzur D, Knox C et al (2007) HMDB: the Human Metabolome Database. Nucleic Acids Res 35:D521–D526
14. Alonso A, Marsal S, Julià A (2015) Analytical methods in untargeted metabolomics: state of the art in 2015. Front Bioeng Biotechnol 3:1–20
15. Feldman AT, Wolfe D (2014) Tissue processing and hematoxylin and eosin staining. In: Day CE (ed) Histopathology: methods and protocols. Springer New York, New York, NY, pp 31–43
16. Vinaixa M, Samino S, Saez I et al (2012) A guideline to univariate statistical analysis for LC/MS-based untargeted metabolomics-derived data. Metabolites 2:775–795
17. Zacharias H, Hochrein J, Klein M et al (2013) Current experimental, bioinformatic and statistical methods used in NMR based metabolomics. Curr Metabolomics 1:253–268
18. van Horn WD, Beel AJ, Kang C et al (2010) The impact of window functions on NMR-based paramagnetic relaxation enhancement measurements in membrane proteins. Biochim Biophys Acta Biomembr 1798:140–149
19. Vu T, Laukens K (2013) Getting your peaks in line: a review of alignment methods for NMR spectral data. Metabolites 3:259–276

Chapter 5

Practical Guidelines for ^{13}C-Based NMR Metabolomics

Arthur S. Edison, Adrien Le Guennec, Frank Delaglio, and Ēriks Kupče

Abstract

We present an overview of ^{13}C-based NMR metabolomics. At first glance, the low sensitivity of ^{13}C relative to ^{1}H NMR might seem like too great an obstacle to use this approach. However, there are several advantages to ^{13}C NMR, whether samples can be isotopically enriched or not. At natural abundance, peaks are sharp and largely resolved, and peak frequencies are more stable to pH and other sample conditions. Statistical approaches can be used to obtain C–C and C–H correlation maps, which greatly aid in compound identification. With ^{13}C isotopic enrichment, other experiments are possible, including both ^{13}C-J-RES and INADEQUATE, which can be used for de novo identification of metabolites not in databases.

NMR instrumentation and software has significantly improved, and probes are now commercially available that can record useful natural abundance 1D ^{13}C spectra from real metabolomics samples in 2 h or less. Probe technology continues to improve, and the next generation should be even better. Combined with new methods of simultaneous data acquisition, which allows for two or more 1D or 2D NMR experiments to be collected using multiple receivers, very rich datasets can be collected in a reasonable amount of time that should improve metabolomics data analysis and compound identification.

Key words ^{13}C NMR detection, Metabolomics, Natural abundance ^{13}C, Isotopic labeling, Multiple receivers, Parallel NMR, Simultaneous acquisition, PANACEA, INADEQUATE, ^{13}C J-RES

1 Introduction

Most NMR-based metabolomics is done by detecting protons (^{1}H). There are several reasons for this: (1) ^{1}H NMR is more sensitive than other biological nuclei, including ^{13}C; (2) ^{1}H atoms are found in most metabolites at nearly 100% natural abundance, so ^{1}H NMR is essentially a universal detector; (3) most biomolecular NMR equipment is optimized for ^{1}H detection; and (4) metabolite libraries and techniques for analysis tend to be focused on ^{1}H NMR, especially one dimensional (1D) ^{1}H NMR.

So why would anyone use ^{13}C detection in NMR metabolomics? The frequency of ^{13}C is 4× lower than ^{1}H. For example, at 14.1 T, ^{1}H resonates at 600 MHz and ^{13}C at 150 MHz. NMR spectroscopy already suffers from relatively low sensitivity, and ^{13}C

G. A. Nagana Gowda and Daniel Raftery (eds.), *NMR-Based Metabolomics: Methods and Protocols*, Methods in Molecular Biology, vol. 2037, https://doi.org/10.1007/978-1-4939-9690-2_5,

is worse than ^{1}H. On top of that, the natural abundance of ^{13}C is only about 1.1%, further lowering the overall sensitivity by 100 fold for samples without isotopic labeling. Fortunately, new approaches using modern NMR spectrometers allow for simultaneous experiments to be collected on the same sample. Therefore, during the time that a ^{13}C experiment is being signal averaged, other types of 1D and 2D data can be obtained with no additional time cost. Before discussing the several advantages of ^{13}C NMR spectroscopy for metabolomics, it is useful to consider two very different scenarios of applications, those at natural abundance ^{13}C and those with isotopic enrichment.

The low natural abundance of ^{13}C hurts overall sensitivity but has a major advantage, namely, almost no homonuclear (e.g., ^{13}C–^{13}C) J-coupling. As a result, ^{13}C resonances at natural abundance are sharp and narrow. This is in contrast to 1D ^{1}H NMR spectroscopy, which has extensive ^{1}H–^{1}H couplings, leading to more complexity and overlap. It is simple to apply heteronuclear decoupling (e.g., ^{1}H from ^{13}C), but homonuclear decoupling is difficult and typically suffers from some degree of sensitivity losses. Finally, the ^{13}C has a considerably broader chemical shift range (~200 ppm for ^{13}C vs ~10 ppm for ^{1}H) that—in combination with the lack of homonuclear couplings at natural abundance—offers advantages in spectral resolution. Sample conditions such as pH and salt cause large changes in some ^{1}H NMR chemical shifts, leading to challenges in spectral alignment, but ^{13}C resonances are less sensitive to these variable sample conditions [1, 2]. Moreover, ^{13}C NMR chemical shifts can be accurately calculated using ab initio [3] or even (to a lesser extent) semiempirical methods, which enables improvements in compound identification and validation [4, 5].

Some samples can be isotopically enriched with ^{13}C, which compensates for low natural abundance. However, adding ^{13}C creates new problems through additional ^{13}C–^{13}C J-couplings. These can be eliminated through a 2D ^{13}C J-RES experiment, which is described in the succeeding text. The ^{13}C–^{13}C J-couplings also offer the possibility of obtaining direct carbon correlation experiments, which can be extremely useful in compound identification and database matching. We describe the use of INADEQUATE experiments for isotopically enriched samples. Finally, isotopic labeling also allows for ^{13}C flux measurements, which can be used to investigate biochemical pathways. Isotopic flux studies are beyond the scope of this chapter and have been recently reviewed [6, 7].

The scope of this chapter is summarized in Table 1. For ^{13}C NMR it is particularly important to consider whether or not the sample is isotopically labeled in picking the right NMR experiments. In Table 1, a single X indicates that that combination can be done and XX indicates that combination is especially useful. If there is no X, that combination is either impossible or impractical.

Table 1
Types of experiments and samples covered in this chapter

Sample	1D ^{13}C	^{13}C-J-RES	INADEQUATE	Multiple receivers
Natural abundance	XX			XX
^{13}C enriched	X	XX	XX	X

The columns represent different types of experiments that we cover in this review and the rows are types of samples (natural abundance or ^{13}C enriched). Not all experiments work equally well with all samples. X indicates that the combination works, XX indicates that it works well, and blank indicates it is an inappropriate combination that will not work or be needed

1.1 1D ^{13}C

Figure 1 compares 1D ^{13}C and 1D ^{1}H NMR spectra collected on the same sample of a mixture of 20 common metabolites at natural abundance ^{13}C [8]. A major advantage of ^{13}C can be seen by comparing, for example, the positions of peaks from isoleucine from both ^{1}H and ^{13}C spectra. The ^{1}H spectrum is more overlapped than the ^{13}C spectrum. This is due to less overall chemical shift dispersion (10 ppm for ^{1}H and 200 ppm for ^{13}C) as well as the absence of J-couplings in the ^{13}C dataset. The spectra shown in Figs. 1 and 2 were collected on a specialized 1.5-mm HTS ^{13}C probe [8] described later in Materials where we discuss various options of instrumentation for ^{13}C NMR.

Figure 2 shows 10 replicates of slightly different mixtures of the same 20 metabolites from Fig. 1, which are meant to represent different biological replicates from a metabolomics study [2]. Using well-established methods to statistically correlate NMR resonances [10], it is possible to obtain carbon correlation networks using just 1D ^{13}C data [2]. The 1D ^{13}C data also allow for improved interpretations of loadings from multivariate scores plots, because the resolution between peaks is greater than the corresponding 1D ^{1}H data.

1.2 ^{13}C-J-RES

As described previously, at natural abundance 1D ^{13}C NMR yields narrow singlets, which significantly improve resolution and interpretation. The reason for these narrow peaks is that the probability of 2 adjacent ^{13}C atoms at natural abundance is 0.011×0.011, so it only occurs once in 8264 molecules, resulting in essentially no ^{13}C–^{13}C couplings. However, with ^{13}C-enriched samples, these couplings are present in the spectrum, making interpretation difficult, especially in the case of complex mixtures. Eliminating these couplings in 1D spectra is very difficult and leads to significant losses in sensitivity, which is especially important in ^{13}C spectroscopy, as mentioned in previous sections.

One of the easiest ways to deal with the ^{13}C–^{13}C couplings is to separate the J-coupling and the chemical shift information in two different dimensions, which is accomplished using the 2D J-RES (J-resolved) pulse sequence [11]. The ^{1}H version of the J-RES

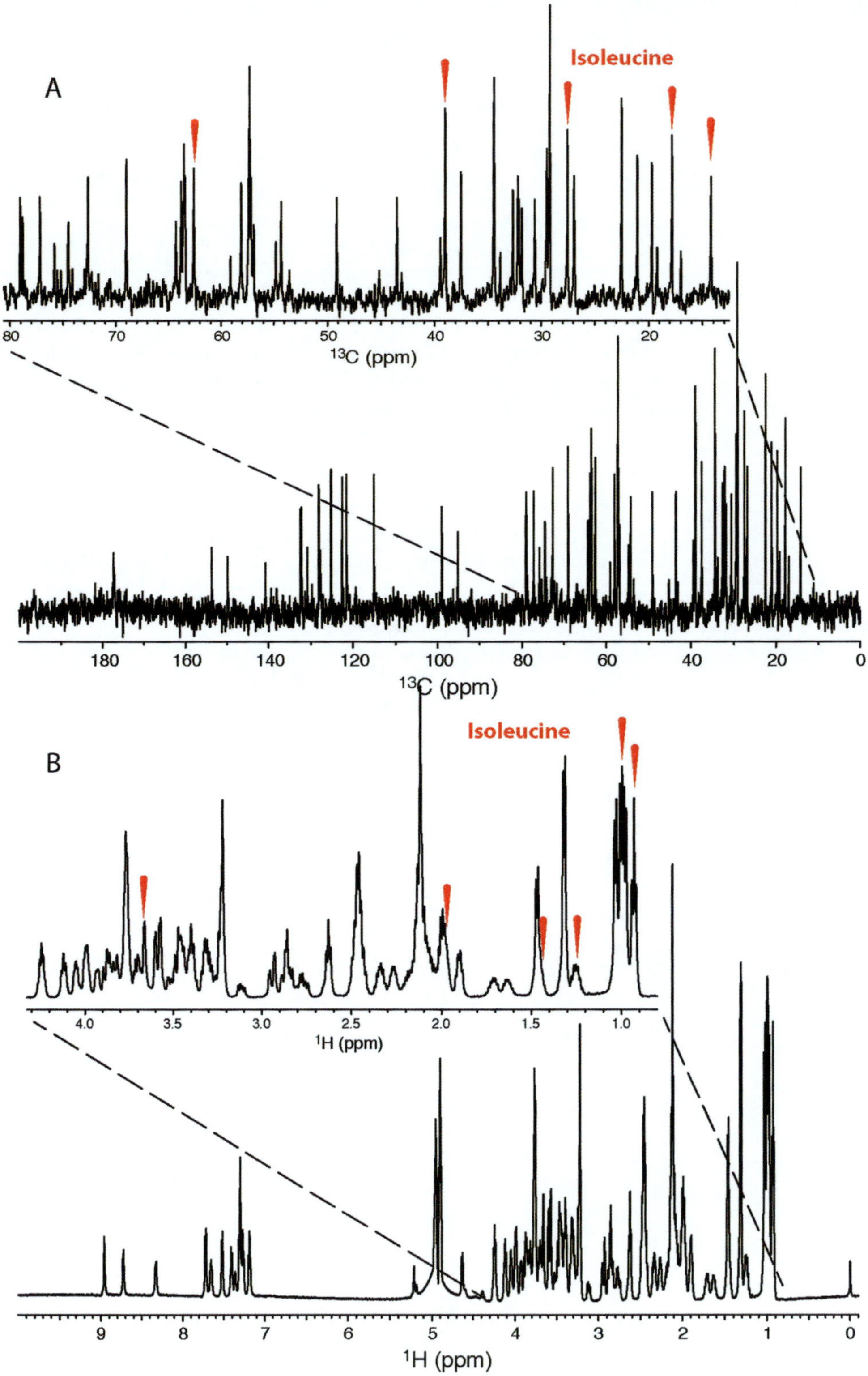

Fig. 1 Synthetic mixture of 20 common metabolites ranging in concentration between 1 and 5 mM. (**a**) 1D ^{13}C NMR spectrum collected in 2 h using a custom 1.5-mm HTS ^{13}C probe with parameters outlined in

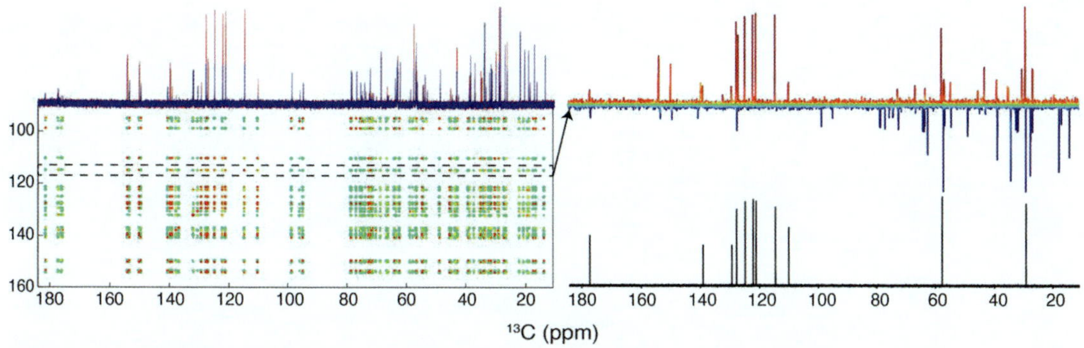

Fig. 2 1D ^{13}C NMR data from the same mixture shown in Fig. 1 was collected on 10 different mixtures representing 2 groups with variation between samples (upper left, red and blue overlaid experimental spectra). These experimental spectra were used to generate the 2D statistical correlation map shown in red and green (lower left). The red dots represent high correlation values between ^{13}C resonances. The trace from the dashed region expanded on the right predominantly corresponds to the carbon network of tryptophan (upper right), and the black spectrum below is tryptophan from the BMRB database [9]. Reproduced with permission from [2]

experiment has been in use for decades and has seen modest use in metabolomics workflows [12]. Normally, J-RES is not needed for ^{13}C, because as described earlier there are no direct ^{13}C–^{13}C couplings at natural abundance. However, the standard ^{1}H J-RES pulse sequence can be easily applied to ^{13}C, but we recommend using adiabatic pulses for more uniform excitation of the large spectral width of ^{13}C and a 180° pulse before the pulse sequence itself in order to allow fast pulsing, if desired. As seen in Fig. 3, a ^{13}C-enriched *C. elegans* extract can be difficult to analyze with a 1D ^{13}C spectrum, even with ^{1}H–^{13}C decoupling. By contrast, the ^{13}C J-RES is less overlapped and easier to read and interpret. Additionally, with partial labeling, it should be possible to measure the fraction of ^{13}C for different sites by comparing the intensity of the multiplet to the singlet at the same chemical shift.

1.3 INADEQUATE

The 2D INADEQUATE experiment (Incredible Natural Abundance DoublE QUAntum Transfer Experiment) was initially developed for natural abundance samples and has been recognized as perhaps the most important NMR experiment for unknown compound identification [13–16]. The problem with this experiment is sensitivity, because it starts with the low probability of neighboring ^{13}C atoms described previously. However, it can be a valuable tool in metabolomics when samples can be isotopically enriched with

Fig. 1 (continued) Subheading 3.3 [8]. (**b**) ^{1}H NMR data collected on the same probe and sample (not simultaneous) in about 3.5 min. The expansions of both spectra roughly correspond to the comparable regions from both spectra, and the peaks from isoleucine (3 mM) are highlighted with red arrows to show different degrees of resolution between the two data types. Reproduced with permission from [8]

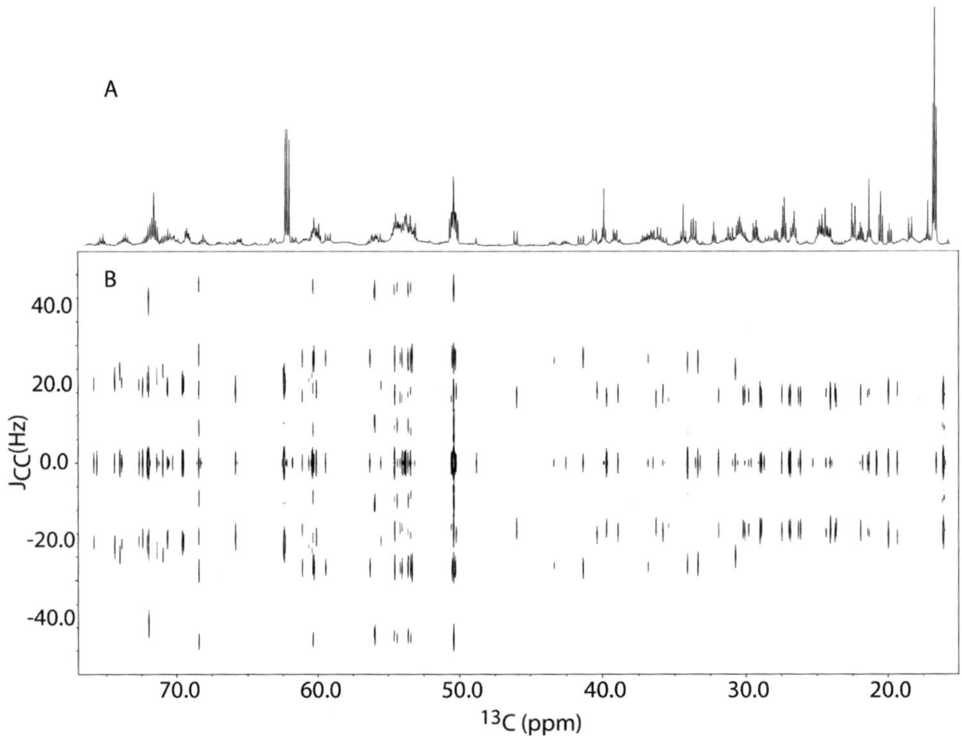

Fig. 3 Expansion of the aliphatic region of uniformly ^{13}C-enriched *C. elegans*. The 1D spectrum in A was obtained with the Bruker *zgig* 1D experiment with ^{13}C detection. The spectral width was 216.57 ppm with the carrier frequency at 100 ppm. The acquisition time was 0.5 s and the recycle delay was 2 s. We acquired 4096 scans, because these datasets were obtained with a conventional (i.e., not ^{13}C optimized) probe. The 2D spectrum in B is a ^{13}C J-RES of the same sample using a modified Bruker pulse sequence called *jresgpprqC13* that is available upon request. B was collected with 32 t_1 points and 32 scans per point. The acquisition SW was 216.57 ppm, and the indirect J dimension was 200 Hz. The ^{13}C carrier frequency was set to 100 ppm. To understand the relationship between the two types of spectra, compare the J-RES data with a particular pattern with the same frequency above in the 1D. If, for example, the J-RES has 2 peaks at a specific ^{13}C frequency (e.g., about 46 ppm), the same frequency in the 1D will be a doublet. Similarly, 3 peaks in J-RES will be a triplet (or doublet of doublets) in the 1D

^{13}C [1]. The advantage of the INADEQUATE experiment is that complete carbon backbones, including quaternary carbons, can be traced with a single experiment. And even though there are no INADEQUATE databases for metabolite matching, it is simple to create a virtual database from assigned 1D ^{13}C spectra, as described [1]. The BMRB contains over 1000 spectra that have been used to construct an INADEQUATE database in our lab for compound matching. Moreover, for compounds not in databases, the networks that can be generated from INADEQUATE data provide an excellent starting point for unknown identification.

1.4 Simultaneous Acquisition

The introduction of multiple receivers in NMR has opened new avenues for simultaneous acquisition of NMR spectra of several nuclear species [17–19]. Simultaneous acquisition is a relatively new area and especially ideal for ^{13}C NMR, because during the relatively long acquisition times that are often required, other more sensitive 1D and 2D experiments can be collected at the same time, enabling more efficient use of spectrometer time and improved datasets for unknown compound identification. We will introduce several ways of doing these experiments in this chapter.

The latest state-of-the-art spectrometers are equipped with a separate receiver on every channel. Up to eight receivers are available on modern NMR consoles allowing recording of up to eight NMR spectra simultaneously. There are several possible ways of recording NMR spectra using multiple receivers. These can be categorized into three main groups: (1) parallel acquisition experiments, (2) sequential acquisition experiments, and (3) interleaved acquisition experiments (*see* Fig. 4).

A simple example of a *parallel acquisition* experiment is the 2D PANSY-COSY experiment that produces the 2D H–H COSY and 2D H–C COSY spectra in a single pass (*see* Fig. 5) [20, 21]. Note that acquisition of the 2D H–C correlation spectra in the second receiver requires no extra time. Therefore, two (or more) spectra can be recorded in the time that is spent to acquire one conventional spectrum, 11 min in this example.

An example of *sequential acquisition* experiment is the TOCSY/HETCOR pulse sequence [20]. In these experiments

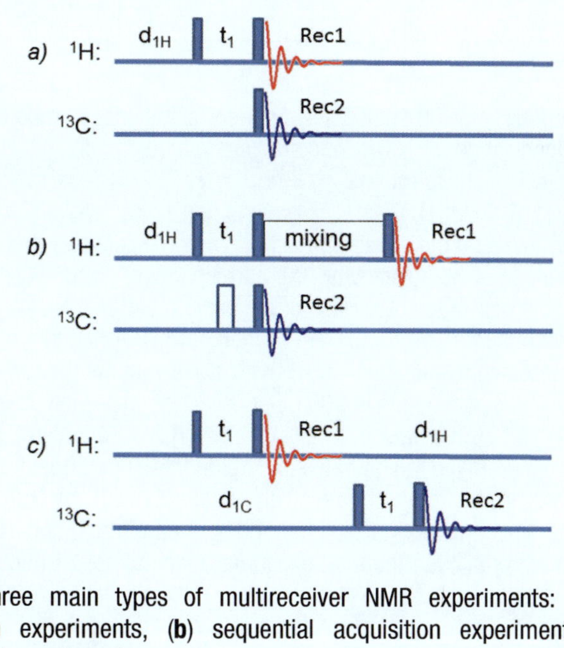

Fig. 4 Three main types of multireceiver NMR experiments: (**a**) parallel acquisition experiments, (**b**) sequential acquisition experiments, and (**c**) interleaved experiments

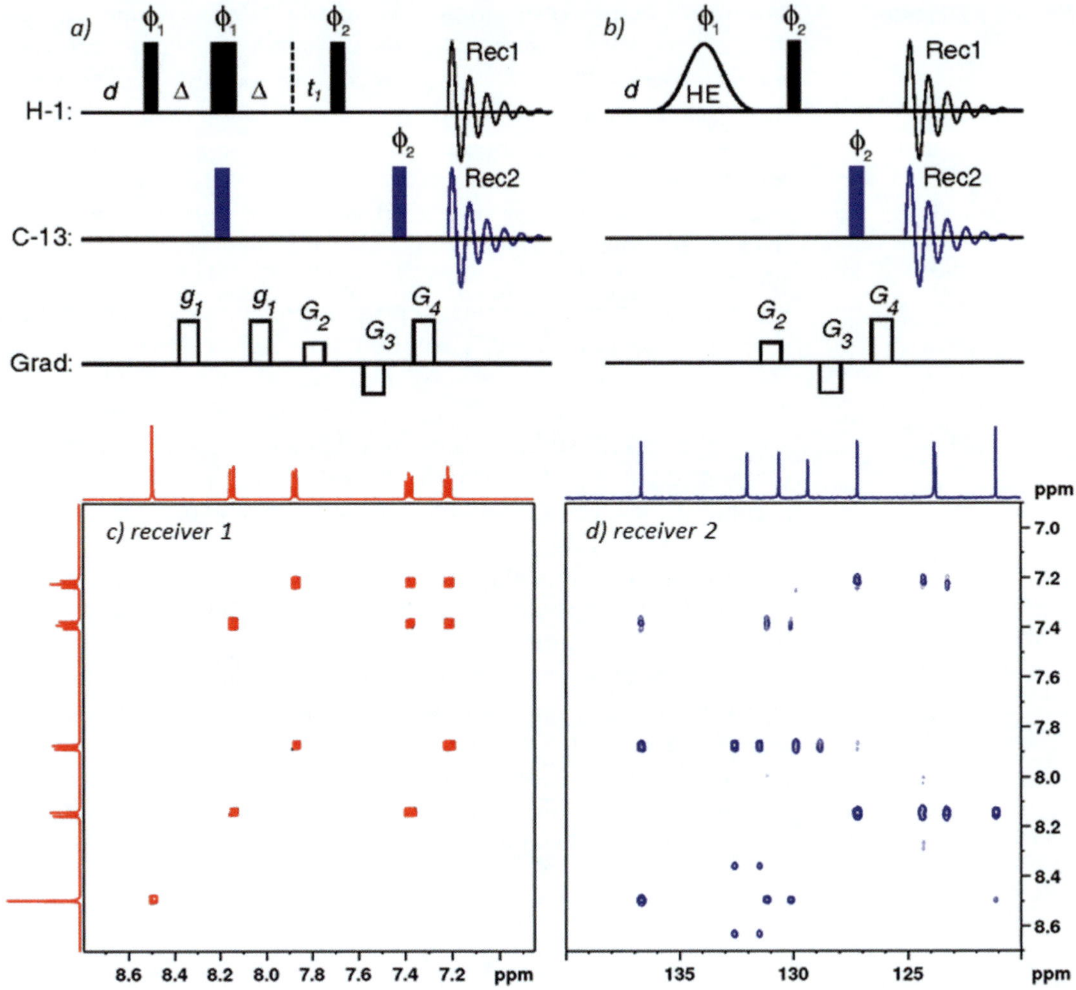

Fig. 5 (**a**) The PANSY-gCOSY pulse sequence for simultaneous recording of 2D H–H COSY and 1J-filtered 2D H–C COSY (long-range HETCOR) spectra (magnitude mode). All pulses are applied with phase x unless indicated otherwise; phase cycling: $\phi_1 = x, -x$, $\phi_2 = 2x, 2(-x)$, rec(^{1}H) = rec(^{13}C) = $x, -x$. Gradients (G/cm, ms): $g_1 = (7, 1)$, $G_2 = (5, 1)$, $G_3 = (15, 1)$, $G_4 = (20, 1)$, $G_3 + G_4 = G_2$, $G_4/G_2 = \gamma C/\gamma H$. Axial peak displacement is achieved by inverting the phase of the 180° proton pulse in every other increment. The optional J-filter can be switched off by simply removing the corresponding (first) ^{13}C pulse. The following phase cycling was employed for the non-gradient version of the experiment: ϕ_1, rec1 and rec2 = $x, -x, -x, x$; $\phi_2 = x, -x$ and phase ϕ_1 is incremented for States-TPPI scheme. (**b**) The Hadamard encoded version of the experiment, HE, denotes the Hadamard encoding pulse, typically a multifrequency Gaussian 90° pulse; (**c**) 2D H–H COSY spectrum of pamoic acid 50 mg in 500 μL of DMSO-d6, recorded using the PANSY-gCOSY pulse sequence; (**d**) ^1J-filtered 2D H–C COSY spectrum (long-range HETCOR) recorded in the same experiment showing an expansion of the aromatic region between 120 and 140 ppm. Spectra were recorded on Avance III HD system operating at 600-MHz proton frequency and equipped with the DCH cryoprobe. The 2D data matrices were recorded using spectral width (F1 × F2) of 6 × 6 ppm (rec1) and 182 × 6 ppm (rec2) and 256 × 4 k data points (rec1 and rec2), with the total experiment time of ca. 11 min

the ^{13}C-detected H–C one-bond correlation spectrum (HETCOR) is recorded during the mixing period of the H–H TOCSY pulse sequence which also provides proton decoupling for the HETCOR spectrum (*see* Fig. 6). While the ^{13}C-detected HETCOR experiment is inferior to the ^{1}H-detected version—HSQC—in terms of sensitivity, it benefits from better resolution, considerably smaller F1 bandwidth, lower decoupling power requirements, and the associated artifacts (heating effects, potential line shape distortions and decoupling sidebands). This is particularly important considering the trends towards higher ^{13}C sensitivity NMR probes and increasingly high magnetic fields in NMR.

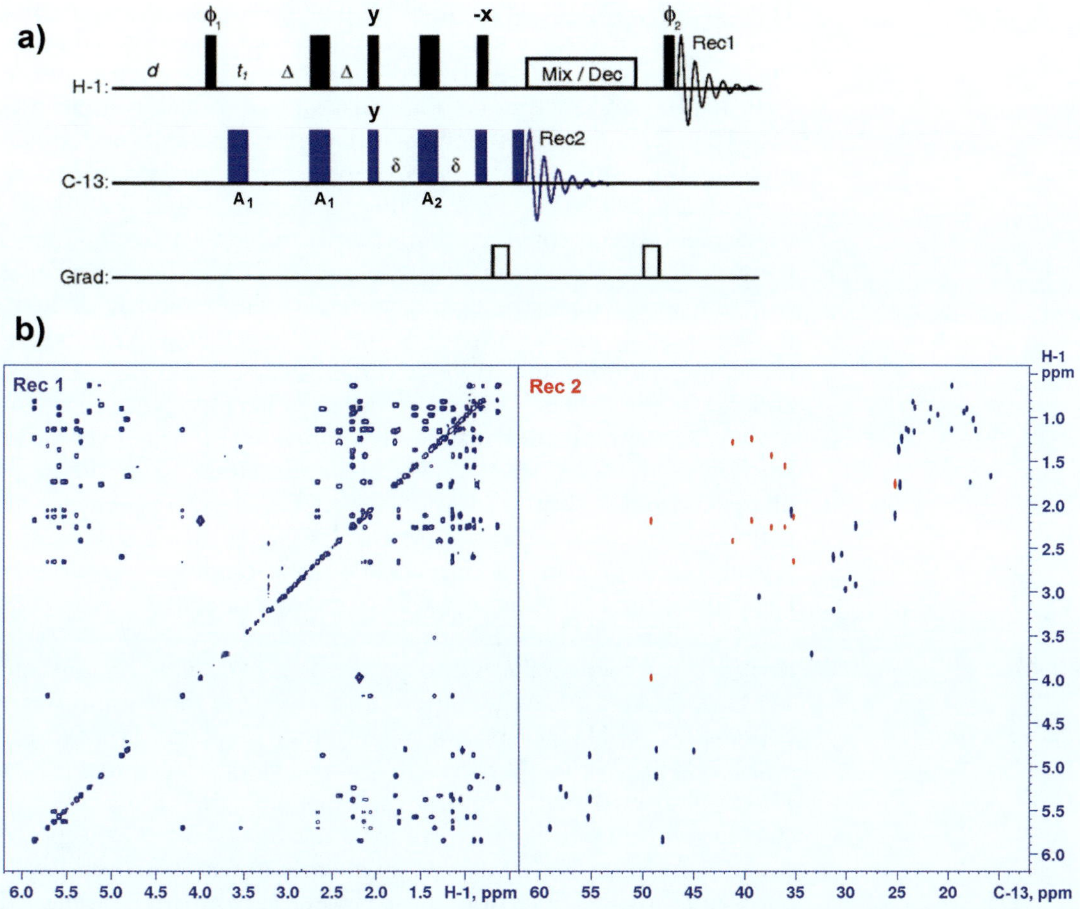

Fig. 6 The dual-receiver PANSY-TOCSY experiment: (**a**) the pulse sequence, (**b**) the 2D H–H TOCSY spectrum (rec1) and the ^{13}C-detected 2D one-bond H–C correlated HETCOR spectrum (rec2) recorded on a 600-MHz Avance-NEO system equipped with the TCI cryoprobe and four receivers; in the HETCOR spectrum the negative (CH_2) peaks are in red and positive (CH and CH_3) peaks are in blue; the DIPSI-2 mixing time was 100 ms; all pulses are applied with phase x unless indicated otherwise; phase cycling: $\phi_1 = x, -x$, $\phi_2 = x, x, -x, -x$, rec1 $= x, -x, -x, x$, rec2 $= x, -x$; A_1 is 500-us-long adiabatic CA WURST-20 pulse, A_2 is 2-ms-long composite adiabatic refocusing pulse. The sample was 50 mM cyclosporine in C_6D_6; with 256 increments and 2 scans per increment the total experiment time was 12 min

The TOCSY/HETCOR-Q experiment [21] combines the time-sharing method [22] with the use of multiple receivers, thus recovering the information about non-protonated carbon sites. Three spectra are recorded simultaneously in this experiment: the 1D ^{13}C spectrum of non-protonated carbon resonances, the 2D H–H TOCSY spectrum, and the ^{13}C-detected one-bond H–C correlation spectrum. Note that time-sharing requires twice as many scans as the conventional version of the experiment, but delivers (by a factor of $\sqrt{2}$) better sensitivity as compared to the two experiments recorded separately.

The PANACEA experiment (Fig. 7a) combines the 2D C–C INADEQUATE pulse sequence with 2D H–C HSQC and 3D H–C HMBC experiments to allow structure elucidation of small molecules from a single measurement [23–26]. As described previously, the INADEQUATE experiment provides unambiguous information about the carbon skeleton of organic molecules [14–16]. It is one of the most reliable experiments for small-molecule structure elucidation. Combined with the HSQC experiment that provides H–C connectivity information and the HMBC experiment that provides connectivities of molecular fragments connected via heteroatoms, the experiment (PANACEA) provides complete molecular structure essentially unambiguously. Due to the intrinsically low sensitivity of the 2D INADEQUATE experiment, several more sensitive experiments, such as 2D H–C HSQC and 2D or 3D HMBC, can be recorded in parallel. Note that 99% of the ^{13}C magnetization is typically suppressed in the conventional versions of the INADEQUATE experiment either by gradients or phase cycling or both. In the PANACEA pulse sequence, this magnetization is refocused, and the ^{13}C frequency encoded and transferred to protons via one-bond or long-range C–H couplings to produce the HSQC and HMBC spectra, respectively.

The basic PANACEA experiment depicted in Fig. 7a is designed for two receivers and produces a 1D ^{13}C spectrum and 2D C–C INADEQUATE in the ^{13}C receiver and three multiplicity edited 2D ^{13}C HSQC spectra and several 2D spectra or a full 3D ^{13}C J-HMBC in the ^{1}H receiver. There is an option to record the HMBC spectra with several settings of the long-range J (CH) evolution period or to record a full 3D J-HMBC spectra to cover a wider range of the C–H couplings. In high-sensitivity samples a shorter 2D version of the HMBC experiment is employed producing the PANACEA spectra in just 23 min [24]. In low-sensitivity samples the full 3D version of the HMBC experiment is used.

The spectra of the hormone melatonin recorded using the PANACEA pulse sequence are shown in Fig. 8 along with the structure elucidation steps. The 1D ^{13}C spectrum (Fig. 8a) provides information about the number of carbon atoms in the molecule. The 2D ^{13}C INADEQUATE spectrum reveals the C–C

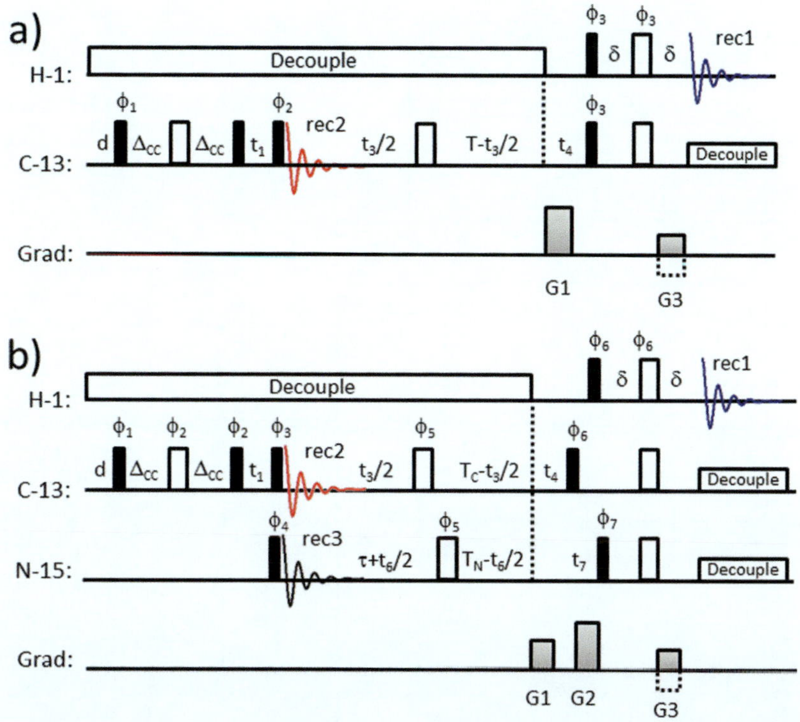

Fig. 7 (**a**) The basic PANACEA pulse sequence designed for two receivers; delays: $\Delta_{CC} = 0.25/^1J_{CC}$, $T = t_2 - t_4$, where t_2 is the acquisition time for receiver *2* (typically *ca* 80 ms) and t_4 is incremented in steps of $0.25/J_{CH}$ to record the multiplicity edited HSQC. Following the first three t_4 increments the ^{13}C decoupling is switched off and the experiment is continued with the t_4 delay set as required for the HMBC experiments; all pulses are applied with the phase x unless indicated otherwise; phase cycling, $\phi_1 = x, -x$ $\phi_2 = x, x, -x, -x;\phi_3 = y, y, -y -y$; rec1 = rec2 = $x, -x, -x, x$; the coherence selection gradients G_1 (40 G/cm) and G_2 (10 G/cm) are 1 ms long; (**b**) the PANACEA sequence designed for three receivers. Delays: T_C, t_2, and t_4 are set as described in (**a**); $T_N = t_2 - t_7$ where t_7 is the J_{NH} evolution delay, t_3 and t_6 are the *FI* evolution delays for the ^{1}H detected experiments; phase cycling, $\phi_1 = 4(x, -x)\,4(-x, x)$; $\phi_2 = x;\phi_3 = 2x, 2(-x);)$; $\phi_5 = x$; $\phi_6 = \phi_7 = y, y, -y -y$; ϕ_4 = rec1 = rec2 = $x, -x, -x, x, -x, x, x, -x$; the coherence selection gradients G_1 (20 G/cm), G_2 (20.5 G/cm) and G_3 (5 G/cm) are 1 ms long. Adopted from [23]

connectivity information (*see* Fig. 8b). Single-bond C–H correlations provided by the HSQC spectra (*see* Fig. 8c) identify the protonated carbon atoms and together with the chemical shift data and the multiplicity information provide indirect evidence about the presence of magnetically inactive heteroatoms in the molecule. Finally, the long-range H–C correlations revealed in the HMBC spectra (*see* Fig. 8d) link the molecular fragments separated by heteroatoms. It has been demonstrated that in the PANACEA experiments, the long-range C–H couplings can be measured with

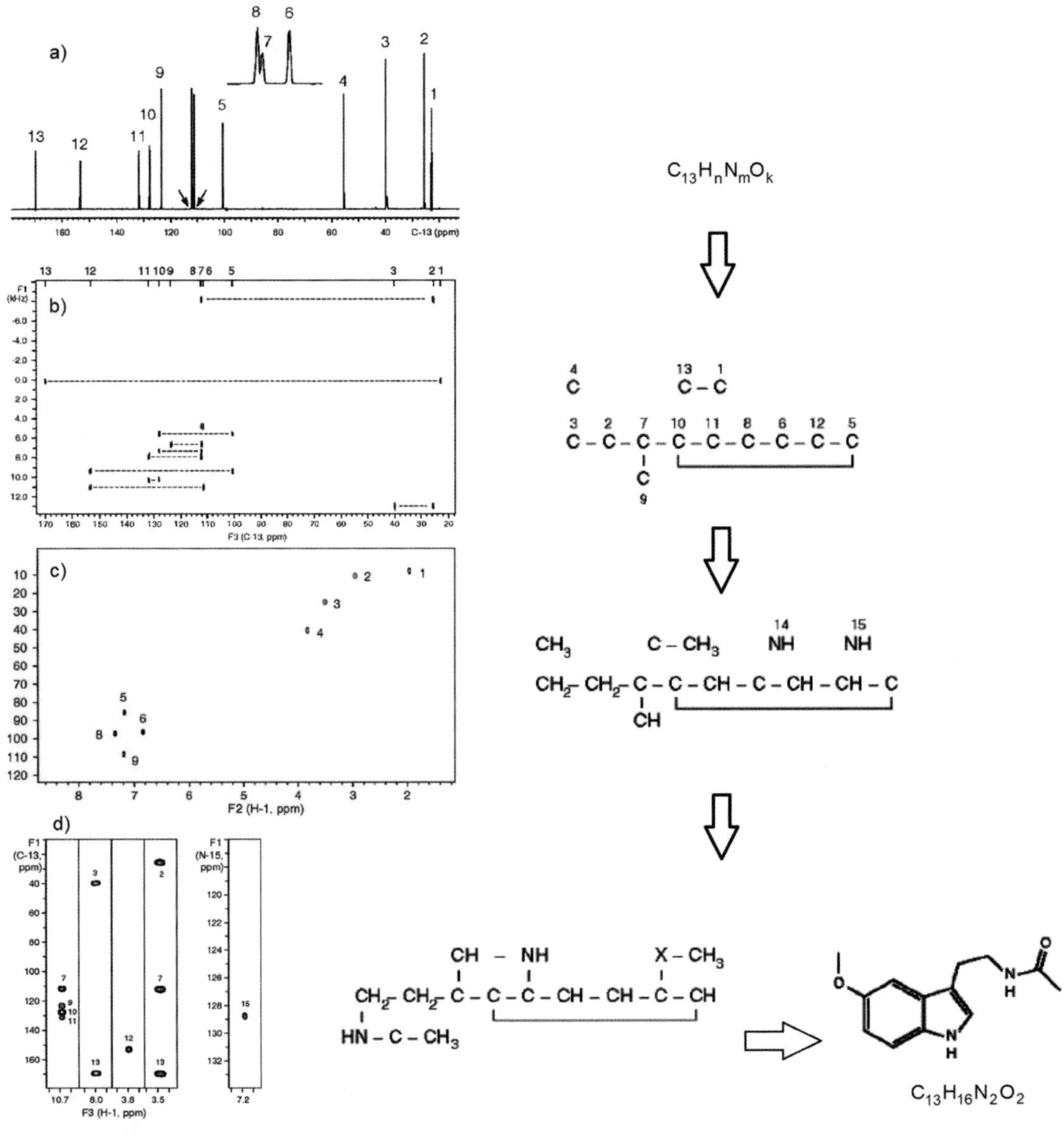

Fig. 8 The PANACEA spectra recorded using the pulse sequence of Fig. 7b are shown along with the structure elucidation steps on the right; (**a**) the 1D ^{13}C spectrum establishes the number of carbon atoms; (**b**) the 2D INADEQUATE spectrum provides the connectivity of the carbon atoms; (**c**) the HSQC spectra indicate the protonated carbon sites and the number of attached protons; (**d**) the long-range H–C and H–N correlations in the HMBC spectra connect the molecular fragments via the heteroatoms

high accuracy providing a wealth of stereospecific information about the structure of small organic molecules (*see* Fig. 9).

In the PANACEA experiments the 1D ^{13}C spectra may be preserved for each individual scan or time increment. It has been shown that such 1D ^{13}C traces can be used as a field/frequency lock signal thus eliminating the necessity for the conventional deuterium lock and opening the possibilities to record the PANACEA

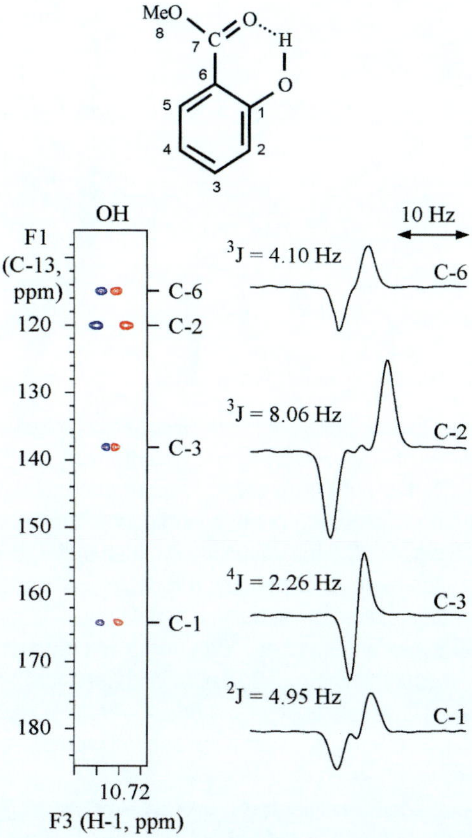

Fig. 9 Long-range C–H couplings (antiphase signals) recorded for methyl salicylate in the 3D HMBC spectra of the high-resolution PANACEA experiment. The timings (ms) are shown for the best F1F3 planes that were chosen used to measure the long-range C–H coupling constants. Adapted from [25]

spectra in pure liquids, such as peanut oil [24]. The INEPT-INADEQUATE version of the PANACEA experiment [25] can provide sensitivity enhancement in proton-rich molecules with only a few non-protonated carbon atoms.

A pulse sequence combining the use of multiple receivers and projection spectroscopy [27, 28] for metabolomics applications at natural abundance has been designed to allow fast data acquisition and resonance assignments (*see* Fig. 10) [29, 30]. The experiment requires two receivers and delivers three 2D spectra in a single measurement: (tilted) HSQC-TOCSY, 2D H–H TOCSY, and 2D H–C HETCOR. The three spectra provide mutually complementary information that is sufficient for reconstruction of the 3D HSQC-TOCSY spectrum from the two orthogonal projections (TOCSY and HETCOR) and the tilted HSQC-TOCSY thus facilitating high-resolution and unambiguous spectral assignments. A mixture of 21 metabolites has been analyzed and assigned using this pulse sequence (*see* Fig. 11).

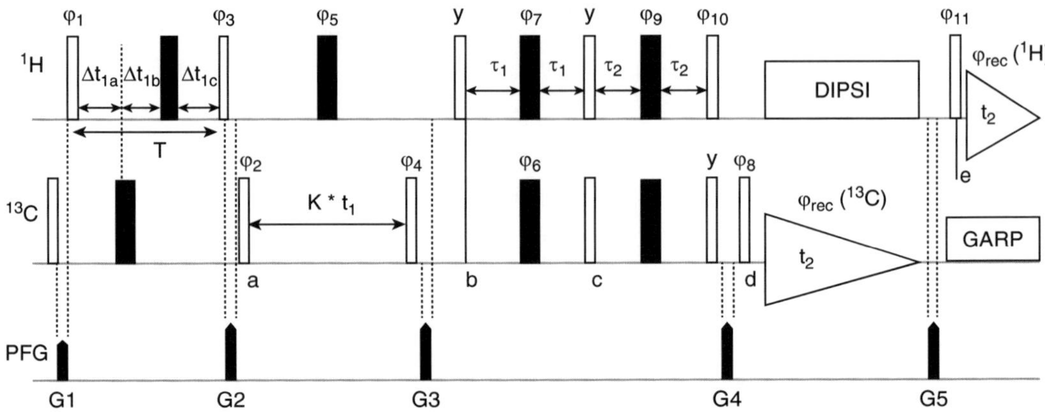

Fig. 10 Dual-receiver HSQC-TOCSY, TOCSY, and HETCOR experiment. The ^{13}C carrier frequency is placed at 40 ppm and the ^{1}H carrier position is set to 4.7 ppm throughout the duration of the experiment. Chemical shift evolution of ^{1}H during t_1 is carried out in a semiconstant time manner. The initial values of $t_{1a}(0)$, $t_{1b}(0)$, and $t_{1c}(0)$ are 0.9 ms, 3 ms, and 0.9 ms, respectively, such that $T = 1.8$ ms. The values of Δt_{1b} and Δt_{1c} are based on the number of increments and the spectral width. For 512 complex points and spectral width of 20 kHz (25 ppm) used for the indirect (^{1}H) dimension, the values of Δt_{1a}, Δt_{1b}, and Δt_{1c} were 25, 23.3, and -1.7 ms. The chemical shift evolution period of ^{13}C during t_1 is co-incremented with that of ^{1}H with a relative scaling factor, κ; and a ^{13}C spectral width of 200 ppm is chosen for the direct dimension in 2D HETCOR and 25 ppm for the indirect (tilted) dimensions. A mixing time of *80 ms* is used for isotropic proton DIPSI mixing. The delays are $\tau_1 = 1.8$ ms, $\tau_2 = 0.9$ ms. All pulses are applied with phase *x* unless indicated otherwise; phase cycling: $\phi_1 = x$; $\phi_2 = x$; $\phi_3 = y, -y$; $\phi_4 = x, -x$; $\phi_5 = x, x, -x, -x$; $\phi_6 = x, x, -x, -x$; $\phi_7 = x, -x$; $\phi_8 = x, x, -x, -x$; $\phi_9 = x, x, -x, -x$; $\phi_{10} = x, -x$; $\phi_{11} = x, x, -x, -x$; $\phi_{rec}(^{13}C) = x, x, -x, -x$; $\phi_{rec}(^{1}H) = x, -x, -x, x$; quadrature detection in t_1 (^{1}H) is carried out in States manner (i.e., $\phi_1 = x, y$); *sinc*-shaped gradients with strengths $G_1 = 19$ G/cm; $G_2 = 19$ G/cm; $G_3 = 19$ G/cm; $G_4 = 19$ G/cm; $G_5 = 11$ G/cm were used. Reproduced from [29]

The experiments involving ***parallel*** and/or ***sequential*** acquisition typically require designing new pulse sequences. The main advantage of this type of multiple acquisition is that additional data are obtained with essentially no time penalty. The repetition time of such experiments is that of the fastest relaxing nuclei (typically ^{1}H) and the polarization of the high gamma nuclei (^{1}H, ^{19}F) can be used for sensitivity enhancement of low gamma nuclei (^{13}C and ^{15}N). The ***interleaved experiments*** typically do not require design of new pulse sequences and only involve minor modifications to account for the availability of several receivers. In dual-receiver interleaved experiments, one of the pulse sequences is placed into the recovery period of the other experiment (*see* Fig. 4c). The drawback of this technique is that the total repetition time is that of the slowest relaxing nuclei [30]. Furthermore, unless the relaxation properties are very different, polarization transfer and decoupling are usually avoided so as not to disturb the nuclei that are in the recovery mode. This limits the applications of interleaved experiments to systems that are not mutually coupled or systems with vanishingly small couplings. These problems may be

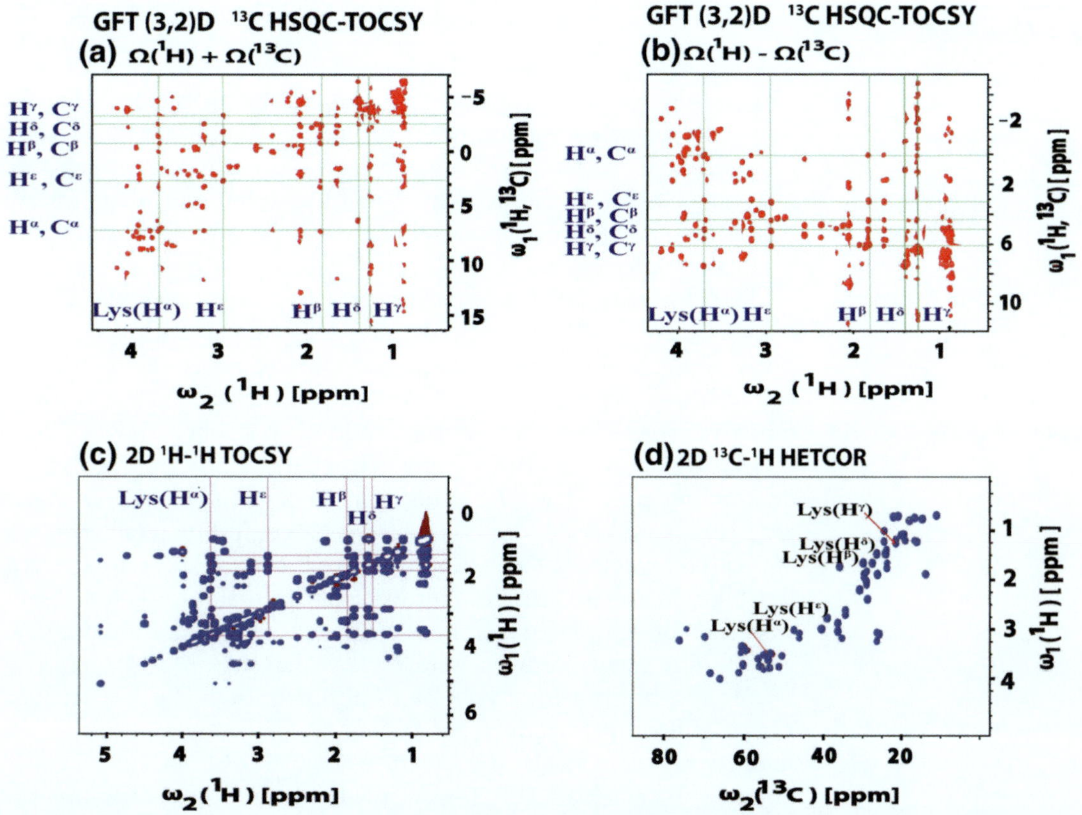

Fig. 11 Spectra acquired for the mixture of 21 natural abundance metabolites in 10.5 h using the pulse sequence shown in Fig. 10 (each metabolite has a concentration of ca. 5–7 mM); (**a**) tilted ($+\alpha$) ^{13}C HSQC-TOCSY, (**b**) tilted ($-\alpha$) ^{13}C HSQC-TOCSY, (**c**) 2D H–H TOCSY, and (**d**) 2D H–C HETCOR. For the purpose of illustration of the peak pattern observed, the spin system corresponding to lysine is shown in the three spectra. The complete assignment of the metabolites was obtained using this set of spectra. Reproduced from [29]

avoided in experiments that are designed for coupling constant measurements. The potential of the interleaved acquisition in metabolomics applications has yet to be explored.

While the majority of the experiments described in this section have not yet been applied in metabolomics research, the individual components of the multiple receiver experiments, such as 1D ^{13}C, TOCSY, COSY, INADEQUATE, HSQC, and HMBC, and the potential applications of the multireceiver experiments are extensive. The experiments described previously may need some adjustments for such applications. So far, the limiting factor has been ^{13}C sensitivity. With progress made in the probe development and the NMR hardware described later, these experiments should significantly increase the efficiency and the information content of the NMR measurements in metabolomics research.

2 Materials

The primary requirement for ^{13}C NMR is a spectrometer and probe that are capable of making the measurements. Current-generation consoles (e.g., Bruker NEO) have multiple receivers incorporated into the design, so implementation of simultaneous experiments described earlier is possible and can easily be added to a metabolomics workflow.

Most modern NMR spectrometers are able to record ^{13}C, but the major factor that impacts the feasibility of these measurements is the NMR probe. The most common way to measure ^{13}C is using a "broadband" probe, which is configured with an inner coil (most sensitive) that can be tuned to a range of frequencies that include ^{13}C and an outer coil (less sensitive) that is dedicated to ^{1}H for decoupling or nuclear Overhauser enhancement of ^{13}C. The coils for broadband probes are copper wire based and generally operate at room temperature (RT). Broadband probes are useful for general flexibility in a shared instrument but not optimal for ^{13}C sensitivity. The size of sample tubes is an important consideration for ^{13}C NMR. The most common sample tube has an outer diameter of 5 mm and holds a total volume of about 550–600 μL of analyte. This size fits most needs in general NMR applications, and the ^{13}C signal-to-noise (S/N) specifications for a 600-MHz (^{1}H frequency) broadband RT 5-mm probe is about 330:1 for the ASTM reference standard (60% C_6D_6 and 40% 1,4-dioxane). NMR probes are available that accommodate tubes in both larger (e.g., 8–10 mm) and smaller sizes (e.g., 1–3 mm). The larger the sample, the greater the overall S/N, but the S/N per mass of sample decreases by about the inverse of the probe diameter. Therefore, the most sensitive measurements for limited material are best done with smaller probes [31–33].

The common RT broadband 5-mm probe described previously is not practical for ^{13}C metabolomics, because the S/N is too low. There are two main ways to improve the S/N for ^{13}C NMR detection, cooling the coils and preamplifiers to reduce noise or making the coils out of material that provides greater signal response (through higher "Q"). Cryogenic ^{13}C-optimized 5-mm probes are available that have a dedicated ^{13}C inner coil with ^{1}H and ^{2}H detected on the less sensitive outer coils. These probes are significantly more sensitive than the RT broadband probe described previously. For example, the Bruker 600-MHz DCH ^{13}C 5-mm cryoprobe has a specification of >3000:1 for the same ASTM standard, an increase of nearly tenfold over the S/N of the RT probe. The drawback to this is that the probe has less overall flexibility and cannot be used for other nuclei except ^{1}H. Also, it costs much more money to purchase and is more difficult and costlier to maintain than a conventional RT probe. However, it makes ^{13}C metabolomics feasible.

The NMR vendor(s) do not typically advertise the material that is used for the coils in the probe, as this information is considered proprietary. However, the type of coil can have a very large effect on the performance. We have worked with instrument vendors and colleagues at the National High Magnetic Field Laboratory to develop very-high-sensitivity NMR coils made from high-temperature superconducting materials (YBCO) [8, 34–38]. A review of these coils and cryogenic systems has been published [33]. These coils are built into standard cryoprobes and use commercial cryogenic systems, but they offer higher sensitivity (at higher cost). We constructed a specialized HTS NMR probe for 1.5-mm samples, which was used to generate data shown in Figs. 1 and 2 [8]. Because the sample volume for this probe is only about 35 μL, its absolute S/N is lower than the DCH cryoprobe described earlier. However, the ASTM measurement for this probe is about 400:1, and if this is scaled to a 5-mm sample tube with a 600 mL volume, the S/N is over 6700:1, more than twofold greater mass sensitivity than the commercial copper wire DCH probe. Some applications using this probe were described previously [1, 2, 5, 39]. We are currently developing a larger-volume HTS ^{13}C cryoprobe that should have even greater absolute S/N.

The supplies needed for ^{13}C NMR are largely the same as for ^{1}H NMR. Samples need to be dissolved in a buffer that has at least 5% deuteration for a lock (e.g., 5% D_2O/95% H_2O or 100% D_2O). For natural abundance ^{13}C studies, nothing else is needed beyond what is used for ^{1}H NMR. For studies using isotopically enriched samples, it is necessary to grow the sample on a labeled carbon source that will lead to either uniform or desired labeling pattern. For metabolomics applications, it is desirable to have an automated sample changer that keeps samples cold when not being measured. Even with the most sensitive ^{13}C probe, the measurement time for a typical sample is rarely less than 2 h for a 1D spectrum, so large-capacity sample changers are not necessary for ^{13}C NMR.

3 Methods

Refer to Table 1 for the combinations of samples and experiments described later.

3.1 Sample Preparation (Natural Abundance ^{13}C)

Samples at natural abundance for ^{13}C NMR metabolomics are prepared exactly like those for ^{1}H NMR metabolomics. Some important general considerations that apply to any NMR metabolomics study are listed in the following:

1. Randomize sample collection, preparation, and data acquisition. This will help avoid batch effects.
2. Establish regular quality assurance/quality control by including solvent and extraction blanks as well as external and internal pooled samples. Our lab adds solvent and extraction blanks at the start, middle, and end of each rack of 96 samples. External and internal pooled controls comprise about 10% of the samples.
3. Barcoding samples is useful in metabolomics to avoid confusion, especially with randomized samples and additional QA/QC samples. Some commercial vendors offer racks of NMR tubes with barcodes on the caps, which helps with large numbers of samples.
4. An advantage of ^{13}C detection is that ^{1}H signals from buffers or solvents do not interfere with the NMR data. However, it is still worth choosing buffers/solvents that will allow for good ^{1}H data, since that is always helpful and can be obtained simultaneously with multiple receivers.
5. Samples should either be extracted from fresh material or material that has been frozen at −80 °C. Always prepare metabolomics samples on ice.
6. If sample quantity is not limited (e.g., human urine or serum), then prepare 600 μL samples for 5-mm tubes. For limited samples (e.g., mammalian cells or small quantities of tissue), dissolving the material in smaller volumes and using 3-mm or 1.7-mm NMR tubes is an advantage, even if the NMR probe is designed for 5-mm tubes.
7. Autosamplers are useful in NMR. If used, it is important that they have sample cooling capabilities to minimize sample degradation, especially with relatively long acquisition times required for 1D ^{13}C measurements.

3.2 Sample Preparation (^{13}C Isotopic Labeling)

Some types of samples can be prepared with ^{13}C stable isotopic labeling. Since each type of sample will be different, specifics are beyond the scope of this chapter. However, a few general considerations can be made that apply to most samples:

1. Samples can be (nearly) uniformly or more specifically ^{13}C labeled. Uniform ^{13}C labeling helps with absolute sensitivity and is useful for ^{13}C–^{13}C correlations from experiments such as INADEQUATE or ^{13}C–^{13}C COSY or TOCSY. These can be helpful for unknown identification or accurate database matching [1, 40] (*see* **Note 1**).
2. For uniform ^{13}C labeling, at least 2 options are available. The most cost effective is to culture material in media with a simple and defined carbon source (e.g., glucose). If this is not possible,

some rich media with uniform ^{13}C is available, but this option is more expensive.

3. For flux studies or to probe specific pathways, an isotopic substrate can often be directly added to a culture, and the products can be detected by NMR. A recent example of this approach was a demonstration using combinations of labeled ketoacids and branched-chain amino acids (BCAAs) that the BCAT1 enzyme produces in leukemia cells [41].
4. To detect the products from a flux or specific labeling protocol, it is possible to either use a standard 1D ^{13}C experiment (succeeding text) or a 1D version of a ^{13}C HSQC experiment, which filters for ^{1}H that are bonded to ^{13}C.

3.3 Standard 1D ^{13}C NMR for Natural Abundance Metabolomics

As described earlier, a simple 1D ^{13}C NMR spectrum can be very valuable for metabolomics.

The following are steps for 1D ^{13}C NMR using natural abundance ^{13}C samples:

1. Obtain a biological sample as described in Subheading 3.1. Attention should be made to concentrate the sample as much as possible to improve the S/N.
2. Insert the sample into the magnet, tune both the ^{13}C and ^{1}H channels, and lock on the ^{2}H signal in the solvent. When using an autosampler, automatic tuning and matching is a desirable feature on the spectrometer. Shim manually or with automation.
3. The ^{13}C NMR needs to be calibrated to find an accurate 90-degree pulse (*see* **Note 2**). This is difficult with samples at natural abundance ^{13}C, so the most practical approach is to use a more concentrated sample in the same solvent that is isotopically labeled. This can be a reference standard or a sample made for this purpose. The analyte should be stable so that the same sample can be used over time. Typically, these calibrations are only required if the ^{13}C coil tuning has been adjusted, e.g., due to variations in the sample salt concentration.
4. Collect the experimental samples using the pulse calibration from the standard. Rather than use a full 90-degree pulse, we find that a 60-degree pulse with a short recovery time gives the best results (*see* **Note 3**). For standard samples, the center frequency should be 100 ppm and the spectral width should be about 210 ppm. Adjust the number of scans empirically using a representative sample from the study. We generally allow about 2 h for each spectrum, but this will vary depending on available hardware and sample concentrations.
5. For 1D ^{13}C NMR it is also important to properly set the ^{1}H channel. There are two considerations, decoupling during

acquisition and the buildup (or lack thereof) of a ^{1}H–^{13}C NOE. There is rarely, if ever, a situation in which ^{1}H decoupling is not an advantage. If it is not done, the beautiful and narrow non-quaternary ^{13}C singlets at natural abundance will be split into 2 (or more) peaks, each with half (or less) of the original amplitude. Even the ^{1}H–^{13}C coupling value is relatively uninformative for small molecules, because it is simply the large 1-bond J-coupling (~140 Hz for aliphatics). We typically acquire 1D ^{13}C data with a NOE, because that will enhance the intensity of ^{1}H-bonded ^{13}C resonances. However, this makes the absolute quantification even worse and further skews the intensities of the quaternary carbons relative to the aliphatic carbons. However, like the consideration discussed previously, ^{13}C data collected with a ^{1}H NOE can still be relatively quantified.

6. The processing of 1D ^{13}C data needs to account for possibilities of initial points that are distorted. This can happen in high-sensitivity probes. If the baseline is not flat after a standard Fourier transform, then follow the steps outlined in **Note 4**.

3.4 ^{13}C J-RES

1. This experiment is not necessary for samples at natural abundance but helps considerably for ^{13}C-labeled samples.
2. Make the sample as described in Subheading 3.2.
3. Insert the sample and tune, lock, shim the sample, and calibrate a 90° ^{13}C pulse as described in Subheading 3.3.
4. The ^{13}C J-RES pulse sequence is very similar to the ^{1}H J-RES pulse sequence. The only differences are the channel being used (^{13}C instead of ^{1}H) and the inclusion of a 180° pulse between the 90° pulse and the spin echo, in order to allow the use of small angles for excitation and therefore compatible for fast pulsing, which is of special interest for the purpose of hyperpolarization techniques. A customized Bruker pulse sequence called *jresgpprqC13* incorporating these adjustments is available from the authors.
5. Adjust the number of t_1 increments according to the required resolution; typically, between 32 and 128 increments provide adequate resolution in F1. The number of scans per increment is then set depending on the sample concentration and a reasonable total experiment time. The number of t_1 points for the spectrum in Fig. 3 was 32, and it gives a very good separation of peaks in the indirect dimension. It should be possible to use as low as 16 t_1 points and still be able to collect a good quality dataset, just as ^{1}H J-RES.
6. The data are typically processed automatically using dedicated processing commands (macros) that involve window multiplication, Fourier transform, baseline correction, and optional

integration and peak picking. The latter is often carried out in third-party software packages, such as NMRPipe [42]. When the quality of the spectra is poor, e.g., due to high salt concentration, misset parameters, or very low concentrations of sample, manual step-by-step processing and parameter adjustment may be required. The processing of ^{13}C J-RES is identical to ^{1}H J-RES; after 2D Fourier transform, the spectrum needs to be tilted and symmetrized along the center of the F2 dimension in order to produce a spectrum where the chemical shift and the J-couplings information are separated along the two dimensions.

3.5 INADEQUATE

1. This experiment is not possible for most metabolomics samples at natural abundance and requires a high-sensitivity ^{13}C NMR probe.
2. Follow **steps 2** and **3** in Subheading 3.4.
3. We typically use the standard Bruker pulse sequence named *inadphsp*, which incorporates a composite adiabatic refocusing pulse to adequately cover the complete ^{13}C bandwidth.
4. In the INADEQUATE experiment, the spectral width in the F1 dimension, sw1, is twice as wide as in the directly detected (F2) dimension. To adequately digitize the large spectral width, we have used more (e.g., 2048) t_1 increments with fewer (e.g., 4) scans per increment [1]. However, to achieve the best resolution in a reasonable time frame, it is customary to allow signal folding and use a significantly reduced sw1 because the chance of spectral overlap in F1 is very small. The number of t_1 increments is then adjusted according to the required resolution; typically, 256 or 512 increments provide adequate resolution in F1, depending on the spectra complexity.
5. For data processing, follow **step 6** in Subheading 3.4 (*see* **Notes 4** and **5**).

3.6 Simultaneous Data Acquisition Experiments

1. These experiments can be applied to both natural abundance and ^{13}C-enriched samples. They are particularly well suited for natural abundance samples that will require long 1D ^{13}C experiments.
2. Prepare the sample and follow steps described previously for tuning, locking, and shimming.
3. The protocols and pulse sequences used will depend on the information desired, the available probe, and the sample concentration.
4. Data from simultaneous data acquisition experiments on Bruker systems are separated on the fly and can be processed and analyzed exactly like their nonsimultaneous counterparts.

4 Notes

1. The problem with uniform ^{13}C labeling is that carbons that are within a few covalent bonds of one another will be split by J-coupling. This can be dealt with using J-RES experiments, but standard 1D ^{13}C experiments are much more complicated to interpret than those at natural abundance ^{13}C. The Markley lab showed that fractional uniform (e.g., 26–45%) ^{13}C labeling provides a compromise between sensitivity gains, the ability to obtain ^{13}C–^{13}C correlations, and cleaner 1D spectra [43]. However, with fractional labeling, the peak shapes are also complicated, so J-RES is often the best solution.
2. The ^{13}C spectral width is quite large, requiring a short pulse for uniform excitation across the spectrum. A manual calibration of the ^{13}C 90° pulse can be obtained by setting the power level for the ^{13}C transmitter to its maximum allowable value (check with the NMR facility manager before doing this step!) and starting with a very short pulse (e.g., 2 μs) that you are sure is less than the 90° pulse. Ideally, the calibration sample would yield a high enough S/N that a single scan can be used for this step. If signal averaging must be used, set the recycle delay time (D1) to a very long value (e.g., >10 s). After the initial spectrum at a low tip angle, apply an exponential window function with small value of line broadening, zero fill, Fourier transform, and phase the spectrum. Increase the value of the pulse to find a maximum value of the signal. At this point, it is close to a 90° pulse. Then double that value and look for a null by adjusting the 180° pulse length. Then double that value again to verify that it is still a null at 360°. These should agree, and the true 90° pulse length is 360/4 or 180/2. Note that it is very important to allow at least 10 s between pulses so that the signal has fully relaxed. These steps are easily automated, but they are useful to at least sometimes perform manually to understand the process.
3. The most unusual setting that we recommend for 1D ^{13}C spectra at natural abundance is a very short relaxation delay (~0.1 s) and an acquisition time of about 0.8 s. This extremely rapid pulsing with a 60-degree pulse does not allow for full relaxation, and slowly relaxing nuclei (e.g., aromatic and quaternary carbons) are attenuated relative to aliphatic nuclei. Under these conditions, absolute quantification is impossible, but relative quantification between replicates in the study still works. An alternative strategy would be to use a relaxation agent that will increase the t_1 relaxation rate, which would be closer to quantitative.

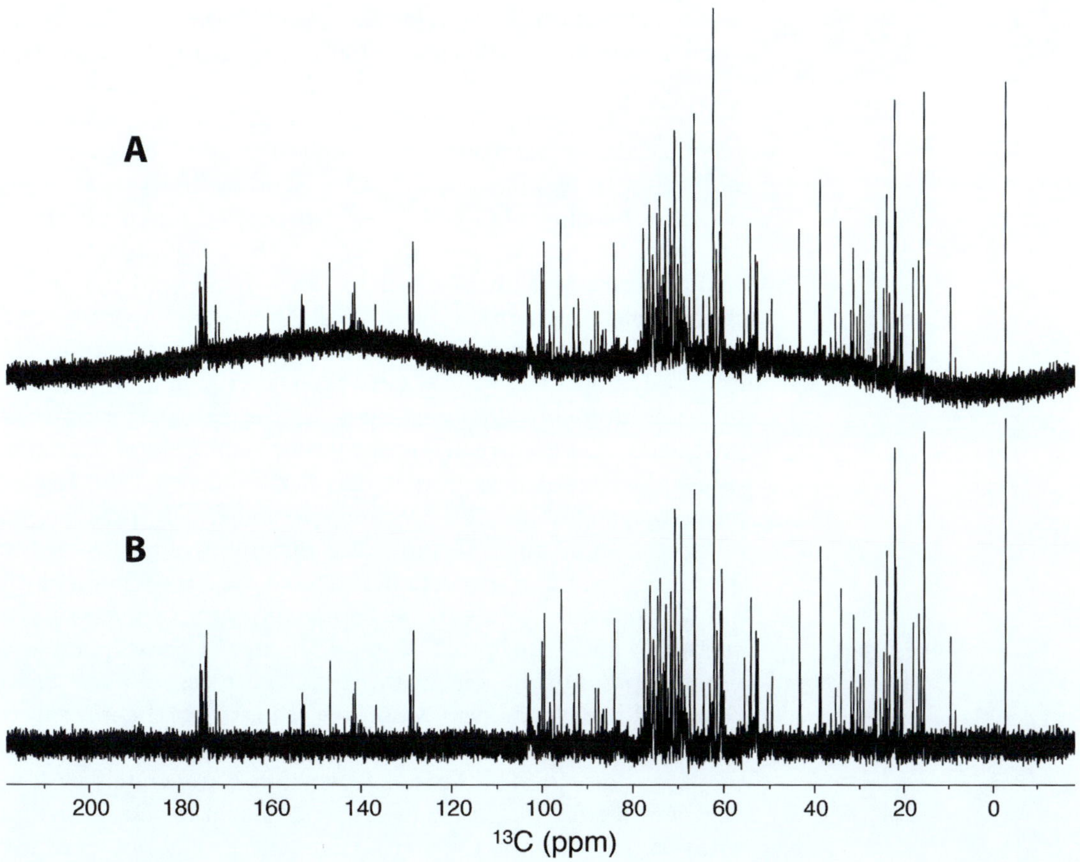

Fig. 12 Illustration of the problem sometimes encountered with 1D ^{13}C spectra. (**a**) and (**b**) are the same dataset, which was obtained in 2 h using a Bruker 5-mm DCH ^{13}C-optimized cryoprobe and Bruker NEO console and parameters outlined in Subheading 3.3. The sample was an extract from *Salmonella enterica* at natural abundance ^{13}C. The spectrum in A was multiplied by an exponential function with 0.5-Hz line broadening, zero filled, Fourier transformed, and phase corrected. The spectrum in (**b**) used the spectrum from (**a**) as input to the correction method in NMRPipe called "tdBaseline1D.com," which uses a variation of iterative soft-thresholding (IST) to achieve baseline correction by replacing initial points in the time-domain version of the data. Here, the initial 10 points of the time domain were replaced, resulting in a flatter baseline

4. Processing 1D ^{13}C data can be surprisingly difficult. The problem lies in the initial points of the time-domain (FID) data, and these can be highly distorted with a high-sensitivity ^{13}C probe. As a first step, try ordinary processing (exponential window function, zero fill, and Fourier transform). If the baseline is bad (e.g., Fig. 12), then you need to do more work. Just removing the first several points will result in a spectrum that is very difficult to phase. We like using NMRPipe [42] for most NMR processing, and this package has several ways of compensating for baseline distortion, including polynomial baseline correction applied in the frequency domain (function POLY), local median baseline correction applied in the frequency

domain (function MED), and linear prediction to achieve baseline correction by replacing initial points in the time domain (function LP [44]). In the case of typical ^{13}C metabolomics spectra, which can have hundreds of signals in the presence of substantial baseline distortions, none of these methods may be sufficient. In response, we developed a variation of iterative soft-thresholding (IST) [45, 46] to replace initial points in the time domain. In our experience, this IST approach works well for spectra with many sharp lines (*see* Fig. 12b) and is currently implemented in NMRPipe as the command "tdBaseline1D.com." The method works by attempting to iteratively improve replacement values for the first *N* points of the time-domain data, starting with the real part of a fully processed, phase-corrected spectrum or region. First, an initial threshold is chosen such that it is low relative to the largest peaks but still above the baseline everywhere. In practice, an initial threshold of 0.33 times the maximum intensity in the spectrum is used as the default. Given a spectrum and threshold, an iteration begins by preparing time-domain versions of "clipped" and "unclipped" data. To generate the clipped version, all parts of the spectrum below the threshold are set to zero, a Hilbert transform is used to reconstruct the imaginary data, and the complex result is inverse Fourier transformed to the time domain. The unclipped version is generated by Hilbert transform and inverse Fourier transform of the spectrum, without the thresholding step. Then, the first N points of the clipped time-domain result are used to replace the first N points of the unclipped time-domain result, this adjusted unclipped result is Fourier transformed back to the frequency domain, the imaginary part is discarded, and the result is used as the input spectrum for the next iteration. At each iteration, the threshold is reduced by a given factor, 0.9 by default. Iterations are terminated when the threshold is less than a selected value, which by default is 3 times the random noise in the frequency domain as determined by automated estimate.

5. As with ^{1}H J-RES spectra, an unshifted SINE window function needs to be used in the indirect (F1) dimension to obtain favorable line shapes, but at the detriment of sensitivity in the ^{13}C J-RES spectra. Similar processing applies to INADEQUATE but in both the acquisition and indirect dimensions. This is because they are detecting antiphase coherence, which has a sine dependency rather than cosine. However, it is possible to mitigate the sensitivity loss by using a backwards linear prediction (the option LPbr for ME_mod in Topspin in the "Procpars" section). A relatively recent publication [47] demonstrated this by setting the number of points predicted (the parameter LPBIN in Topspin) and back-predicted (the

parameter TDoff in Topspin) to the positive and negative values of the t_1 increments, respectively. For example, if 32 t_1 increments are used, the value of points back-predicted should be −32. For the number of coefficients used for predicting the next points of the FID (the parameter NCOEFF in Topspin), we used 32 for the spectrum in Fig. 3, but that number may be increased depending on the complexity of the dataset. Of course, the number of points in the spectrum in the indirect dimension (the SI parameter in Topspin) should be at least equal to twice the number of t_1 increments in order for the back-predicted point to be included in the FID before Fourier transform.

Acknowledgements

We thank John Glushka in the CCRC NMR facility for his help and support. Members of the Edison lab provided the enriched *C. elegans* sample used in Fig. 3. Diana Downs kindly provided the *S. enterica* sample in Fig. 12. Bill Brey, Vijay Ramaswamy, Nicolas Freytag, and Ghoncheh Amouzandeh are instrumental in the ^{13}C probe development project. Chaevien Clendinen, Ricardo Borges, and Rahil Taujale all contributed to INADEQUATE metabolomics experiments. ASE was supported by NIH R01 GM120151 and U2C ES030167, and the Georgia Research Alliance generously provides support to ASE and the CCRC NMR facility. The ^{13}C DCH NMR probe and Bruker NEO console was from a shared instrumentation grant to Jim Prestegard (NIH S10 OD021623).

References

1. Clendinen CS, Pasquel C, Ajredini R, Edison AS (2015) 13-C NMR metabolomics: INADEQUATE network analysis. Anal Chem 87 (11):5698–5706. https://doi.org/10.1021/acs.analchem.5b00867
2. Clendinen CS, Lee-McMullen B, Williams CM, Stupp GS, Vandenborne K, Hahn DA et al (2014) 13-C NMR metabolomics: applications at natural abundance. Anal Chem 86 (18):9242–9250. https://doi.org/10.1021/ac502346h
3. Wang B, Dossey AT, Walse SS, Edison AS, Merz KM Jr (2009) Relative configuration of natural products using NMR chemical shifts. J Nat Prod 72(4):709–713. https://doi.org/10.1021/np8005056
4. Bingol K, Bruschweiler-Li L, Yu C, Somogyi A, Zhang F, Bruschweiler R (2015) Metabolomics beyond spectroscopic databases: a combined MS/NMR strategy for the rapid identification of new metabolites in complex mixtures. Anal Chem 87(7):3864–3870. https://doi.org/10.1021/ac504633z
5. Clendinen CS, Stupp GS, Wang B, Garrett TJ, Edison AS (2016) 13-C metabolomics: NMR and IROA for unknown identification. Curr Metabolomics 4(2):116–120. https://doi.org/10.2174/2213235X04666160407212156
6. von Morze C, Merritt ME (2018) Cancer in the crosshairs: targeting cancer metabolism with hyperpolarized carbon-13 MRI technology. NMR Biomed e3937. https://doi.org/10.1002/nbm.3937
7. Lane AN, Fan TW (2017) NMR-based stable isotope resolved metabolomics in systems biochemistry. Arch Biochem Biophys 628:123–131. https://doi.org/10.1016/j.abb.2017.02.009

8. Ramaswamy V, Hooker JW, Withers RS, Nast RE, Brey WW, Edison AS (2013) Development of a (1)(3)C-optimized 1.5-mm high temperature superconducting NMR probe. J Magn Reson 235:58–65. https://doi.org/10.1016/j.jmr.2013.07.012
9. Ulrich EL, Akutsu H, Doreleijers JF, Harano Y, Ioannidis YE, Lin J et al (2008) BioMagResBank. Nucleic Acids Res 36(Database issue):D402–D408. https://doi.org/10.1093/nar/gkm957
10. Cloarec O, Dumas ME, Craig A, Barton RH, Trygg J, Hudson J et al (2005) Statistical total correlation spectroscopy: an exploratory approach for latent biomarker identification from metabolic 1H NMR data sets. Anal Chem 77(5):1282–1289. https://doi.org/10.1021/ac048630x
11. Aue WP, Karhan J, Ernst RR (1976) Homonuclear broad-band decoupling and 2-dimensional J-resolved NMR-spectroscopy. J Chem Phys 64(10):4226–4227. https://doi.org/10.1063/1.431994
12. Ludwig C, Viant MR (2010) Two-dimensional J-resolved NMR spectroscopy: review of a key methodology in the metabolomics toolbox. Phytochem Anal 21(1):22–32. https://doi.org/10.1002/pca.1186
13. Bax A, Freeman R, Frenkiel TA (1981) An NMR technique for tracing out the carbon skeleton of an organic-molecule. J Am Chem Soc 103(8):2102–2104
14. Bax A, Freeman R, Kempsell SP (1980) Investigation of C-13-C-13 long-range couplings in natural-abundance samples. J Magn Reson 41 (2):349–353
15. Bax A, Freeman R, Kempsell SP (1980) Natural abundance carbon-13-carbon-13 coupling observed via double-quantum coherence. J Am Chem Soc 102(14):4849–4851. https://doi.org/10.1021/ja00534a056
16. Bax A, Freeman R, Frenkiel TA, Levitt MH (1981) Assignment of C-13 NMR-spectra via double-quantum coherence. J Magn Reson 43 (3):478–483
17. Freeman R, Kupče Ē (2016) NMR spectroscopy using several parallel receivers. In: Williams A, Martin G, Rovnyak D (eds) Modern NMR approaches to the structure elucidation of natural products. Royal Society of Chemistry, Cambridge, pp 119–145. https://doi.org/10.1039/9781849735186-00119
18. Kupče Ē, Freeman R (2011) Parallel receivers and sparse sampling in multidimensional NMR. J Magn Reson 213(1):1–13. https://doi.org/10.1016/j.jmr.2011.08.027
19. Kupče Ē (2015) NMR with multiple receivers. eMagRes 4:721–732. https://doi.org/10.1002/9780470034590.emrstm1404
20. Kupče Ē, Freeman R, John BK (2006) Parallel acquisition of two-dimensional NMR spectra of several nuclear species. J Am Chem Soc 128(30):9606–9607. https://doi.org/10.1021/ja0634876
21. Gierth P, Codina A, Schumann F, Kovacs H, Kupče Ē (2015) Fast experiments for structure elucidation of small molecules: Hadamard NMR with multiple receivers. Magn Reson Chem 53(11):940–944. https://doi.org/10.1002/mrc.4292
22. Parella T, Nolis P (2010) Time shared NMR experiments. Concepts Magn Reson 36A:1–23
23. Kupče Ē, Freeman R (2008) Molecular structure from a single NMR experiment. J Am Chem Soc 130(32):10788–10792
24. Kupče Ē, Freeman R (2010) Molecular structure from a single NMR sequence (fast-PANACEA). J Magn Reson 206(1):147–153. https://doi.org/10.1016/j.jmr.2010.06.018
25. Kupče Ē, Freeman R (2010) High-resolution NMR correlation experiments in a single measurement (HR-PANACEA). Magn Reson Chem 48(5):333–336. https://doi.org/10.1002/mrc.2595
26. Kupče Ē, Wrackmeyer B (2010) Multiple receiver experiments for NMR spectroscopy of organosilicon compounds. Appl Organomet Chem 24:837–841
27. Freeman R, Kupče Ē (2011) Concepts in projection-reconstruction. Top Curr Chem 316:1–20
28. Kupče Ē, Kay LE, Freeman R (2010) Detecting the "afterglow" of 13C NMR in proteins using multiple receivers. J Am Chem Soc 132 (51):18008–18011. https://doi.org/10.1021/ja1080025
29. Pudakalakatti SM, Dubey A, Jaipuria G, Shubhashree U, Adiga SK, Moskau D et al (2014) A fast NMR method for resonance assignments: application to metabolomics. J Biomol NMR 58(3):165–173. https://doi.org/10.1007/s10858-014-9814-6
30. Kovacs H, Kupče Ē (2016) Parallel NMR spectroscopy with simultaneous detection of 1H and 19F nuclei. Magn Reson Chem 54:544–560
31. Ardenkjaer-Larsen JH, Boebinger GS, Comment A, Duckett S, Edison AS, Engelke F et al (2015) Facing and overcoming sensitivity challenges in biomolecular NMR spectroscopy. Angew Chem Int Ed Engl 54 (32):9162–9185. https://doi.org/10.1002/anie.201410653
32. Li Y, Logan TM, Edison AS, Webb A (2003) Design of small volume HX and triple-resonance probes for improved limits of detection in protein NMR experiments. J Magn Reson 164(1):128–135

33. Ramaswamy V, Hooker JW, Withers RS, Nast RE, Edison AS, Brey WW (2013) Microsample cryogenic probes: technology and applications. eMagRes 2:215–228. https://doi.org/10.1002/9780470034590

34. Ramaswamy V, Edison AS, Brey WW (2017) Inductively-coupled frequency tuning and impedance matching in HTS-based NMR probes. IEEE Trans Appl Supercond 27 (4):1–5. https://doi.org/10.1109/tasc.2017.2672718

35. Ramaswamy V, Hooker JW, Withers RS, Nast RE, Edison AS, Brey WW (2016) Development of a 1H-13C dual-optimized NMR probe based on double-tuned high temperature superconducting resonators. IEEE Trans Appl Supercond 26(3):1–5. https://doi.org/10.1109/TASC.2016.2522302

36. Hooker JW, Ramaswamy V, Arora RK, Edison AS, Brey WW (2016) Effects of dielectric substrates and ground planes on resonance frequency of Archimedean spirals. IEEE Trans Appl Supercond 26(3):1. https://doi.org/10.1109/TASC.2016.2531007

37. Hooker JW, Ramaswamy V, Arora RK, Edison AS, Withers RS, Nast RE et al (2015) An empirical expression to predict the resonant frequencies of Archimedean spirals. IEEE Trans Microwave Theory Tech 63 (7):2107–2114. https://doi.org/10.1109/TMTT.2015.2434814

38. Brey WW, Edison AS, Nast RE, Rocca JR, Saha S, Withers RS (2006) Design, construction, and validation of a 1-mm triple-resonance high-temperature-superconducting probe for NMR. J Magn Reson 179(2):290–293. https://doi.org/10.1016/j.jmr.2005.12.008

39. Clendinen CS, Stupp GS, Ajredini R, Lee-McMullen B, Beecher C, Edison AS (2015) An overview of methods using (13)C for improved compound identification in metabolomics and natural products. Front Plant Sci 6:611. https://doi.org/10.3389/fpls.2015.00611

40. Bingol K, Zhang F, Bruschweiler-Li L, Brüschweiler R (2012) TOCCATA: a customized carbon total correlation spectroscopy NMR metabolomics database. Anal Chem 84 (21):9395–9401. https://doi.org/10.1021/ac302197e

41. Hattori A, Tsunoda M, Konuma T, Kobayashi M, Nagy T, Glushka J et al (2017) Cancer progression by reprogrammed BCAA metabolism in myeloid leukaemia. Nature 545 (7655):500–504. https://doi.org/10.1038/nature22314

42. Delaglio F, Grzesiek S, Vuister GW, Zhu G, Pfeifer J, Bax A (1995) Nmrpipe - a multidimensional spectral processing system based on Unix pipes. J Biomol NMR 6(3):277–293

43. Oh BH, Westler WM, Darba P, Markley JL (1988) Protein carbon-13 spin systems by a single two-dimensional nuclear magnetic resonance experiment. Science 240 (4854):908–911

44. Marion D, Bax A (1989) Baseline correction of 2d Ft Nmr-spectra using a simple linear prediction extrapolation of the time-domain data. J Magn Reson 83(1):205–211. https://doi.org/10.1016/0022-2364(89)90307-7

45. Stern AS, Donoho DL, Hoch JC (2007) NMR data processing using iterative thresholding and minimum l(1)-norm reconstruction. J Magn Reson 188(2):295–300. https://doi.org/10.1016/j.jmr.2007.07.008

46. Hyberts SG, Milbradt AG, Wagner AB, Arthanari H, Wagner G (2012) Application of iterative soft thresholding for fast reconstruction of NMR data non-uniformly sampled with multidimensional Poisson Gap scheduling. J Biomol NMR 52(4):315–327. https://doi.org/10.1007/s10858-012-9611-z

47. Sakhaii P, Bermel W (2014) Improving the sensitivity of conventional spin echo spectra by preservation of initial signal-to-noise ratio. J Magn Reson 242:220–223. https://doi.org/10.1016/j.jmr.2014.02.024

Chapter 6

Analysis of Coenzymes and Antioxidants in Tissue and Blood Using 1D ^{1}H NMR Spectroscopy

G. A. Nagana Gowda and Daniel Raftery

Abstract

Cellular coenzymes including coenzyme A (CoA), acetyl coenzyme A (acetyl-CoA), coenzymes of redox reactions and of energy, and antioxidants mediate biochemical reactions fundamental to the functioning of all living cells. The redox coenzymes include NAD^+ (oxidized nicotinamide adenine dinucleotide), NADH (reduced nicotinamide adenine dinucleotide), $NADP^+$ (oxidized nicotinamide adenine dinucleotide phosphate), and NADPH (reduced nicotinamide adenine dinucleotide phosphate); the energy coenzymes include ATP (adenosine triphosphate), ADP (adenosine diphosphate), and AMP (adenosine monophosphate); and the antioxidants include GSSG (oxidized glutathione) and GSH (reduced glutathione). Their measurement is important to better understand cellular metabolism. Recent advances have pushed the limit of metabolite quantitation using NMR methods to an unprecedented level, which offer a new avenue for analysis of the coenzymes and antioxidants. Unlike the conventional enzyme assays, which need separate protocols for analysis, a simple 1D ^{1}H NMR experiment enables analysis of all these molecular species in one step. In this chapter, we describe protocols for their identification and quantitation in tissue and whole blood using NMR spectroscopy.

Key words 1D NMR, Quantitation, Tissue, Blood, Coenzymes, Antioxidants, CoA, Acetyl-CoA, NAD^+, NADH, $NADP^+$, NADPH, ATP, ADP, AMP, GSH, GSSG

1 Introduction

Cellular coenzymes including coenzyme A (CoA), acetyl coenzyme A (acetyl-CoA), coenzymes of redox reactions and coenzymes of energy, and antioxidants mediate biochemical reactions fundamental to the functioning of all living cells. Major cellular redox coenzymes include NAD^+ (oxidized nicotinamide adenine dinucleotide), NADH (reduced nicotinamide adenine dinucleotide), $NADP^+$ (oxidized nicotinamide adenine dinucleotide phosphate), and NADPH (reduced nicotinamide adenine dinucleotide phosphate); major energy coenzymes include ATP (adenosine triphosphate), ADP (adenosine diphosphate), and AMP (adenosine monophosphate); and major antioxidants include GSSG (oxidized glutathione) and GSH (reduced glutathione). The concentration

G. A. Nagana Gowda and Daniel Raftery (eds.), *NMR-Based Metabolomics: Methods and Protocols*, Methods in Molecular Biology, vol. 2037, https://doi.org/10.1007/978-1-4939-9690-2_6, © Springer Science+Business Media, LLC, part of Springer Nature 2019

ratio between reduced and oxidized forms of redox coenzymes indicates important cellular functions including the overall redox status and regulation of ion channels, cell signaling, cell survival, and death [1–4]. The redox balance also indicates pathological conditions of major diseases including heart disease, diabetes, and cancer. Cellular levels of the antioxidants GSH and GSSH report on the measure of oxidative stress as well as redox signaling and are involved in the regulation of critical metabolic functions [5]. The coenzyme ATP represents the energy currency of living cells; ADP and AMP are closely associated with ATP as its precursors and products. Reliable and high-throughput quantitative measurements of these molecules is therefore important for investigations focused on the mechanistic understanding of cellular functions in health and disease [6].

Conventional methods currently used for the analysis of coenzymes typically involve enzymatic assays. Enzymatic methods are suboptimal as separate protocols are used for analysis of each coenzyme or their ratios [7–12]. In addition, these assays are challenged by the interference from sample matrix and the finite linear range of the assays, which necessitates repeated recalibrations of the solution concentration to match to standard curves. The deleterious effect of the separate analysis of these molecules is that errors associated with such analysis potentially outweigh biological changes. A number of efforts have focused on developing methods for simultaneous analysis of the coenzymes based on chromatographic separation and detection using UV-vis absorption [13] or mass spectrometry (MS) [14, 15]. Mass spectrometry is most promising owing to its ability to analyze the coenzymes in one step and with high sensitivity. Its challenges, however, are ion suppression and interference due to the unit mass difference for many coenzymes, as well as similar retention times in chromatography. For example, NAD^+ and NADH have very similar structures and differ by a unit mass, and the same is true for $NADP^+$ and NADPH. In addition, ATP and ADP can undergo in-source fragmentation to AMP, before detection, which can lead to inaccurate values. Owing to such challenges, the analysis methods have so far been far from optimal.

NMR spectroscopy represents a simple method for analysis of the coenzymes and antioxidants owing to its unique characteristics. However, the challenges for their analysis by NMR are numerous, including the (1) relatively low sensitivity of NMR, (2) complexity of biological mixtures, (3) virtually identical structures for many coenzymes, (4) very low concentration for some of the coenzymes ($\leq$1 μM), and (5) unstable nature of many coenzymes due to enzyme activity and oxidation. These challenges have so far prevented analysis of these molecules by NMR. The instability of many coenzymes evades their detection partly or wholly and hence NMR spectra under such conditions were not reproducible. Until recently, the inability to obtain reproducible NMR spectra posed a major challenge for unambiguous identification of the coenzymes. It may be noted, however, that the unstable nature of the coenzymes is a challenge not only for NMR analysis but also for other methods including mass spectrometry, UV-vis, and the conventional enzymatic assays.

Recent advances in NMR spectroscopy have enabled routine quantification of metabolites down to sub-micromolar concentrations using a single internal reference [16] and opened new avenues for quantitative analysis of major coenzymes and antioxidants in one step using a single internal standard [17–19]. An important outcome is a realization of the importance of optimal sample harvesting and extraction for the detection of all the coenzymes by NMR. As an example, Fig. 1 shows spectra for the same type of tissue (mouse heart) obtained using different harvesting and extraction methods. It can be seen that depending on the method used, ATP is converted into ADP and AMP, and the NADH and NADPH peaks are missing altogether. The method described in Fig. 1a retains the integrity of the coenzymes. Utilizing such new methods, in this chapter, we provide protocols for analysis of major coenzymes and antioxidants in tissue and whole blood specimens using simple 1D NMR spectroscopy methods. The ability to visualize important molecules fundamental to cellular functions in one step represents an important avenue to interrogate the mechanistic details of cellular function in health and disease.

2 Materials

2.1 Samples

1. Mouse tissue used for analysis: heart, kidney, brain, liver, and skeletal muscle tissues (*see* **Note 1**).
2. Whole human blood.

2.2 Chemicals and Solvents

1. Chemicals: monosodium phosphate (NaH_2PO_4), disodium phosphate (Na_2HPO_4), sodium salt of 3-(trimethylsilyl) propionic acid-2,2,3,3-d_4 (TSP).
2. Solvents: methanol, chloroform, deionized (DI) water, deuterium oxide (D_2O).
3. Phosphate buffer: buffer solution (100 mM; pH = 7.4) made by dissolving 1124.0 mg anhydrous Na_2HPO_4 and 249.9 mg anhydrous NaH_2PO_4 in 100 g D_2O. Add a solution of TSP to achieve a final concentration of 25 μM (*see* **Note 2**).

2.3 Lab Supplies and Equipment

1. Pipette (1 mL, 200 μL).
2. Eppendorf tubes (1.5 or 2 mL).
3. Ice bucket and ice.

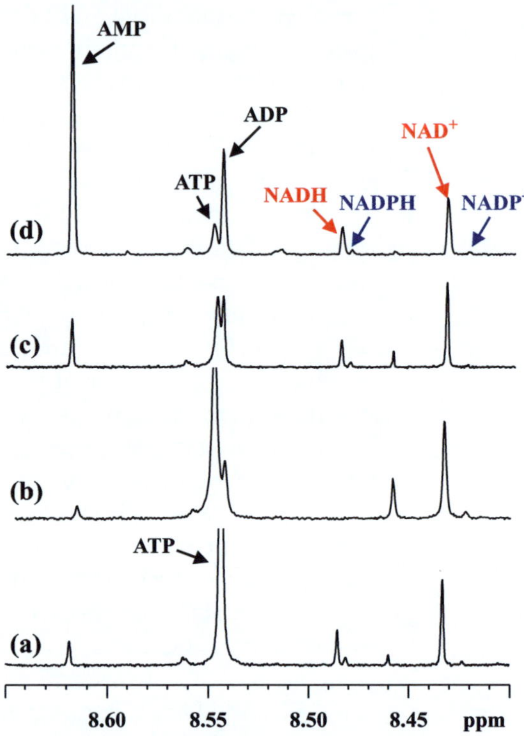

Fig. 1 Portions of 800-MHz ^{1}H NMR spectra obtained using 3-mm sample tubes highlighting the sensitivity of coenzyme levels to tissue harvesting/extraction protocols: (**a**) mouse heart harvested, washed with glucose/pyruvate solution, and freeze clamped, followed by extraction using methanol-chloroform mixture; (**b**) mouse heart harvested, perfused with glucose/pyruvate solution, and freeze clamped, followed by extraction using methanol-chloroform mixture; (**c**) mouse heart freeze clamped in vivo, followed by extraction using methanol-chloroform mixture; and (**d**) mouse heart harvested, washed with PBS, and freeze clamped, followed by extraction using methanol-water mixture. The most prominent changes are the missing NADH/NADPH peaks in (**b**) and diminished ATP and enhanced ADP/AMP in (**b–d**) (reproduced with permission from ref. 17)

4. Tissue blender.
5. Vortex mixer.
6. −20 °C refrigerator, −80 °C freezer.
7. Vacuum concentrator/freeze drier.
8. Ultrasound sonicator.

2.4 NMR Instrument and Accessories

1. High-resolution NMR spectrometer (*see* **Note 3**).
2. NMR data processing software (*see* **Note 4**).
3. 5-mm or 3-mm NMR tubes (*see* **Note 5**).

3 Methods

3.1 Sample Collection

1. Mouse tissue harvesting: Anesthetize mice and separate heart, kidney, brain, liver, and skeletal muscle tissues, rinse in glucose/pyruvate solution, and snap freeze in liquid nitrogen (*see* **Note 6**). The resulting tissue can then be used for metabolite extraction or stored at −80 °C until used for analyses.
2. Whole blood: collect human blood (~4 mL) in a heparin tube (*see* **Notes** 7 and **8**).

3.2 Coenzyme and Antioxidant Extraction from Tissue (See Note 9)

1. Weigh tissue specimen (~5 to 80 mg).
2. Mix with a 1-mL mixture of cold methanol and chloroform (1:2 v/v; 4 °C) in 2-mL Eppendorf vials.
3. Homogenize using tissue blender.
4. Sonicate for 20 s.
5. Add 800-μL cold chloroform/DI water mixture (1:1 v/v).
6. Vortex the sample and set aside for 30 min on ice to separate the solvent layers.
7. After centrifugation at 13,400 rcf for 30 min, collect the aqueous (top) layer, and freeze dry.
8. Mix dried samples with 200 or 600 μL phosphate buffer containing 25 μM TSP.
9. Spin to sediment any residue.
10. Transfer the supernatants to NMR tubes for analysis (*see* **Notes 10** and **11** and Fig. 2).

3.3 Coenzyme and Antioxidant Extraction from Whole Blood (See Note 12)

1. Mix 200–400 μL whole blood with cold methanol and chloroform in a 1:2:2 sample:methanol:chloroform (v/v/v) ratio.
2. Vortex for 30 s.
3. Sonicate for 2 min at 4 °C.
4. Incubate at −20 °C for 20 min.
5. Centrifuge the mixture at 13,400 rcf for 30 min to pellet proteins and cell debris.
6. Transfer the clear aqueous solution to a fresh vial (*see* **Note 13**).
7. Dry using a vacuum concentrator for 5 h.
8. Mix dried samples with 200 or 600 μL phosphate buffer containing 25 μM TSP.
9. Spin to sediment any residue.
10. Transfer the supernatants to 3- or 5-mm NMR tubes for analysis (*see* **Notes 10** and **11**).

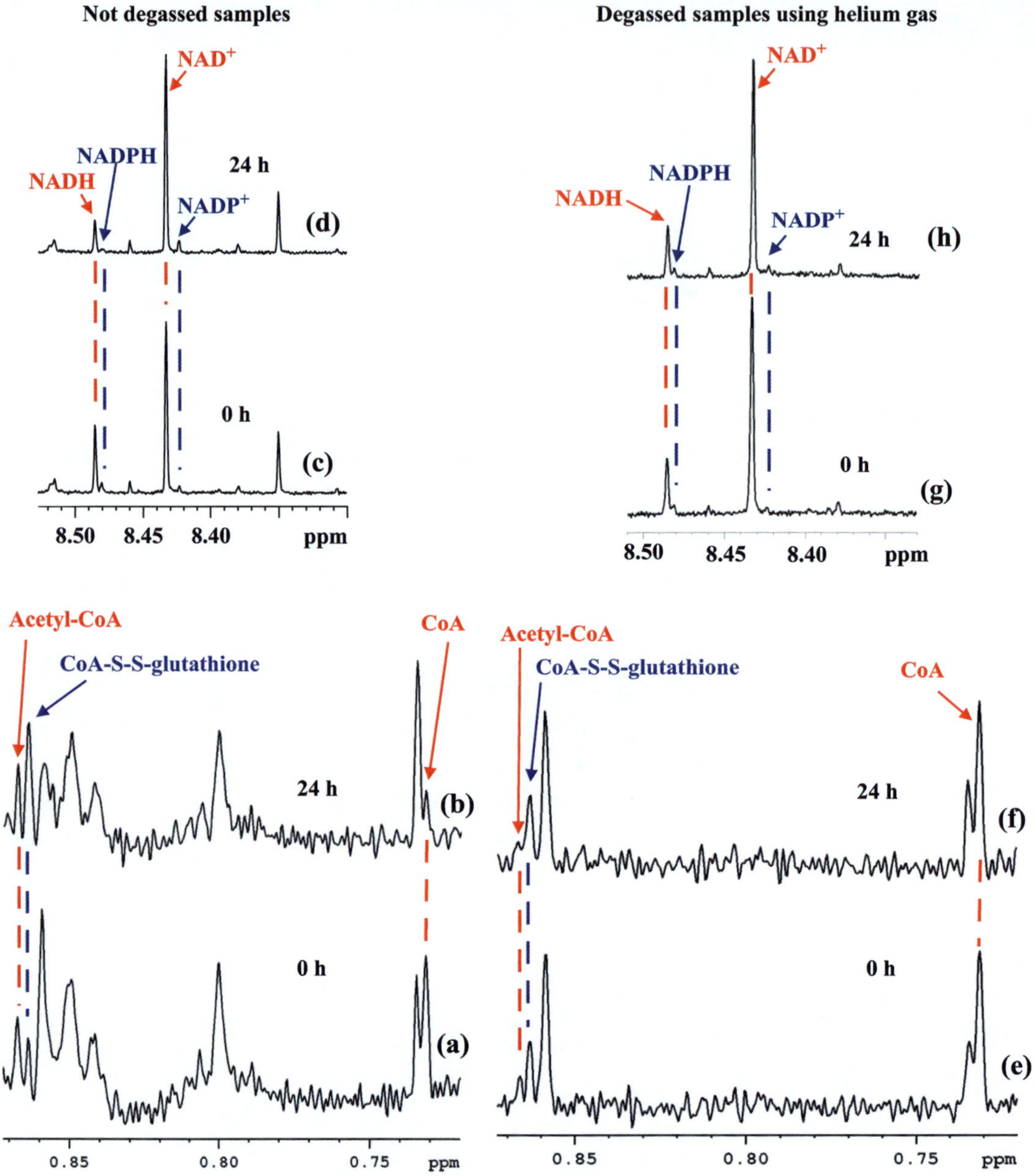

Fig. 2 Portions of 800-MHz ^{1}H NMR spectra of mouse heart tissue extracts obtained without degassing, immediately after sample preparation (Figs. 2a and 2c) and 24 h after preparation (Figs. 2b and 2d). Note, that while the acetyl-CoA level is unaltered, the CoA, NADH, and NADPH levels are reduced drastically with a concomitant increase of the CoA-S-S-glutathione, NAD^+, and $NADP^+$ levels, respectively, 24 h after sample preparation. Portions of spectra of mouse heart tissue extracts with degassing of both the NMR tubes and D_2O buffer using helium gas obtained immediately after sample preparation (Figs. 2e and 2g) and 24 h after preparation (Figs. 2f and 2h). Note that the levels of the CoA, acetyl-CoA, and CoA-S-S-glutathione, NADH, NADPH, NAD^+, and $NADP^+$ are all unaltered even 24 h after sample preparation, which indicates no oxidation of CoA, NADH, and NADPH

3.4 NMR Experiments: Shimming Using Standard Sample

1. Set NMR probe temperature to 298 K.
2. Insert a standard sample consisting of a mixture of H_2O and D_2O (90:10 v/v).
3. Tune the probe.
4. Read most recent shim file for the current probe.
5. Perform field-frequency locking using D_2O signal.
6. Perform 1D gradient shimming (*see* **Note 14**).
7. Perform automated 3D gradient shimming first and then 1D gradient shimming.
8. Save shims as the most recent shim values.
9. Remove sample from the magnet.

3.5 NMR Experiment for Tissue and Whole Blood Extracts

1. Set pulse sequence: one-pulse or 1D NOESY pulse sequence with residual water suppression using presaturation (*see* **Note 15**).
2. Set parameters: temperature, 298 K; spectral width, 10,000 Hz; recycle delay, 6 s; number of transients, 128 transients; time domain points (TD), 32 or 64 K (*see* **Note 16**).
3. Tune the probe, perform locking and shimming.
4. Calibrate 90-degree pulse.
5. Acquire data.
6. Perform Fourier transformation after multiplying the FID (free induction decay) by exponential multiplication with a line broadening of 0.3 or 0.5 Hz.
7. Perform phase correction and baseline correction and set the reference (TSP) signal to 0 ppm.

3.6 Coenzyme and Antioxidant Identification and Quantitation

1. Identify the characteristic peaks for coenzymes and antioxidants based on the chemical shifts (*see* Table 1) and spectral features as shown in Fig. 3 (*see* **Note 17**) and Fig. 4 (*see* **Note 18**) for tissue and whole blood, respectively.
2. Determine area for the characteristic peaks of coenzymes and antioxidants relative to the area of the reference (TSP) peak (*see* **Note 19** and Fig. 5).
3. Using the peak areas and the following equation, obtain coenzyme concentration:

$$C_m = \frac{I_m n_r C_r}{I_r n_m}$$

where C_m and C_r are concentrations of the metabolite and reference compound (TSP) in solution, I_m and I_r are integrated peak areas of the coenzyme/antioxidant and reference (TSP), and n_m and n_r are number of protons that the peaks from coenzyme/antioxidant and reference represent, respectively (*see* **Note 20**).

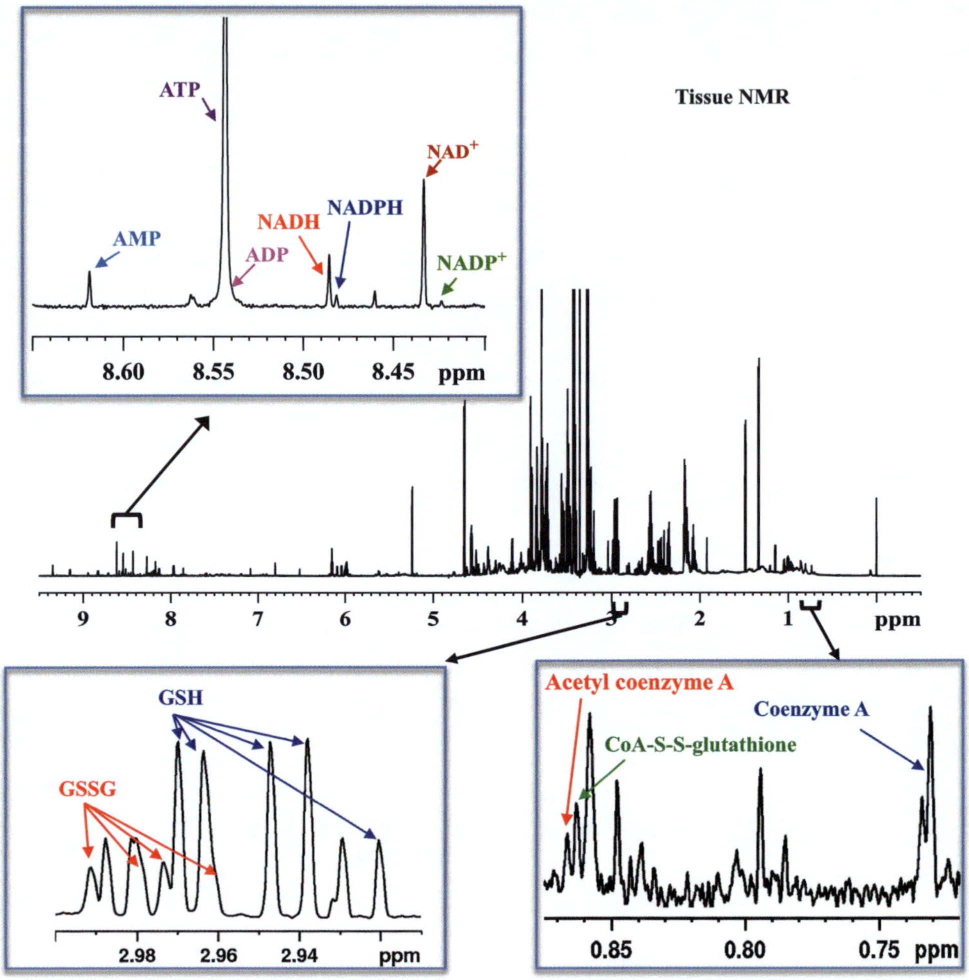

Fig. 3 Typical 800-MHz ^{1}H NMR spectrum of a mouse heart tissue extract with expanded regions labeled with characteristic peaks of coenzymes: coenzyme A (CoA), acetyl coenzyme A (acetyl-CoA), coenzyme A glutathione disulfide (CoA-S-S-glutathione), oxidized nicotinamide adenine dinucleotide (NAD^{+}), oxidized nicotinamide adenine dinucleotide phosphate ($NADP^{+}$), reduced nicotinamide adenine dinucleotide (NADH), reduced nicotinamide adenine dinucleotide phosphate (NADPH), adenosine triphosphate (ATP), adenosine diphosphate (ADP) and adenosine monophosphate (AMP), and antioxidants: reduced glutathione (GSH) and oxidized glutathione (GSSG). Note: ADP peak is too small to be distinguished from ATP peak

4 Notes

1. We describe analysis of coenzymes using mouse tissue; however, a similar procedure can be extended to other types of tissue.
2. The pH of the buffer solution should be 7.4. A variation of pH by about 0.1 from 7.4, however, does not alter the chemical shifts appreciably and hence such solutions may be used without further pH correction.

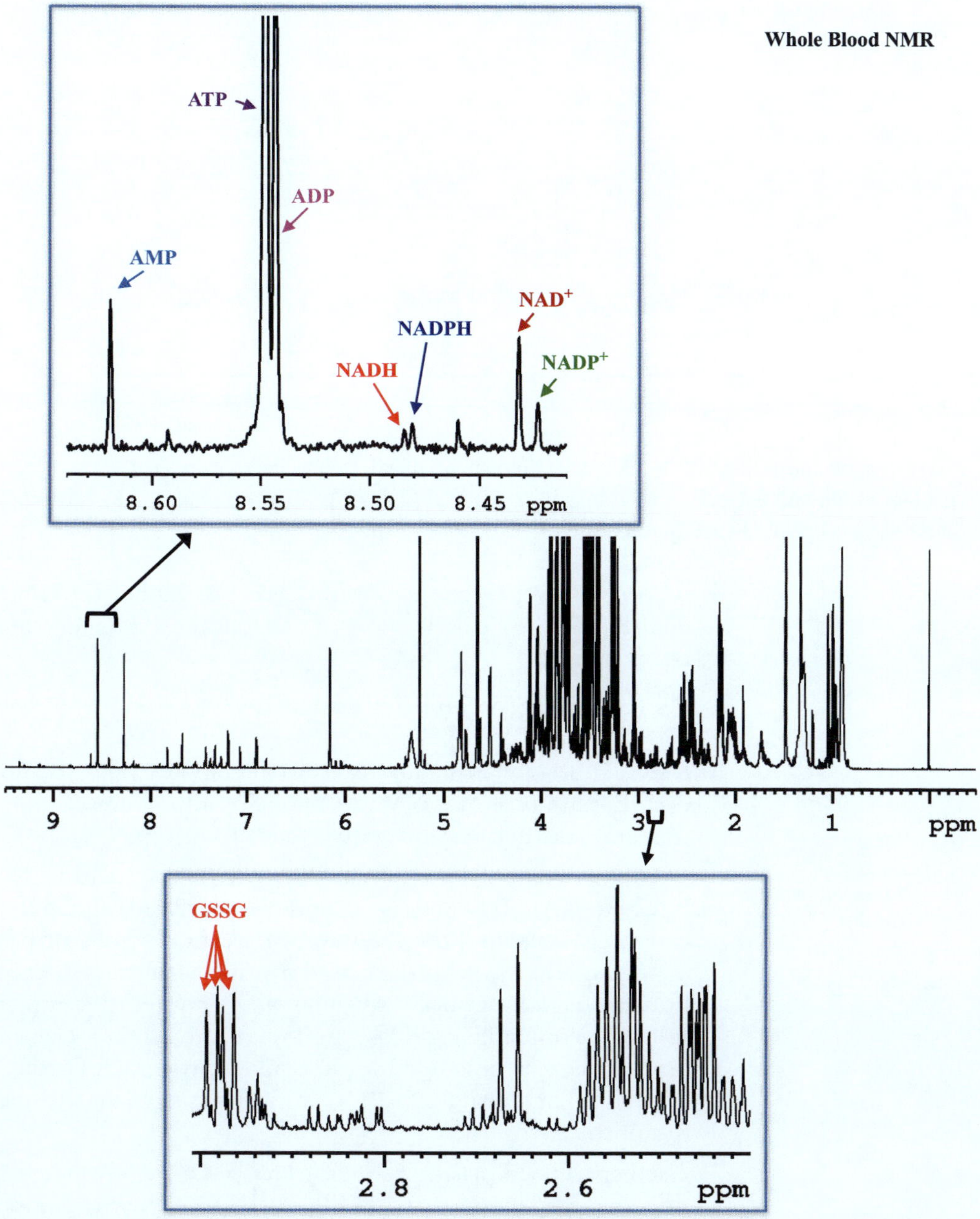

Fig. 4 Typical 800-MHz ^1H NMR spectrum of human whole blood extract with expanded regions labeled with characteristic peaks of coenzymes: oxidized nicotinamide adenine dinucleotide (NAD^+), oxidized nicotinamide adenine dinucleotide phosphate ($NADP^+$), reduced nicotinamide adenine dinucleotide (NADH), reduced nicotinamide adenine dinucleotide phosphate (NADPH), adenosine triphosphate (ATP), adenosine diphosphate (ADP) and adenosine monophosphate (AMP), and antioxidant: oxidized glutathione (GSSG)

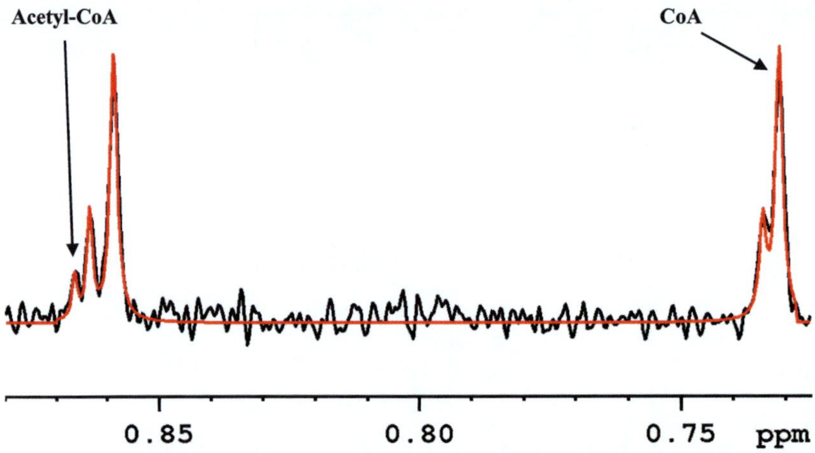

Fig. 5 A portion of the 800-MHz ^{1}H NMR spectrum of a typical mouse heart tissue extract highlighting deconvolution of the characteristic peaks of CoA and acetyl-CoA for peak integration: black, experimental spectrum; red, deconvoluted spectrum using Bruker TopSpin software

3. A spectrometer frequency of 600 MHz or higher is preferred owing to the closely spaced chemical shifts for many coenzymes.
4. NMR instrument vendor, Bruker, provides TopSpin NMR data processing software free of cost for academic use.
5. 5-mm tubes are generally used for metabolite analysis; however, the use of 3-mm NMR tubes helps improve both resolution and sensitivity significantly even with 5-mm NMR probes.
6. It is important to note that the coenzymes such as CoA, NADH, NADPH, and ATP and the antioxidant GSH are extremely labile and they can evade detection wholly or partly depending on the procedure used for tissue harvesting/extraction (*see* Fig. 1). Quickly washing the harvested tissue with a solution containing glucose and pyruvate, freeze clamping using tweezers precooled with liquid nitrogen, followed by extraction, or storing at −80 °C until analysis retains the integrity of the coenzymes.
7. Use heparinized plasma tubes for blood collection since heparin does not interfere with the analysis of whole blood or plasma metabolites. Avoid citrate and EDTA plasma tubes for blood collection as citrate and EDTA interfere with the analysis of whole blood or plasma metabolites.
8. Whole blood can be used to analyze other metabolites apart from the coenzymes. Red blood cells constitute more than 99% of the blood cells and are metabolically active even at 4 °C. Glycolysis is the major metabolic pathway in RBC and under hypoxic conditions RBCs exhibit accelerated glucose consumption leading to accumulation of lactate [20, 21]. Hence, to

avoid metabolic activity, it is important to analyze the samples as soon as the blood is obtained or store immediately at −80 °C until analysis.

9. This procedure is optimized for mouse heart tissue extraction. Further standardization may be required for other types of tissue to optimize the extraction of the coenzymes.
10. If the sample is dissolved in 200 μL buffer, use a 3-mm NMR tube. However, if the sample is dissolved in 600 μL buffer, transfer the solution to a 5-mm NMR tube. Use of a 3-mm NMR tube is recommended, since it provides significantly higher sensitivity as well as better resolution.
11. In solution, the coenzymes CoA, NADH, and NADPH decrease with time and the levels of their oxidized forms, CoA-S-S-glutathione, NAD^+, and $NADP^+$, increase proportionately. However, degassing the NMR solvent and flushing the sample tube with helium gas followed by sealing the tube with parafilm prevented such changes (*see* Fig. 2). Therefore, either the solutions should be analyzed as soon as they are made or the solvent and the sample tube should be degassed using an inert gas such as helium.
12. This procedure is optimized for human whole blood. However, the protocol can be extended for blood from other sources.
13. Cell debris and precipitated protein lie in between the top aqueous layer and bottom chloroform layer.
14. If the result of 1D gradient shimming is good (final standard deviation of B_o of typically ~0.2 Hz), there is no need for 3D shimming (skip Subheading 3.4, **step** 7) from the magnet and proceed to the next section.
15. We recommend using the 1D NOESY pulse sequence with presaturation of the signal from residual water.
16. These are typical parameters. The parameters can be altered depending on the sample amount and the requirement of specific resolution and quantitation accuracy. For accurate quantitation, use a recycle delay 5 times T_1 of the signal with the longest relaxation time or perform relaxation correction to compensate for a shorter recycle delay [22].
17. Chemical shifts of the characteristic peaks that can be used for quantitation are given in Table 1.
18. CoA and acetyl-CoA are not detected in whole blood.
19. Due to weak and closely spaced coenzyme peaks, we recommend peak deconvolution as shown in Fig. 5 to obtain accurate peak areas.

Table 1
^{1}H NMR chemical shifts (in ppm with reference to TSP) and *J* couplings (in Hz) for the coenzymes/metabolites of redox reaction and cellular energy. Chemical shifts for characteristic peaks of metabolites that may be used for identification and quantification using 1D ^{1}H NMR are shown in bold (modified with permission from ref. 17)

Coenzyme/metabolite	Mouse heart/kidney/brain/liver/ skeletal muscle	Authentic compounds[a]
Coenzyme A (CoA)	8.5641 (1H, s); 2.6089 (2H, t); **0.7314 (3H, s)**; 0.8594 (3H, s)	8.5641 (s); 8.2796 (s); 6.1795 (d, $^{3}J_{H,H}$ = 7.17); 4.8517 (m); 4.7648 (m); 4.5771 (m); 4.2331 (m); 4.0036 (s); 3.5352 (dd, $^{2}J_{H,H}$ = 9.62, $^{3}J_{H,31P}$ = 4.82); 3.8055 (dd, $^{3}J_{H,31P}$ = 4.69); 3.4658 (t, $^{3}J_{H,H}$ = 6.58); 3.3161 (t,$^{3}J_{H,H}$ = 6.65); 2.4653 (t, $^{3}J_{H,H}$ = 6.58); 2.6089 (t, $^{3}J_{H,H}$ = 6.65); 0.7315(s); 0.8594 (s)
Acetyl-coenzyme A (acetyl-CoA)	8.5626 (1H, s); 0.7340 (3H, s); **0.8674 (3H, s)**	8.5626 (s); 8.2729 (s); 6.1755 (d, $^{3}J_{H,H}$ = 7.16); 4.8519 (m); 4.7632 (m); 4.5758 (m); 4.2328 (m); 4.0107 (s); 3.5408 (dd,$^{2}J_{H,H}$ = 9.77, $^{3}J_{H,31P}$ = 4.82); 3.8173 (dd,$^{3}J_{H,31P}$ = 4.78); 3.4589 (m, $^{3}J_{H,H}$ = 6.59); 3.4413 (m,$^{2}J_{H,H}$ = 13.42, $^{3}J_{H,H}$ = 6.59); 3.3190 (t,$^{3}J_{H,H}$ = 6.40); 2.9642 (t, $^{3}J_{H,H}$ = 6.40); 2.4273 (t, $^{3}J_{H,H}$ = 6.59); 2.3495 (s); 0.7340 (s); 0.8674 (s)
Nicotinamide adenine dinucleotide, oxidized (NAD^{+})	4.371 (m); 4.387 (m); 4.429 (m); 4.487 (t, *J* = 5.213); 4.515 (m); 4.545 (m); 6.041 (1H, d, *J* = 6.042); 6.093 (1H, d, *J* = 5.482); 8.174 (1H, s); 8.193 (1H, dd); **8.434 (1H, s)**; 8.832 (1H, d, *J* = 7.992); 9.147 (1H, d, *J* = 6.158); 9.344 (1H, s)	4.191–4.291 (m); 4.369 (m); 4.386 (m); 4.430 (m); 4.488 (t; *J* = 5.196); 4.514 (m); 4.546 (m); 6.042 (d, *J* = 6.089); 6.092 (d, *J* = 5.510); 8.176 (s); 8.193 (dd); 8.435 (s); 8.833 (d; *J* = 8.032); 9.148 (d; *J* = 6.138); 9.344 (s)
Nicotinamide adenine dinucleotide phosphate, oxidized ($NADP^{+}$)	6.101 (d); 8.146 (1H, s), **8.424 (1H, s)**; 9.104 (1H, d); 9.300 (1H, s)	4.181–4.236 (m); 4.284–4.342 (m); 4.377 (m); 4.411 (m); 4.461 (t; *J* = 5.286); 4.503 (m); 4.617 (t, *J* = 4.990); 4.968 (m); 6.037 (d; *J* = 5.560); 6.101 (d, *J* = 4.969); 8.146 (s); 8.168–8.199 (m); 8.423 (s); 8.818 (d, *J* = 8.008); 9.106 (d; *J* = 6.259); 9.299 (s)
Nicotinamide adenine dinucleotide, reduced (NADH)	6.137 (d); 6.942 (1H, s); 8.247 (1H, s); **8.486 (1H, s)**	2.669 (br. d, *J* = 17.976); 2.793 (br. d, *J* = 17.976); 4.082 (br. m); 4.097 (br. m) 4.169–4.289 (br. m); 4.389 (br. m); 4.512 (t, *J* = 4.448); 4.709 (t, *J* = 5.272); 5.980 (d,

(continued)

Table 1 (continued)

Coenzyme/metabolite	Mouse heart/kidney/brain/liver/ skeletal muscle	Authentic compounds[a]
		$J = 8.424$); 6.138 (d, $J = 5.479$); 6.943 (s); 8.248 (s); 8.487 (s)
Nicotinamide adenine dinucleotide phosphate, reduced (NADPH)	6.216 (d); 6.936 (1H, s); **8.481** (1H, s)	2.737 (br. d, $J = 18.093$); 2.838 (br. d, $J = 18.093$); 4.044 (br. m); 4.066 (br. m); 4.154–4.236 (m); 4.285–4.325 (br. m); 4.390 (br. m); 4.597 (t, $J = 5.049$); 4.950 (m); 5.963 (d, $J = 8.363$); 6.216 (d, $J = 4.597$); 6.935 (s); 8.246 (s); 8.481 (s)
Adenine triphosphate (ATP)	4.410 (m); 4.621(m); 6.155 (d); 8.274 (s); **8.549** (1H, s)	4.192–4.233 (m); 4.278–4.318 (m); 4.392–4.421 (br. m); 4.620–4.648 (m); 6.153 (d, $J = 5.974$); 8.274 (s); 8.557 (s)
Adenine diphosphate (ADP)	4.387 (m); 4.621(m); 6.155 (d); 8.273 (s); **8.544** (1H, s)	4.187–4.219 (m); 4.244–4.283 (m); 4.377–4.398 (br. m); 4.611–4.634 (m); 6.156 (d, $J = 5.383$); 8.273 (s); 8.547 (s)
Adenine monophosphate (AMP)	4.014 (m); 4.372 (m); 4.515 (m); 6.147 (d); 8.271 9(s); **8.619** (1H, s)	3.995–4.026 (m); 4.359–4.383 (br. m); 4.501–4.526 (m); 6.148 (d, $J = 5.965$); 8.272 (s); 8.620 (s)
Glutathione, reduced (GSH)	4.5713 (dd); **2.9551** (2H, abx); 2.170 (m); 2.554 (m)	4.5707 (dd, $^3J_{HH} = 5.02$, $^3J_{HH} = 7.13$); 3.783 (m); 2.9551 (abx, $^2J_{HH} = 14.20$, $^3J_{HH} = 7.28$ $^3J_{HH} = 5.05$); 2.5616 (m); 2.1707 (m)
Glutathione, oxidized (GSSG)	**2.9763** (1H, dd); 2.554 (m); 2.170 (m)	4.7657 (dd, $^3J_{HH} = 4.26$, $^3J_{HH} = 9.58$); 3.7813 (m); 3.3148 (dd, $^3J_{HH} = 4.40$, $^2J_{HH} = 14.21$); 2.9763 (dd, $^3J_{HH} = 9.57$, $^2J_{HH} = 14.21$); 2.5438 (m); 2.1684 (m)

Chemical shifts for authentic compounds are shown for comparison

br. broad, *s* singlet, *d* doublet, *dd* doublet of doublets, *t* triplet, *m* multiplet, *abx* AB part of ABX pattern

[a]Spectra for the authentic compounds were obtained near their physiological concentrations (100 μM) in D_2O buffer at pH 7.45 at 298 K

20. The number of protons in the characteristic peaks are represented as follows: 9 for the reference TSP (0 ppm), 3 for CoA (0.7315 ppm) and acetyl-CoA (0.8674 ppm), 2 for GSH (2.955 ppm), and 1 for NAD^+ (8.434 ppm), $NADP^+$ (8.424 ppm), NADH (8.486 ppm), NADPH (8.481 ppm), ATP (8.549 ppm), ADP (8.544 ppm), AMP (8.619 ppm), and GSSG (2.976 ppm) (*see* Table 1).

Acknowledgements

We acknowledge financial support from the NIH (National Institute of General Medical Sciences 2R01GM085291).

References

1. Kilfoil PJ, Tipparaju SM, Barski OA, Bhatnagar A (2013) Regulation of ion channels by pyridine nucleotides. Circ Res 112(4):721–741
2. Koch-Nolte F, Haag F, Guse AH, Lund F, Ziegler M (2009) Emerging roles of NAD+ and its metabolites in cell signaling. Sci Signal 2(57):mr1. https://doi.org/10.1126/scisignal.257mr1
3. Ying W (2008) NAD+/NADH and NADP+/NADPH in cellular functions and cell death: regulation and biological consequences. Antioxid Redox Signal 10(2):179–206
4. Engel PC (2014) Glutamate dehydrogenases: the why and how of coenzyme specificity. Neurochem Res 39(3):426–432
5. Aquilano K, Baldelli S, Ciriolo MR (2014) Glutathione: new roles in redox signaling for an old antioxidant. Front Pharmacol 5:196. https://doi.org/10.3389/fphar.2014.00196
6. Bar-Or D, Bar-Or R, Rael LT, Brody EN (2015) Oxidative stress in severe acute illness. Redox Biol 4:340–345
7. Du J, Cleghorn W, Contreras L, Linton JD, Chan GC, Chertov AO et al (2013) Cytosolic reducing power preserves glutamate in retina. Proc Natl Acad Sci U S A 110 (46):18501–18506
8. Smith JS, Brachmann CB, Celic I, Kenna MA, Muhammad S, Starai VJ et al (2000) A phylogenetically conserved NAD+-dependent protein deacetylase activity in the Sir2 protein family. Proc Natl Acad Sci U S A 97:6658–6663
9. Allred JB, Guy DG (1969) Determination of coenzyme A and acetyl CoA in tissue extracts. Anal Biochem 29:293–299
10. Kato T (1975) CoA cycling: an enzymatic amplification method for determination of CoASH and acetyl CoA. Anal Biochem 66:373–392
11. Szutowicz A, Bielarczyk H (1987) Elimination of CoASH interference from acetyl-CoA cycling assay by maleic anhydride. Anal Biochem 164:292–296
12. Takamura Y, Kitayama Y, Arakawa A, Yamanaka S, Tosaki M, Ogawa Y (1985) Malonyl-CoA: acetyl-CoA cycling. A new micromethod for determination of acyl-CoAs with malonate decarboxylase. Biochim Biophys Acta 834:1–7
13. Sporty JL, Kabir MM, Turteltaub KW, Ognibene T, Lin SJ, Bench G (2008) Single sample extraction protocol for the quantification of NAD and NADH redox states in Saccharomyces cerevisiae. J Sep Sci 31 (18):3202–3211
14. Evans C, Bogan KL, Song P, Burant CF, Kennedy RT, Brenner C (2010) NAD+ metabolite levels as a function of vitamins and calorie restriction: evidence for different mechanisms of longevity. BMC Chem Biol 10:2. https://doi.org/10.1186/1472-6769-10-2
15. Trammell SA, Brennera C (2013) Targeted, LCMS-based metabolomics for quantitative measurement of NAD(+) metabolites. Comput Struct Biotechnol J 4:e201301012. https://doi.org/10.5936/csbj.201301012
16. Nagana Gowda GA, Gowda YN, Raftery D (2015) Expanding the limits of human blood metabolite quantitation using NMR spectroscopy. Anal Chem 87(1):706–715
17. Nagana Gowda GA, Abell L, Lee CF, Tian R, Raftery D (2016) Simultaneous analysis of major coenzymes of cellular redox reactions and energy using ex vivo (1)H NMR spectroscopy. Anal Chem 88(9):4817–4824
18. Nagana Gowda GA, Raftery D (2017) Whole blood metabolomics by 1H NMR spectroscopy provides a new opportunity to evaluate coenzymes and antioxidants. Anal Chem 89 (8):4620–4627
19. Nagana Gowda GA, Abell L, Tian R (2019) Extending the scope of 1H NMR spectroscopy for the analysis of cellular coenzyme A and acetyl coenzyme A. Anal Chem 91:2464–2471. https://doi.org/10.1021/acs.analchem.8b05286
20. Murphy JR (1960) Erythrocyte metabolism. II. Glucose metabolism and pathways. J Lab Clin Med 55:286–302
21. Hamasaki N, Asakura T, Minakami SJ (1970) Effect of oxygen tension on glycolysis in human erythrocytes. Biochemist 68 (2):157–161
22. Lane AN (2012) NMR applications in metabolomics. In: Fan TW-M, Lane AN, Higashi RM (eds) Handbook of metabolomics. Humana

Part III

Unknown Resonance Identification

Chapter 7

UHPLC-QTOF-MS/MS-SPE-NMR: A Solution to the Metabolomics Grand Challenge of Higher-Throughput, Confident Metabolite Identifications

Anil Bhatia, Saurav J. Sarma, Zhentian Lei, and Lloyd W. Sumner

Abstract

Metabolomics represents a powerful, complementary approach for studying biological system responses to various biotic and abiotic stimuli. A major challenge in metabolomics is the lack of reliable annotations for all metabolites detected in complex MS and/or NMR data. To meet this challenge, we have developed an integrated UHPLC-QTOF-MS/MS-SPE-NMR system for higher-throughput metabolite identifications, which provides advanced biological context and enhances the scientific value of metabolomics data for understanding systems biology. This integrated instrumental method is less labor-intensive and more cost-effective than conventional individual methods (LC; MS; SPE; NMR). It enables the simultaneous purification and identification of primary and secondary metabolites present in biological samples. In this chapter, we describe the configuration and use of UHPLC-MS/MS-SPE-NMR in metabolite analyses ranging from sample extraction to higher-throughput metabolite annotation. With the integrated UHPLC-QTOF-MS/MS-SPE-NMR method, we have purified and confidently identified more than 100 previously known as well as unknown triterpene and flavonoid glycosides while noting that most of the identified compounds are not commercially available.

Key words Ultrahigh-performance liquid chromatography (UHPLC), Mass spectrometry (MS), Solid-phase extraction (SPE), Online and offline sample transfer, Nuclear magnetic resonance (NMR) spectroscopy

1 Introduction

Metabolomics is a well-defined domain within systems biology and has undergone exponential growth with extensive applications encompassing areas from agriculture to precision medicine and even space exploration [1–6]. Much of this development has resulted from significant advancements during the last two decades associated with two major analytical techniques, mass spectrometry (MS), and nuclear magnetic resonance (NMR) spectroscopy. The number one grand challenge of metabolomics is to unambiguously identify all the metabolites observed in a biological sample, and this

G. A. Nagana Gowda and Daniel Raftery (eds.), *NMR-Based Metabolomics: Methods and Protocols*, Methods in Molecular Biology, vol. 2037, https://doi.org/10.1007/978-1-4939-9690-2_7,

typically involves high-resolution separations and purifications followed by structure elucidation. Although multiple analytical techniques are available for identification and structure elucidation of novel metabolites, MS and NMR are the two most widely used techniques with unparalleled preeminence [7–10]. In fact, most organic chemistry journals require evidence of purity and accurate mass and NMR data for confident structural assignments.

The "metabolome" is a heterogeneous mixture of molecules that vary significantly in physical and chemical properties including sensitivity toward MS and NMR. Both MS and NMR have unique advantages as well as limitations as stand-alone techniques. However, MS has dominated the area of metabolomics primarily due to its very high specificity and sensitivity, i.e., ability to detect very low levels of metabolites with high resolution [11]. Application of tandem mass spectrometry (MS/MS) and computational predictive tools have enabled the identification of many molecular substructures with relatively high confidence through spectral matching with authentic compounds compiled in several large MS databases [12–17]. However, the success of MS is often dependent on the ionization efficiency of the metabolites and eliminating suppression from interfering ions. Therefore, MS is often coupled with separation techniques such as GC, LC, or CE for relative purification prior to detection to minimize ion interferences and spectral overlap, but this can require additional sample preparation steps such as metabolite derivatization for GC [11, 18]. Also, MS/MS spectra can differ significantly from instrument to instrument, and most of the metabolome is still not represented in existing MS databases used for spectral matching and unknown metabolite identification. This suggests that MS alone is not sufficient to provide absolute identification for all metabolites, and authentic standards are only available for a modest number of specialized metabolites. Thus, MS alone is not likely to yield confident metabolite identifications in the absence of authentic or synthetic standards, and in these situations, MS can and should be complimented with other technologies to acquire orthogonal data.

NMR is the gold standard for absolute structure elucidation of metabolites by identifying connections among nuclei within a molecule based upon their different electronic environments and magnetic properties with a very high degree of reproducibility as well as quantitative accuracy [19]. NMR does not require any intricate derivatization and allows data collection for molecules with diverse functionalities. In addition, metabolites can be identified in a nondestructive manner, allowing complete sample recovery. However, NMR still suffers from lower relative sensitivity compared to MS. This lower sensitivity is the result of very low population difference between the energy states responsible for generating the signal of detection and in certain cases the low natural abundance of the NMR-sensitive isotope of a particular element, such as

^{13}C [20]. Another limitation of NMR, especially in metabolomics, is the lack of sufficiently large spectral databases for spectral matching in order to identify metabolites. This limitation is more pronounced when compared to the data in MS/MS databases. Thus, it will be essential to build and expand NMR databases in the future. The concentrations of metabolites in biological systems are typically small, and metabolites typically require purification and concentration to acquire sufficient amounts to facilitate quality NMR data collection. Innovations in NMR hardware in recent years, including the design of NMR spectrometers with higher magnetic field strengths and NMR probes with cryogenically cooled receiver coils, have helped to mitigate these limitations and significantly enhance the signal-to-noise (S/N) ratios [21, 22]. Further, the introduction of 1.7 mm MicroCryoProbe with a 14-fold higher sensitivity than a conventional 5 mm room temperature probe has immensely benefited the area of metabolomics wherein identification of minute amounts/micrograms of metabolites is now possible.

Efforts to maximize the benefits of NMR applications in the analysis of complex mixtures of molecules led to the emergence of hyphenated systems through online coupling of NMR with other analytical techniques. An early example of such efforts was the introduction of LC-NMR where a flow probe was used. In this system, chromatographically separated fractions from a mixture of compounds were directly transported to the NMR probe, eliminating the need for a separate sample preparation method and NMR sample tubes [23]. Although well received in the pharmaceutical industry, the needed use of deuterated solvents during LC led to substantial cost increases and limited the manipulation of LC conditions and optimized separations. Continuous flow mode allowed only enough time to collect ^{1}H NMR data, limiting spectral information content. Collection of 2D NMR data was made possible by introducing stopped flow and loop collection methods, but cross contamination of fractions posed a challenge for this method [24]. Also, inevitable sample diffusion, especially for those with lower solubility and chromatographic peak broadening in all forms of LC-NMR methods, continued as other shortcomings [25].

In 1995, scientists from Pfizer first demonstrated the online integration of LC, NMR, and MS together where the separated fractions from LC could be simultaneously introduced to NMR and MS for maximum structural information [26]. Parallel coupling of MS and NMR facilitated the use of either method to screen the LC fractions for selection of the fraction of interest which could be subsequently analyzed by the other method. However, as long as LC and NMR modules were coupled directly, limitations such as sample dilution and need of a dedicated NMR flow probe persisted. Accordingly, the concept of separating the LC-MS and NMR

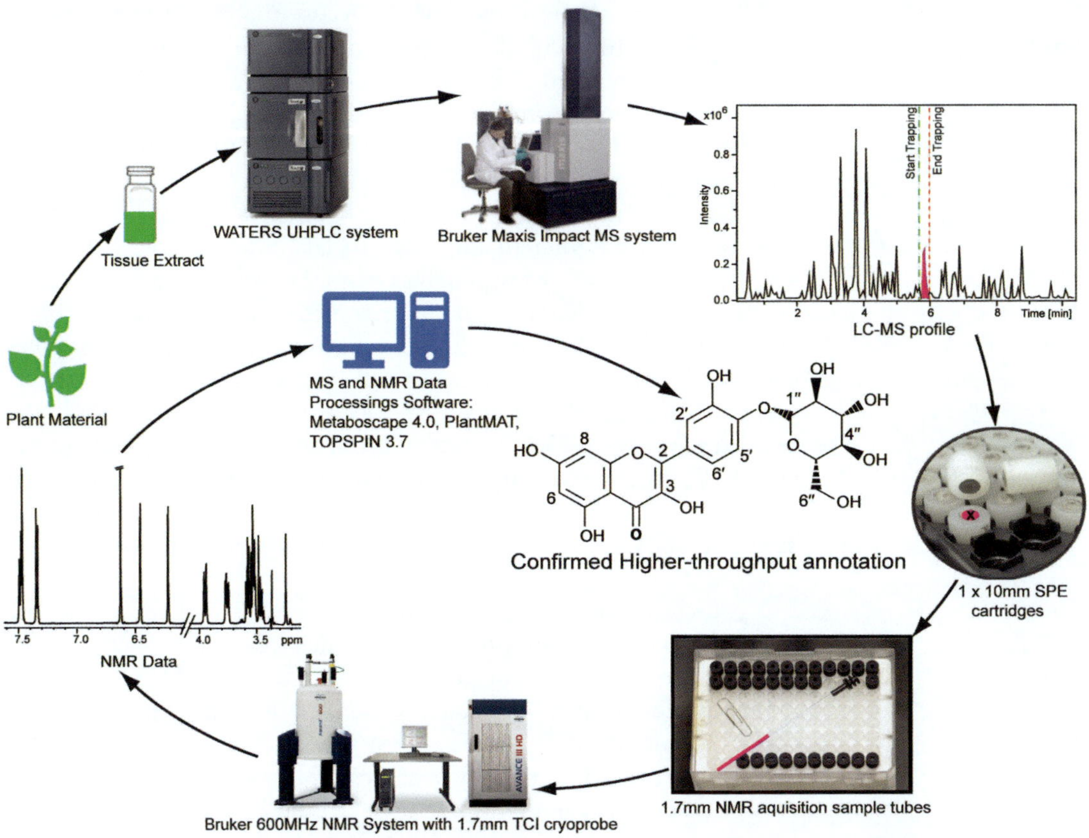

Fig. 1 Schematic representation of a UHPLC-MS-SPE-NMR analytical workflow for automated purification and concentration of targeted metabolites followed by structure elucidation based upon retention time, accurate mass MS, MS/MS, 1D ^{1}H NMR, and 2D NMR data

components gained traction that culminated in the insertion of a solid-phase extraction (SPE) module between the LC-MS and NMR platforms [27, 28]. This intervention allowed the monitoring of eluting compounds of interest from the LC from a small proportion (5%) of the eluent with the MS while simultaneously diverting most of the eluent (95%) onto a SPE cartridge designed to retain the analyte without the use of costly deuterated solvents (*see* Fig. 1). Repetitive injections of the sample enabled the accumulation and concentration of the desired metabolite in the SPE cartridge from which the metabolite could be subsequently eluted using a minimal amount of a preferred deuterated solvent for NMR analysis in either an online or offline fashion. This method has not only minimized the use of costly deuterated solvents but also has allowed optimization of LC-MS conditions without affecting NMR signal quality. In addition, this method has facilitated the use of diverse types of NMR probes and experiments needed for observing different nuclei, without being restricted to dedicated flow probes. Currently, LC-MS-SPE-NMR has emerged as a very powerful and less labor-intensive analytical platform to selectively purify and concentrate metabolites from complex biological samples prior to structure elucidation using a multitude of 1D and 2D NMR techniques in a cost-effective and higher-throughput workflow [23, 29–34].

2 Materials

2.1 Chemicals and Consumables

1. Use LC-MS grade solvents to prepare the following solutions (*see* **Note 1**):
   (a) 80% methanol in water.
   (b) UHPLC-MS gradient solvents: Solvent (A) Water with 0.1% formic acid and solvent (B) 100% acetonitrile (*see* **Note 2**).
   (c) Mass spectrometer tuning and mass calibration solution using sodium formate cluster ions (9.9 mL of 50% isopropanol +0.2% formic acid and 100 μL of 1 M NaOH) (*see* **Note 3**).
   (d) Lock mass calibration solution: 1 mg hexakis-(2,2-difluoroethoxy)-phosphazene in 1 mL 50% isopropanol; negative ion *m/z* 556.0020, add to adsorber surface (*see* **Note 4**).
   (e) Makeup solvent for SPE system: Water with 0.05% formic acid.
   (f) Solvent for analyte elution after SPE collection: 75% acetonitrile in water (*see* **Note 5**).
   (g) Extraction solution: 80% methanol containing 18 μg/mL umbelliferone as an internal standard.
2. Deuterated methanol-d_4 for NMR acquisition (*see* **Note 6**).
3. SampleJet 1.7 × 103.5 mm (4 in.) NMR capillary tube with a hole on top and 2.5 mm ball for sealing the top (*see* **Note 7**).
4. Flat bottom, 2 mL glass vial, 12 × 32 mm with 9 mm screw neck vial with the cap.
5. Powder-free nitrile gloves and cryogenic gloves for handling liquid nitrogen.
6. Syringes: 5 in needle length, zero dead volume, 50 μL syringe volume, removable needle, and bevel tip syringe.
7. Glass 5.75 in Pasteur pipettes.
8. Texpad Tape Head Cleaner Pads (*see* **Note 8**).
9. Tubing: PEEK Red 1/16 in. OD × 0.005 in. ID × 5 ft; PEEK Nat 1/16 in. OD × 25 μm ID × 2 in.; fused silica 20 μm × 350 μm × 500 cm.

10. One tee, 1/16″ zero dead volume splitter.
11. Kimwipes, 4.4 × 8.4 in.
12. Solvent reservoir bottle caps.
13. Autosampler 300 μL vial inserts.

2.2 Processing Plant Material

1. Root tissue of *Medicago truncatula* (*see* **Note 9**).
2. Liquid nitrogen for freezing and quenching internal metabolism in its current state (*see* **Note 10**).
3. Grind sample tissue with pre-solvent rinsed, oven-dried, and precooled with liquid nitrogen pestle and mortar for homogenization (*see* **Note 11**).
4. Pre-solvent rinsed clean steel spatula.

2.3 Ultrahigh-Performance Liquid Chromatography-Mass Spectrometry (UHPLC-MS)

1. UHPLC system with PDA detector, column heater with active preheating (CH-A), autosampler, and binary solvent pump, such as a Waters UPLC ACQUITY I-Class.
2. Waters BEH C_{18} column, 130 Å, 1.7μm, 2.1 mm × 150 mm (*see* **Note 12**).
3. Bruker maXis Impact QTOF mass spectrometer or similar.
4. Acquisition and data processing software (e.g., Waters ACQUITY UPLC console, Bruker's HyStar acquisition, QTOF control and Compass DataAnalysis, and MetaboScape 4.0).
5. KD Scientific syringe pump (or similar) for continuously pumping the mass calibrant solution at 500 nL/min (*see* **Note 13**).

2.4 Solid-Phase Extraction (SPE)

1. Spark Holland Prospekt-2 SPE system coupled with UHPLC-MS system.
2. Waters OASIS online SPE, Symbiosis/Prospekt-2 cartridges, HLB, 10 × 1 mm, 96/pk.
3. Knauer pump to dilute UHPLC column eluent with 0.05% aqueous formic acid to improve analyte absorption on SPE cartridge.
4. New or highly clean glass bottle as a solvent reservoir.
5. 1/16″ zero dead volume splitter.
6. Fraction collector for SPE trapped elutions.
7. Use 96 vial rack for the offline SPE elution method. For the online SPE elution method, use 192 position rack of 1.7 mm NMR tubes (*see* **Note 14**).

2.5 Nuclear Magnetic Resonance (NMR)

1. 600 MHz NMR Ascend UltraShield spectrometer.
2. 1.7 mm TCI MicroCryoProbe.
3. SampleJet 480 positions autosampler system.
4. Acquisition and data processing software (e.g., TopSpin 3.5 or equivalent).

3 Methods

3.1 Tissue Extractions

1. After sample collection, arrest metabolism and enzymatic activity in the tissue samples by immersion in liquid nitrogen (*see* **Note 15**).
2. Lyophilize frozen tissue until dry; i.e., for 2–3 days or until the constant dried mass is obtained. The dried tissue should be stored at −20 °C until further analysis.
3. Grind dried tissue using clean mortar and pestle while cold.
4. Approximately, 10 ± 0.06 mg of the dry tissue is weighed and transferred into a 1-dram glass vial (3.8 mL) with a clean spatula.
5. Add 1 mL of 80% methanol containing 18 μg/mL umbelliferone as an internal standard (IS) (*see* **Note 16**).
6. Sonicate the tissue sample for 10 s to burst the tissue cells followed by vortexing for 20 s.
7. Gently shake the sample containing vial on an orbital shaker for 2 h at room temperature.
8. Centrifuge sample vial for 30 min at 3000 × *g* at 4 °C.
9. Transfer 500 μL of supernatant to an autosampler vial (*see* **Note 17**).
10. Store sample at −20 °C until sample analysis.

3.2 UHPLC-MS-SPE Analysis

1. We are working on the isolation and higher-throughput annotation of glycosylated compounds usually flavonoid and triterpene glycosides present in root and aerial tissues of *Medicago truncatula*. Thus, we developed an in-house method for the separation of glycosylated mixtures using UHPLC and a BEH C_{18} reverse phase column due to variations in the polarities of the different positional isomers.
2. Place the zero dead volume splitter on the column outlet as shown in Fig. 2 (*see* **Note 18**).
3. For UHPLC analysis, prime solvents in LC reservoir for 1 min followed by the equilibrating UHPLC system with A:B (95:5) for 10 min.
4. Prepare the UHPLC gradient method for the analysis with an initial mobile phase ratio of (percent A:B) 95:5, then change

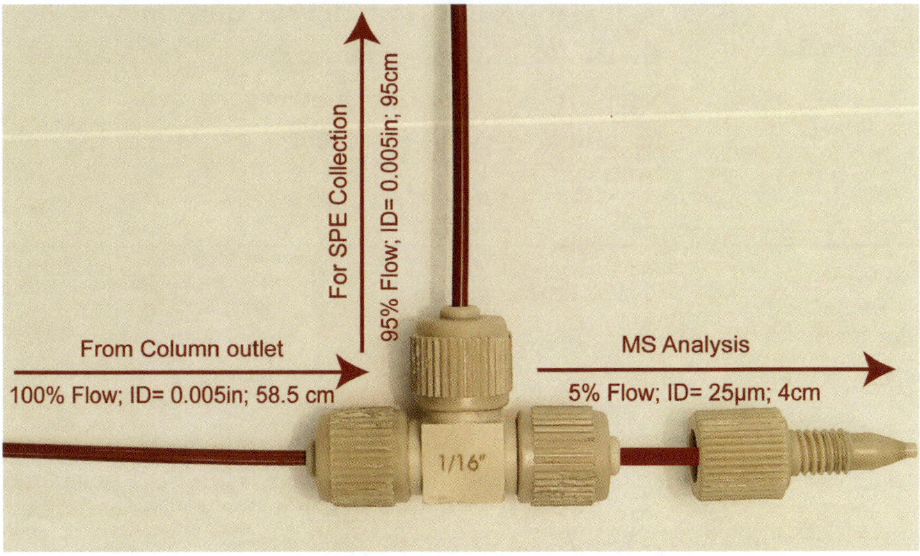

Fig. 2 Flow diagram of the UHPLC column eluent using a passive tee splitter which diverts a proportionate flow amount to MS and SPE using differential tubing lengths and internal diameters. The PEEK tubing length for the SPE collection was 95 cm with an internal diameter (ID) = 0.005 in, while the PEEK tubing length to the mass spectrometer was much less at 4 cm with an ID = 25 μm. The PEEK tubing length from the column to the splitting tee was 58.5 cm with an ID = 0.005 in.

linearly to 30:70 over 30 min, then to 5:95 over 3 min, hold at 5:95 for 3 min and return to 95:5 over 3 min, and hold at 95:5 over 1 min for column equilibration. The flow rate is 0.56 mL/min throughout the analysis, while the column temperature is maintained at 60 °C. Set the PDA detector wavelength to 280 nm for the analysis. The injection needle is washed for 10 s before and after every injection with methanol-acetonitrile-isopropanol-water; each 25%.

5. Set up the MS and MS^2 parameters on the QTOF instrument. The optimum MS and MS^2 conditions for Bruker Impact QTOF-MS were as follows: Mass range 50–1500 *m/z* with a spectral acquisition time of 0.25 s and absolute threshold 12 cts. Ion source settings were end plate offset 500 V, capillary 4000 V, nebulizer 43.5 psi, dry gas 12.0 L/min, and dry temp 250 °C. To tune the system, the transfer parameters for funnel 1RF and 2RF were 300.0 Vpp, CID 0.0 eV, and hexapole RF 150.0 Vpp. For the quadrupole, ion energy 8.0 eV and low mass 100.00 *m/z*. For the collision cell: collision energy 8.0 eV, collision RF 720.0 Vpp, transfer time 100.0 μs, and prepulse storage 5.0 μs (*see* **Note 19**).
6. Place the aqueous methanol extract (obtained at Subheading 3.1) in the autosampler chamber at 4 °C. Program the UHPLC-MS system for the test sample analysis for peak-picking information or the selection for the SPE collection with 2 μL of injection volume (*see* **Note 20**).

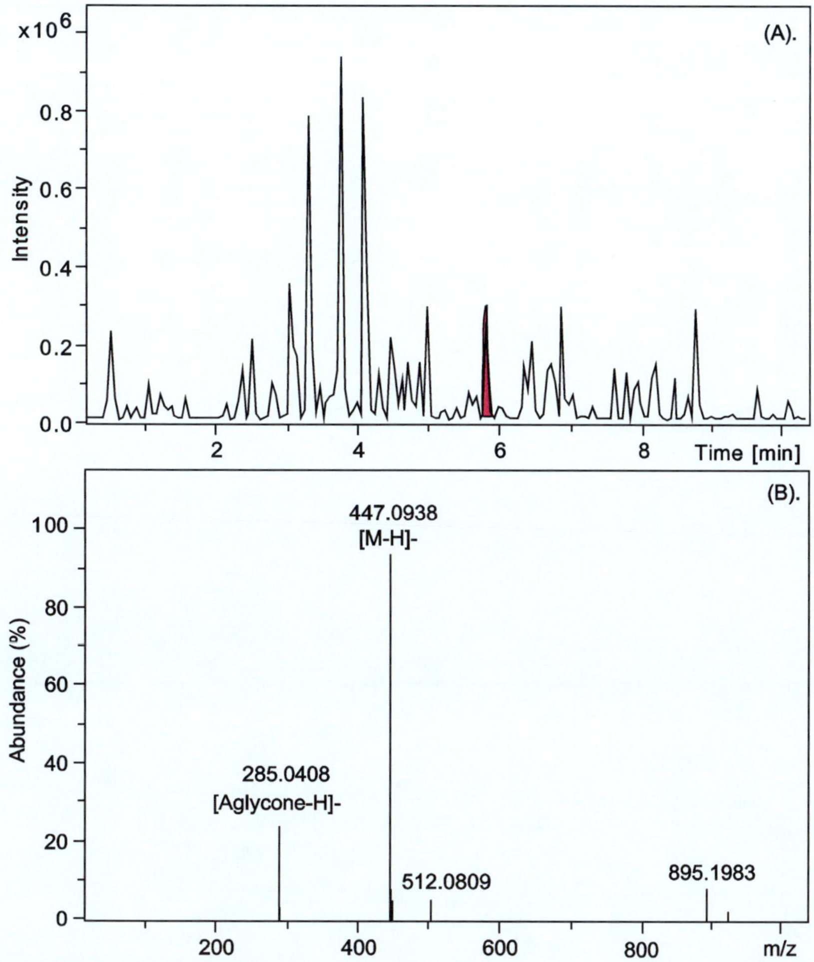

Fig. 3 (**a**) UHPLC Negative electrospray ionization MS chromatogram and (**b**) mass spectrum of the highlighted compound targeted for SPE purification. The peak at 5.85 min (highlighted in red) was selected from UHPLC-MS chromatogram of a *Medicago truncatula* root extract and targeted for solid-phase extraction (SPE) and concentration

7. Set up a method for SPE collection of targeted peaks after initialing determining their retention times and *m/z* values (*see* Fig. 3). The parameters used were as follows: number of injections 30; volume for each injection 5 μL; runtime for Knauer makeup pump from 20 s before the trapping of the first peak until 4 min after the retention time of the last peak, at the flow rate of 1.2 mL/min. The parameters for trapping peaks include the parent ion or targeted ion *m/z* value or retention time window (*see* Fig. 4). Multiple trappings of the same targeted analyte onto the same SPE cartridge can yield sufficient amounts (1–10 μg) for NMR data acquisition (*see* Fig. 5) (*see* **Note 21**).

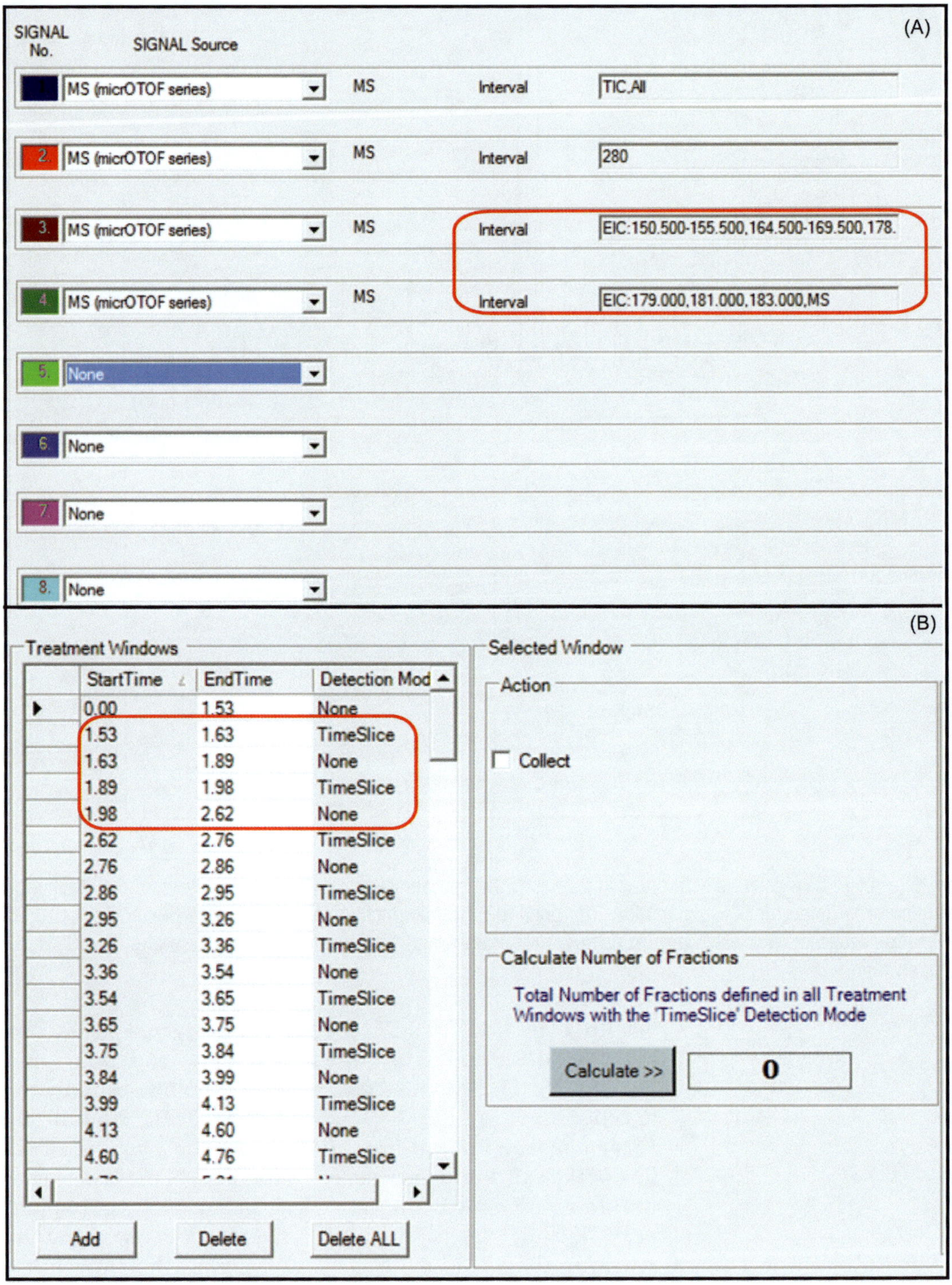

Fig. 4 Parameters used to setup SPE peak trappings in Bruker's HyStar acquisition software on the basis of parent ion or targeted ion observed during negative ESI-MS (**a**) and start and end retention time ranges for peak collection (**b**)

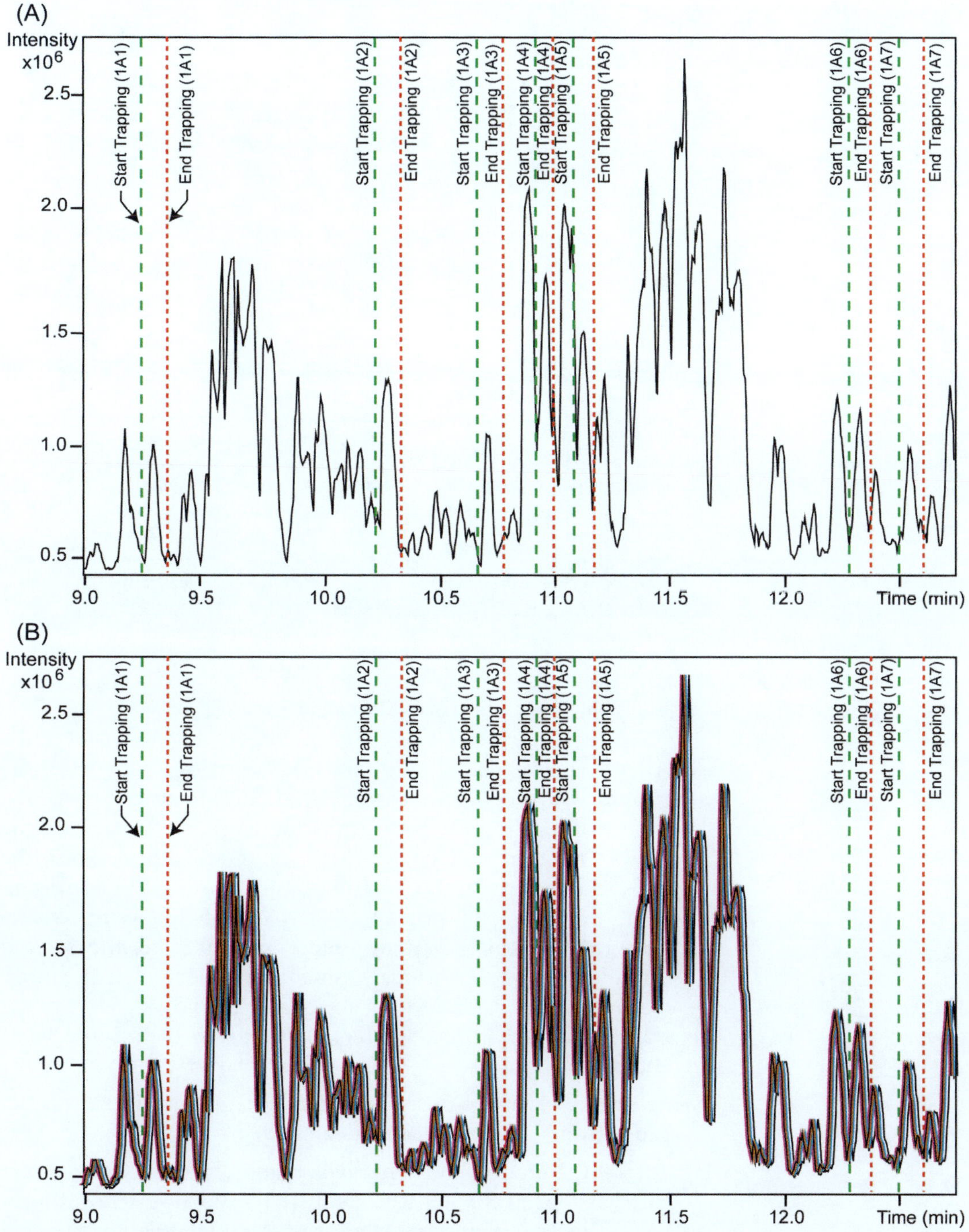

Fig. 5 Collection of peaks targeted for SPE purification: UHPLC-MS chromatogram illustrating seven eluting peaks targeted for SPE purification with start and stop intervals noted with vertical markers (**a**); overlay of multiple chromatograms from multiple injections of the same extract for repetitive SPE concentration to obtain microgram quantities of material that will produce quality NMR spectra (**b**)

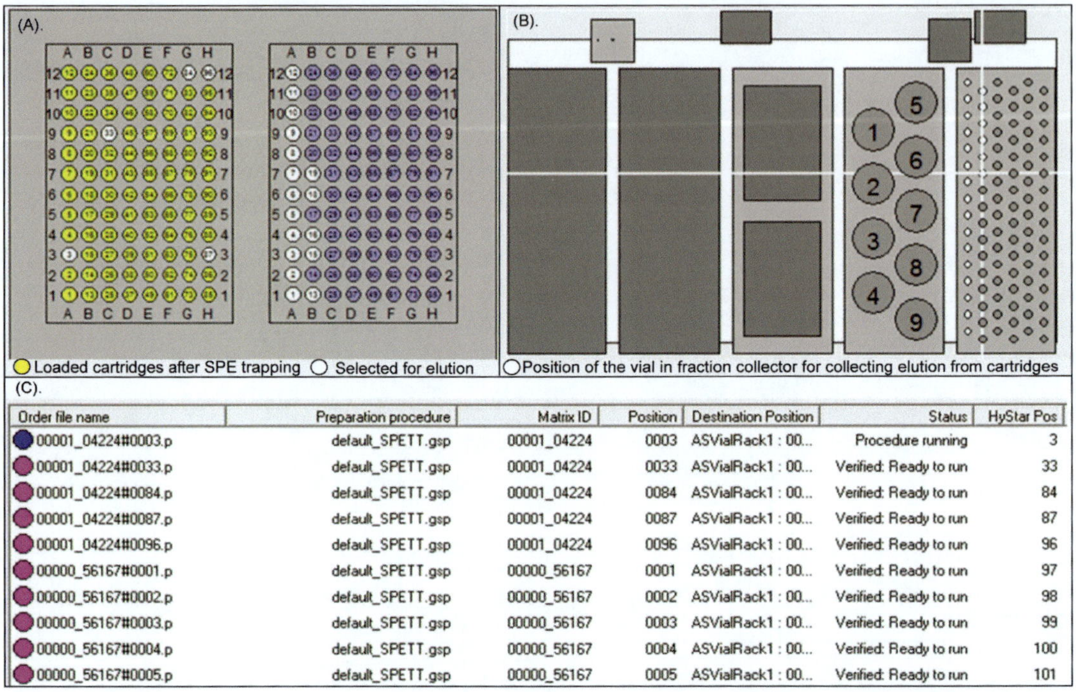

Fig. 6 Screenshot image for the peak elution from the SPE cartridge software. Illustration of loaded cartridges in the tray of Bruker's SPE Prospekt-2 system (**a**); elution selection vials position for elution in Gilson fraction collector (**b**); job order created for elution of loaded peaks from SPE cartridges to vial (**c**)

8. After trapping peaks onto the HLB cartridges, dry the cartridges with nitrogen gas for 1 min. It is necessary to prime the needle, injection port, and diluter purge system of the Gilson fraction collector before the first elution from the Prospekt-2 system. Elute compounds from SPE cartridges using 830 μL a solution of 70% ACN and 30% water with a flow rate of 250 μL/min. Before switching to the second cartridge, it is necessary to wash and dry the line after elution from the first cartridge to avoid any carryover. Set up the washing and drying parameters as 600 μL with elution solvent and dry for 2 min.
9. Select the cartridge position and destination vial in the tray of the Gilson fraction collector (*see* Fig. 6).
10. After SPE elution into the vial, evaporate to dryness under nitrogen gas at 30 °C. Reconstitute the fraction with 200 μL of 80% methanol and transfer into the 300 μL vial inserts, and measure the purity and intensity of the peak with a 2 μL injection volume using UHPLC-MS with similar parameters as before, prior to moving onto NMR data acquisition.
11. The SPE collected sample is then redried under a stream of nitrogen gas for the preparation of 1D (one-dimensional) and

2D (two-dimensional) NMR analyses to confirm the putatively assigned structure or de novo structural elucidation.

12. Try to identify metabolites through UHPLC-MS/MS spectral matching prior to NMR. We use Bruker Compass DataAnalysis, MetaboScape 4.0 with Sumner's Bruker MetaboBASE Plant Library https://www.bruker. com/products/mass-spectrometry-and-separations/ms-software/metabolomics-spectral-libraries/overview.html.
13. If UHPLC-MS/MS identification is unsuccessful via spectral matching, predict metabolite identity prior to NMR data acquisition using commercial or custom software such as our in-house developed Plant Metabolite Annotation Toolbox (PlantMAT) https://sumnerlab.missouri.edu/plantmat/. PlantMAT [16] is used to generate putative structural information for metabolites based upon MS, MS/MS, and retention times of analytes. The software will provide information regarding the aglycone and degree of glycosylation present for flavonoid and triterpene conjugates (*see* **Note 22**).

3.3 Nuclear Magnetic Resonance Analysis

1. Reconstitute the sample with 30 μL of deuterated methanol and transfer into 1.7 mm NMR tube with a 5" needle length syringe, and then seal the tube from the top with ball (*see* **Note 23**).
2. Set the NMR parameters for the acquisition of ^{1}H pulse sequence, e.g., ZG30 or WET, with 65 K data points, 64 dummy scans (DS), 1024 number of scans (NS), 9 ppm spectral width (SW), and 4.5 ppm as an offset value (O1) (*see* **Note 24**).
3. Select the spectral window (*see* **Note 25**).
4. For processing 1D ^{1}H spectra, select line broadening (LB) as 0.30 Hz and perform automatic or manual phase correction (*see* Fig. 7).
5. 2D experiments such as correlation spectroscopy (COSY), heteronuclear multiple quantum coherence (HMQC), heteronuclear multiple bond correlation (HMBC), and total correlation spectroscopy (TOCSY) use the same SW and O1 as for proton 1D spectra. Parameters for carbon are 0–160 ppm SW with 80 ppm O1 in HSQC and 0–220 ppm SW with 110 O1 for HMBC experiments, with 256 NS, 16 DS with FID size of 2048 points in F2, and 256 points in F1 (*see* Figs. 8, 9, and 10) (*see* **Note 26**).
6. Selective (Sel) 1D pulse sequences such as Sel-TOCSY, Sel--ROESY (Rotating-frame Overhauser SpectroscoPY) and Sel-NOESY (Nuclear Overhauser SpectroscopY) use NS = 512, DS = 64, and 65 K data points with respective mixing time (D9 for TOCSY = 2–4 s, P15 for

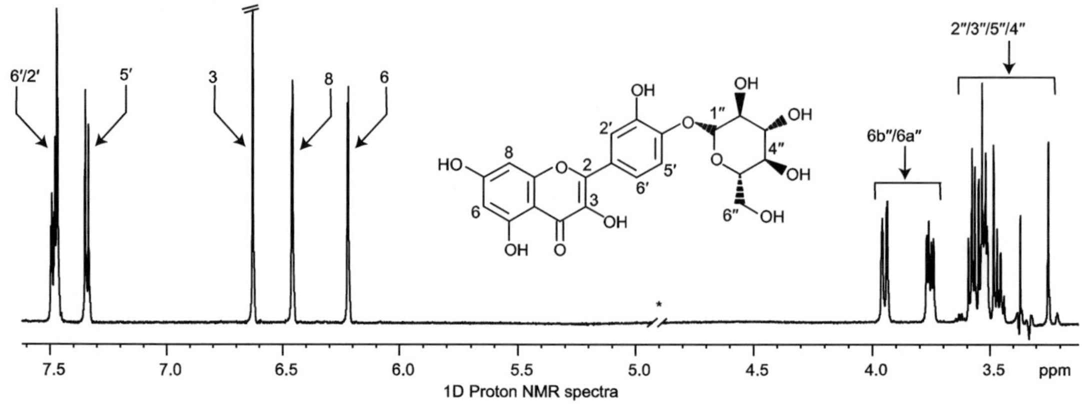

Fig. 7 Example of a one-dimensional ^{1}H NMR spectrum of a UHPLC-MS-SPE purified metabolite (peak at 5.85 min) with peak assignments. This peak has been identified as luteolin-4′-glucoside based upon MS, MS/MS, 1D NMR, and 2D NMR

ROESY = 0.4–2.0 s, D8 for NOESY = 0.4–2.0 s) (*see* **Note 27**).

7. After combining all the information from 1D and 2D NMR experiments, the metabolites are identified as exemplified for luteolin-4′-glucoside in Figs. 7–10.

4 Notes

1. Many solvents are toxic and potentially flammable and/or explosive. Their contact and/or vapors can irritate or damage the skin or internal organs when inhaled. Use in a fume hood and handle with solvent resistant nitrile gloves. Extra high purity solvents such as LC-MS grade solvents are really needed as solvent impurities can erode MS and NMR spectral quality when working with microgram quantities of sample.
2. Cover solvent bottle tops with solvent reservoir bottle caps and placed under an exhaust snorkel. The purpose of the cap is to avoid airborne contamination of the solvent(s).
3. It is necessary to calibrate the MS instrument before all acquisitions. Be sure the calibration covers wide ranges in the negative (112.9856–1404.7467 *m/z*) and positive (90.9766–1382.7377 *m/z*) electrospray ionization modes. Calibration score ≥95% is good for data processing after the acquisition.
4. Use the lock mass calibration within each of the test sample acquisitions and in the validation of peak purity after elution from SPE cartridges.

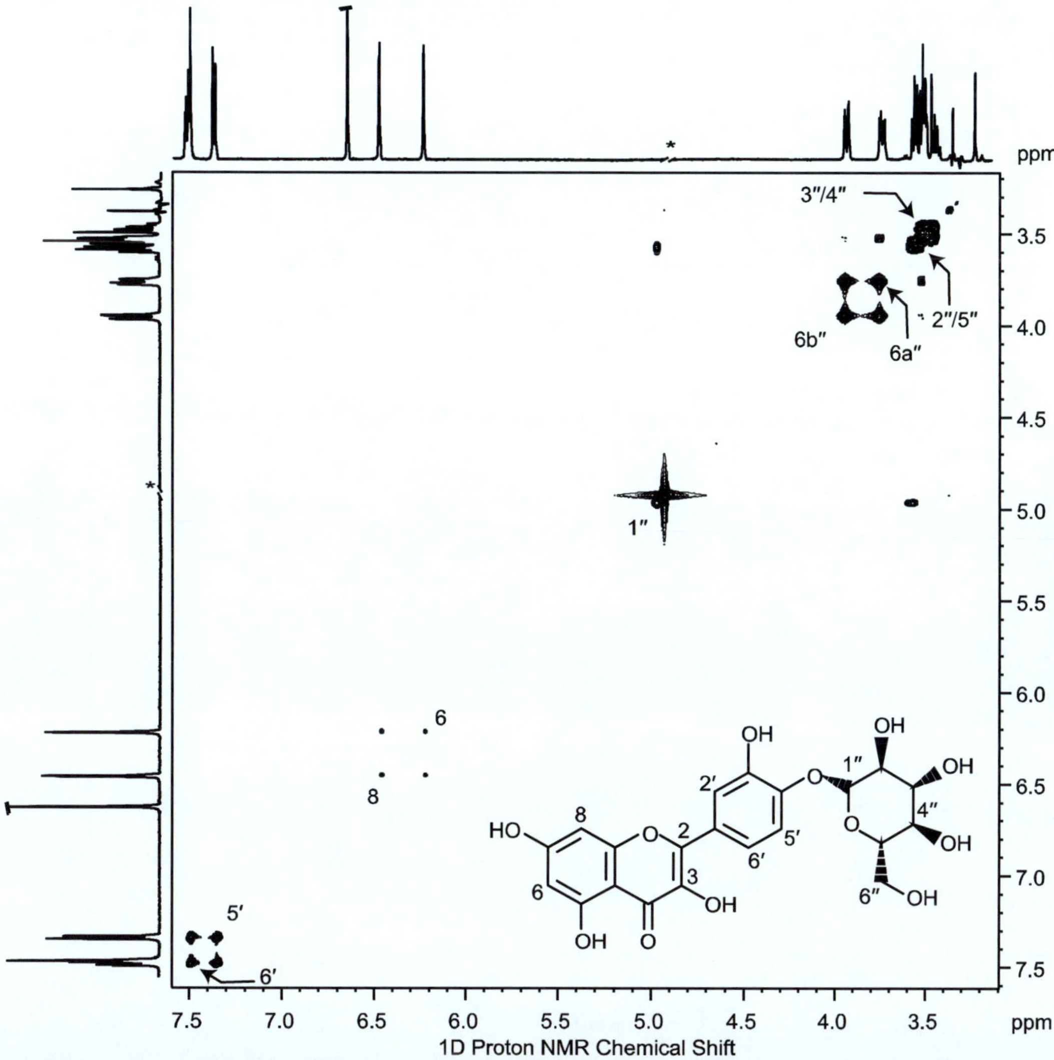

Fig. 8 Example of a two-dimensional $^1H-^1H$ COSY NMR spectrum of a UHPLC-MS-SPE purified metabolite (peak at 5.85 min) with the peak assignments. This targeted compound has been identified as luteolin-4′-glucoside based upon MS, MS/MS, 1D NMR, and 2D NMR

5. We do not recommend eluting the analyte fraction with any 100% organic solvent in an attempt to help avoid co-elution of lipid contaminants that are typically observed with the analytes on the SPE cartridges. Instead, we recommend eluting with generically 70% organic or with a percent organic calculated based upon the elution solvent composition plus 5–10% organic observed for each individual metabolite.

6. Keep deuterated solvent(s) in an airtight container or desiccator due to their hygroscopic and hydrogen-deuterium exchange properties.

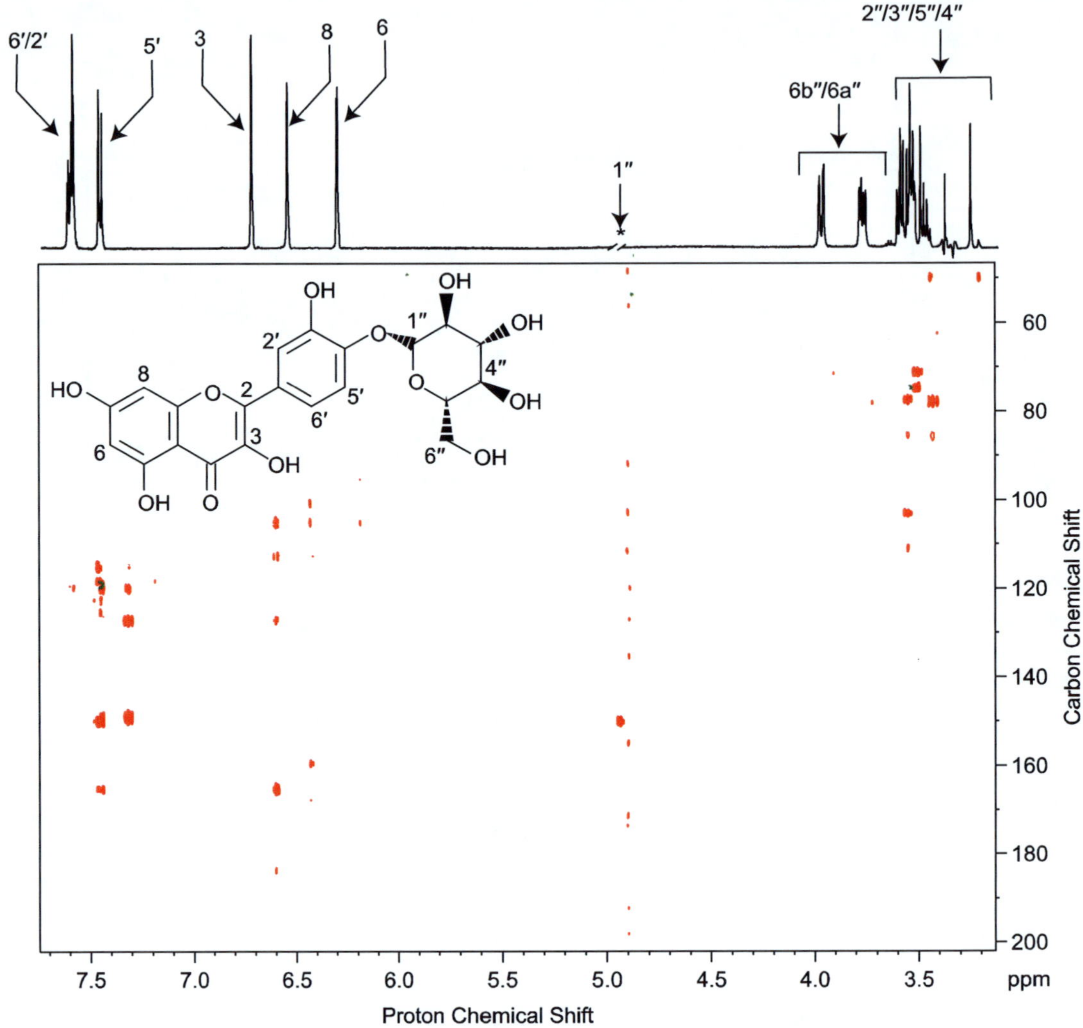

Fig. 9 Example of a two-dimensional ^{1}H–^{13}C HMBC NMR spectrum of a UHPLC-MS-SPE purified metabolite (peak at 5.85 min) with the complete assignment. This targeted compound has been identified as luteolin-4′-glucoside based upon MS, MS/MS, 1D NMR, and 2D NMR

7. 1.7 mm NMR acquisition tubes are open at the top. Thus, store them in an airtight container, dehumidifier, oven or in a closed cabinet to avoid moisture, dust particles or any other contaminants.
8. Bruker's 1.7 mm TCI cryoprobe is a highly sensitive probe. Thus, clean the external glass surface of the NMR tube using Texpad Tape Head Cleaner Pads before putting it into the SampleJet autosampler.
9. We are focusing on the isolation and identification of metabolites from *Medicago truncatula* root tissue. You can use this or a similar UHPLC-MS-SPE-NMR protocol for other biological systems.

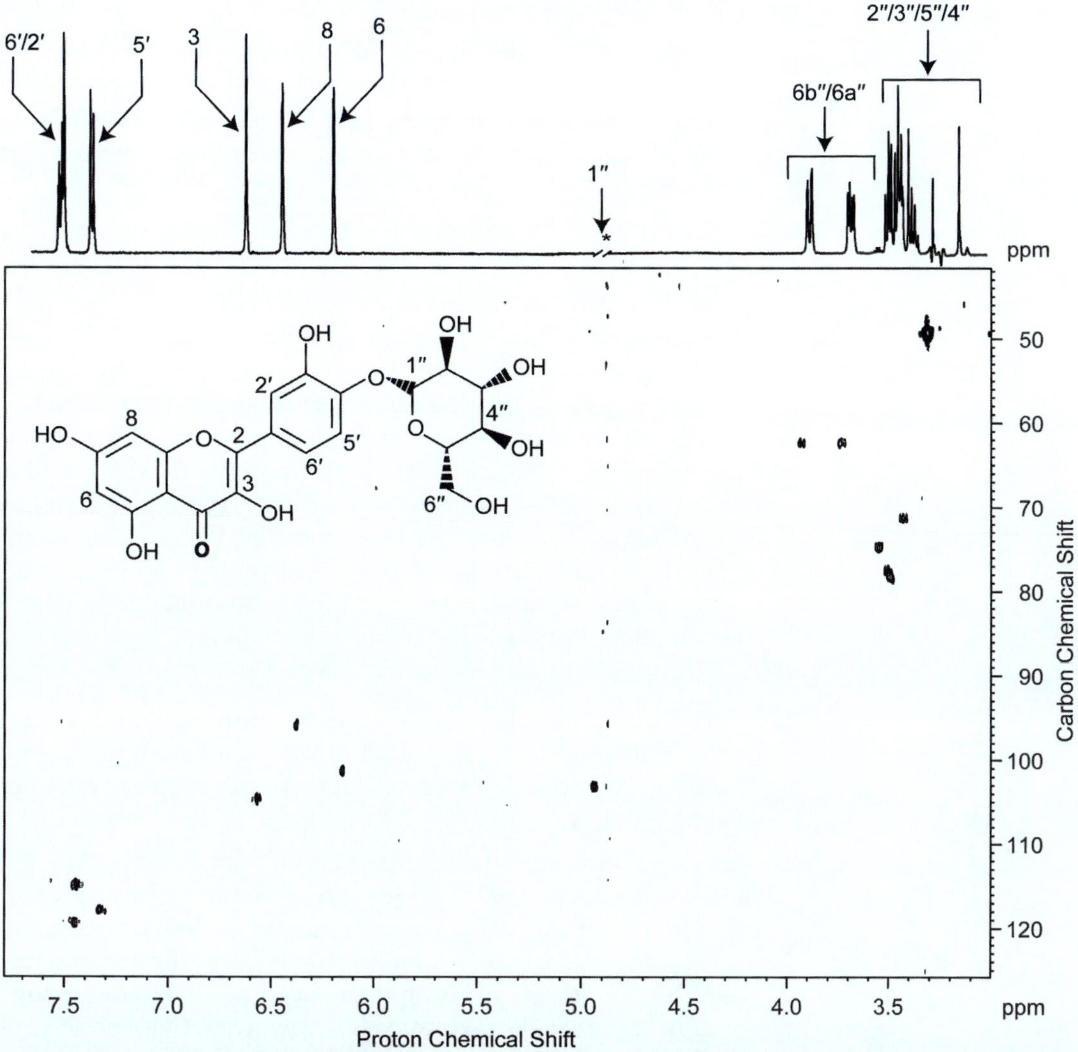

Fig. 10 Example of a two-dimensional $^{1}H–^{13}C$ HMQC NMR spectrum of UHPLC-MS-SPE purified metabolite (peak at 5.85 min) with peak assignments. This targeted compound has been identified as luteolin-4-′-glucoside based upon MS, MS/MS, 1D NMR, and 2D NMR

10. After plant harvesting or collecting sample tissues, it is recommended to limit ongoing metabolism and metabolite degradation which may alter the real information of the sample. Thus, transport your samples from the site of collection to the lab in liquid nitrogen and store at −80 °C until further analyses.
11. If your sample is liquid (urine, plasma, etc.) there is no need for the tissue grinding process.
12. You can use alternative reverse-phase columns that have similar chemistry or better specificity or resolution with respect to your preferred analyte.

13. Check the volume and internal diameter of the syringe before setting the syringe pump flow rate settings. This will help you to maintain accurate flow rates.
14. There are two modes of fraction elution from Prospekt-2-loaded cartridges, and these depend on the objectives as follows.

    Online method: Fraction is eluted from the cartridge using deuterated solvent directly into a 1.7 mm NMR tube using the Gilson fraction collector.

    Offline method: Target fractions are eluted from the cartridges using a non-deuterated LC-MS grade solvent into a flat bottom screw cap vial using the Gilson fraction collector. We recommend the offline method as the SPE elution volume for maximum recovery from 1 mm × 1 cm SPE cartridges is approximately 300 μL. This volume will not fit into the 1.7 mm NMR tubes. We also recommend UHPLC-MS analysis of SPE eluent prior to NMR acquisition to validate the purity of the eluted peak fraction from the cartridges. This approach is also more economical and no hydrogen-deuterium exchange property exists.
15. Use nitrile gloves throughout the experiment to avoid lipid contamination. Use clean glass tip or glass syringes. Avoid parafilm or plastic pipette tips to minimize polymer contamination.
16. Selection of an internal standard (IS) is a challenging task for LC-MS analysis. The physical and chemical properties of IS should be known prior to analysis, e.g., solubility with the mobile phase, inert with sample mixture, not produced naturally by sample tissue and not overlapping in retention time with other molecules. Nowadays, many researchers prefer stable isotope-labeled metabolites, but these are typically more costly. The addition of an internal standard will help you calculate how much quantity you can inject for SPE collection with respect to column absorption properties. An overloaded injection may shift the retention times and might result in the trapping of impurities associated with both the targeted and untargeted peaks in the profile. Moderate injection quantities also help in avoiding carryover from the column.
17. To collect a soluble fraction, use a beveled-tip syringe and put beveled-edge against the side of glass vial to avoid uptake of insoluble particles present in the extract.
18. The tee splitter should be close to the MS spray chamber to avoid delays for peak trapping.
19. For compounds of 300–500 m/z use collision energy 20–35 eV, for 500–800 m/z use 35–50 eV, and for 800–1500 m/z use 50–80 eV.

20. For a single batch, always make fresh solvents dependent on the volume you need.
21. SPE trapping from 30 injections of a peak of moderate intensity, i.e., >20% of base peak, is usually sufficient to yield reasonable NMR data. More abundant peaks may require less, while less abundant peaks might require more trappings.
22. As we are targeting specific classes of metabolites to understand the systems biology of roots, we are using multiple software packages and databases for assisting in identification. You can establish your own in-house database or use public databases for generating putative information.
23. Do not use other material to seal the top of the 1.7 mm NMR tube. It can increase the weight of the tube which may break the tube inside the probe during insert or exit movements.
24. The number of scans should be appropriate because as you increase the number of scans, you will also increase the signal-to-noise ratio.
25. We use a spectral window range 0.5–6.5 ppm with offset of 3.5 ppm for triterpene saponins and 1.0–9.0 ppm with offset of 5.0 ppm for flavonoid saponins. You can select the range of the chemical shift according to the chemical environment of your targeted molecule.
26. After obtaining 1D data, it is beneficial to acquire specific 2D data for complete structural elucidation: e.g., for proton-proton correlation, COSY; proton with directly attached carbon, HMQC; proton with two- or three-bond range carbon, HMBC and 2–4 range proton–proton correlations, or TOCSY. HMBC experiments also yield information on quaternary carbons present in the metabolite, which reduce the need for 1D carbon experiments.
27. For the confirmation of overlapping peaks (sugar region in saponins) or spatial orientation of the metabolite, we suggest the use of selective TOCSY, ROESY, and NOESY experiments instead of their regular experiments. In such experiments, you can irradiate specific chemical shift regions or peaks to better observe their coupling through space.

Acknowledgments

The Sumner lab has been graciously supported by several entities over the years for the development of metabolite identifications, natural products profiling, and plant metabolomics. These specifically include support from the University of Missouri, the Samuel Roberts Noble Foundation, Bruker BioSpin and Daltonics Gmbh, Agilent Technologies, NSF-JST Metabolomics for a Low Carbon

Society #1139489, NSF-MRI-DBI #1126719, NSF-RCN #1340058, and NSF-MCB #1024976. We thank Drs. Ulrich Braumann (Bruker BioSpin), Aiko Barsch (Bruker Daltonics), Daniel J. Wherritt (Noble Foundation), and Dennis Fine (Noble Foundation) for their operational and technical guidance in assembling the UHPLC-MS-SPE-NMR ensemble.

References

1. Beckles DM, Roessner U (2012) 5—Plant metabolomics: applications and opportunities for agricultural biotechnology. In: Altman A, Hasegawa PM (eds) Plant biotechnology and agriculture. Academic Press, San Diego, pp 67–81
2. Dixon RA, Gang DR, Charlton AJ, Fiehn O, Kuiper HA, Reynolds TL et al (2006) Applications of metabolomics in agriculture. J Agric Food Chem 54(24):8984–8994
3. Resham S, Akhter F, Ashraf M, Kazi AG (2014) Chapter 2—Metabolomics role in crop improvement. In: Ahmad P, Rasool S (eds) Emerging technologies and management of crop stress tolerance. Academic Press, San Diego, pp 39–55
4. Wishart DS (2016) Emerging applications of metabolomics in drug discovery and precision medicine. Nat Rev Drug Discov 15:473
5. Armitage EG, Southam AD (2016) Monitoring cancer prognosis, diagnosis and treatment efficacy using metabolomics and lipidomics. Metabolomics 12(9):146
6. Schmidt MA, Goodwin TJ (2013) Personalized medicine in human space flight: using omics based analyses to develop individualized countermeasures that enhance astronaut safety and performance. Metabolomics 9 (6):1134–1156
7. Dunn WB, Broadhurst DI, Atherton HJ, Goodacre R, Griffin JL (2011) Systems level studies of mammalian metabolomes: the roles of mass spectrometry and nuclear magnetic resonance spectroscopy. Chem Soc Rev 40 (1):387–426
8. Lei Z, Huhman DV, Sumner LW (2011) Mass spectrometry strategies in metabolomics. J Biol Chem 286(29):25435–25442
9. Nicholson JK, Lindon JC (2008) Metabonomics. Nature 455:1054
10. Boiteau MR, Hoyt WD, Nicora DC, Kinmonth-Schultz AH, Ward KJ, Bingol K (2018) Structure elucidation of unknown metabolites in metabolomics by combined NMR and MS/MS prediction. Metabolites 8 (1):8
11. Aretz I, Meierhofer D (2016) Advantages and pitfalls of mass spectrometry based metabolome profiling in systems biology. Int J Mol Sci 17(5):632
12. Guijas C, Montenegro-Burke JR, Domingo-Almenara X, Palermo A, Warth B, Hermann G et al (2018) METLIN: a technology platform for identifying knowns and unknowns. Anal Chem 90(5):3156–3164
13. Horai H, Arita M, Kanaya S, Nihei Y, Ikeda T, Suwa K et al (2010) MassBank: a public repository for sharing mass spectral data for life sciences. J Mass Spectrom 45(7):703–714
14. Lai Z, Tsugawa H, Wohlgemuth G, Mehta S, Mueller M, Zheng Y et al (2017) Identifying metabolites by integrating metabolome databases with mass spectrometry cheminformatics. Nat Methods 15:53
15. Vinaixa M, Schymanski EL, Neumann S, Navarro M, Salek RM, Yanes O (2016) Mass spectral databases for LC/MS- and GC/MS-based metabolomics: state of the field and future prospects. Trends Anal Chem 78:23–35
16. Qiu F, Fine DD, Wherritt DJ, Lei Z, Sumner LW (2016) PlantMAT: a metabolomics tool for predicting the specialized metabolic potential of a system and for large-scale metabolite identifications. Anal Chem 88 (23):11373–11383
17. Wishart DS, Tzur D, Knox C, Eisner R, Guo AC, Young N et al (2007) HMDB: the human metabolome database. Nucleic Acids Res 35 (suppl_1):D521–D526
18. Kuehnbaum NL, Britz-McKibbin P (2013) New advances in separation science for metabolomics: resolving chemical diversity in a post-genomic era. Chem Rev 113 (4):2437–2468
19. Reynolds WF (2017) Chapter 29—Natural product structure elucidation by NMR spectroscopy. In: Badal S, Delgoda R (eds) Pharmacognosy. Academic Press, Boston, pp 567–596
20. Emwas A-HM (2015) The strengths and weaknesses of NMR spectroscopy and mass spectrometry with particular focus on metabolomics research. In: Bjerrum JT

(ed) Metabonomics: methods and protocols. Springer New York, New York, NY, pp 161–193

21. Molinski TF (2010) NMR of natural products at the 'nanomole-scale'. Nat Prod Rep 27 (3):321–329
22. Molinski TF (2010) Microscale methodology for structure elucidation of natural products. Curr Opin Biotechnol 21(6):819–826
23. Hooft Justin JJ, Mihaleva V, Vos Ric CH, Bino Raoul J, Vervoort J (2012) A strategy for fast structural elucidation of metabolites in small volume plant extracts using automated MS-guided LC-MS-SPE-NMR. Magn Reson Chem 49(S1):S55–S60
24. Exarchou V, Krucker M, van Beek Teris A, Vervoort J, Gerothanassis Ioannis P, Albert K (2005) LC–NMR coupling technology: recent advancements and applications in natural products analysis. Magn Reson Chem 43 (9):681–687
25. Sharman GJ, Jones IC (2003) Critical investigation of coupled liquid chromatography–NMR spectroscopy in pharmaceutical impurity identification. Magn Reson Chem 41 (6):448–454
26. Pullen FS, Swanson AG, Newman MJ, Richards DS (1995) 'On-line' liquid chromatography/nuclear magnetic resonance mass spectrometry—a powerful spectroscopic tool for the analysis of mixtures of pharmaceutical interest. Rapid Commun Mass Spectrom 9 (11):1003–1006
27. Schlotterbeck G, Ceccarelli SM (2009) LC–SPE–NMR–MS: a total analysis system for bioanalysis. Bioanalysis 1(3):549–559
28. Godejohann M, Tseng L-H, Braumann U, Fuchser J, Spraul M (2004) Characterization of a paracetamol metabolite using on-line LC-SPE-NMR-MS and a cryogenic NMR probe. J Chromatogr A 1058(1):191–196
29. Clarkson C, Hansen SH, Jaroszewski JW (2005) Hyphenation of solid-phase extraction with liquid chromatography and nuclear magnetic resonance: application of HPLC-DAD-SPE-NMR to identification of constituents of Kanahia laniflora. Anal Chem 77 (11):3547–3553
30. Sprogøe K, Stærk D, Jäger AK, Adsersen A, Hansen SH, Witt M et al (2007) Targeted natural product isolation guided by HPLC–SPE–NMR: constituents of Hubertia species. J Nat Prod 70(9):1472–1477
31. Khakimov B, Tseng HL, Godejohann M, Bak S, Engelsen BS (2016) Screening for triterpenoid Saponins in plants using hyphenated analytical platforms. Molecules 21(12)
32. Kenny O, Smyth TJ, Hewage CM, Brunton NP, McLoughlin P (2014) 4-Hydroxyphenylacetic acid derivatives of inositol from dandelion (Taraxacum officinale) root characterised using LC–SPE–NMR and LC–MS techniques. Phytochemistry 98:197–203
33. Kongstad KT, Wubshet SG, Johannesen A, Kjellerup L, Winther A-ML, Jäger AK et al (2014) High-resolution screening combined with HPLC-HRMS-SPE-NMR for identification of fungal plasma membrane H+-ATPase inhibitors from plants. J Agric Food Chem 62 (24):5595–5602
34. Liu F, Wang Y-N, Li Y, Ma S-G, Qu J, Liu Y-B et al (2017) Rhodoterpenoids A–C, three new rearranged triterpenoids from Rhododendron latoucheae by HPLC–MS–SPE–NMR. Sci Rep 7(1):7944

Chapter 8

Fast NMR Methods for Identification of Resonances and Metabolic Pathways

Abhinav Dubey, Shivanand M. Pudakalakatti, and Hanudatta S. Atreya

Abstract

High-throughput analysis of NMR data in metabolomics involves both rapid data acquisition and analysis. We describe here a data collection and analysis protocol, which enables fast multidimensional NMR data acquisition and automated analysis of NMR spectra to rapidly identify the metabolites and assign them to active metabolic pathways in the system.

Key words Fast NMR methods, Multiple receivers, Pattern recognition, Metabolite assignments, Metabolic pathways

1 Introduction

NMR-based metabolomics has witnessed rapid advances in recent years due to the development of new methods for data collection and analysis [1, 2]. There are three important points to be considered when it comes to a high-throughput pipeline in metabolomics: (1) the speed with which data are collected, (2) the sensitivity/resolution provided by the NMR experiment, and (3) how easily and rapidly the NMR spectra can be analyzed. Toward each of these requirements, several methods have been developed during the past decade [1, 2].

Typically, a one-dimensional (1D) ^{1}H NMR spectrum is preferred owing to its advantages of rapid acquisition and relatively high sensitivity. However, the 1D NMR spectrum suffers from the problem of overlap of peaks, because the sample (urine, blood, cell lysate) usually contains a mixture of large numbers of metabolites. This can be overcome by recording a 2D NMR spectrum, which resolves the peaks in the second dimension. A 2D [^{1}H, ^{1}H] total correlation spectroscopy (TOCSY) spectrum provides ^{1}H–^{1}H chemical shift correlations within a spin system, which helps to identify the metabolites. A 2D [^{13}C, ^{1}H] heteronuclear single-

G. A. Nagana Gowda and Daniel Raftery (eds.), *NMR-Based Metabolomics: Methods and Protocols*, Methods in Molecular Biology, vol. 2037, https://doi.org/10.1007/978-1-4939-9690-2_8,

quantum coherence (HSQC) experiment and/or 2D [^{13}C, ^{1}H] HSQC-TOCSY provides information on ^{13}C and ^{1}H chemical shift correlations for spin system identification, with the added benefits of high chemical shift dispersion in the ^{13}C dimension [3]. The ^{1}H–^{13}C correlations thus provide invaluable additional information for spectral annotation/quantitation. The advantages of 2D NMR for metabolomics have been recently reviewed along with the evaluation of methods, which assist in rapid data acquisition [4]. Once the data are acquired, the next important step is to analyze the 2D spectra to identify the metabolites present in the sample. Several automated/semiautomated methods have been proposed recently to achieve this objective [1, 2].

We describe here a data collection and analysis protocol developed in our laboratory, which enables both fast data acquisition and identification of metabolites or metabolic pathways active in the system being studied. The method described for data acquisition involves multiple NMR receivers [5], which are nowadays built into all high-field NMR spectrometers and have been shown to be highly effective for fast data collection [6]. The three 2D NMR spectra alluded to above are collected in a single data set. Following acquisition, the automated spectral analysis methods—PROMEB and ChemSMP [7, 8]—utilize the acquired 2D NMR data to identify and assign the metabolites or metabolic pathways that are active in the biological system. These steps are depicted in Fig. 1. For each of the methods, we first describe the principle involved, followed by a set of instructions to run the experiment/program and analyze the data.

2 Materials

1. The amount/type of sample required in this protocol is similar to that required for other NMR-based metabolomics experiments. That is, 600 μL of the sample in a 5 mm NMR tube or 300 μL taken in a 3 mm or a Shigemi tube. The concentration of the metabolites present is an important consideration (and also the level of ^{13}C isotopic enrichment; *see* **Note 1**). We have observed a lower limit of 50 μM for 2D NMR detection on our Bruker AVANCE III-HD 800 MHz NMR spectrometer equipped with a TCI cryoprobe (cooled by helium gas) using the parameters and measurement time mentioned below (*see* **Note 2**). Typically, below this concentration, not all [^{13}C, ^{1}H] correlations from a given metabolite are observed in the 2D [^{13}C, ^{1}H] HSQC spectrum. This leads to an incomplete spin system in the 2D TOCSY spectrum. However, the automated methods—PROMEB and ChemSMP—have been designed to handle the missing peaks.

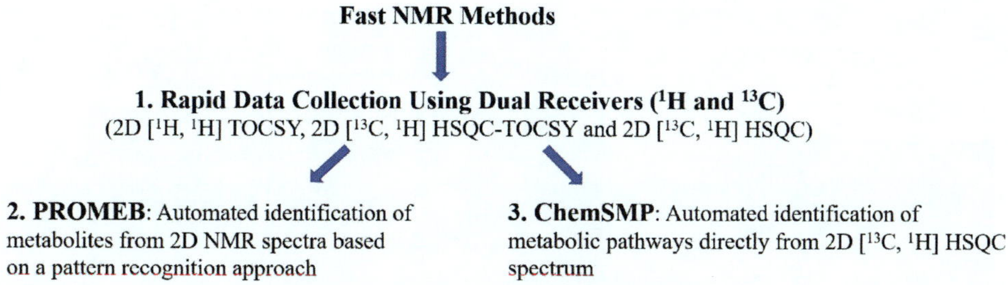

Fig. 1 Schematic illustration of the data collection and analysis protocol

2. For acquiring and processing the 2D NMR data on a Bruker NMR spectrometer, the TOPSPIN software available on the spectrometer is sufficient (*see* **Note 3**). The data acquisition and processing parameters are discussed below. The spectra processed in TOPSPIN can be directly used or can be processed in NMRPipe and/or exported to other formats for peak picking (such as CARA, Sparky, XEASY, etc.). Both PROMEB and ChemSMP consist of Python scripts, which can be run on a regular personal desktop/laptop computer with Windows or Linux. No special graphics or processing capabilities are needed.
3. The complete parameter set for the dual receiver experiment and the software PROMEB and ChemSMP, together with instructions to run the experiments/programs, are available online at http://nrc.iisc.ernet.in/hsa/software.htm

3 Methods

3.1 Fast NMR Data Acquisition Using Dual Receivers

3.1.1 Principle of the Method

1. In this experiment, the following three 2D spectra are acquired simultaneously in a single data set: (1) 2D [^{13}C, ^{1}H] HSQC, which gives ^{13}C, ^{1}H correlations for every CH/CH_2/CH_3 moiety but with ^{13}C in the direct dimension, as in a 2D HETCOR experiment; (2) 2D [^{1}H, ^{1}H] TOCSY which correlates the chemical shift of two protons within a network of scalar-coupled spins; and (3) 2D [^{13}C, ^{1}H] HSQC-TOCSY, which correlates the ^{13}C and its directly attached ^{1}H chemical shifts to all the protons within a spin system. The 2D HSQC-TOCSY thus combines the information from 2D [^{13}C, ^{1}H] HSQC and 2D [^{1}H, ^{1}H] TOCSY.
2. The three 2D spectra can be visualized as the three different 2D projections of a 3D [^{13}C, ^{1}H] HSQC-TOCSY (as depicted in Fig. 2). The three projections together provide complete 3D spectral information. However, instead of a single 3D spectrum, which would require long minimal measurement times to achieve sufficient digital resolution, it is much faster to

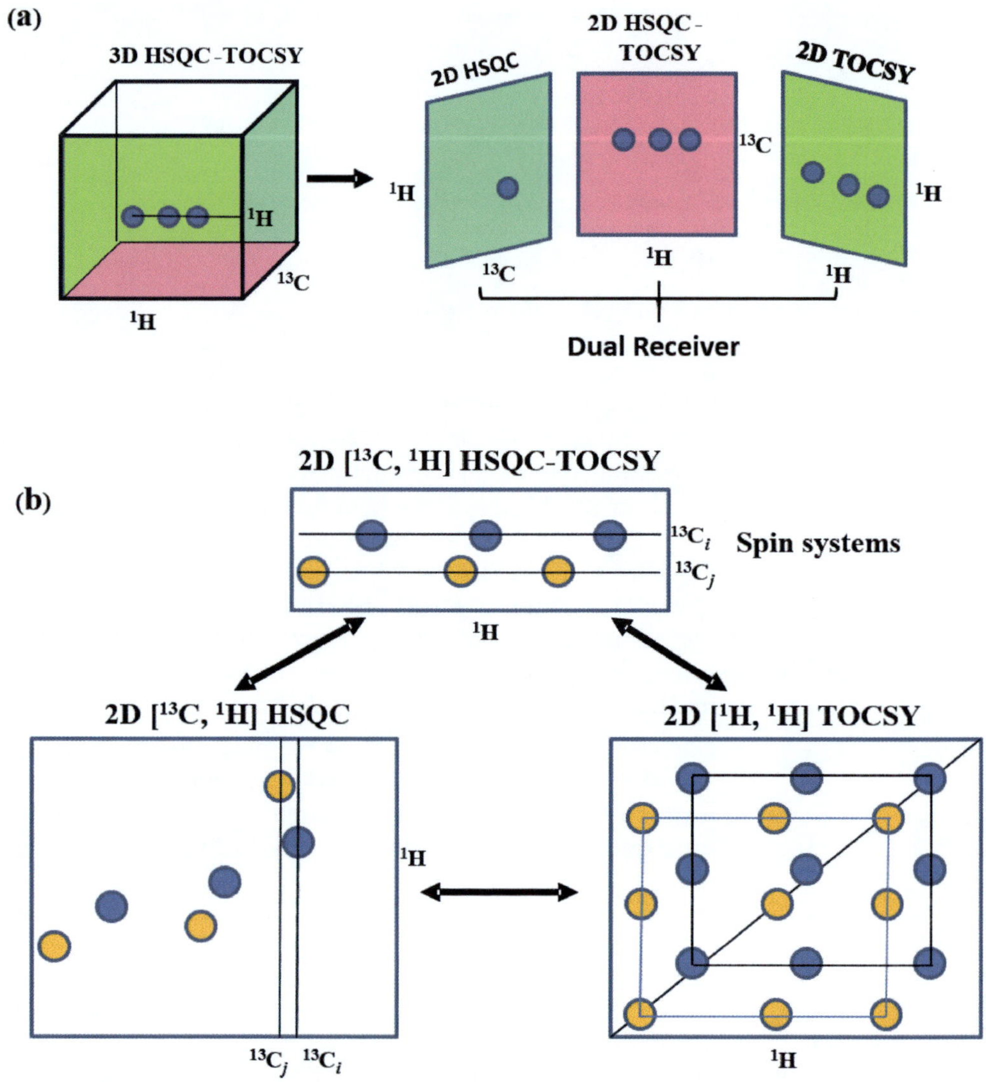

Fig. 2 (**a**) The [^{13}C, ^{1}H] 3D HSQC-TOCSY spectrum can be visualized as a combination of three 2D experiments: 2D [^{13}C, ^{1}H] HSQC, 2D [^{13}C, ^{1}H] HSQC-TOCSY, and 2D [^{1}H, ^{1}H] TOCSY, which can be simultaneously acquired in the dual receiver mode. (**b**) The peak pattern observed in the three 2D spectra, illustrated for two spin systems

record the individual 2D spectra with high resolution. This can be further accelerated using the dual receiver approach used here, wherein all the three 2D spectra are acquired simultaneously. A similar but different idea is involved in the technique of projection-reconstruction NMR techniques (PR-NMR) proposed previously [9]. In PR-NMR the goal is to construct a *N*D spectrum from its lower-dimensional projections. However, the projections that need to be acquired involve the joint sampling and relative scaling of two chemical shifts as in

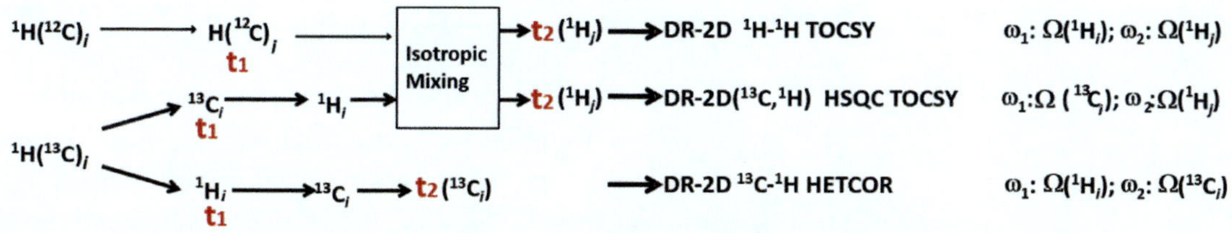

Fig. 3 A schematic depiction of the flow of magnetization through the dual receiver experiment

Table 1
Acquisition parameters

	Spectral width (ppm) and nucleus detected				Acquisition time (t_{max}) (ms)		No. of complex points	
Experiment name	**Nuc.**	**F1**	**Nuc**	**F2**	**F1**	**F2**	**F1**	**F2**
2D ^{13}C–^1H HSQC-TOCSY	^{13}C	72	^1H	16	10.4	160	150	2048
2D ^1H–^1H TOCSY	^1H	18	^1H	16	10.4	160	150	2048
2D ^{13}C–^1H HSQC/HETCOR	^1H	18	^{13}C	200	10.4	78	150	3072

reduced-dimensionality experiments [9]. No such joint sampling of chemical shifts is involved here.

3. A schematic diagram of how the magnetization flows during the experiment is shown in Fig. 3. The magnetization of ^1H attached to ^{13}C and ^{12}C is channeled through different pathways. The experiments are implemented in a manner such that the free induction decays (FIDs) from each of the three different experiments are acquired simultaneously.

3.1.2 Steps to Acquire and Process the Data

1. Table 1 shows acquisition parameters used for data collection.
   The total measurement time using the above parameters along with a relaxation delay of 1–1.5 s between scans and 16 scans for single averaging is about 4–5 h.
2. The spectra require preprocessing before proceeding for peak picking. Briefly, the free induction decays (FIDs) of 2D HSQC-TOCSY and 2D [^1H–^1H] TOCSY both of which are proton detected are acquired in a single data set and that of 2D HSQC/HETCOR data (which involves direct detection of ^{13}C) is stored separately as the second receiver data. The spectra are acquired in an interleaved manner such that in one set of transients, cross peaks of 2D HSQC-TOCSY and 2D [^1H–^1H] TOCSY have the same phase and in the next set of transients, opposite phase. After completion of the experiment, the two interleaved data sets are subjected to addition/subtraction to

separate the FIDs of 2D HSQC-TOCSY and 2D [^{1}H–^{1}H] TOCSY. The 2D HSQC data stored separately does not require any preprocessing and is directly processed as such.

3. The three spectra obtained from the dual receiver experiment provide complementary information. Assignment begins by picking arbitrarily a peak, "C_i,H_i," in the 2D HSQC spectrum. Using the proton chemical shift of "C_i,H_i," the diagonal peak is identified in 2D [^{1}H, ^{1}H] TOCSY. Considering the cross peaks "H_j" in the 2D [^{1}H, ^{1}H] TOCSY corresponding to the diagonal peak "H_i," the spin system can be identified. The peak pattern of spin system is further validated by finding "C_i,H_i" correlation peaks in ^{13}C-edited 2D HSQC-TOCSY, where along the ω_2 axis the same "H_j" peaks as observed in the 2D TOCSY are found along ω_2 axis (*see* Fig. 2). After confirming the spin system, all "C_i,H_i" correlations corresponding to a given spin system in the 2D [^{13}C, ^{1}H] HSQC spectrum are noted. The protocol is repeated until all peaks in the HSQC spectrum and hence all the spin systems in the sample are identified (*see* **Note 4**).

3.2 Assignment of Metabolites Using PROMEB

3.2.1 Principle of the Method

The conventional approach for assignment of metabolites involves a database search, wherein chemical shifts are assigned to specific metabolites by use of a tolerance limit. This is inefficient because deviation in chemical shifts associated with pH or temperature variations, as well as missing peaks, impairs a robust comparison with the database. The method, PROMEB (*P*attern *R*ec*O*gnition based assignment for *ME*ta*B*olomics), is a novel database search approach that addresses these problems and takes care of both systematic and non-systematic deviations in the observed chemical shifts of metabolites relative to those in the database and is also robust to missing peaks (*see* **Note 5**). The idea is to match the pattern of chemical shifts observed for a given spin system rather than its absolute chemical shift values, which may differ from the database. This approach leverages the mathematical technique of geometric hashing, which is extensively used in the area of image processing [10, 11].

The idea is to take the archived NMR data corresponding to each metabolite from the database and generate chemical shift patterns via the geometric hashing technique. PROMEB uses the HMDB database (http://www.hmdb.ca/). Subsequently, when a spin system observed in the NMR spectra (as described in Subheading 3.2.2) is taken as a query, it is also converted to a set of patterns by geometric hashing and matched with the database-derived patterns. The matching metabolites are then ranked in priority. Figure 4 schematically depicts the algorithm.

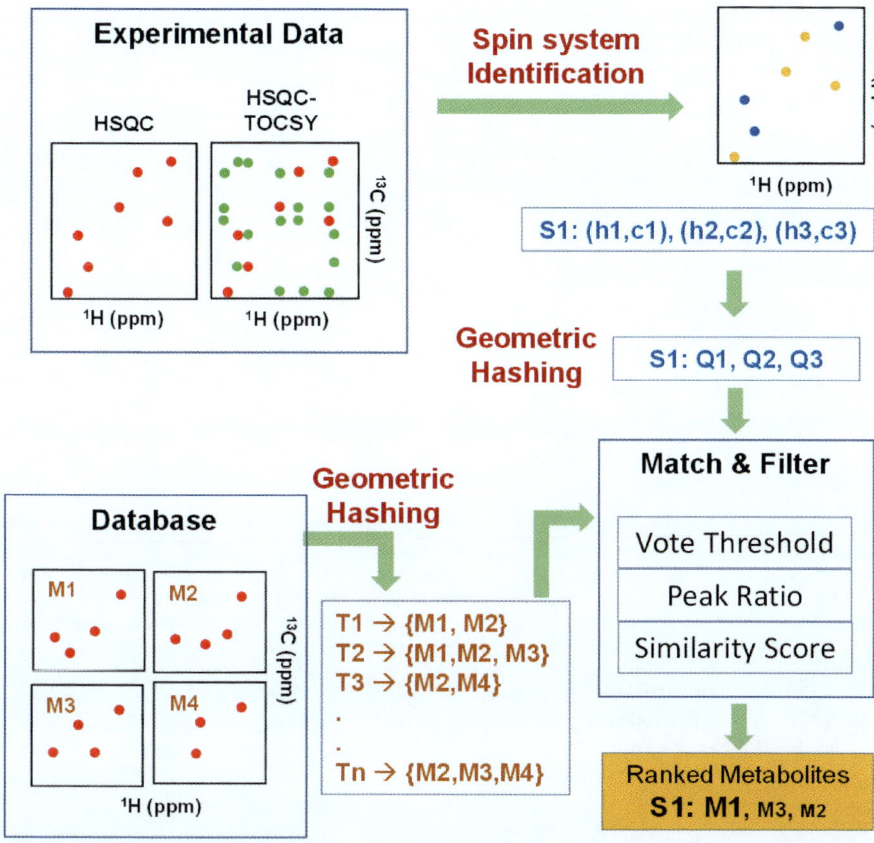

Fig. 4 An illustration of PROMEB with the steps from experimental data to obtaining output. Chemical shifts belonging to a given spin system are first identified using a combination of HSQC and HSQC-TOCSY spectra. The chemical shifts of these spin systems are transformed using geometric hashing to set of keys (Q_1, Q_2, and Q_3). Similar operations are applied to chemical shifts for all spin systems from the database to transform chemical shifts to keys (T_1, T_2, … T_n) which point to the metabolite they belong. The keys generated from experimental chemical shifts and database chemical shifts are then matched, and metabolites not qualifying with respect to the vote threshold and peak ratio criteria are removed. The remaining metabolites are ranked based on similarity score

3.2.2 Steps to Input and Analyze PROMEB Data

The input-output data format of the program is illustrated in Fig. 5.

1. Each line of input file is a spin system in the following format:

$$h_1\ c_1\ h_2\ c_2\ h_3\ c_3 \ldots h_n\ c_n$$

   where $h_1, \ldots h_n$ are the ^{1}H chemical shift, $c_1, \ldots c_n$ are the ^{13}C chemical shifts, and (h_i, c_i) is a peak from the spectrum. These chemical shifts correspond to one-spin system identified from the 2D TOCSY spectrum as described above.

2. The output is a list of 30 metabolites (with HMDB IDs) ranked according to relevance along with the similarity score and metabolite name. The lowest rank represents the most probable match for the input spin system. Other metabolites should

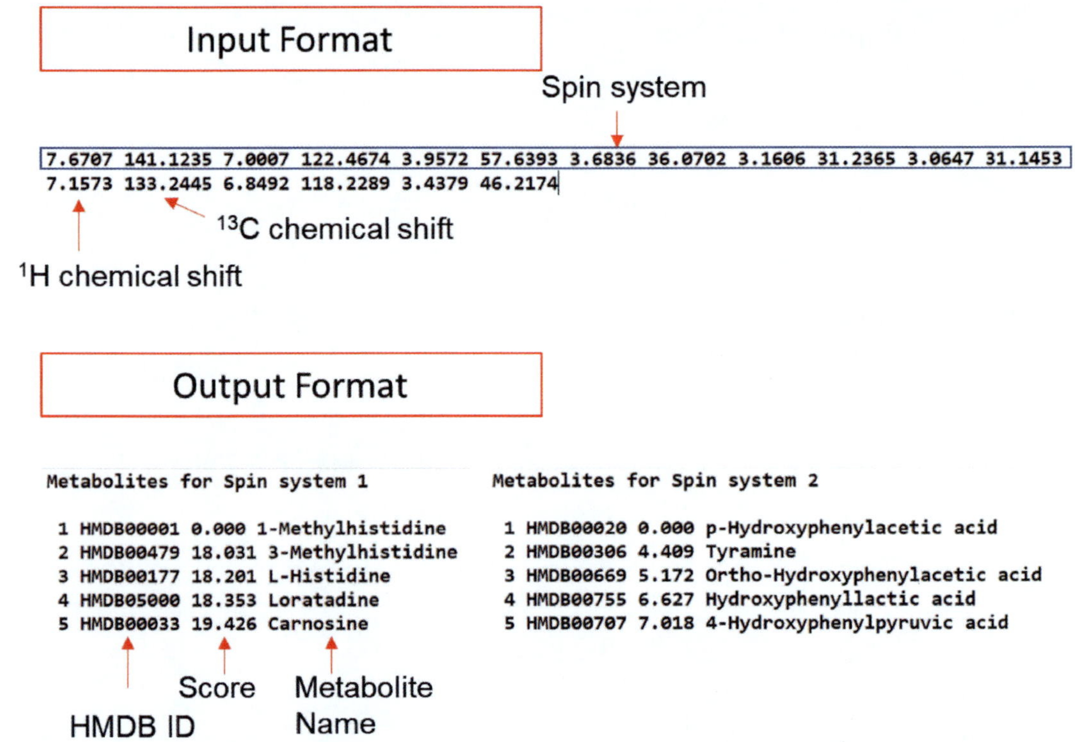

Fig. 5 The input-output file format of PROMEB

also be considered if they have a similarity score close to the lowest ranked metabolite. However, the top 10 ranked metabolites almost always (~75%) contain the correct metabolite corresponding to the query spin system (*see* **Note 5**).

3.3 ChemSMP: From 2D [^{13}C, ^{1}H] HSQC Spectrum to Metabolic Pathways

3.3.1 Principle of the Method

One of the important goals in metabolomics is to investigate the active metabolic pathways in the system being studied. For this purpose, in the conventional approach, the metabolites need to be first identified in the spectra. However, identification of metabolites from chemical shifts has several challenges. First, under steady-state conditions, certain metabolites in a metabolic pathway accumulate, whereas others are present below the detection limit of NMR, giving rise to missing peaks in the spectrum. Second, the chemical shift database currently has nonuniform coverage of chemical shifts data for metabolic pathways. Both of these bottlenecks impede rapid metabolite assignment and the consequent mapping to metabolic pathways.

To overcome these limitations, ChemSMP (*Chem*ical *S*hifts to *M*etabolic *P*athways) identifies putative metabolic pathways active in the system directly from a 2D [^{13}C, ^{1}H] HSQC spectrum, without recourse to individual metabolite assignments. Thus, ChemSMP offers the distinct advantage of facilitating the rapid analysis of spectra in the context of metabolic pathways skipping

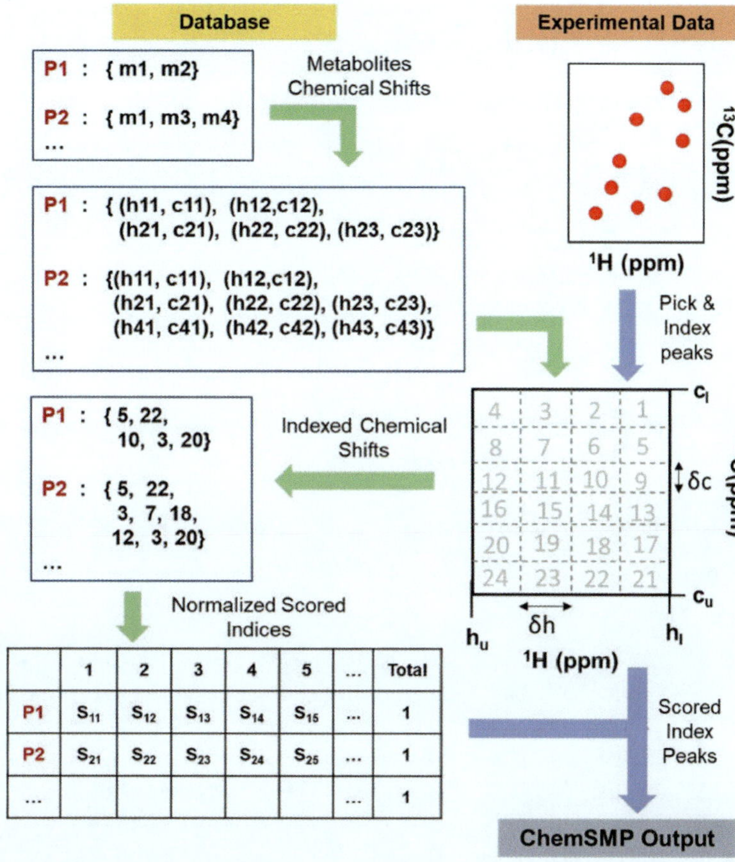

Fig. 6 Illustration of ChemSMP. Metabolites present in given pathway (denoted as P1, P2, etc.) are converted into their [^{1}H,^{13}C] chemical shifts based on the assignment of metabolites in those pathways available in the database. These chemical shifts are indexed using user-defined parameters cl, cu, hl, hu, dh, and dc. The pathways are then represented as collection of indices. These indices for a given pathway are assigned a score. The cumulative sum of these scores for a given pathway is normalized to 1. The peaks from experimental data are indexed using same parameters and are scored for the presence of pathway using a score table generated using the database

the rate-limiting step of metabolite assignments. The underlying principle behind the workings of ChemSMP is to first assume that presence of all metabolic pathways is equally likely for metabolites that are shared between these pathways and then ranking them in the order of the amount of evidence available for each pathway from the 2D spectrum. This approach avoids the time-consuming process of identifying individual metabolites, thereby accelerating the analysis. Figure 6 depicts schematically the principle and steps used in ChemSMP for identification of metabolic pathways.

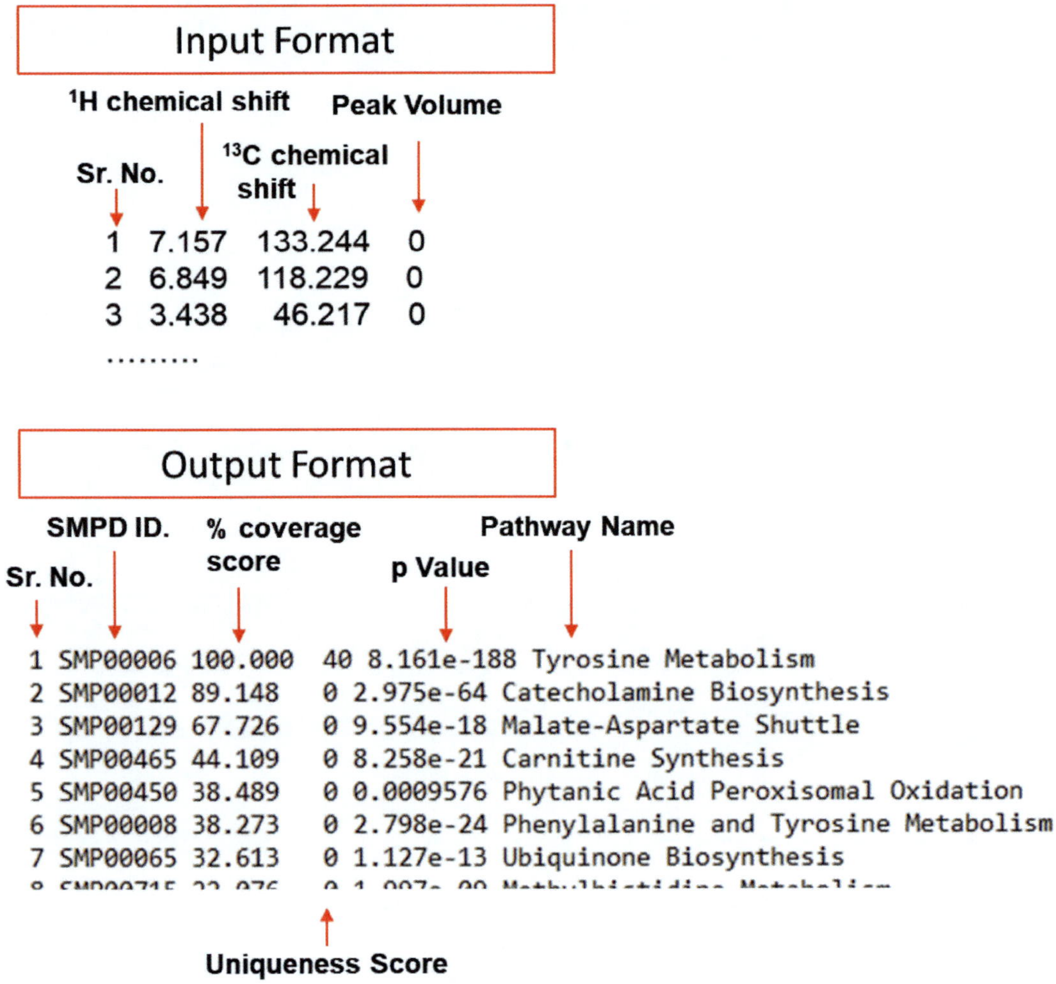

Fig. 7 The input-output file format of ChemSMP

3.3.2 Steps to Input Data and Analyze ChemSMP Output

The input-output data format of the program is illustrated in Fig. 7.

1. Each line of input file appears in a four-column format separated by spaces:

$$\text{sr } h_1 \ c_1 \ v$$

where "sr" is the serial number, "h_1" is the ^{1}H chemical shift, "c_1" is the ^{13}C chemical shift, and "v" is the peak volume (0 if not available). The (h_1, c_1) pair is a peak in the 2D [^{13}C, ^{1}H] HSQC spectrum, which can also be picked from the spectrum in an automated manner (*see* **Notes 6** and 7).

2. The output is a list of metabolic pathways sorted in order of their coverage score along with the uniqueness score, statistical significance, and metabolic pathway name, followed by detailed information about the putative metabolites found in the spectra.

4 Notes

1. The sensitivity of the three spectra obtained using the dual receiver experiment depends on the isotopic enrichment of the metabolites. This is due to the fact that the experiment relies on protons bound to both ^{12}C and ^{13}C. At natural abundance (non-enriched) state, the percentage of ^{13}C is almost always 1.1%. In such cases, the sensitivity of the 2D HSQC and 2D HSQC-TOCSY spectra would be relatively low compared to 2D [^{1}H–^{1}H] TOCSY, which is derived from the 99% of ^{1}H attached to ^{12}C. On the other hand, in a ^{13}C-enriched molecule, the percentage of ^{1}H bound to ^{12}C would be low, resulting in decreased sensitivity of the 2D [^{1}H–^{1}H] TOCSY experiment but good sensitivity for the other two spectra.
2. At lower field strengths and/or spectrometers equipped with a room temperature probe, the lower limit of detection would increase. Hence, it is advisable to run these experiments on the highest field strengths available, equipped with a cryogenically (helium) cooled probe that is optimized for both ^{1}H and ^{13}C detection (i.e., it has a cold ^{13}C preamplifier in addition to cold ^{1}H radiofrequency coil/preamplifier).
3. For acquiring data using multiple receivers, it is necessary to refer to the spectrometer manual for the relevant commands. On the Bruker NMR spectrometer running TOPSPIN 3.2, the dual receiver experiment is first set up in a given experiment folder, which contains only the acquisition parameters for ^{1}H and ^{13}C, where ^{1}H is in the direct dimension. An additional experiment folder is created where the FID with ^{13}C detection is to be stored. The command "mulexpno" is used to link the two experiments, and they are run simultaneously using the "multizg" command. The raw data (FID) gets stored simultaneously in the two data sets.
4. At this stage, only the spin system has been identified and not assigned to any particular metabolite. The assignment is carried out using PROMEB, which is described below.
5. PROMEB is robust to missing peaks in a given spin system. Even in the case of 50% missing peaks (e.g., if only 4 out of 8 (^{1}H, ^{13}C) chemical shift correlation for a given metabolite are observed in the 2D HSQC or HSQC-TOCSY spectrum), most of the metabolites are correctly identified within rank 10.
6. In ChemSMP, an important parameter is the grid size in the 2D [^{13}C, ^{1}H] HSQC spectrum, which is specified in both ^{1}H and ^{13}C dimensions. The default values for these are set to 0.03 and 0.3 ppm, respectively. But these can be varied to obtain optimal

results, because the peaks may not fall in a particular grid when matched with the database due to variations in temperature and pH.

7. In the ChemSMP package, a Python script "ChemSMPBruker.py" is available to generate a ChemSMP compatible input peak list directly from a 2D [^{13}C, ^{1}H] HSQC spectrum acquired on a Bruker spectrometer. For this, you need to install the packages "nmrglue," "numpy," "pylab," and "tkinter" in Python 3.x and execute the script.

   On executing ChemSMPBruker.py, it asks for the directory containing the "2rr" file, which corresponds to the processed 2D spectrum. Two parameters are to be adjusted interactively to obtain peaks: (a) the lowest contour level so as to visualize all peaks but not noise and (b) the relative ratio to maximum peak intensity, which is used as a cutoff to pick peaks. Once satisfied with peak picking, it asks for additional parameters: (a) spectral width in ^{13}C and ^{1}H dimension (in ppm) and (b) spectral offset in the ^{13}C and ^{1}H dimensions (in ppm). With this information, it will create and save the peak list at a desired location in the required four column format, which is required as input for ChemSMP.

Acknowledgments

The high-field NMR facility at NMR Research Centre, Indian Institute of Science, supported by the Department of Science and Technology (India) is gratefully acknowledged. HSA acknowledges research grant from SERB (EMR/2016/003181).

References

1. Nagana Gowda GA, Raftery D (2017) Recent advances in NMR-based metabolomics. Anal Chem 89:490–510
2. Markley JL, Bruschweiler R, Edison AS, Eghbalnia HR, Powers R, Raftery D, Wishart DS (2017) The future of NMR-based metabolomics. Curr Opin Biotechnol 43:34–40
3. Ross A, Schlortterbeck G, Dieterle F, Senn H (2007) NMR spectroscopy techniques for application to metabolomics. In: Lindon JC, Nicholson JK, Holmes E (eds) The handbook of metabolomics, 1st edn. Elsevier, Amsterdam
4. Guennec AL, Giraudeau P, Caldarelli S (2014) Evaluation of fast 2D NMR for metabolomics. Anal Chem 86:5946–5954
5. Pudakalakatti SM, Dubey A, Jaipuria G, Shubhashree U, Adiga SK, Moskau D, Atreya HS (2014) A fast NMR method for resonance assignments: application to metabolomics. J Biomol NMR 58:165–173
6. Kupce E (2013) NMR with multiple receivers. Top Curr Chem 335:71–96
7. Dubey A, Rangarajan A, Pal D, Atreya HS (2015) Pattern recognition based approach for identifying metabolites in nuclear magnetic resonance-based metabolomics. Anal Chem 87:7148–7155
8. Dubey A, Rangarajan A, Pal D, Atreya HS (2015) Chemical shifts to metabolic pathways: identifying metabolic pathways directly from a

single 2D NMR spectrum. Anal Chem 87:7148–7155

9. Atreya HS, Szyperski T (2005) Rapid NMR data collection. Methods Enzymol 394:78–108
10. Wolfson HJ (1990) Model-based object recognition by geometric hashing. In: Faugeras O Computer vision-ECCV 90: first European conference on computer vision antibes, France, April 23-27, 1990 proceedings. Springer, Berlin
11. Wolfson HJ, Rigoutsos I (1997) Geometric hashing: an overview. IEEE Comput Sci Eng 4(4):10–21

Part IV

Stable Isotope-Resolved NMR

Chapter 9

Stable Isotope-Resolved Metabolomics by NMR

Penghui Lin, Andrew N. Lane, and Teresa W.-M. Fan

Abstract

Stable isotopes enable the tracing of atoms from precursors to products via metabolic transformations in situ and in crude extracts. NMR has some unique capabilities that are especially suited for metabolic analysis, including atom position-resolved isotope analysis and isotope editing. In this chapter we describe NMR-based analysis of positional isotopomer distribution using metabolic tracers.

Key words Stable isotope-resolved metabolomics, Isotopomers, Spectral editing, Quantification

Abbreviations

DSS	2,2-Dimethyl-2-silapentane-5-sulfonate
FID	Free induction decay
rf	Radiofrequency
RoI	Region of interest
SIRM	Stable isotope-resolved metabolomics
SNR	Signal-to-noise ratio
TMS	Tetramethylsilane

1 Introduction

Metabolic activity is both quantitatively and qualitatively specific to tissue and cell types and is generally highly responsive to changes in external (environmental) or internal (e.g., genetic) perturbations. Thus, cellular metabolism provides a very rich and molecular read-out of the functional state of cells [1]. In metabolic research, stable isotope tracing provides much more rigorous metabolic pathway details than total metabolite profiling, as it can help resolve which intersecting pathways are active and be used for estimating network fluxes [1–4].

G. A. Nagana Gowda and Daniel Raftery (eds.), *NMR-Based Metabolomics: Methods and Protocols*, Methods in Molecular Biology, vol. 2037, https://doi.org/10.1007/978-1-4939-9690-2_9,

For stable isotope tracing, there are two main analytical approaches that are widely applied, mass spectrometry (MS) and NMR. Although NMR is much less sensitive than most MS detection methods, it has several unique advantages for metabolite analysis. These are (1) quantification without the need for authentic standards; (2) reliable quantitation with high dynamic range; (3) ability to elucidate molecular structures with multidimensional and multinuclear capabilities; (4) ability to determine distributions of labeled atom position(s), i.e., isotopomers; and (5) isotope editing capabilities for spectral simplification in complex mixtures.

The goal of this chapter is to provide detailed experimental procedures for NMR identification and quantification of metabolites and their isotopomer distributions in stable isotope-resolved metabolomics (SIRM) studies. It should be noted that the experiments and analyses can be used for those with or without stable isotope enrichment.

The sensitivity of an NMR experiment is determined primarily by the amount (concentration) of each analyte and the nature of the detector system, primarily the probe type and magnetic field strength. Sensitivity theoretically scales as ${B_0}^{1.5}$, though in practice it is closer to linear in B_0 especially for the now prevalent cryoprobes. Cryoprobes may yield a sensitivity enhancement of three- to fourfold compared with an RT probe counterpart, which corresponds to about ≥tenfold reduction in experimental time for a given signal-to-noise ratio (SNR). However, this may be optimistic for salty samples, and the loss of sensitivity due to the presence of mobile ions increases with increasing magnetic field strength. The salt-associated losses, however, are somewhat offset by using smaller diameter microprobes, such as 3 mm or 1.7 mm compared with the "standard" 5 mm probe. When samples are severely limited, such as for tissue biopsy, microprobes outperform standard diameter probes. Routine high-resolution 5 mm cryoprobes at 16.45–18.8 T typically have an SNR of 6000–8000:1 for 0.1% ethyl benzene in $CDCl_3$. This translates to an SNR of 10:1 in 10 min acquisition (for a 6 s recycle time) in 1D ^1H NMR experiments for a volume of 300 μL at 0.5 nmol (i.e., ~1.7 μM) for each analyte in a Shigemi tube with optimal shimming and probe tuning.

2 Materials

1. *Cell or tissue extracts, biofluids.* Cells and tissue can be extracted simultaneously into fractions containing polar and nonpolar (lipids) metabolites, respectively [5, 6]. Polar extracts are best lyophilized to minimize metabolite degradation during the drying process. Lipid extracts can be evaporated under vacuum centrifugation or N_2 purge. Lipid extracts can be lyophilized if they contain excess water.

2. After lyophilization, the polar residue is best redissolved in 99.9% D_2O (or higher grade) or in a buffer in D_2O, preferably low in paramagnetic impurities. An external standard such as DSS-d_6 (4,4-dimethyl-4-silapentane-1-sulfonic acid, e.g., 0.5 mM) is added to serve both as a chemical shift reference standard and as a concentration reference, which should be prepared as accurately as possible by weighing. For optimal referencing and quantification, the DSS concentration should be sufficiently high but not excessively higher than those of the metabolites of interests, due to consideration for the dynamic range of signal digitization. We have been using 18–27.5 nmoles of DSS-d_6 per 55 μL of D_2O for our extracts.

   We find that dissolving extracts in D_2O will suffice in providing NMR spectra with consistent chemical shifts for most biological samples, as the extract pHs do not vary significantly. This avoids adding buffer salts to the extract. A common aqueous buffer for reconstituting polar extracts is 50 mM sodium phosphate in D_2O buffered to pD 7. For tissue extracts or extracts with high levels of paramagnetic ions such as Fe^{3+}, it is advisable to include 1 mM or higher EDTA-d_{12} to minimize line-broadening effects.

3. The nonpolar extracts after drying can be dissolved in either d_4-methanol or $CDCl_3$. The chemical shift references can be the residual proton resonance of the solvent. For quantification, an internal standard (e.g., tetramethylsilane or TMS) must be added at a known amount.

4. For biofluids such as blood plasma, CSF, or urine, it is possible to perform NMR analysis without processing other than dilution with a strong buffer to normalize the pH. Biofluids are typically very salty, which can compromise the sensitivity of cryoprobes at high magnetic field strengths. Biofluids also contain high concentrations of protein and lipid, which interfere with quantification, particularly for ^1H spectroscopy, as they produce broad overlapping peaks that make accurate peak integration problematic. Techniques for suppressing the broad resonances are available, but they also differentially affect the intensities of the sharper metabolite resonances. Furthermore, the intense water resonance needs to be suppressed, and several suppression techniques either obliterate a large region of the spectrum, leading to loss of information, or produce spectral artefacts (*see* below).

5. An alternative is to extract biofluids using a suitable solvent (e.g., 80% acetone) [7] to remove proteins and lipids, followed by reconstitution as described above for cells and tissues in **item 1**.

6. Samples should be centrifuged before use to remove particulates that may negatively impact shims and thus sensitivity and resolution.

3 Methods

3.1 General Setup

1. Add an appropriate amount of DSS-containing buffer to lyophilized samples depending on the length of the sensing region in the NMR probe. A typical amount is 35–55 μL for 1.7 mm tubes, 120–200 μL for 3 mm tubes, and 300–500 μL for 5 mm Shigemi tubes. Make sure that the samples are completely dissolved by rigorous vortexing and removal of particulates by centrifugation at ≥20,000 × *g* for 5 min before transferring to the NMR tubes.
2. Load sample into the magnet from the top opening or via an autosampler. If possible, samples in the autosampler should be cooled to ≤10 °C to better maintain sample integrity during prolonged runs.
3. Regulate the sample temperature using the instrument controls. Allow at least 2 min for temperature to stabilize before any NMR experiment. We find that 15 °C is a good compromise between optimal spectral quality and reduced sample deterioration.
4. Tune the probe by switching the "tune" and "match" knobs under the probe or through the autotune and automatch feature (if available). The tune and match capacitors are not independent; thus, both must be adjusted iteratively for optimal result. Make sure that the resonance curve is centered on the reference frequency with least reflection (deepest curve).
5. Select appropriate solvent and lock the sample. Adjust the lock power, gain, and phase for optimal lock signal. Typically, the lock level should be set around 50–60%.
6. Shim the sample. Typically, use a representative sample of the whole set for shim optimization. Start by loading the last best shim file from the system; adjust z1, z2, and radial shims for the best lock signal. Then apply gradient shimming routine with an appropriate shim map for optimal gradient shims. In the ^{1}H spectrum of a well-shimmed sample, the linewidth at half height of DSS methyl resonance should be less than 1 Hz, and the silicon satellite peaks should be well resolved (i.e., <12 Hz at 2.3% height, since $^{2}J_{H\text{-}Si}$ is ca. 6 Hz and the ^{29}Si natural abundance is 4.67%, which gives a pair of satellites of about 2.3% on either side of the methyl resonance).
7. Set the carrier frequency for proton to be on resonance with water peak, if water suppression is needed [8]. This is achieved by obtaining a test ^{1}H spectrum without water suppression (typically one scan with very small flip angle or excitation pulse width and receiver gain set at very low level to avoid

signal saturation). Locate the water peak center and read the offset frequency value. Set this value as the transmitter carrier frequency offset for the presaturation experiment.

8. All experiments are run without sample spinning.

3.2 Specific Experimental Setup

3.2.1 PRESAT (^{1}H) Experiment

1. Calibrate the 90° excitation pulse width of proton precisely for each set of samples for best quantification. Typically setting the transmitter power to the lowest, while keeping the ^{1}H pulse width under 10 μs, is optimal.
2. Select the water presaturation pulse sequence with the correct transmitter carrier frequency offset set from **step 7** in Subheading 3.1 and the presaturation time set to 4 s for optimal water peak saturation. For a 99.9% D_2O sample, the water concentration is still about 55 mM, which is much higher than those of metabolites. Weak field presaturation with a field strength of ca. 20 Hz is sufficient to suppress the water signals. However, for samples with higher water content, a stronger rf field will be required (50 Hz or more).
3. Set the acquisition time to 2 s for achieving 0.5 Hz/point resolution in the raw free induction decay (FID) data. In this case, the total recycle time is 6 s, sufficient ($> 5T_1$) to recover most of the metabolites' magnetization (>95%), since the T_1 of most small molecules are in the range of 1 s (*see* **Notes 1 and 2** and Fig. 3).
4. Use at least 8 dummy transients to achieve steady state. 256 + 8 dummy transients take 26.4 min of acquisition.
5. Process the FIDs with linear prediction once and zero-filling to four times of the original size. For example, a 2 s acquisition with a spectral width of 12 ppm on a 14.1 T spectrometer results in 14,368 complex points in the raw FID. Linear prediction once results in 32k points, followed by zero-filling to give 128k points for a final resolution <0.1 Hz/point.
6. Apodize the FIDs with an exponential function that matches the natural half-height linewidth (e.g., 1 Hz) before Fourier transformation.
7. Adjust phase carefully; correct baseline using a third-order Bernstein polynomial for accurate peak integration or deconvolution.
8. Reference the spectrum to DSS at 0 ppm

3.2.2 1D ^{1}H {^{13}C}-HSQC

Compared to 1D ^{1}H experiments, which detect all observable protons, HSQC detects the protons directly bonded to ^{13}C. It is a much more sensitive method for ^{13}C isotope detection than direct ^{13}C observation, which requires much longer acquisition times due to the low gyromagnetic ratio of ^{13}C.

1. Tune the ^{13}C channel and carefully calibrate the ^{13}C pulse width using the first increment of the 2D HSQC experiment by maximizing the signal as a function of the INEPT delay.
2. Set the proton carrier frequency on the HOD resonance the same way as **step** 7 in Subheading 3.1 if presaturation is desired.
3. The HSQC acquisition time is limited by the maximal ^{1}H decoupling time that the probe is allowed to prevent overheating, typically less than 250 ms.
4. Set spectral width to 12 ppm for the proton dimension and 200 ppm for the carbon dimension.
5. Apply adiabatic decoupling for broadband decoupling of carbons from protons. Set the coupling constant to the middle of the typical one-bond H–C coupling constant range (ca. 146 Hz), corresponding to a transfer delay of 3.425 ms (1/2 J). Presaturation may be applied during the relaxation time of 1.75 s giving a total cycle time of 2 s per transient. We typically acquire 1024 transients, which would need ca. 35 min of acquisition (*see* **Note 3**).
6. The FID is zero-filled to 16k data points, apodized with an unshifted half Gaussian (or cosine squared) function plus a 4 Hz exponential function, and Fourier-transformed.
7. Adjust the phase and correct the baseline, and reference the spectrum as described in **steps** 7 and **8** in Subheading 3.2.1.

3.2.3 2D ^{1}H {^{13}C}-HSQC

All setups are the same as the 1D ^{1}H {^{13}C}-HSQC. Typical 2D HSQC setup includes a total of 64 real increments (128 total) in the indirect carbon dimension with 32 transients per increment. Under these conditions, the experiment would take 2.3 h.

3.2.4 High-Resolution 2D ^{1}H {^{13}C}-HSQC

The maximal spectral resolution in the proton dimension is set by the acquisition time, but in the indirect carbon dimension it can be adjusted.

1. Set the carrier frequency at 50 ppm and a spectral width of 120 ppm in the carbon dimension to span the entire aliphatic carbon region. Such spectral width reduction optimizes signal detection via efficient ^{1}H decoupling and resolution of ^{13}C–^{13}C couplings in ^{13}C-enriched compounds via better digital resolution.
2. Set the increment in the indirect dimension to be 512 or more (real) with 32 transients per increment. Total experiment time is around 18 h.
3. For 2D data processing, linear predicts the FIDs once to 4096 and 1024 points in the ^{1}H and ^{13}C dimensions, followed by zero-filling to 8192 and 2048 data points, respectively. Then,

apodization with an unshifted half Gaussian (or cosine squared) function plus an exponential line-broadening function of 1 Hz in t_1 and 4 Hz in t_2 is followed by Fourier transformation. The final resolution in the indirect carbon dimension is 8.8 Hz/point, which is sufficient to resolve the one-bond ^{13}C–^{13}C couplings of about 40 Hz.

4. Adjust the phases of the spectrum in F_1 and F_2, correct the baseline, and reference both dimensions to DSS at 0 ppm or to a known resonance at the proper chemical shift (e.g., the glucose 1-α resonance at 5.22 ppm for ^1H and 94.61 ppm for ^{13}C).

3.2.5 1D $^1H\{^{15}N\}$-HSQC

Since ^{15}N has a natural abundance of 0.37% and a gyromagnetic ratio that is 1/10th of the proton, it is not suitable for direct detection and therefore not a routine experiment for metabolite analysis. However, if metabolites are enriched in ^{15}N from labeled tracer precursors (e.g., ^{15}N-glutamine, ^{15}N-tryptophan, etc.), this experiment can provide direct evidence of nitrogen incorporation into metabolites at specific positions [2, 4] or for detection of ^{15}N-labeled metabolites via chemoselective derivatization using ^{15}N-labeled reagents [9, 10].

Typical one-bond NH coupling constants are around 90–95 Hz. However, most NHs are not detectable in proton spectra due to chemical exchange with water. If the proton to be detected is directly attached to ^{15}N such as in amides and peptides, the experiment must be carried out in ^1H_2O and for optimal signals at acidic pH (e.g., pH 4–5) and lower temperature (e.g., 10 °C) to minimize proton exchange. Multiple-bond NH coupling such as in a molecular fragment NCH may also be detected. Two-bond to three-bond NH coupling constants are in the range 1–5 Hz [9].

1. Set carrier frequencies at water resonance for proton dimension and 120 ppm for the nitrogen dimension (to cover both amide and non-amide resonances) with spectral widths of 10 ppm and 50–60 ppm for proton and nitrogen dimensions, respectively.
2. Tune the ^{15}N channel and carefully calibrate the ^{15}N pulse width using the first increment of the HSQC experiment (cf. **step 1** in Subheading 3.2.2).
3. Set the acquisition time in the proton dimension to be the same as the broadband decoupling duration, typically less than 250 ms. GARP decoupling is applied on the nitrogen dimension during acquisition. If carbon decoupling is also desired, it can be applied simultaneously during acquisition.
4. Apply a very weak, soft pulse presaturation during the relaxation time if needed. For 1024 + 8 steady state transients, the experimental time is 34 min.

5. Zero-fill the final FID to 16k data points, apodize with an unshifted half Gaussian (or cosine squared) function plus 4 Hz exponential line-broadening before Fourier transformation.
6. Adjust phase, reference, and correct baseline as in **step 7**, Subheading 3.2.1.

3.2.6 2D $^1H\{^{15}N\}$-HSQC

All setups are identical to the 1D $^1H\{^{15}N\}$ HSQC experiment.

1. A typical setup includes 64 increments in the indirect dimension with 32 transients per increment, which amounts to approximately 2.3 h total acquisition time.
2. Process the 2D data, adjust phase, baseline correct, and reference the spectrum as described in **step 4**, Subheading 3.2.4.

3.2.7 High-Resolution 2D $^1H\{^{15}N\}$-HSQC

Similar to the 2D $^1H\{^{13}C\}$ HSQC experiment, resolution in the proton dimension is limited by the decoupling time allowed by the probe, but the resolution in the indirect ^{15}N dimension can be adjusted as desired.

1. Set the total number of increments in the indirect dimension to 512 over a 60 ppm spectral width. With 32 transients per increment, the total experiment time is around 9 h.
2. Linear predict the FIDs once with one zero-filling before apodization and Fourier transformation as in **step 3**, Subheading 3.2.4. This provides the digital resolution in the ^{15}N dimension to be around 1.7 Hz/point, sufficient to resolve ^1H–^{15}N one- and ^{15}N–^{13}C one and two-bond couplings of 8–16 Hz [11, 12].
3. Process the 2D data, adjust phase, baseline correct, and reference the spectrum as described in **step 4**, Subheading 3.2.4. ^{15}N shifts can be referenced indirectly according to [13].

3.2.8 TOCSY (Total Correlation Spectroscopy)

This experiment correlates scalar-coupled spins in the entire spin systems.

1. Set the offset of proton carrier frequency at the HOD resonance determined by the 1D PRESAT experiment (ca. 4.8 ppm at 20 °C).
2. Set the spectral widths to 10 ppm (e.g., −0.2 to 9.8 ppm) in both dimensions for samples in D_2O unless the 1D spectrum shows resonances outside this range.
3. Set the acquisition and relaxation delay times in the direct dimension (t_2) to 1 s.
4. Apply solvent suppression the same way as in the PRESAT 1D ^1H experiment in Subheading 3.2, and apply soft pulse presaturation during relaxation time.

5. Set the isotropic mixing spinlock time to 50 ms via the MLEV17 or DIPSI-2 pulse sequence. This duration enables magnetization relay for up to four bonds, thereby providing up to four covalent bond correlations in the molecular structure, which is sufficient for small metabolites.
6. Calibrate the B_1 field strength to deliver 8–10 kHz for the spin lock.
7. Apply magnetic gradients with a zero-quantum filter to obtain better quality spectra.
8. Set 4096 real points and 256 real increments in the direct (t_2) and indirect (t_1) dimensions, respectively. Determine the number of transients according to the sample concentration. Typical samples use 16 transients per increment with 256 real increments, which amounts to a total experimental time of 4.6 h.
9. Zero-fill FIDs to a final matrix of 8192 × 1024 real data points, apodize with an unshifted half Gaussian (or cosine squared) function plus a 1 Hz exponential function, and Fourier transform in both dimensions.
10. Process the 2D data, adjust phase, baseline correct, and reference the spectrum as described in **step 4**, Subheading 3.2.4, except that both spectral dimensions are referenced to DSS at 0 ppm.

3.2.9 Double-Quantum-Filtered COSY (DQ-COSY)

This is a pure phase COSY experiment that detects scalar coupling between protons 2–4 bonds apart.

1. Set offset of proton carrier frequency to the HOD resonance determined by the 1D PRESAT experiment.
2. Set the spectral width to 10 ppm in both dimensions, acquisition times of 1 s in t_2, 256 real (512 total) increments in t_1, and a relaxation delay of 1 s with solvent suppression as needed. For 16 transients per increment, this requires 4.6 h of experimental time (*see* **Note 4**).
3. Process the 2D data, adjust phase, baseline correct, and reference the spectrum as described in **step 10**, Subheading 3.2.8.

3.2.10 Heteronuclear Multiple-Bond Correlation Spectroscopy (HMBC)

This experiment complements HSQC as it detects X–^{1}H correlations over 2–4 bonds, with coupling constants typically in the range of 1–4 Hz. It can therefore detect non-protonated atoms such as quaternary ^{13}C or ^{15}N or ^{13}C carbonyl groups.

1. Set proton carrier frequency for the HOD resonance determined by the 1D PRESAT experiment and the ^{13}C carrier offset to 100 ppm.

2. Set the spectral widths to 10 ppm in the ^1H and 200 ppm in the ^{13}C dimensions.
3. Optimize the HMBC delay in a 1D experiment to assess sensitivity over the range of couplings present (1–4 Hz). This will determine the effective *J* values used for the evolution time.
4. Set 4096 real points and 128 real increments (256 total) in the proton and carbon dimensions, respectively, with a recycle time of 1.5 s. For 64 transients per increment, the experimental time is 6.8 h (*see* **Note 5**).
5. Process the 2D data, adjust phase, baseline correct, and reference the spectrum as described in **step 4**, Subheading 3.2.4.

3.2.11 Other Editing Experiments

NMR is exceptionally rich in the variety of experiments that can be performed, especially when NMR-active stable isotopes are present [4, 14–16]. In addition to the common experiments described above, other less utilized isotope-edited experiments are also valuable in providing complementary and cross-validating information on isotopomer distributions in labeled tracer studies.

1. *2D HNCO*: This experiment must be carried out in ^1H_2O and requires both ^{15}N and ^{13}C enrichment in metabolites to detect protons attached to ^{15}N alpha to a ^{13}C carbonyl group. ^{15}N must be enriched to a high level. For amides, this is best carried out at pH 4 and lower temperature (e.g., 10 °C) to minimize proton exchange. The experiment can be carried out either as $^1H^{13}C$ plane or $^1H^{15}N$ plane. The HC plane is especially useful, which requires only a small number of increments due to a narrow spectral width (ca. 20 ppm centered at 180 ppm). For 32 real increments with 32 transients per increment, the experimental time per plane is ca. 68 min.
2. *HCCH-TOCSY*: This experiment detects protons attached to consecutively ^{13}C enriched carbons. The experimental setup is similar to TOCSY with spectral widths of 10 ppm in both ^1H dimensions, real increments of 256, and recycle time of 2 s. Typically, with 32 transients per increment the total experiment time is 9 h.
3. *HSQC TOCSY*: This experiment detects the coupling of an *X* nucleus such as ^{13}C or ^{15}N to directly bonded protons and protons that are 2–4 bonds away. The acquisition conditions are otherwise the same as individual HSQC and TOCSY experiments as described above.

3.3 Assessing Sensitivity to Determine Experimental Time Needed

Most of the 2D experiments have maximal sensitivity in the first increment (DQ-COSY is an exception as it has no intensity in the first increment). Thus, it is useful to test spectral sensitivity by performing the corresponding 1D experiment using the first increment.

1. Execute a 1D experiment with all of the optimized parameters for the first increment (for DQ-COSY, set the initial delay to 40 ms to assess sensitivity).
2. Assess the intensity of peaks that are desired for analysis, e.g., nucleotides, lactate, Ala, Glu, glutathione, etc. Although there is a 2D advantage in signal detection for the full 2D experiment, if the resonances are not visible in the 1D spectra acquired with the first increment, they are unlikely to be visible in the 2D spectra (*see* **Note 6**).

3.4 NMR Assignments

Processed NMR spectra can be assigned based on chemical shifts, covalent connectivity, and other NMR parameters as in [17–20]. Detailed assignment strategies are beyond the scope of this chapter.

3.5 Concentration Determination

3.5.1 1D NMR Methods

Concentrations are determined from integrated peak areas, *A* as follows [2, 6]:

$$A = k \times c \times n \times f$$

where c is the concentration of the compound, n is the number of equivalent protons in the resonance (e.g., 3 for a methyl, 2 for a methylene and 1 for a methine), f is the saturation factor (*see* **Note 7**), and k is an undetermined proportionality constant that depends on the spectrometer and settings.

Suppose there are two compounds, 1 and 2, then.

$$A_1/A_2 = c_1 f_1 n_1/c_2 f_2 n_2$$

$$c_1 = c_2(A_1/A_2)(f_2 \times n_2/f_1 \times n_1)$$

If c_2 is known, then c_1 can be determined. The expected error in c_1 is then (assuming no covariance):

$$| \mathrm{d}c_1/c_1 | = \mathrm{d}c_2/c_2 + \mathrm{d}A_2/A_2 + \mathrm{d}f_2/f_2 + \mathrm{d}A_2/A_2 + \mathrm{d}f_1/f_1$$

For well-resolved peaks, with good SNR (>10:1), the errors in area determination should be <10%.

The procedure for concentration determination by 1D NMR is as follows.

1. Determine the peak areas by integration or peak deconvolution including the methyl protons of the DSS standard and their ^{29}Si satellites (*see* **Note 8**).
2. Determine the number of protons for a given resonance to be quantified, e.g., 9 for DSS methyl protons; 1–3 for CH, CH_2, CH_3, respectively, for most metabolites).
3. Determine the f values based on the experimental setup. R_1 can be determined by using the fast inversion recovery pulse sequence for the particular samples or set of samples in the same matrix (*see* **Note 7**).

3.5.2 2D NMR Methods

Volume integration can be performed using the 3D version of the trapezoidal rule, which has shown to be reliable when the digitization in both dimensions is adequate [21].

1. Volume integration using the VNMRJ software:
   (a) Linear predict, zero-fill, apodize, and Fourier transform 2D data; phase, reference, and baseline correct, as described above.
   (b) Set ins2ref = 1, ins2 = 1e6 to 1e8.
   (c) Choose region of interest (RoI).
   ll2d('clear'),
   ll2d('peak','pos')—selects positive peaks in the RoI,
   ll2d('adjust','volume'),
   ll2d('writetext','filename').
   (d) The text file includes peak volumes (v) corresponding to picked peaks and RoI selected.
   (e) Paste peak volumes with the assignments into an excel file (*see* **Note 9**).
2. Volume integration using the MestReNova software:
   (a) Load FID data into the MestReNova software. Process each spectrum as described above including linear prediction, zero-filling, phase adjustment, referencing, and baseline correction if applicable.
   (b) For volume quantification of 2D NMR spectra, MestReNova provides two different tools: peak picking and integration.
   (c) For 2D spectral peak picking, two different options can be used, "manual threshold" and "peak by peak." In manual threshold mode, drag the mouse over the area to be picked, the MestReNova algorithm will automatically detect the contours under a predefined threshold, and fit the contour with deconvolution. In "peak by peak" mode, the mouse drag will jump from point to point on the highest point of each contour. Clicking the mouse will pick the contour where the peak point sits. However, in this mode the deconvolution will not be applied.
   (d) In the peak-picking mode, all peak information can be directly exported to an Excel file including chemical shifts, height, linewidth, volume, etc.
   (e) For 2D integration, MestReNova also provides several tools: automatic integration, predefined region integration, and manual integration. In the automatic integration mode, a 2D peak-picking algorithm will be applied in the background, so it is essentially the same as using the 2D

peak-picking methods with deconvolution. In the predefined region integration mode, the user can set up the regions of interest for the software to integrate instead of the whole spectrum. More importantly, these regions can be saved and applied to the next spectrum, which is very useful for metabolomics analysis as the same metabolites are usually analyzed across a large number of samples. In manual integration mode, the user can click and drag with the mouse to select the region to be integrated. The first integral will be normalized to 1.0, and all following integrals will be referenced to the first one. The absolute values can be found in the integral tables generated by MestReNova. These tables are also readily exported into excel files (*see* **Note 10**).

3.6 Isotopomer Distribution Analysis

3.6.1 1D ^1H NMR Analysis

Isotopomers of given metabolites are always present due to the natural abundances of ^{13}C (1.1%) and ^{15}N (0.37%). These can be observed (with low sensitivity) by direct detection methods or by the appearance of satellite peaks in the proton spectrum. In the case of ^{13}C satellites, there are two doublet resonances spaced equally on either side of the resonance of ^1H attached to ^{12}C (*see* Fig. 1, top spectrum), with intensities of about 0.55%. In $^{13}C_6$-glucose labeling experiments, the satellite peaks are more complex and increase in relative intensity (*see* Fig. 1, bottom spectrum). The absolute and fractional enrichment can be readily quantified from such spectrum as follows.

1. Process 1D ^1H spectra as described in Subheading 3.2.1.
2. Integrate (*see* **Note 8**) the ^{12}C–^1H peak and its two satellites.
3. Correct for partial saturation using measured T_1 values (*see* **Note 7**), if needed.
4. Calculate the fraction ^{13}C at this site as:

$$F = A(^{13}C)/[A(^{13}C) + A(^{12}C)]$$

5. Assign the isotopomer from the fine splitting pattern (e.g., $^{13}C_3$ lactate or ^{13}C1,2,3-lactate).

3.6.2 2D ^1H NMR Analysis

2D ^1H TOCSY spectra are recorded and processed as described above (*see* Subheading 3.2.8). Figure 2 shows the spectral region where the cross-peaks for Asp and Glu and their ^{13}C satellites occur. The isotopomer distribution can be analyzed as follows.

1. Volume integrate all of the components of the cross-peaks, including their fine structure [21].
2. Identify the $^{12}C^{12}C$ cross-peaks.
3. Identify the $^{13}C^{12}C$ and $^{13}C^{13}C$ cross-peaks and add volumes of all related peaks (*see* Fig. 2) to obtain v_{sat}.

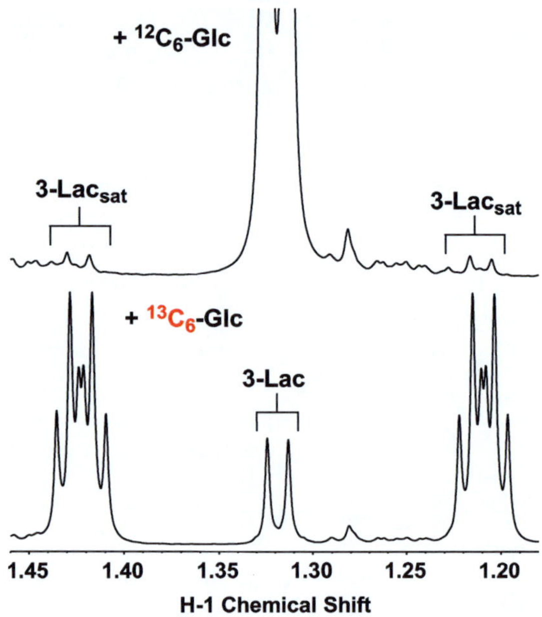

Fig. 1 1D PRESAT spectrum of medium extracts from human macrophage cultures. Polarized human macrophages were grown in the presence of unlabeled ($^{12}C_6$-Glc) or $^{13}C_6$-glucose for 24 h. The lactate (Lac) methyl resonance of protons attached to ^{12}C occurs at 1.32 ppm in both spectra. The two ^{13}C satellite (Lac$_{sat}$) peaks appear as a pair of doublets in the natural abundance spectrum (top) and as a pair of multiplets in the labeled spectrum (bottom). The multiplets comprise 6 peaks each due to coupling to the ^{13}C2H, ^{13}C2, and ^{13}C3, showing that this lactate isotopomer is ^{13}C1,2,3 [2]. The fraction (*F*) of ^{13}C enrichment is determined to be 0.90

4. Sum the volumes of all of the cross-peak components to obtain ν_{tot}.
5. Calculate the mole fraction of each component as $F = \nu_{sat}/\nu_{tot}$.

4 Notes

1. NOESY1D with water presaturation [22] and excitation sculpting [23] methods may also be applied for proton spectroscopy where the solvent is predominantly ^1H_2O. The advantage of using NOESY1D is that it eliminates most of the macromolecule signals to provide a cleaner baseline whereas water sculpting method suppresses the water peak better. However, NOESY1D tends to suppress a larger bandwidth in which metabolite signals close to water are also suppressed (e.g., glucose and oxidized glutathione), leading to inaccurate quantification. Excitation sculpting results in phase distortion in the peaks that is impractical to phase, which also lead to inaccurate quantification.

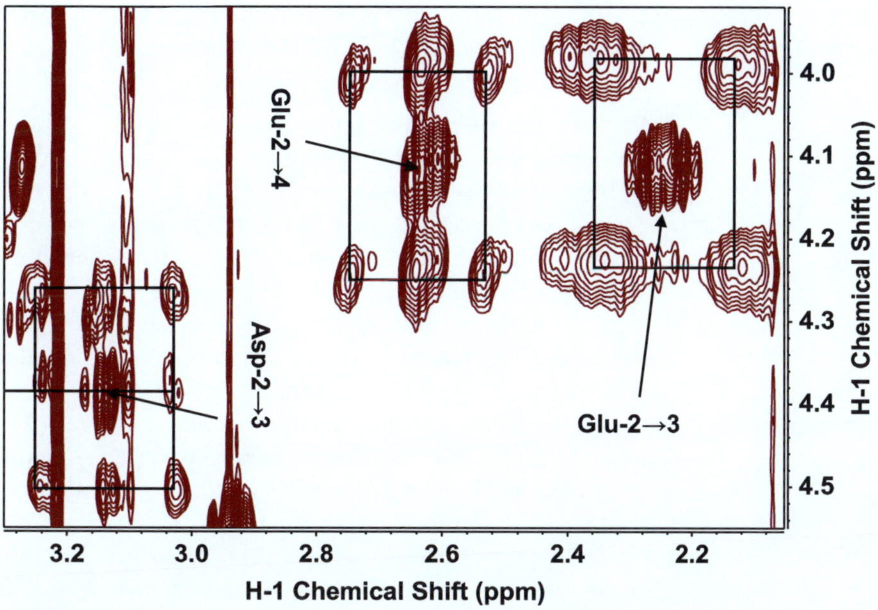

Fig. 2 2D ^{1}H TOCSY spectrum of a cancer cell extract. MCF7 breast cancer cells were grown in the $^{13}C_5$-Gln and unlabeled glucose for 24 h. The ^{1}H TOCSY spectrum was acquired at 14.1 T for the cell extract. The spectral region where cross-peaks from Asp and Glu occur is shown. The annotated central cross-peaks correspond to protons attached to ^{12}C. The surrounding satellite peaks (delineated by boxes) arise from protons attached to ^{13}C. For the Asp 2 → 3 cross-peak, there are four satellite peaks corresponding to the isotopomers $^{13}C2^{13}C3$, two peaks for $^{13}C2^{12}C3$, and two for $^{12}C2^{13}C3$. The Glu2 → 4 and Glu2 → 3 cross-peaks show only four satellites, respectively, for $^{13}C2^{13}C4$ and $^{13}C2^{13}C3$

2. The number of transients or scans can be determined by the signals from the peaks of interests. In general, 512 transients would result in 52 min acquisition, which provides sufficient SNR for most peaks while limiting the experimental time to a reasonable scale. If the SNR is not adequate within 1 h of data acquisition, it may not be worth investing more experimental time for detection. As SNR increases with the square root of the number of acquisitions, e.g., to increase the SNR from 5:1 to 10:1 would increase the acquisition time from 1 h to 4 h.
3. Under these settings, most protons will be partially saturated, and saturation correction needs to be made based on T_1 values of each proton group.
4. For cell extracts with ca. 1 mg total solutes, this is usually the minimum time for analysis. As the cross-peak multiplets are in antiphase, the peak volumes can be underrepresented, so DQ-COSY is not well-suited for quantitative isotopomer analysis.
5. The HMBC is significantly less sensitive than the HSQC experiment due to T_2 relaxation during the longer delays and the spreading of signals over multiple resonances.

6. If sufficient sensitivity is not reached in 64–96 transients, then one should increase the sample concentration or seek measurement on a higher sensitivity instrument.
7. Partially saturated peak intensity can be corrected by obtaining the saturation factor (f) where $f = 1/(1 - \exp\text{-}R_1D)$. Here, R_1 is the spin-lattice relaxation rate constant ($=1/T_1$), and D is the experimental recycle time, i.e., acquisition time + relaxation delay for a PRESAT experiment. f values for different relaxation rate constants for three different recycle times are shown in Fig. 3. The DSS methyl resonance has a short R_1 value (ca. 0.3 s^{-1}), which corresponds to an underestimated peak area by up to 20% without correction, and thus a corresponding underestimated concentration. The other source of error arises from uncertainty in c_2. Note that the analysis of peak areas, saturation factors, and reference concentration should be accurate to better than 1% to achieve an estimate error of 1–2% for c_2 determination. Errors accruing from inaccuracies in T_1 determination have been treated in [24]. Also note that protons attached to ^{13}C will relax faster than those attached to ^{12}C, and thus the T_1 values for the satellite peak must also be measured [25].
8. There are several methods to integrate peaks in NMR spectra, and all require accurate phase adjustment and good baselines. The standard integration method, based on the trapezoidal

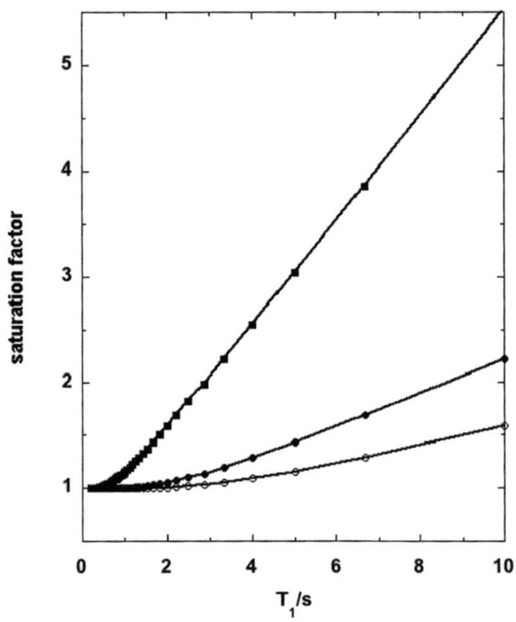

Fig. 3 Saturation factors versus T_1 values for three different recycle times. Saturation factors were calculated using $f = 1/(1 - \exp\text{-}R_1D)$ with $D = 2$ s (■), 6 s (•), and 10 s (0)

rule, is very accurate if the baseline is properly adjusted and there is no significant overlap. It is tolerant of unresolved couplings, imprecise shimming, and thus lineshape distortions (which may occur in automated runs). An alternative method is lineshape fitting, which performs an iterative nonlinear regression to a mixture of Lorentzian and Gaussian lineshapes, and can thus deconvolute overlapping peaks. In software packages like MestReNova (http://mestrelab.com/software/mnova/), lineshape fitting can be semiautomated and run in a batch mode. However, it is sensitive to deviations form Lorentzian/Gaussian lineshapes and unresolved couplings. The Chenomx software (https://www.chenomx.com/) uses proprietary libraries of experimentally derived peak shapes of metabolites for sequential curve fitting and spectral subtraction, which can detect obscured metabolite signals but is very time-consuming.

9. Integration of empty regions of the spectrum should also be done and subtracted from the estimated volumes.

10. Peak volume integration in MestReNova is defined as the sum of all digital intensities within the footprint (integration area or RoI). The footprint shape can be set to rectangular or elliptical. In low-resolution spectra with no significant fine peak structure, the elliptical RoI is superior. However, in higher-resolution spectra where multiplet fine structures are resolved and need to be independently integrated, the rectangular footprint may be more accurate.

Acknowledgments

This work was supported in part by NIH grants 1U24DK097215-01A1, CA163223-01A1, 5P20GM121327 (COBRE) the Carmen L. Buck endowment (to ANL), and the Edith D. Gardner Chair in Cancer Research (TWMF).

References

1. Lane AN, Fan TWM (2017) NMR-based stable isotope resolved metabolomics in systems biochemistry. Arch Biochem Biophys 628:123–131
2. Fan TWM, Lane AN (2016) Applications of NMR to systems biochemistry. Prog NMR Spectrosc 92:18–53
3. Bruntz RC, Higashi RM, Lane AN, Fan TWM (2017) Exploring cancer metabolism using stable isotope resolved metabolomics (SIRM). J Biol Chem 292:11601–11609
4. Fan TWM, Lane AN (2011) NMR-based stable isotope resolved metabolomics in systems biochemistry. J Biomol NMR 49 (3-4):267–280
5. Fan TWM (2012) Considerations of sample preparation for metabolomics investigation. Handbook of metabolomics, Human New York
6. Fan TWM, Lane AN (2013) Assignment strategies for NMR resonances in metabolomics research. In: Lutz N, Sweedler JV, Weevers RA (eds) Methodologies for metabolomics: experimental strategies and techniques. Cambridge University Press, Cambridge
7. Fan TWM, WarmoesMO SQ, Song H, Turchan-Cholewo J, Martin JT et al (2016) Distinctly perturbed metabolic networks underlie differential tumor tissue damages induced by immune modulator beta-glucan in

a two-case ex vivo non-small-cell lung cancer study. Cold Spring Harb Mol Case Stud 2(4): a000893
8. Giraudeau P, Silvestre V, Akoka S (2015) Optimizing water suppression for quantitative NMR-based metabolomics: a tutorial review. Metabolomics 11:1041–1055
9. Lane AN, Arumugam S, Lorkiewicz PK, Higashi RM, Laulhe S, Nantz MH et al (2014) Chemoselective detection of carbonyl compounds in metabolite mixtures by NMR. Magn Reson Chem 53:337–343
10. Tayyari F, Nagan Gowda GA, Gu H, Raftery D (2013) 15N-cholamine--a smart isotope tag for combining NMR- and MS-based metabolite profiling. Anal Chem 85:8715–8721
11. Heikkinen S, Permi P, Kilpela¨inen I (2001) Methods for the measurement of ^{1}JNCa and ^{2}JNCa from a simplified 2D ^{13}Ca-coupled ^{15}N SE–HSQC spectrum. J Magn Reson 148:53–60
12. Niu C, Bertrand RD, Shindo H, Cohen JS (1079) Cross-peptide bond ^{13}C--^{15}N coupling constants by ^{13}C and J cross-polarization ^{15}N NMR. J Biochem Biophys Methods 1 (3):135–143
13. Wishart DS, Bigam CG, Yao J, Abildgaard F, Dyson HJ, Oldfield E et al (1995) 1H, 13C and 15N chemical shift referencing in biomolecular NMR. J Biomol NMR 6(2):135–140
14. Clore GM, Gronenborn AM (1998) NMR structure determination of proteins and protein complexes larger than 20 kDa. Curr Opin Chem Biol 2(5):564–570
15. Otting G, Wuthrich K (1990) Heteronuclear filters In 2-dimensional H-1, H-1 NMR-spectroscopy—combined use with isotope labeling for studies of macromolecular conformation and intermolecular interactions. Q Rev Biophys 23(1):39–96
16. Gardner KH, Kay LE (1998) The use of H-2, C-13, N-15 multidimensional NMR to study the structure and dynamics of proteins. Annu Rev Biophys Biomol Struct 27:357–406
17. Fan TWM, Lane AN (2013) Assignment strategies for NMR resonances in metabolomics research. In: Lutz N, Sweedler JV, Wevers RA (eds) Methodologies for metabolomics: experimental strategies and techniques. Cambridge University Press, Cambridge, p 525
18. Fan TWM (1996) Metabolite profiling by one- and two-dimensional NMR analysis of complex mixtures. Prog Nucl. Magn Reson Spectro 28:161–219
19. Fan TWM, Lane AN (2008) Structure-based profiling of metabolites and isotopomers by NMR. Prog NMR Spectrosc 52:69–117
20. Fan TW, Lane AN (2016) Applications of NMR spectroscopy to systems biochemistry. Prog Nucl Magn Reson Spectrosc 92–93:18–53
21. Lane AN, Fan TW (2007) Quantification and identification of isotopomer distributions of metabolites in crude cell extracts using ^{1}H TOCSY. Metabolomics 3:79–86
22. Le Guennec A, Tayyari F, Edison AS (2017) Alternatives to nuclear overhauser enhancement spectroscopy presat and Carr–Purcell–Meiboom–Gill presat for NMR-based metabolomics. Anal Chem 89:8582–8588
23. Hwang TL, Shaka AJ (1995) Water suppression that works. excitation sculpting using arbitrary wave-forms and pulsed-field gradients. J Magn Reson Ser A 112:275–279
24. Lane AN (2012) In: Fan TW-M, Lane AN, Higashi RM (eds) NMR applications in metabolomics in Handbook of metabolomics. Humana, New York
25. Lane AN, Fan TW, Higashi RM (2008) Isotopomer-based metabolomic analysis by NMR and mass spectrometry. Methods Cell Biol 84:541–588

Chapter 10

Tracking Metabolic Rewiring of Cancer Metabolism in Humans Using Isotope-Resolved NMR

Kumar Pichumani

Abstract

Altered metabolism is considered one of the hallmarks of cancer. The findings that malignant brain tumors and brain metastases utilize acetate as an alternative nutrient are relatively recent and offer new avenues for investigation of altered metabolism in human cancers. Here, we describe comprehensively the details of the ^{13}C NMR-based isotopomer methodology to measure in vivo acetate utilization in brain tumor patients, including the contribution from acetate metabolism of peripheral tissues. Methods described in this chapter can be readily extended to other cancer types.

Key words ^{13}C NMR, Isotopomer, Acetyl-CoA, Citric acid cycle, ^{13}C–^{13}C spin–spin coupling, [1,2-^{13}C]acetate, Pyruvate recycling

1 Introduction

Metabolomics provides a global assessment of the cellular state of complex biological systems. It represents the investigation of the endpoint of the chain of biological events involving the upstream cellular events including gene transcription, translation, post-translational modifications, and enzymatic reactions. Metabolite concentrations can be accurately measured using either mass spectrometry or nuclear magnetic resonance (NMR) spectroscopy. Although metabolite levels can be linked to increased production or decreased utilization during metabolism, they do not provide information about the activity or flux of the metabolic pathways. To alleviate such challenges in the metabolomics field, carbon tracers have long been used to measure flux in the metabolic pathways. The basic principle behind the use of carbon tracers is the unique distribution of the tracers in metabolites, which provides specific activities of the underlying biochemical pathways. Traditionally, radio-labeled carbon (^{14}C) tracer molecules were used to obtain activities of metabolic pathways in isolated biochemical systems; it is less attractive to study in vivo metabolism in humans since it

G. A. Nagana Gowda and Daniel Raftery (eds.), *NMR-Based Metabolomics: Methods and Protocols*, Methods in Molecular Biology, vol. 2037, https://doi.org/10.1007/978-1-4939-9690-2_10, © Springer Science+Business Media, LLC, part of Springer Nature 2019

involves very minimal interaction between various metabolic pathways [1, 2, 3, 4]. Therefore, the use of ^{13}C tracer molecules (non-radioactive, stable isotope of carbon) has been preferred in the tracer studies to probe complex metabolic networks. Probing metabolic phenotypes in intact tumors is critical to understand how tumor cells grow in the human body under the complex tumor microenvironment. To date, the vast majority of knowledge on cancer metabolism is derived from cultured cell lines [5, 6]. Unfortunately, cancer cell lines are not ideal for understanding in vivo metabolism in patients since the cell lines have diverged genetically and epigenetically from patient tumors to adapt to the artificial culture conditions. Analysis of cell lines in a dish also misses the influence of tumor microenvironment in the growing tumors in patients [7, 8]. Furthermore, cell line studies lack the information about the input from the metabolism of peripheral organs, such as liver, skeletal muscle, and kidney to accurately model dynamic changes in the metabolism in a human body. While it is generally accepted that best models for human disease are human patients, most experimental approaches cannot be implemented easily for human studies.

^{13}C NMR spectroscopy offers unique advantages for investigations of altered metabolic pathways through ^{13}C labeling patterns that are generated via ^{13}C–^{13}C spin–spin (*J*) couplings. This method has long been used to investigate metabolic pathway activities; in 1983 Chance et al. first demonstrated the use of ^{13}C distributions in glutamate carbons to determine citric acid cycle flux in the isolated heart [9]. The technology of probing metabolic pathways based on ^{13}C–^{13}C spin-coupled multiplets from the ^{13}C NMR spectra is widely known as isotopomer analysis [10].

It is well known that altered energy metabolism via metabolic reprogramming is one of the hallmarks of cancer [11]. Recently, it was demonstrated that human brain tumors were capable of oxidizing glucose in the citric acid cycle in addition to the Warburg effect that involves producing lactate through glycolysis [12]. This study involved intravenous infusion of ^{13}C-labeled glucose to brain tumor patients followed by surgical removal of the tumors and the ^{13}C NMR spectroscopy of the tumor extracts. The outcome of this study led to the conclusion that only a small portion of acetyl-CoA was derived from the glucose in the brain tumor patients and therefore these tumors depended on alternative energy sources in order to generate the remaining acetyl-CoA pool in these tumors. While searching for the alternative fuels, we have identified acetate as a major energy source in malignant brain tumor patients and shown that acetate contributed significantly to the acetyl-CoA production in these patient tumors [13]. Subsequently, based on a number of studies, we established the feasibility of safely studying metabolic pathways in situ in human gliomas and brain metastases using ^{13}C NMR spectroscopy, which involves intravenous infusion of nontoxic nutrients labeled with nonradioactive stable isotopes of carbon (^{13}C) in patients who were undergoing surgical resection of tumors [13–18]. In this chapter, we will summarize the details of ^{13}C NMR-based isotopomer analysis to investigate in vivo acetate metabolism in brain tumor patients including the contributions from acetate metabolism of peripheral tissues.

2 Materials

1. Sterile and pyrogen-free [1,2-^{13}C]acetate can be purchased from Cambridge Isotope Laboratories or similar vendors (*see* **Notes 1–4**).
2. D_2O (NMR solvent), sodium deuteroxide, and DSS-d_6 (internal chemical shift and concentration standard) can be purchased from Sigma Aldrich, Cambridge Isotope Laboratories, or similar vendors.
3. Chloroform and tri-*n*-octylamine can be purchased from Sigma Aldrich.

3 Methods

3.1 Isotope Introduction and Sample Processing

1. Prior to the start of the study, approval must be obtained from the local Institutional Review Board (IRB) to perform infusion of [1,2-^{13}C]acetate (*see* **Notes 2–4**).
2. Intravenous infusion of sterile and pyrogen-free [1,2-^{13}C]acetate has to be performed according to the following parameters [13]: 6.0 mg/kg/min for the first 5 min, followed by 3.0 mg/kg/min for 2 h.
3. Grind the snap-frozen tumor tissue samples (that were surgically removed from patients) in a mortar under liquid nitrogen with 4.0% percholoric acid.
4. Allow the mixture to thaw and centrifuge at 25,000 × *g* for 20 min.
5. Collect the supernatant and add chloroform/tri-*n*-octylamine to increase the pH to close to 6.
6. Centrifuge the mixture again at 3500 × *g* for 15 min.
7. Carefully remove the aqueous phase to a new tube, and dry the sample using a speed-vac.
8. Reconstitute the lyophilized powder in 200 μL D_2O containing 1 mM of DSS-d_6.

9. Adjust the final pH to 7.0 using sodium deuteroxide and centrifuge for 1 min (*see* **Note 5**).
10. Finally, carefully remove the supernatant and transfer to a 3 mm NMR tube for NMR data acquisition.

3.2 NMR Acquisition

1. The following acquisition parameters can be used for ^{13}C NMR experiments:
   (a) 1D Proton decoupled ^{13}C spectra of tumor extracts and plasma can be acquired at 150 MHz or similar (for ^{13}C).
   (b) 10-mm broadband cryogenically cooled probe (*see* **Note 6**).
   (c) Flip angle of the nonselective RF pulse, 45°.
   (d) Spectral width, 225 ppm.
   (e) Acquisition time, 2.5 s.
   (f) Relaxation delay, 1.5 s.
   (g) ^{1}H decoupling sequence, WALTZ-16.
   (h) Number of scans, 15,000.
2. Perform Fourier transformation, phase and baseline correction.
3. Set lactate C3 carbon signal at 20.8 ppm as internal chemical shift reference (*see* **Note 7**).

3.3 Acetate Utilization in Brain Tumor Patients via [1,2-^{13}C]Acetate Intravenous Infusion Prior to Surgical Resection of the Tumor Mass

1. To understand the basics behind the method to determine the fate of [1,2-^{13}C]acetate in tumor cells, *see* Fig. 1. The figure shows the details of the flow of ^{13}C carbons from the infused [1,2-^{13}C]acetate. [1,2-^{13}C]acetate is converted to [1,2-^{13}C] acetyl-CoA in the tumor cell mitochondria and enters the citric acid cycle, which leads to ^{13}C labeling of carbons 4 and 5 of α-ketoglutarate (α-KG), glutamate and glutamine during the first turn of the cycle. Subsequently, carbons 3, 4, and 5 of α-KG, glutamate, and glutamine are labeled after multiple turns of the cycle.
2. ^{13}C-labeled carbons 4 and 5 produce a ^{13}C–^{13}C spin-coupled doublet (D45) at the carbon C4 position, whereas ^{13}C-labeled carbons 3, 4, and 5 produce a ^{13}C–^{13}C–^{13}C spin-coupled quartet (Q) in the ^{13}C NMR spectrum of C4 glutamate (*see* Fig. 2). The C4 carbon signals of glutamate represent the ^{13}C pool of acetyl-CoA and oxaloacetate (OAA). Thus, the C4 glutamate multiplet represents a direct readout of the ^{13}C-labeled acetyl-CoA pool in the tumor [13].
3. To determine the amount of [1,2-^{13}C]labeled acetyl-CoA generated from [1,2-^{13}C]acetate metabolism under non-steady-state conditions, measure peak areas of quartet signals (Q) and multiply it by the ratio of C4 and C3 glutamate signals [13, 19].

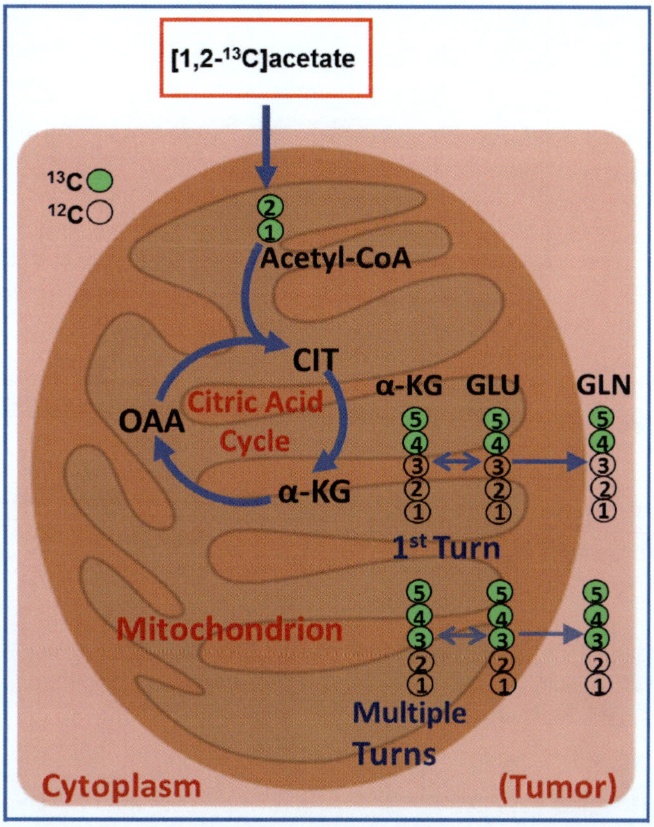

Fig. 1 A schematic depiction of the metabolic fate of carbons from [1,2-^{13}C] acetate during its oxidation in the citric acid cycle. *CIT* citrate, *α-KG* α-ketoglutarate, *GLU* glutamate, *2-HG* 2-hydroxyglutarate, *GLN* glutamine, *OAA* oxaloacetate

4. The [1,2-^{13}C]acetate level in the plasma, which is required to produce [1,2-^{13}C]acetyl-CoA in the tumor cells, needs to be measured as follows: obtain a ^{1}H NMR spectrum from the blood sample collected at the end of the [1,2-^{13}C]acetate infusion period (*see* **Note 8**) [17]. Well-resolved ^{13}C satellites (arising from ^{13}C–^{1}H J coupling) from methyl protons of the acetate in ^{1}H NMR spectrum confirm the presence of [1,2-^{13}C]acetate in the blood (*see* Fig. 3). ^{13}C enrichment of acetate in the plasma is determined from the ratio, [($^{13}CH_3$)/($^{12}CH_3$ + $^{13}CH_3$)], based on the peak areas from the $^{12}CH_3$ and $^{13}CH_3$ proton signals (*see* Fig. 3).

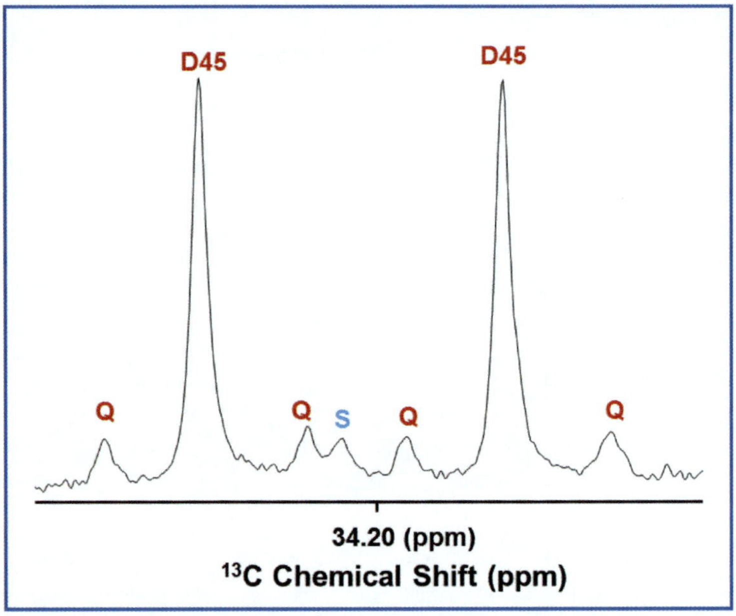

Fig. 2 A portion of ^{13}C NMR spectrum showing C4 glutamate from tumor extract from a GBM patient after the infusion of [1,2-^{13}C]acetate. ^{13}C-multiplets D45 and Q arise from the robust oxidation of acetate in the citric acid cycle of the tumor cells. S is the singlet due to ^{13}C natural abundance

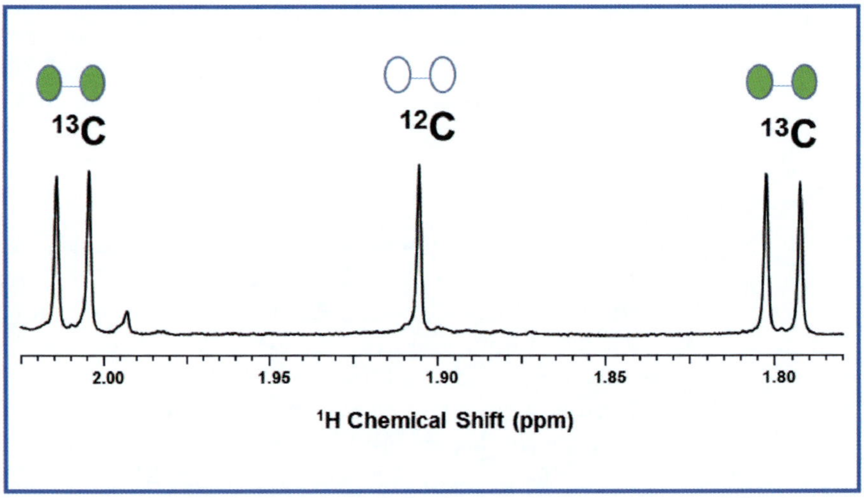

Fig. 3 A portion of ^{1}H NMR spectrum showing methyl proton signals of [1,2-^{13}C]acetate from the plasma of a GBM patient, who had the infusion of [1,2-^{13}C]acetate. $^{12}CH_3$ (open circle) refers to the methyl proton signals that are covalently bonded to ^{12}C; and $^{13}CH_3$ (filled circle) refers to the methyl protons that are covalently bonded to ^{13}C carbon, which appear as two sets of doublets separated by $^{1}J_{C\text{–}H} = 127.10$ Hz. The splitting observed within each set of the doublets (5.94 Hz) is due to $^{2}J_{C\text{–}H}$

3.4 The Role of [1,2-^{13}C]Acetate Metabolism in Peripheral Tissues on Brain Tumor Energy Metabolism

3.4.1 Blood-Borne ^{13}C-Labeled Lactate and Glucose

1. Peripheral tissues play significant roles during acetate metabolism of brain tumors. A portion of the ^{13}C-acetate can be metabolized by the liver, skeletal muscle, and renal tissues, and lead to the generation of ^{13}C-labeled glucose, lactate, and glutamine. These ^{13}C-labeled metabolites in the peripheral organs can be detected in the circulation, which can also enter the brain tumor and serve as nutrients along with ^{13}C-acetate, and contribute to the ^{13}C-acetyl-CoA pool in the brain tumor. Probing such peripheral metabolism of acetate enables assessing flux through pyruvate cycling and also the role of glutamine and lactate in energy metabolism in brain tumors in vivo.
2. As a proof of the contribution of peripheral metabolism, previously unexpected, [2,3-^{13}C] and [1,2-^{13}C]lactate were detected in the ^{13}C NMR spectra of brain tumor patients, who were given a [1,2-^{13}C]acetate intravenous infusion prior to surgery (*see* Fig. 4b, c). There is no direct metabolic route for the conversion of [1,2-^{13}C]acetate to ^{13}C-labeled lactate in the brain tumor cells.
3. Hepatic gluconeogenesis in the tumor patients leads to the following biological processes that help quantify the contribution of the ^{13}C labeled nutrients, from peripheral metabolism to brain tumor metabolism.
   (a) Glucose is not synthesized directly from [1,2-^{13}C]acetate in the mammalian liver. However, in the citric acid cycle of the liver [1,2-^{13}C]acetate transfers the ^{13}C labels to oxaloacetate, which is a key precursor for gluconeogenesis and eventually leads to the generation of ^{13}C-labeled glucose isotopomers [1,2-^{13}C] and [2,3-^{13}C]. These glucose isotopomers can be readily detected in the plasma of the tumor patient (*see*, Fig. 4d, e). Further, in the tumor cells, these ^{13}C-labeled glucose isotopomers produce [2,3-^{13}C] and [1,2-^{13}C]lactate isotopomers (*see* Fig. 4b, c).
   (b) ^{13}C-enriched pyruvate or lactate can also arise from [1,2-^{13}C]acetate through [1,2-^{13}C]acetyl-CoA, which upon entering citric acid cycle generates [1,2-^{13}C] and [3,4-^{13}C] labeled oxaloacetate (OAA) (*see* Fig. 4a). An equimolar mixture of [1,2-^{13}C] and [3-^{13}C]pyruvate and lactate isotopomers can be produced by either the malic enzyme or the combined activity of the enzymes phosphoenolpyruvate carboxykinase (PEPCK) and pyruvate kinase (PK) decarboxylation from [1,2-^{13}C]/[3,4-^{13}C]OAA via the process known as pyruvate cycling (*see* Fig. 4a) [20].
   (c) [1,2-^{13}C]pyruvate formed from [1,2-^{13}C]acetate could produce mono-labeled [1-^{13}C]acetyl-CoA through the

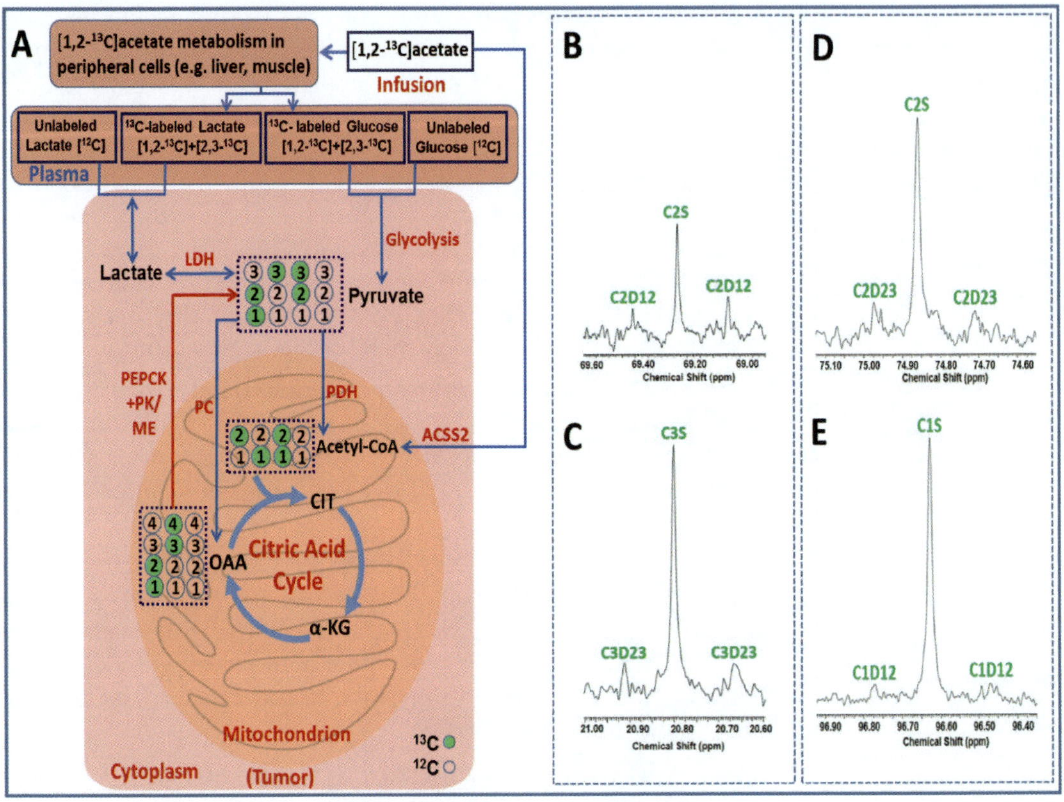

Fig. 4 (**a**) A schematic depiction of the multiple sources of ^{13}C label entering the tumor pyruvate/lactate pool from circulation. (**b**, **c**) Lactate ^{13}C labeling in a GBM patient's tumor. (**d**, **e**) Glucose ^{13}C labeling in the plasma of the same patient. Abbreviations: *PC,* pyruvate carboxylase; *PDH,* pyruvate dehydrogenase; *ME,* malic enzyme; *PEPCK,* phosphoenolpyruvate carboxykinase; *PK,* pyruvate kinase; *ACSS2,* acetyl-CoA synthetase 2; *C1D12 or C2D12,* ^{13}C–^{13}C *J* coupling between the carbons C1 and C2; *C2D23 or C3D23,* ^{13}C–^{13}C *J* coupling between the carbons C2 and C3; *C1S,* singlet of C1 carbon signal; *C2S,* singlet of C2 carbon signal; *C3S,* singlet of C3 carbon signal

enzymatic action of pyruvate dehydrogenase (PDH). To determine the amount of [1-^{13}C]acetyl-CoA, multiply the ratio between C5 glutamate singlet and doublet (C5D) by the total [1,2-^{13}C]acetyl-CoA [15, 17].

3.4.2 Blood-Borne ^{13}C-Labeled Glutamine

1. Glutamine is a conditionally essential amino acid as a cellular precursor for biomolecular synthesis. Higher glutamine utilization by the tumor supports its growth through nucleotide de novo synthesis and other anabolic processes [21, 22]. Gliomas show higher uptake of glutamine than the normal brain as observed through PET imaging using ^{18}F–4F-glutamine (^{18}F-FGln) [23].
2. In human brain tumors, glutamine can form from [1,2-^{13}C] acetate through α-ketoglutarate (α-KG) (*see* Fig. 1) and

glutamate [13]; in brain tumors glutamine synthetase (GS) is expressed, which converts glutamate into glutamine [17]. Glutamine can also be formed from acetate metabolism in skeletal muscle, liver and renal tissues [24, 25].

3. [1,2-^{13}C]acetate influences hepatic gluconeogenesis pathways that lead to ^{13}C label incorporation into pyruvate, lactate, and glucose in the liver citric acid cycle, which upon entering the citric acid cycle in the tumor contributes (via α-ketoglutarate) to the glutamate and glutamine pools.
4. The description below provides a comprehensive analysis of ^{13}C turnover of the tumor glutamate and glutamine pools using C4 glutamate/glutamine isotopomer analysis in the presence of blood-borne ^{13}C-labeled glutamine, ^{13}C-lactate, and ^{13}C-glucose formed by the infused [1,2-^{13}C]acetate.
   (a) ^{13}C isotopomers that contribute to C4 glutamate/glutamine ^{13}C NMR multiplet signal (*see* Fig. 5) are:

   a = the fraction of glutamine or glutamate that does not have ^{13}C in C4 position.

   b = the fraction of glutamine or glutamate that has ^{13}C only in C4 position.

   c = the fraction of glutamine or glutamate, which is a [3,4-^{13}C]labeled isotopomer.

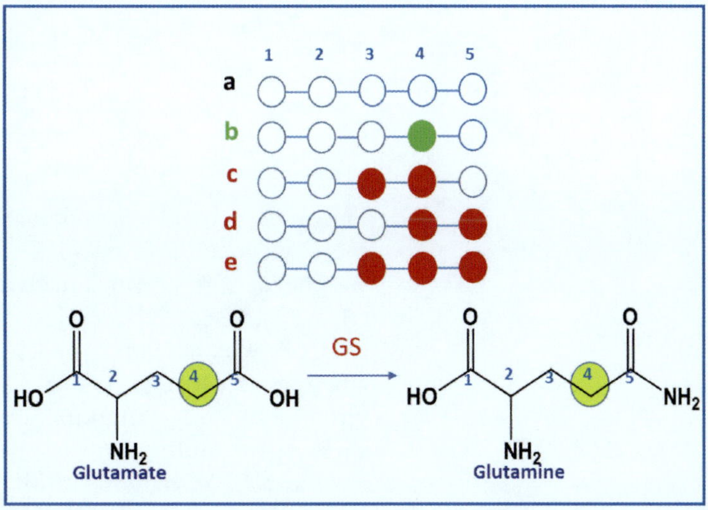

Fig. 5 Illustration of various ^{13}C isotopomers (**a–e**) of C4 carbon of glutamate or glutamine used in this analysis. (**a**) ^{12}C fraction, (**b**) ^{13}C fraction that has glutamate or glutamine only at C4 either due to natural abundance or generated from [2-^{13}C]acetyl-CoA, (**c**) [3,4-^{13}C]glutamate or glutamine, (**d**) [4,5-^{13}C]glutamate or glutamine, (**e**) [3,4,5-^{13}C]glutamate or glutamine. The isotopomers (*c, d,* and *e*) that are purely generated from the metabolism of [1,2-^{13}C]acetate are shown in red. *GS,* glutamine synthetase

d = the fraction of glutamine or glutamate, which is a [4,5-^{13}C]labeled isotopomer.

e = the fraction of glutamine or glutamate, which is a [3,4,5-^{13}C]labeled isotopomer.

By definition, $(a + b + c + d + e) = 1$.

(b) ^{13}C NMR signals from C4 glutamine or glutamate provide peak areas of the ^{13}C–^{13}C spin–spin-coupled doublets (D45 and D34), quartet (Q), and singlet (S). The singlet (S) can arise from the natural abundance of ^{13}C (*see* Fig. 2); further, [2-^{13}C]acetyl-CoA (Fc2, fraction of acetyl-CoA enriched only in C2 methyl carbon) can also contribute to the singlet C4 (S) signal through the generation of [4-^{13}C]glutamate on the first turn citric acid cycle. In the absence of any blood-borne ^{13}C-enriched nutrients due to peripheral metabolism of [1,2-^{13}C]acetate, Fc2 should be at the level of the natural abundance of ^{13}C (1.1%).

(c) Measure peak areas of all the multiplets in the ^{13}C NMR signal of C4 glutamate/glutamine (*see* **Notes 9** and **10**). The solutions of the following four equations provide fractional contributions of C4glutamate and glutamine isotopomers*: a, b, c, d*, and *e.*

$$b/(a+b) = \mathrm{Fc2} \quad (1)$$

$$(c/b) = \mathrm{D34/S} \quad (2)$$

$$(d/b) = \mathrm{D45/S} \quad (3)$$

$$(e/b) = \mathrm{Q/S} \quad (4)$$

(d) Figure 6 presents a schematic diagram illustrating the metabolism of infused [1,2-^{13}C]acetate in a tumor cell. By the action of nucleocytosolic acetyl-CoA synthetase (ACSS2), [1,2-^{13}C]acetate is converted to [1,2-^{13}C]acetyl-CoA which enters the citric acid cycle yielding ^{13}C-labeling at carbons 4 and 5 of α-ketoglutarate (α-KG), glutamate, and glutamine during the first turn of the cycle. After multiple turns of the cycle, carbons 3, 4, and 5 of α-KG, glutamate, and glutamine are labeled with ^{13}C. Glutamine synthetase (GS), the key enzyme responsible for the conversion of glutamate to glutamine, is highly expressed in human brain tumors and supports in situ glutamine synthesis from glutamate [13, 17].

(e) We have observed similar levels of doublets (D45) and quartets (Q) for both glutamate and glutamine C4 carbon signals in ^{13}C NMR spectra of tumor extracts (*see*

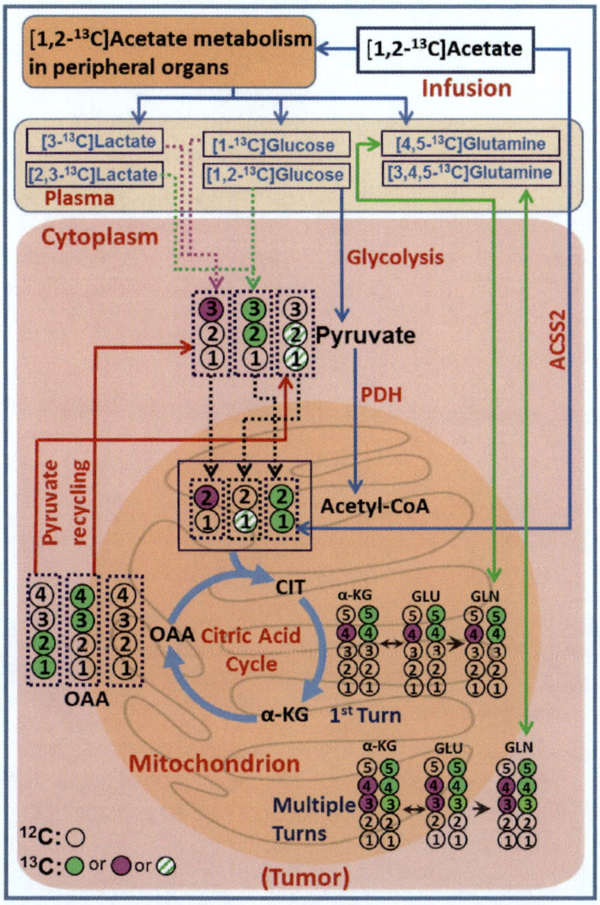

Fig. 6 Comprehensive metabolic model including the contributions from the peripheral metabolism of [1,2-^{13}C]acetate to the brain tumor energy metabolism. ^{13}C labeling patterns in α-KG, GLU, and GLN are shown along with the numbers representing the carbon atom positions. [4,5-^{13}C] and [3,4,5-^{13}C]GLU/GLN (green filled circles) were generated from acetate-derived [1,2-^{13}C]acetyl-CoA. [4-^{13}C] and [3,4-^{13}C]GLU/GLN (purple filled circles) were mainly generated from blood-borne ^{13}C-labeled glucose/lactate that originated from [1,2-^{13}C]acetate metabolism of peripheral organs. Abbreviations: *α-KG,* α-ketoglutarate; *GLU,* glutamate; *GLN,* glutamine; *OAA*, oxaloacetate; *PDH*, pyruvate dehydrogenase; *ACSS2,* nucleocytosolic acetyl-CoA synthetase 2. Filled circle (green or purple) refers to ^{13}C and open circle refers to ^{12}C carbon nuclei. Green pattern-filled circles correspond to ^{13}C isotopomers that generate ^{13}C-labeled C5 glutamate isotopomer and do not influence C4 ^{13}C fractional enrichment

Fig. 7a–c), which provides evidence for de novo glutamine synthesis from glutamate in the tumor cells [17]. Measure these peak areas to determine fractional contributions of various C4 ^{13}C isotopomers of glutamate and glutamine. An example of the results obtained from such an analysis is summarized in Tables 1 and 2 [17].

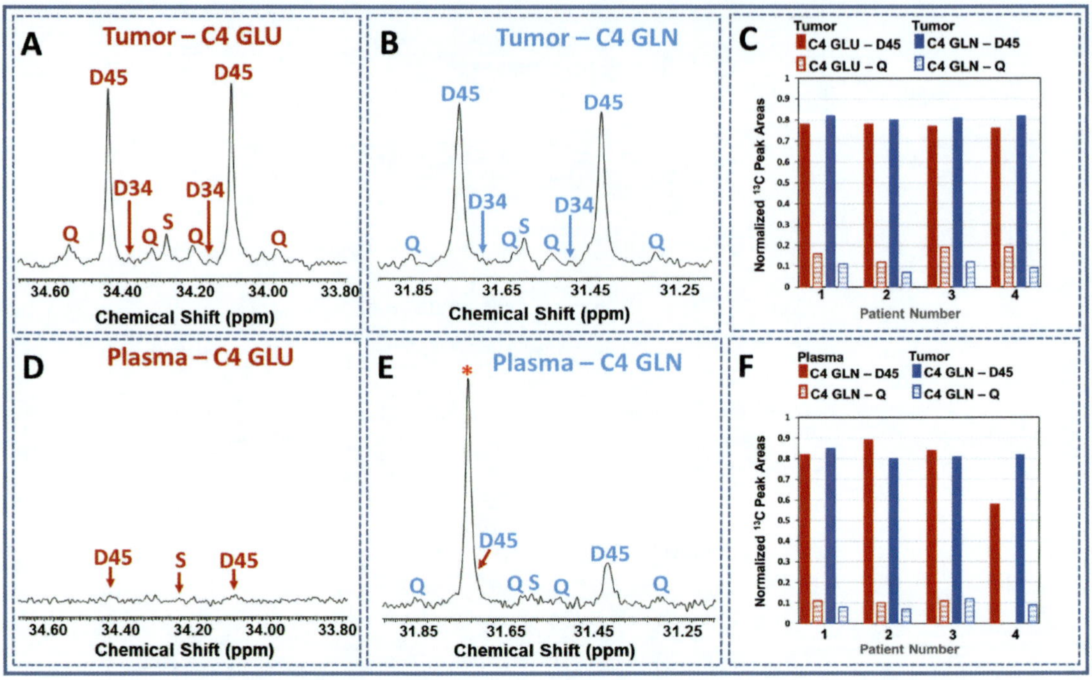

Fig. 7 ^{13}C NMR spectral data. (**a**, **b**) and (**d**, **e**) show portions of the ^{13}C NMR spectra showing peaks for C4 carbon of glutamate and glutamine in tumor and plasma, respectively, from a GBM (glioblastoma) patient. In **e**, $*$ refers to an overlapping peak from unknown carbon signal with one of the D45 GLN doublet peaks. The patient was infused with [1,2-^{13}C]acetate prior to the resection of the tumor. Charts **c** and **f** show normalized ^{13}C peak areas of C4 carbon multiplets of glutamine and glutamate from four brain tumor patients participated in the study. Chart **c** compares the peak areas between glutamate and glutamine C4 carbon multiplets in the tumor. Chart **f** shows the comparison of C4 GLN peak areas between plasma and tumor. *S* singlet is due to both natural abundance of ^{13}C and Fc2, *D34* doublet is due to ^{13}C–^{13}C *J* coupling between the carbons 3 and 4, *D45* doublet is due to ^{13}C–^{13}C *J* coupling between the carbons 4 and 5, *Q* quartet is due to ^{13}C–^{13}C *J* couplings between the carbons 3, 4, and 5

Table 1
^{13}C fractional enrichments of GLU C4 in tumors

Tumor type	*a*	*b*	*c*	*d*	*e*	FE	(*b* + *c*)/FE	(*d* + *e*)/FE
GBM	0.758	0.014	0.002	0.188	0.038	0.242	0.066	0.934
GBM	0.717	0.024	0.003	0.222	0.035	0.284	0.095	0.905
Breast metastasis	0.537	0.017	0.006	0.354	0.087	0.464	0.050	0.950
Lung metastasis	0.554	0.019	0.006	0.339	0.083	0.446	0.056	0.944

a–*e* were determined from ^{13}C NMR spectra of C4 GLU as described in Subheading 3 (*see* Figs. 5 and 6). FE is equal to (*b* + *c* + *d* + *e*) and corresponds to fractional enrichment of GLU C4 in the tumor. (b+c) represents the contributions from the [1,2-^{13}C]acetate metabolism by periphery to tumor C4 GLU pool

Table 2
C4 GLN ^{13}C fractional enrichments in tumors

Tumor type	*a*	*b*	*c*	*d*	*e*	FE	(*b* + *c*)/FE	(*d* + *e*)/FE
GBM	0.762	0.014	0.002	0.196	0.027	0.239	0.067	0.933
GBM	0.787	0.026	0.002	0.171	0.015	0.213	0.131	0.869
Breast metastasis	0.659	0.021	0.003	0.276	0.041	0.341	0.070	0.930
Lung metastasis	0.669	0.023	0.009	0.271	0.028	0.331	0.097	0.903

a–e were determined from ^{13}C NMR spectra of C4 GLN as described in Subheading 3 (*see* Figs. 5 and 6). FE is equal to ($b + c + d + e$) and corresponds to fractional enrichment of GLN C4 in the tumor. (b+c) represents the contributions from the [1,2-^{13}C]acetate metabolism by periphery to tumor C4 GLN pool

(f) In addition to the tumor, plasma of the patients shows ^{13}C-labeled glutamine arising from the [1,2-^{13}C]acetate metabolism in peripheral organs. Figure 7d, e show ^{13}C NMR spectra of C4 carbon multiplet signals of plasma glutamate and glutamine of a GBM (glioblastoma) patient. Detection of the prominent doubly labeled 4 and 5 carbons of glutamine ^{13}C isotopomer (D45) in the plasma of a GBM patient (Fig. 7e) indicates the presence of ^{13}C-labeled glutamine in the circulating blood. Plasma C4 glutamine ^{13}C-multiplet peak areas are shown in Fig. 7e and are dominated by D45 with weaker quartet (Q) signals. As illustrated in Fig. 7d, the C4 glutamate signals are very weak in the plasma [17].

(g) Summary of the results:

- Some of the blood-borne ^{13}C-labeled glutamine from the metabolism of [1,2-^{13}C]acetate by peripheral tissues may have entered the tumor, and both glutamine pools show very similar D45 and *Q* values (*see* Fig. 7f) [17].
- The ^{13}C fractional enrichment of C4 glutamate pool in the tumors was 35.90 ± 11.25% (($b + c + d + e$) of tumor C4 GLU in Table 1); whereas, the value in C4 glutamine was only 28.10 ± 6.44% (($b + c + d + e$) of tumor C4 GLN in Table 2) [17]. If the ^{13}C-glutamine in the tumor was mainly synthesized from ^{13}C-glutamate (Tables 1 and 2), then the total ^{13}C fractional enrichment of the C4 carbon of glutamine is expected to be less than or equal to that of its precursor, glutamate C4 carbon, assuming similar pool sizes. Our current results of lower ^{13}C fractional enrichment of C4 glutamine (Table 2) compared to C4 glutamate in the tumor suggests that ^{13}C labeling from infused [1,2-^{13}C]acetate entered the glutamate pool in the tumor before it reached glutamine. ($d + e$)/FE in Tables 1 and

2 corresponds to the portion of C4 ^{13}C fractional enrichment of glutamate and glutamine that was derived from the direct utilization of [1,2-^{13}C]acetate by tumor cells. (b+c) represents the contributions from the [1,2-^{13}C]acetate metabolism by periphery to tumor C4 GLU/GLN pools. Similarly, measure peak areas of C4 glutamate/glutamine from the plasma and follow the above analysis to quantify the amount of blood-borne ^{13}C-glutamine.

(h) As illustrated in Fig. 6, the amount of [2-^{13}C]acetyl-CoA gives the readout of the contributions from peripheral metabolism of the infused [1,2-^{13}C]acetate to the tumor. Through the following mechanisms, [2-^{13}C]acetyl-CoA may have attributed to ^{13}C labeling at C4 of glutamate and glutamine (C4S), in addition to background signal due to the natural abundance of ^{13}C (1.1%):

- As described earlier, ^{13}C label from [1,2-^{13}C]acetate entering the liver citric acid cycle may enter the gluconeogenesis pathways via the oxaloacetate pool, which led to the detection of multiple ^{13}C-labeled isotopomers including [1-^{13}C]/[1,2-^{13}C]glucose in the plasma [15]. The [3-^{13}C]lactate isotopomer was also detected in the plasma in these patients due to the systemic metabolism of infused acetate. The ^{13}C-labeled glucose and lactate enters the tumor and generate [2-^{13}C]acetyl-CoA, which then yields [4-^{13}C]glutamate/glutamine and [3,4-^{13}C]glutamate/glutamine after the first and multiple turns of the cycle respectively (*see* Fig. 4a) [15].
- Increased flux through the oxidative branch of the pentose phosphate pathway (PPP) was reported in human orthotopic mouse models of GBM and renal cell carcinoma [26]. The following is a method to measure PPP flux relative to glycolysis using peripherally derived [1,2-^{13}C]glucose in the plasma: C3 lactate singlet (C3S) is produced through the combination of flux through the oxidative branch of PPP and a minor contribution from the natural abundance ^{13}C signal of C3 lactate carbon (1.1%). The ratio between the singlet (C3S) to doublet (D23) of the lactate C3 multiplet provides the measure of PPP flux relative to glycolysis. The activity of PPP may contribute to ^{13}C labeling at C3 lactate, which produces [2-^{13}C]acetyl-CoA via [3-^{13}C]pyruvate. As described earlier, the [2-^{13}C]acetyl-CoA yields [4-^{13}C]glutamate/glutamine and [3,4-^{13}C]glutamate/glutamine after the first and multiple turns of the cycle, respectively (*see* Fig. 8) [15, 17].

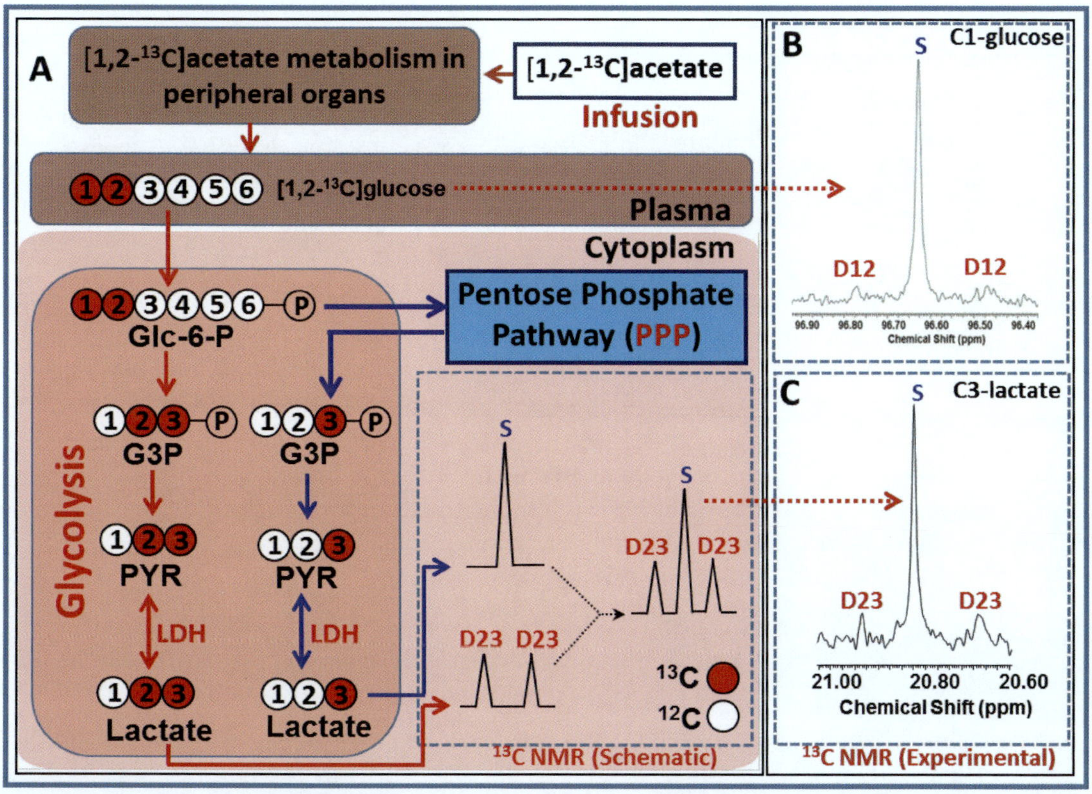

Fig. 8 (**a**) Schema showing acetate-derived [1,2-^{13}C]glucose metabolism via PPP. ^{13}C NMR spectrum of (**b**) C1 glucose region in the plasma showing the presence of D12 doublet and (**c**) tumor C3 lactate signal showing higher singlet (S) to doublet (D23) ratio. Abbreviations: *D12,* ^{13}C–^{13}C *J* coupling between the carbons C1 and C2; *D23,* ^{13}C–^{13}C *J* coupling between the carbons C2 and C3; *S,* singlet; *Glc-6-P,* glucose-6-phosphate; *G3P,* glycerol-3-phosphate; *PYR,* pyruvate; *LDH,* lactate dehydrogenase

- Pyruvate recycling in the tumor cells generates [3-^{13}C] and [1,2-^{13}C]pyruvate isotopomers from [3,4-^{13}C] and [1,2-^{13}C]OAA isotopomers, respectively, after the first turn of the citric acid cycle during the metabolism of [1,2-^{13}C]acetate [3, 9, 10, 18]. [3-^{13}C]pyruvate further metabolizes to generate [2-^{13}C]acetyl-CoA via pyruvate dehydrogenase (PDH) and yields [4-^{13}C]glutamate/glutamine and [3,4-^{13}C]glutamate/glutamine in the citric acid cycle (*see* Fig. 6). [1,2-^{13}C]pyruvate generates [1-^{13}C]acetyl-CoA, which upon entering the citric acid cycle yields [5-^{13}C]glutamate and glutamine isotopomers. Together, $(b + c)$/FE (*see* Tables 1 and 2) represents the contribution from the peripheral metabolism of acetate to yield the overall C4 ^{13}C fractional enrichment of glutamate and glutamine, respectively [15, 17].

4 Notes

1. The [1,2-^{13}C]acetate that we used was purchased from Cambridge Isotope Laboratories Inc., Tewksbury, MA.
2. This study involves human subjects and therefore an IRB protocol needs to be approved prior to beginning this study.
3. Make sure the Investigational Drug Service (IDS) at your Institution performs its own tests for sterility and pyrogenicity of [1,2-^{13}C]acetate prior to administering to patients.
4. Although^{13}C-acetate is commercially purchased, the ^{13}C enrichment in the stock solution will have to be determined in order to confirm the ^{13}C enrichment in the [1,2-^{13}C]acetate, before it is administered intravenously.
5. The pH of the aqueous extracts needs to be precisely adjusted to 7.0 as the ^{13}C chemical shifts are sensitive to pH changes.
6. We have used 600 MHz Avance Bruker NMR Spectrometer (Bruker Biospin, Billerica, MA) equipped with a 10-mm broadband cryogenically cooled probe.
7. For processing, we used the Advanced Chemistry Development (Toronto, Canada) software program.
8. Always obtain a ^{1}H NMR spectrum prior to acquiring the ^{13}C NMR spectrum.
9. Assumptions of isotopic or metabolic steady state conditions are not required. The analysis described here is fully applicable for non-steady-state conditions.
10. Excellent shimming of the samples during NMR experiments is critical to obtain well-resolved C4 glutamate multiplets. Poor shimming leads to broad lines and often masks the ^{13}C–^{13}C spin-coupled multiplets (Q) that are required for accurate ^{13}C isotopomer analysis.

Acknowledgments

This work was supported by Donna and Kenneth Peak, The Kenneth R. Peak Foundation, The Kenneth R. Peak Brain and Pituitary Tumor Treatment Center at Houston Methodist Hospital, The Houston Methodist Foundation, The Taub Foundation, The Pauline Sterne Wolff Foundation, The Veralan Foundation, The Marilee A. and Gary M. Schwarz Foundation, The John S. Dunn

Foundation and The McKone Family Foundation. We are grateful to the many patients and their families who have participated in our studies, and who are dedicated to join us in our fight against brain cancer.

References

1. Strisower EH, Kohler GD, Chaikoff IL (1952) Incorporation of acetate carbon into glucose by liver slices from normal and alloxan-diabetic rats. J Biol Chem 198:115–126
2. Weinman EO, Strisower EH, Chaikoff IL (1957) Conversion of fatty acids to carbohydrate: applications of isotopes to this problem and the role of Krebs cycle as a synthetic pathway. Physiol Rev 37:252–272
3. Coxon RV, Robinson RJ (1959) Movements of radioactive carbon dioxide within the animal body during oxidation of ^{14}C-labeled substances. J Physiol 147:487–570
4. Coxon RV, Robinson RJ (1959) The transport of radioactive carbon dioxide in the blood stream of the dog after administration of radioactive biocarbonate. J Physiol 147:469–486
5. Ramanathan A, Wang C, Schreiber SL (2005) Peturbational profiling of a cell-line model of tumorigenesis by using metabolic maesurements. Proc Natl Acad Sci U S A 102:5992–5997
6. Forbes NS, Meadows AL, Clark DS et al (2006) Estradiol stimulates the biosynthetic pathways of breast cancer cells: detection by metabolic flux analysis. Metab Eng 8:639–652
7. Lee J, Kotliarova S, Kotliarov Y et al (2006) Tumor stem cells derived from glioblastoma cultured in bFGF and EGF more closely mirror the phenotyope and genotype of primary tumors than do serum-cultured cell lines. Cancer Cell 9:391–403
8. Marian CO, Cho SK, McEllin BM et al (2010) The telomerase antagonist, imetelstat, effectively targets glioblastoma tumor-initiating cells leading to decreased proliferation and tumor growth. Clin Cancer Res 16:154–163
9. Chance EM, Seeholzer SH, Kobayashi K et al (1983) Mathematical analysis of isotope labeling in the citric acid cycle with applications to 13C NMR studies in perfused rat hearts. J Biol Chem 258:13785–13794
10. London RE (1987) ^{13}C labeling in studies of metabolic regulation. Prog NMR Spectrosc 20:337–383
11. Ward PS, Thompson CB (2012) Metabolic reprogramming: a cancer Hallmark even Warburg did not anticipate. Cancer Cell 21:297–308
12. Maher EA, Marin-Valencia I, Bachoo RM et al (2012) Metabolism of [U-^{13}C]glucose in human brain tumors in vivo. NMR Biomed 25:1234–1244
13. Mashimo T, Pichumani K, Vemireddy V et al (2014) Acetate is a bioenergetics substrate for human glioblastoma and brain metastases. Cell 159:1603–1614
14. Pichumani K, Mashimo T, Baek H-M et al (2015) Conditions for 13C NMR detection of 2-Hydroxyglutarate in tissue extracts from isocitrate dehydrogenase-mutated gliomas. Anal Biochem 481:4–6
15. Pichumani K, Mashimo T, Vemireddy V et al (2016) Hepatic gluconeogenesis influences ^{13}C enrichment in lactate in human brain tumors during metabolism of [1,2-^{13}C]acetate. Neurochem Int 97:133–136
16. Maher EA, Pichumani K, Sarode V et al (2016) Differential metabolism of glucose and acetate in mitochondria of early stage breast cancer in vivo. Proc Intl Soc Mag Reson Med 24:841
17. Pichumani K, Mashimo T, Vemireddy V et al (2017) Measurement of ^{13}C turnover into glutamate and glutamine pools in brain tumor patients. FEBS Lett 591:3548–3554
18. Courtney KD, Bezwada D, Mashimo T et al (2018) Isotopic tracing of human clear cell renal cell carcinomas demonstrates suppressed glucose oxidation *In Vivo.* Cell Metab 28:793–800
19. Malloy CR, Thompson JR, Jeffrey FH et al (1990) Contributions of exogenous substrates to acetyl coenzyme a: measurement by ^{13}C NMR under non-steady-state conditions. Biochemistry 29:6756–6761
20. Cerdan S, Kunnecke B, Seelig J et al (1990) Cerebral metabolism of [1,2-$^{13}C_2$]acetate as detected by *in Vivo* and *in Vitro* ^{13}C NMR. J Biol Chem 265:12916–12926
21. DeBerardinis RJ, Mancuso A, Daikhin E et al (2007) Beyond aerobic glycolysis: transformed cells can engage in glutamine metabolism that exceeds the requirement for protein and nucleotide synthesis. Proc Natl Acad Sci U S A 104:345–350
22. Tardito S, Oudin A, Ahmed SU et al (2015) Glutamine synthetase activity fuels nucleotide biosynthesis and supports growth of

glutamine-restricted glioblastoma. Nat Chem Biol 17:1556–1568

23. Venneti S, Dunphy MP, Zhang H et al (2015) Glutamine-based PET imaging facilitates enhanced metabolic evaluation of gliomas in vivo. Sci Transl Med 7:274ra17
24. Szczepaniak L, Babcock EE, Malloy CR et al (1996) Oxidation of acetate in rabbit skeletal muscle: detection by ^{13}C NMR spectroscopy *in vivo*. Magn Reson Med 36:451–457
25. Befroy DE, Perry RJ, Jain N et al (2014) Direct assessment of hepatic mitochondrial oxidative and anaplerotic fluxes in humans using dynamic ^{13}C magnetic resonance spectroscopy. Nat Med 20:98–102
26. Valencia IM, Cho SK, Rakjeja D et al (2012) Glucose metabolism via the pentose phosphate pathway, glycolysis and Krebs cycle in an Orthotopic mouse model of human brain tumors. NMR Biomed 25:1177–1186

Part V

Biomarker Discovery

Chapter 11

NMR-Based Metabolomics in Cardiac Research

Julian L. Griffin and Cecilia Castro

Abstract

Proton (^{1}H) nuclear magnetic resonance (NMR) spectroscopy can measure a range of metabolites in both cardiac tissue and blood plasma for following cardiovascular disease. For solution-state NMR spectroscopy, it is necessary to create a tissue extract, with perchloric acid, acetonitrile/water, and chloroform/methanol being popular extraction media. Alternatively, high-resolution magic angle spinning (HRMAS) ^{1}H NMR spectroscopy can be used to derive a metabolic profile directly from intact cardiac tissue. This chapter will discuss the practical methods used for ^{1}H NMR spectroscopy to follow cardiovascular diseases both in terms of metabolic changes in cardiac tissue and changes in blood plasma.

Key words Cardiovascular disease, High-resolution ^{1}H NMR spectroscopy, Blood plasma

1 Introduction

Nuclear magnetic resonance (NMR) spectroscopy has long been used for assessing the metabolic status of the heart [1], particularly in the use of ^{31}P NMR to assess its energetic status by measuring ATP, inorganic phosphate, and phosphocreatine both in vivo and in tissue extracts [2]. In addition, ^{13}C NMR spectroscopy has been used to measure carbon flux in the heart, including measuring the citric acid cycle in the working heart [3] and the application of hyperpolarization techniques. With the advent of metabolomics, high-resolution ^{1}H NMR spectroscopy has increasingly been used to profile the high-concentration metabolites in cardiac tissue extracts. This was the basis of one of the first metabolomics papers describing changes in cardiac metabolism in mouse models of cardiac and cardiovascular disease [4].

Since then, a variety of cardiovascular studies have used ^{1}H NMR spectroscopy to profile the major metabolites in cardiac tissue and blood plasma, including studies in conjunction with other omics, such as proteomics [5, 6] and transcriptomics [7]. While one-dimensional ^{1}H NMR spectroscopy is the main NMR technique applied within metabolomics, a variety of two-dimensional

G. A. Nagana Gowda and Daniel Raftery (eds.), *NMR-Based Metabolomics: Methods and Protocols*, Methods in Molecular Biology, vol. 2037, https://doi.org/10.1007/978-1-4939-9690-2_11,

approaches are also used [8], particularly with respect to metabolite identification. In some rarer cases, two-dimensional NMR spectroscopy has also been used to generate multivariate models to increase metabolite coverage and dispersion. In addition, solution-state ^{1}H NMR spectroscopy and high-resolution magic angle spinning (HRMAS) ^{1}H NMR spectroscopy have also been used to profile intact tissue [8]. With an increase in interest in the use of stable isotopes for flux analysis, there has also been a resurgence in the use of ^{13}C NMR spectroscopy to follow flux through metabolic pathways either in tissue extracts or in situ in the working heart. In this chapter, we will focus on the use of ^{1}H NMR spectroscopy to profile metabolites in cardiac tissue extracts, intact tissue, and blood plasma.

2 Materials

1. High-resolution NMR spectrometer: For metabolomics the commonest field strengths have been 400–900 MHz ^{1}H frequency (9.4–22.2 T), with 600 MHz NMR spectrometers (14.1 T) being favored. All the work performed in the authors' laboratory has been on a 500 MHz NMR ADVANCE II+ spectrometer (Bruker, Karlsruhe, Germany).
2. The NMR spectrometer is interfaced with an NMR probe to acquire spectra. For blood plasma and tissue extracts, we use a 5 mm inverse-detect triple-resonance TXI probe (Bruker, Karlsruhe, Germany), which is designed to be more sensitive in terms of detection of ^{1}H NMR spectra compared with ^{13}C and ^{15}N spectra. However, a variety of probes could be used to give comparable results.
3. For HRMAS, a HRMAS probe is interfaced with the magnet and samples packed into zirconium oxide 3 mm rotors.
4. All solutions were prepared using analytical grade reagents and stored at room temperature.
5. Sodium-3-(trimethylsilyl)-2,2,3,3-tetradeuteriopropionate (TSP) (Cambridge Isotope Laboratories, MA, USA) was used as a chemical shift reference.
6. [^{13}C]formate (Isotec-Sigma Aldrich, St. Louis, MO, USA) was used as a chemical shift reference for ^{13}C NMR spectroscopy.
7. Deuterated water (D_2O, 99.9%) (Goss Scientific, UK) was used as a lock reference.
8. Software: For acquiring data and processing NMR spectra on the NMR spectrometer, we used TopSpin (version 3.2; Bruker, Karlsruhe, Germany), Chenomx NMR Suite 7.7 (Chenomx, Alberta, Canada), and ACD one-dimensional NMR processor (version 12, ACD, Toronto, Canada).

3 Methods

3.1 Solution-State ^1H NMR Spectroscopy

1. Metabolites were extracted from cardiac tissue (30–100 mg wet weight) using a methanol–chloroform–water (2:2:1) procedure. 600 μL of methanol–chloroform mix (2:1 v:v) were added to the cells, and samples were sonicated for 15 min at room temperature (*see* **Note 1**).
2. 200 μL each of chloroform and water were added, the samples centrifuged, and the separated aqueous and lipid phases collected. The procedure was repeated twice, and the aqueous and lipid fractions from each extraction pooled.
3. The aqueous layer was dried overnight in an evacuated centrifuge.
4. The dried aqueous fractions were rehydrated in 600 μL D_2O, containing 0.05 mM TSP as an internal chemical shift reference and pipetted into 5 mm high-resolution NMR tubes. The samples were inserted on a stream of nitrogen gas into a 11.8 T superconducting magnet interfaced with an AVANCE II+ NMR spectrometer operating at 500.13 MHz for the ^1H frequency and 125.72 MHz for the ^{13}C frequency (Bruker, Germany) using a 5 mm inverse triple-resonance TXI probe.
5. Spectra were collected using a solvent suppression pulse sequence based on a one-dimensional nuclear Overhauser effect spectroscopy (NOESY) pulse sequence to saturate the residual ^1H water signal and improve the flatness of the baseline (relaxation delay = 2 s, t_1 increment = 3 μs, mixing time = 150 ms, solvent presaturation applied during the relaxation time and the mixing time). 128 transients were collected into 16k data points over a spectral width of 12 ppm at 27 °C. Spectra were collected using the software TopSpin (version 2.3, Bruker, Germany).
6. In addition, representative samples of each data set were also examined by two-dimensional correlation spectroscopy (COSY), using a standard pulse sequence (cosygpprqf) and 0.5 s water presaturation during the relaxation delay, 8 kHz spectral width, 2048 data points, 32 scans per increment, and 512 increments (*see* **Note 2**).
7. Assignment of the peaks was done using the COSY spectra in conjunction with reference to previous literature and databases and the Chenomx spectral database contained in Chenomx NMR Suite 7.7 (Chenomx, Alberta, Canada).
8. NMR spectra were processed using the ACD one-dimensional NMR processor (version 12, ACD, Toronto, Canada). Free induction decays were Fourier transformed following multiplication by a line broadening of 1 Hz and referenced to TSP at

0.0 ppm. Spectra were phased and baseline corrected, manually.

9. Each spectrum was integrated using 0.02 ppm integral regions between 0.5 and 4.3 and 4.7–8.5 ppm. The spectral region for each sample was scaled such that the total sum of integrals for each sample was equal. Alternatively, the integrals of the different metabolites were obtained using Chenomx. The data is then ready for subsequent processing using univariate or multivariate statistics.

3.2 Intact Tissue HRMAS ^1H NMR Spectroscopy

1. Samples of cardiac tissue (5–10 mg in wet weight) were washed in D_2O to remove any blood on the tissue and placed into a zirconium oxide MAS rotor along with 10 μL of D_2O (deuterium lock reference) containing 10 mM TSP (chemical shift reference). The rotor was placed in a HRMAS probe and spun at 5 kHz (*see* **Note 3**).
2. HRMAS ^1H NMR spectra were acquired using a 11.8 T superconducting magnet interfaced with an AVANCE II+ NMR spectrometer operating at 500.13 MHz for ^1H frequency, interfaced with a high-resolution MAS probe (Bruker, Rheinstetten, Germany) at 4 °C. The probe was chilled using nitrogen gas passed through a cooling unit.
3. Spectra were acquired using a conventional solvent suppressed pulse/acquire sequence based on the nuclear Overhauser spectroscopy pulse sequence to suppress baseline artefacts resulting from B_0 and B_1 field inhomogeneities (relaxation delay—$\pi/2$—t_1—$\pi/2$—t_m—$\pi/2$ TR, 2 s; SW, 10 kHz; 32,000 data points, solvent suppression applied during the preparation time (t_1) of 4 μs and mixing time of 150 ms).
4. Spectra were also acquired with a Carr–Purcell–Meiboom–Gill (CPMG) pulse sequence using a TE of 40 ms (spin-echo delay, 500 μs; total number of spin echoes, 40; other parameters identical to those described for the nuclear Overhauser spectroscopy presaturation pulse sequence) (*see* **Note 4**).
5. Spectra were processed using the software and details described in Subheading 3.1 above.

3.3 High-Resolution ^1H NMR Spectroscopy of Blood Plasma

1. Blood plasma (200 μL) was diluted in D_2O containing 0.9% NaCl and 0.05 mM TSP as an internal chemical shift reference. The samples were vortex mixed and spun using a minicentrifuge.
2. Samples were pipetted into 5 mm high-resolution NMR tubes and inserted into a 11.8 T superconducting magnet interfaced with an AVANCE II+ NMR spectrometer operating at 500.13 MHz for ^1H frequency, and an inverse triple-resonance TXI probe (Bruker, Rheinstetten, Germany) at 27 °C.

3. Spectra were collected using a solvent suppression pulse sequence based on a one-dimensional nuclear Overhauser effect spectroscopy (NOESY) pulse sequence to saturate the residual ^{1}H water signal and improve the flatness of the baseline (relaxation delay = 2 s, t_1 increment = 3 μs, mixing time = 150 ms, solvent presaturation applied during the relaxation time and the mixing time). 128 transients were collected into 16k data points over a spectral width of 12 ppm at 27 °C. Spectra were collected using the software TopSpin (version 2.3, Bruker, Germany).
4. Spectra were also acquired with a CPMG pulse sequence using a TE of 40 ms (spin-echo delay, 500 μs; total number of spin echoes, 40; other parameters were identical to those described for the nuclear Overhauser spectroscopy presaturation pulse sequence).
5. Spectra were processed using the software and details described in Subheading 3.1 above.

4 Notes

1. The extraction procedure used in this chapter is based on the Folch and Bligh-Dyer extraction protocols. It is particularly useful if one wants to use the organic fraction for lipid analysis either by gas chromatography or liquid chromatography mass spectrometry.
2. To aid metabolite identification or to examine metabolite labeling, ^{13}C NMR spectroscopy can be performed. For inverse probes, where the ^{1}H coil is closest to the sample, spectra with better signal to noise are often collected using inverse spectroscopy (i.e., the ^{13}C coupling is detected through the ^{1}H signal). One such pulse sequence is the heteronuclear single-quantum coherence (HSQC) two-dimensional pulse sequence. For this experiment, the dried aqueous fractions were resuspended in 600 μL D_2O, containing 0.05 mM TSP as an internal standard for ^{1}H-NMR and 0.5 mM [^{13}C]formate (Isotec-Sigma Aldrich, St. Louis, MO, USA) for ^{13}C-NMR. Spectra were acquired using the Bruker pulse sequence hsqcetgpsp.3 (phase-sensitive gradient edited-2D HSQC, using adiabatic pulses for inversion and refocusing and background signals effects). 2048 data points were acquired over 12 ppm spectral width (acquisition time 170 ms) in the F2 dimension, using 64 scans with 2 s delay. For spectral width of 180 ppm, 256 increments were acquired in F1 (acquisition time 5.6 ms) that resulted in a total experimental time of ~10 h per sample.

3. For HRMAS ^{1}H NMR spectroscopy, a useful "rule of thumb" is to spin the samples at ten times the ^{1}H frequency of the magnet (ignoring the MHz units). So, for a 500 MHz NMR magnet, a spin rate of 5000 Hz would ensure the spinning side bands are outside the region of the NMR spectrum containing metabolite resonances.
4. For HRMAS ^{1}H NMR spectroscopy, the CPMG pulse sequence is commonly used to attenuate resonances from lipids (which often give broad resonances) and improve the detection of resonances from aqueous metabolites by virtue of the difference in T_2 relaxation values.

Acknowledgments

This work was funded by the Medical Research Council (Lipid Profiling and Signaling, MC UP A90 1006, and Lipid Dynamics and Regulation, MC PC 13030).

References

1. Heather LC, Wang X, West JA, Griffin JL (2012) A practical guide to metabolomic profiling as a discovery tool for human heart disease. J Mol Cell Cardiol 55:2–11
2. Cross HR, Radda GK, Clarke K (1995) The role of Na^{+}/K^{+} ATPase activity during low flow ischemia in preventing myocardial injury: a ^{31}P, ^{23}Na and ^{87}Rb NMR spectroscopic study. Magn Reson Med 34(5):673–685
3. Yu X, White LT, Alpert NM, Lewandowski ED (1996) Subcellular metabolite transport and carbon isotope kinetics in the intramyocardial glutamate pool. Biochemistry 35(21):6963–6968
4. Griffin JL, Williams HJ, Sang E, Clarke K, Rae C, Nicholson JK (2001) Metabolic profiling of genetic disorders: a multitissue (1)H nuclear magnetic resonance spectroscopic and pattern recognition study into dystrophic tissue. Anal Biochem 293(1):16–21
5. Mayr M, Yusuf S, Weir G, Chung YL, Mayr U, Yin X et al (2008) Combined metabolomic and proteomic analysis of human atrial fibrillation. J Am Coll Cardiol 51(5):585–594
6. Gulston MK, Rubtsov DV, Atherton HJ, Clarke K, Davies KE, Lilley KS et al (2008) A combined metabolomic and proteomic investigation of the effects of a failure to express dystrophin in the mouse heart. J Proteome Res 7 (5):2069–2077
7. Atherton HJ, Gulston MK, Bailey NJ, Cheng KK, Zhang W, Clarke K et al (2009) Metabolomics of the interaction between PPAR-alpha and age in the PPAR-alpha-null mouse. Mol Syst Biol 5:259. https://doi.org/10.1038/msb.2009
8. Griffin JL, Williams HJ, Sang E, Nicholson JK (2001) Abnormal lipid profile of dystrophic cardiac tissue as demonstrated by one- and two-dimensional magic-angle spinning (1)H NMR spectroscopy. Magn Reson Med 46 (2):249–255

Chapter 12

Breast Cancer Metabolomics Using NMR

Uma Sharma and Naranamangalam R. Jagannathan

Abstract

Continued progress is being made in understanding the breast cancer metabolism using analytical magnetic resonance (MR)-based methods like nuclear magnetic resonance (NMR) and in-vivo MR spectroscopy (MRS). Analyses using these methods have enhanced the knowledge of altered biochemical pathways associated with breast cancer progression, regression, and pathogenesis. Comprehensive metabolic profiling of biological samples like tissues, cell lines, fine needle aspirate, and biofluids such as sera and urine enables identification of new biomarkers and abnormalities in biochemical pathways. These methods are not only useful for diagnosis, therapy monitoring, disease progression, and staging of cancer but also for the identification of new therapeutic targets and designing new treatment strategies. Additionally, in-vivo MRS studies have established choline-containing compounds (tCho) as biomarkers of malignancy, which is useful for enhancing the diagnostic specificity of magnetic resonance imaging (MRI). Recent technological developments related to in-vivo MRS such as increased magnetic field strength, multichannel phased array breast coils, and absolute quantification of tCho have provided a better understanding of the tumor heterogeneity, metabolism, and pathogenesis. This chapter focuses on providing the experimental aspects of in-vitro, ex-vivo, and in-vivo MR spectroscopy methods used for metabolomics studies of breast cancer.

Key words In-vitro NMR, Ex-vivo NMR, In-vivo MRS, Metabolomics, Breast cancer, Choline, Perchloric acid extraction, Acetonitrile extraction, Chloroform-methanol extraction

1 Introduction

Despite significant progress in the diagnosis and assessment of breast cancer, it remains a major health care challenge for women worldwide. Early diagnosis of breast cancer is critical for improving the quality of life and survival of patients. Over the last few years, several studies have applied NMR-based metabolomics approaches using malignant tissue, sera, nipple aspirate, and cell lines to discover biomarkers for the diagnosis and therapy evaluation in breast cancer [1–9]. Metabolomics provides a detailed description of the metabolic profile of a living system, which reflects the changes due to pathological, physiological, genetic, and environmental stresses.

Commonly used techniques for metabolomics studies are NMR spectroscopy and mass spectroscopy. Mass spectroscopy is

G. A. Nagana Gowda and Daniel Raftery (eds.), *NMR-Based Metabolomics: Methods and Protocols*, Methods in Molecular Biology, vol. 2037, https://doi.org/10.1007/978-1-4939-9690-2_12, © Springer Science+Business Media, LLC, part of Springer Nature 2019

generally used in combination with chromatography techniques like liquid chromatography or gas chromatography. NMR spectroscopy provides insights into the metabolic profiles of a wide variety of biological samples such as tissue specimens, fine needle aspirates (FNA), fine needle biopsy (FNB), and biofluids such as blood plasma/sera, urine, saliva, cerebrospinal fluid, etc. With the availability of high-field automated NMR spectrometers equipped with cryo-probes and nano-probe technologies, it is now possible to acquire data from samples with micromolar or even lower concentration metabolites in a short time, of the order of a few minutes. NMR also enables the determination of absolute concentrations of metabolites. An ex-vivo NMR technique, high-resolution magic angle spinning (HRMAS)-based NMR has the potential to provide metabolic profiles of intact tissue and cell lines [6–9]. A flowchart describing the details of the methodology of in-vitro and ex-vivo NMR spectroscopy is shown in Fig. 1. Figure 2 presents the representative one-dimensional (1D) ^{1}H NMR spectrum acquired from the perchloric acid extract of a malignant tissue obtained surgically from a breast cancer patient. Resonances corresponding to several metabolites including amino acids, organic acids, sugars, energy metabolites, and membrane components are assigned in the spectrum.

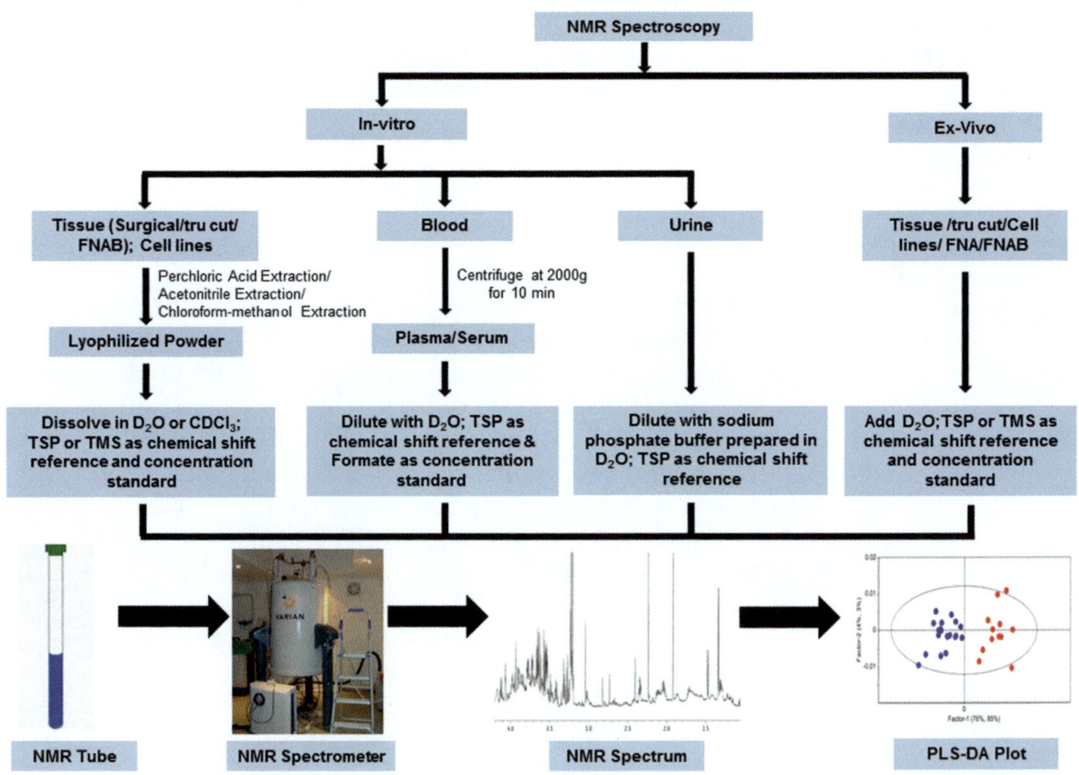

Fig. 1 Flowchart depicting major steps involved in NMR spectroscopy-based metabolomics of breast cancer

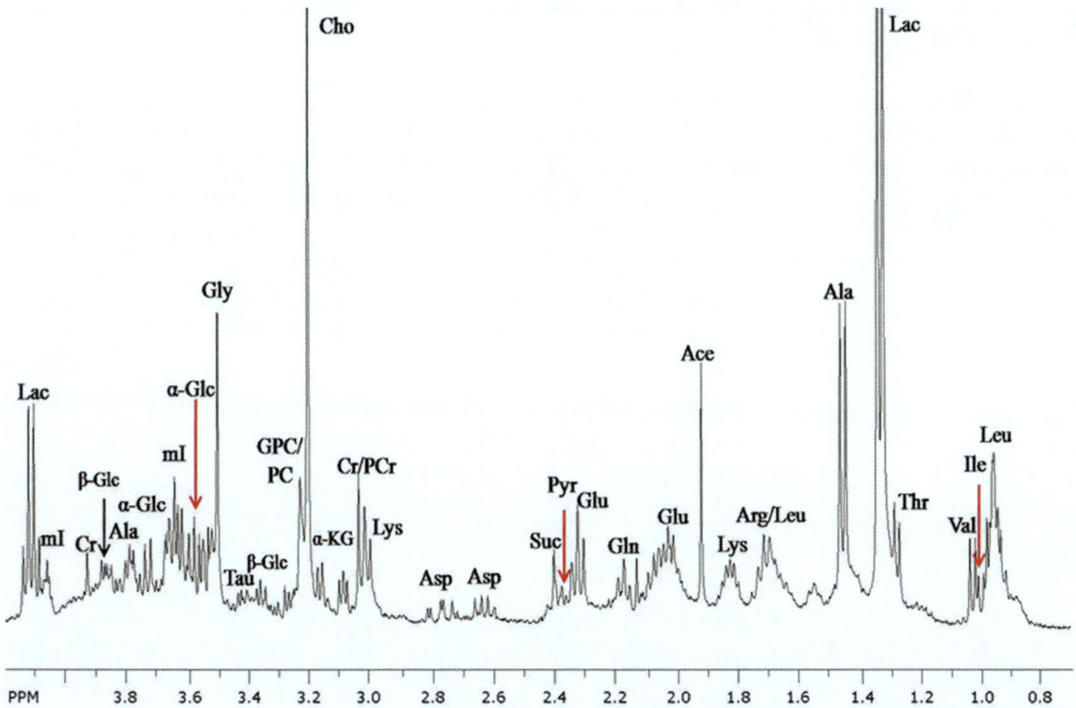

Fig. 2 A portion of the in-vitro ^{1}H NMR spectrum of a breast cancer tissue extract recorded at 400 MHz. *Ala* Alanine, *Ace* Acetate, *Arg* Arginine, *Asp* Aspartate, *Cho* Choline, *Cr* Creatine, *Glc* Glucose, *Glu* Glutamate, *Gln* Glutamine, *GPC* Glycerophosphocholine, *Gly* Glycine, *Ile* Isoleucine, α-*KG* α-Ketogultarate, *Lac* Lactate, *Leu* Leucine, *Lys* Lysine, *mI* Myo-inositol, *PCr* Phosphocreatine, *PC* Phosphocholine, *Pyr* Pyruvate, *Suc* Succinate, *Tau* Taurine, *Thr* Threonine, *Val* Valine (Reproduced from ref. 1)

Localized in-vivo ^{1}H MRS is another approach that provides information on biochemicals from a selected volume of interest (VOI) or voxel from any organ of the human body noninvasively. Several in-vivo MRS studies have documented that malignant breast tumors are characterized by high levels of choline-containing metabolites (tCho) [10–19]. The in-vivo observation of a unique peak at 3.22 ppm corresponding to tCho has been documented as a biomarker for the discrimination of malignancy from benign lesions and normal breast tissues [10–19]. Recent developments include the quantitative estimation of tCho concentrations in-vivo and cut-off values determined for the diagnosis of breast cancer [12]. In-vivo monitoring of the tCho signal following therapy in breast cancer patients has been demonstrated to have the potential for predicting the response of the tumor to various therapeutic regimens [11, 17–19]. Although in-vivo breast MRS has shown promising results, the complexity of acquisition procedures necessitates further advances for routine clinical use.

This chapter describes the methodology of in-vitro, ex-vivo, and in-vivo studies of breast cancer metabolomics using NMR. The readers may also refer to the article describing the experimental procedure for NMR spectroscopy of tissues and biofluids [20].

2 Materials

2.1 Materials for In-Vitro NMR Spectroscopy

2.1.1 Chemicals

1. Deuterium oxide (D_2O) (99.99%), 3-trimethylsilyl propionic-2,2,3,3-d4 acid sodium salt 98% D (TSP), deuterated tetramethylsilane (TMS), perchloric acid ($HClO_4$), formate (HCOOH), chloroform ($CHCl_3$), methanol (CH_3OH), acetonitrile (CH_3CN), deuterated chloroform ($CDCl_3$, 99.96 atom% D), deuterated methanol-d4 99.8% (CD_3OD), sodium azide (NaN_3), sodium hydrogen phosphate (Na_2HPO_4), sodium dihydrogen phosphate (NaH_2PO_4). These chemicals should be of analytical grade and may be purchased from standard companies like Sigma-Aldrich, Merck, etc.
2. Phosphate buffer: Prepare phosphate buffer (pH = 7.4) by adding 288.5 mg Na_2HPO_4 with 52.5 mg NaH_2PO_4 in 10 mL of D_2O.

2.1.2 Biological Samples

1. Tissue (tru cut biopsy or surgical).
2. Cell lines.
3. Fine needle aspiration (FNA).
4. Fine needle biopsy (FNB).
5. Blood plasma/sera.
6. Urine.

2.1.3 Equipment

1. 400 MHz or higher field NMR spectrometer (Bruker, Agilent Technologies, Jeol, etc.).
2. NMR probe (5 mm broad-band inverse (BBI) or equivalent, dual $^{1}H/^{13}C$, cryo-probe; nano-probe) or flow-injection probe for automated sampling (*see* **Note 1**).
3. 5 mm NMR tubes. Use 2 mm NMR tubes for low volume (*see* **Note 2**).
4. Tissue homogenizer.
5. Deep freezer (−40 °C or lower).
6. 4 °C refrigerator.
7. Refrigerated centrifuge.
8. pH meter.
9. Vortexer.
10. Vacuum concentrator.
11. Lyophilizer.

2.1.4 Accessories for Automation

1. Autosampler (*see* **Note 3**).
2. Gilson or similar flow-injection system.
3. Eppendorf tubes.
4. Well plate (96 wells or similar).

2.2 Materials for Ex-Vivo NMR Spectroscopy

1. Sample rotor.
2. NMR Spectrometer (400 MHz or higher field) with HR-MAS probe.
3. Deep freezer (−40 °C or lower).
4. 4 °C refrigerator.

2.2.1 Biological Samples

1. Tissue (tru cut biopsy/surgical).
2. Cell lines.
3. FNB.

2.3 Materials for In-Vivo MR Spectroscopy

2.3.1 Subjects

1. Breast cancer patients.
2. Patients with benign lesion.
3. Healthy women volunteers.

2.3.2 Equipment

1. Human MR Scanner (1.5 T, 3 T, 4 T, 7 T) from Siemens, Phillips, GE, etc.
2. Breast coil (double breast multichannel phased array or single coil).
3. Phantoms (for optimization of parameters).

3 Methods

3.1 Sample Preparation for In-Vitro NMR Spectroscopy

3.1.1 Tissue Sample Collection

Tissue samples are collected (tru cut biopsy, surgery or FNB), immediately snap frozen in liquid nitrogen to arrest further metabolism and stored frozen at (−40° or lower) until further analysis (*see* **Note 4**).

3.1.2 Tissue Extraction

Choose the appropriate method based on the requirement.

1. Water-soluble metabolites are extracted using perchloric acid or acetonitrile.
2. Both lipid-soluble and hydrophilic metabolites are extracted using a chloroform and methanol extraction procedure.

3.1.3 Perchloric Acid Extraction of Water-Soluble Metabolites

1. Weigh frozen tissue and homogenize in 6% $HClO_4$ (5 mL/gm) using ultrasonicator. Alternatively, the tissue sample can be pulverized in a mortar pestle with liquid N_2 and then homogenized in ice-cold 6% $HClO_4$ to a paste consistency (*see* **Note 4**).
2. Centrifuge the homogenate at 10,000 × *g* at 4 °C.

3. Collect the supernatant and adjust its pH to 7.0 by using ice-cold 3 M potassium hydroxide. Alternatively, ice-cold 2 M K_2CO_3 can be used for neutralization of the supernatant.
4. After neutralization (pH adjustment), remove the precipitate containing perchlorate salts by centrifuging at 10,000 × *g* at 4 °C for 10 min.
5. Collect the supernatant and lyophilize.
6. Store the lyophilized tissue powder in a freezer (−40 °C or lower) until NMR analysis.

3.1.4 Acetonitrile Extraction of Water-Soluble Metabolites

1. Prepare ice-cold acetonitrile and water solvent.
2. Homogenize the weighed frozen tissue in ice-cold CH_3CN/H_2O 1:1, v/v (5 mL/g tissue) using ultrasonicator or mortar pestle.
3. Centrifuge the homogenate at 12,000 × *g* for 10 min at 4 °C.
4. Collect the supernatant and lyophilize.
5. Store the lyophilized powder at −40 °C or lower until used for NMR spectroscopy.

3.1.5 Chloroform Methanol Extraction of Both Lipids and Water-Soluble Metabolites

1. Weigh frozen tissue and homogenize with ice-cold solvents, CH_3OH (4 mL/g tissue) and water (0.85 mL/g tissue).
2. Vortex the homogenate and add $CHCl_3$ (2 mL/g tissue).
3. Add 2 mL of $CHCl_3$ and 2 mL of H_2O per g tissue weight and vortex the homogenate again.
4. Keep homogenate on ice or in the refrigerator for 15 min.
5. Centrifuge the homogenate at 1000 × *g* for 15 min at 4 °C.
6. Two separate phases are formed: the lower $CHCl_3$ phase containing lipids and the upper CH_3OH/water phase containing water-soluble metabolites.
7. Transfer these two layers into separate glass vials.
8. Remove solvents by either using a nitrogen gas stream or vacuum concentrator. Store dried samples at low temperature (−40 °C or lower) until used for NMR.

3.1.6 Solutions for NMR Spectroscopy

Dissolve the lyophilized tissue powder in 600 μL of suitable deuterated solvent and transfer this solution into the 5 mm NMR tubes.

1. For ^{1}H NMR studies of water-soluble molecules add D_2O as a solvent and TSP (0.5 mM) as a reference for both chemical shift calibration and concentration determination (*see* **Note 5**).
2. For hydrophobic molecules, use $CDCl_3$ containing TMS as the solvent. TMS is used as a chemical shift reference.

3.2 Extraction of Metabolites from Cell Lines

3.2.1 Perchloric Acid Extraction

1. Wash the cells three times with phosphate buffer and add 12% ice-cold $HClO_4$.
2. Gently scrape the cells using rubber scrapers, collect the cells and sonicate on the ice twice for 30 s each.
3. Centrifuge cell lysates at 3000 × *g* at 4 °C for 20 min.
4. Collect supernatant in ice-cold tubes.
5. Neutralize the supernatant using ice-cold 3 M KOH and adjust pH to 7.0.
6. Centrifuge this mixture to remove perchlorate salt at 3000 × *g* at 4 °C for 20 min.
7. Collect the supernatant, lyophilize and store at −40 °C or lower until NMR spectroscopy.

3.2.2 Chloroform-Methanol Extraction

1. Wash cells with ice-cold water and add a mixture of ice-cold $CHCl_3/CH_3OH$ (9:1, v/v).
2. Lyse the cells by keeping the mixture at −40 °C or lower for 5 min and thaw at room temperature.
3. Centrifuge the lysates at 10,000 × *g* for 5 min.
4. Collect the supernatant and dry using vacuum concentrator.
5. Add 600 μL of D_2O and TSP (0.5 mM) to the dried sample for NMR spectroscopy.

3.3 Preparation of FNA Sample

1. Collect fine needle aspirate samples in polypropylene vial containing phosphate buffer saline prepared in D_2O and immediately freeze at −40 °C or lower until NMR spectroscopy.
2. Before NMR spectroscopy, thaw the FNA solution and transfer to an NMR tube and add 0.5 mM TSP for chemical shift referencing.

3.4 Preparation of Biofluids

3.4.1 Blood Plasma/Serum

1. Use heparin vacutainer for collection of blood plasma.
2. Collect the blood sample in a pre-chilled vacutainer during the morning pre-prandial period after overnight fasting. Collection of 1.0 mL of blood is sufficient for NMR studies.
3. Separate blood plasma by centrifugation of the blood at 2000 × *g* for 15 min at 4 °C and freeze (−40 °C or lower) until NMR analyses.
4. Use serum separating tubes for collection of serum. After collection, keep the blood at room temperature for 30–45 min to clot.
5. Collect supernatant serum obtained after centrifugation of the coagulate at 2000 × *g* for 15 min at 4 °C and freeze at −40 °C or lower until NMR analyses.

6. To perform ^{1}H NMR spectroscopy, add D_2O (400 μL), TSP (0.5 mM) and sodium formate (0.5 mM) to the blood plasma (200 μL) or serum sample (200 μL) and transfer to 5 mm NMR tube. Alternatively, blood samples can be transferred to 96-well plates, if using autosampler acquisition robotic technology.
7. TSP (δ 0.0 ppm) is used as a chemical shift reference while formate (δ 8.46 ppm) is used as a standard for determination of metabolite concentrations.

3.4.2 Urine

1. Collect urine under sterile conditions in the morning pre-prandial period after overnight fasting.
2. Place on ice and immediately freeze in liquid nitrogen and store at −40 °C or lower.
3. To perform ^{1}H NMR experiments, add 200 μL sodium phosphate buffer prepared in D_2O to the 400 μL of urine and adjust the pH to 7.4 (*see* **Note 6**).
4. To inhibit bacterial growth in the solution, add sodium azide (3 mM).
5. Add TSP (0.5 mM) for chemical shift referencing and quantification of concentration.

3.5 ^{1}H NMR Spectroscopy

3.5.1 Tissue Extract, Cell Extracts, FNA, and Biofluids

1. NMR spectrometer operating at frequencies of 400 MHz or higher can be used depending on the availability.
2. Transfer sample (600 μL) into a 5 mm NMR tube, keep the tube inside the probe of NMR instrument and allow it to equilibrate for 2–3 min.
3. Set the temperature of the probe to 25 °C.
4. Tune and match the NMR probe for each sample, carry out shimming and check the line width of the water resonance (*see* **Note 7**).
5. Perform 90° pulse calibration for each type of sample, separately.
6. For tissue/cell extracts, optimize power level for suppression of water signal and acquire 1D spectrum (*see* **Note 8**) using the one pulse (90° pulse) sequence with pre-saturation of water. Typical parameters used for 1D ^{1}H NMR at 700 MHz are: spectral width of 9000 Hz; 32 K data points, 32–48 scans depending on the concentration of the sample. Water suppression is used to enable detection of low concentration metabolites; usually pre-saturation is used to suppress the water signal.
   (a) Optimized relaxation delay should be used for each type of sample for the determination of absolute calculations. We recommend a relaxation delay of 14 s, which was

optimized for 700 MHz. Most metabolomics studies in the literature have used a relaxation delay of 2–5 s and such NMR data were used for multivariate analysis after binning.

7. For plasma/serum, ^1H NMR spectroscopy is performed using the 1D Carr–Purcell–Meiboom–Gill (CPMG) sequence with pre-saturation. The acquisition parameters for the 1D CPMG experiment used at 700 MHz spectrometer are: spectral width: 9000 Hz; data points: 32 K; number of scans: 64; spin echo delay $\tau = 15$ ms and a relaxation delay = 70 s.
   (a) Use longer relaxation delay for plasma/serum samples when formate is used as a concentration standard, since it typically has a long T_1 relaxation time. At 700 MHz, we recommend 70 s.
8. For urine samples, generally 1D spectrum with water suppression is acquired with a single 90° pulse with the following parameters: 32–64 scans, 9000 Hz spectral width, and 32 K data point.
9. 1D NOESY pulse sequence with pre-saturation is another useful experiment for the blood plasma and urine samples.
10. If necessary, perform two-dimensional correlation spectroscopy (COSY) and total correlation spectroscopy (TOCSY) using the following parameters (for 700 MHz spectrometer): 2 K data points, 9000 Hz spectral width, 2 s relaxation delay, 256 experiments (time domain points) collected in t_1 dimension, and 32–64 scans (*see* **Note 8**). For TOCSY experiments a mixing time of 80 ms is used.
11. The data are usually processed on a workstation using the software provided by the instrument manufacturer or other software such as Chenomx and MNOVA. For processing of the NMR data, multiply the FID by an exponentially decaying function using a line broadening of 0.3 Hz, zero-fill to 64 K data points and perform Fourier transformation. Chemical shifts of metabolites resonances are referenced to TSP/TMS at 0.0 ppm.

3.5.2 Ex-Vivo NMR Spectroscopy

1. Place the tissue (surgical/tru cut/biopsy/cell line) in the HR-MAS sample rotor.
2. Insert rotor into the probe and align rotor with the magic angle (54.7°).
3. Spin the sample at the magic angle.
4. Perform NMR experiments using the pulse sequences used for in-vitro NMR studies (*see* above Subheading 3.5.1).

3.6 Data Analysis

3.6.1 Quantification of Metabolites

The concentration of a metabolite is directly proportional to the peak intensity and the number of protons that the peak represents in ^{1}H NMR spectrum. A comparison of the integrated intensity of the resonance peak from a metabolite with the integrated intensity of the standard compound provides the concentration of the metabolite using the following formula:

$$\text{For tissue sample, } C_A = C_B \times (N_B/N_A) \times (A_A/A_B) \times (V_S/W_S)$$

$$\text{For blood and urine, } C_A = C_B \times (N_B/N_A) \times (A_A/A_B) \times \text{Dilution factor}$$

where C_A = metabolite concentration to be calculated; C_B = concentration of reference standard; N_B = number of protons that the standard peak represents; N_A = number of protons contributing to the peak of the metabolite to be determined; A_B = integrated peak area of the standard; A_A = integrated peak area of the metabolite; V_S = volume of the sample; W_S = wet weight of the tissue sample.

3.6.2 Statistical Analysis: Univariate and Multivariate Methods

Proton NMR spectra of biological samples contain hundreds of resonances due to the presence of a large number of metabolites. Both univariate and multivariate statistical methods are used to obtain biological information from these datasets [21].

1. Univariate analyses: This is the simplest statistical method for data analyses, where a single variable across samples is compared at a time. Usually parametric methods like Student's *t*-test for two groups and analyses of variance (ANOVA) for comparison among multiple groups are employed to determine the significance of findings. Similarly, nonparametric versions like the Wilcoxon rank sum test for two samples and the Kruskal–Wallis test are used with appropriate correction methods for multiple testing.
2. Multivariate analyses: This method analyzes the spectral dataset that contains large numbers of variables by applying methods such as data reduction, pattern recognition, and clustering. Two major types of multivariate analyses methods used are unsupervised and supervised:
   (a) Unsupervised algorithms include principal component analyses (PCA). Here, the number of variables is reduced by the transformation of the high dimension dataset into a low-dimensional set of principal components.
   (b) Partial least squares regression and discriminant analyses (PLS-DA), multiple regressions, support vector machines, etc., are supervised methods. Cluster analysis is used for classifying the data with the same features so that they tend to form the same cluster. Further details are available in the review on the multivariate data analyses applied in metabolomics studies [21]. PLS-DA can have good

predictive power and it predicts a set of dependent variables derived from a set of factors. A typical PLS-DA score plot demonstrating separate clusters of two groups of samples is shown in Fig. 1.

3.7 In-Vivo MR Spectroscopy

3.7.1 Single Voxel MR Spectroscopy (SVS)

1. In-vivo MRS described here is performed at 1.5 T using a 4-channel phased array breast matrix receive coil (*see* **Note 9**).
2. Patient positioning: Usually the patient is positioned prone in the human MR scanner with the breast being placed in each cup of the double volume coil (Fig. 3). A single breast coil can also be used (*see* **Note 10**).
3. Use soft cushions for compressing the breast to reduce motion.
4. The body coil of the MRI scanner is used as the transmitter coil while the dedicated double breast multichannel phased array volume coil serves as a receiving coil.
5. Lesion localization using MR images: Perform localizer imaging followed by T_1- and T_2-weighted imaging with fat suppression in three orthogonal planes (axial, sagittal, and coronal). Use these images for visualization of the lesion and accurate

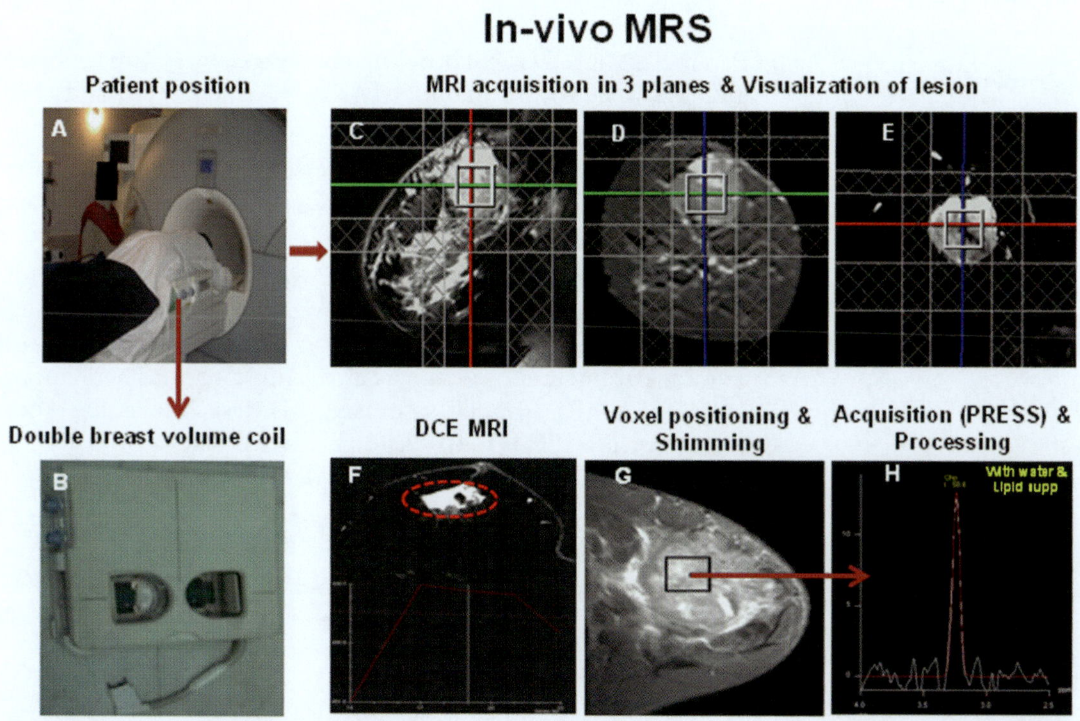

Fig. 3 Flowchart depicting the various steps of the methodology of in-vivo experiments on breast cancer patients, patient position in MR scanner (**A**), double breast volume coil (**B**), MR images in three orthogonal planes showing voxel location (**C–E**), DCE MRI (**F**), voxel positioning and shimming (**G**) and acquisition and processing of ^{1}H MR spectrum showing the tCho peak (**H**)

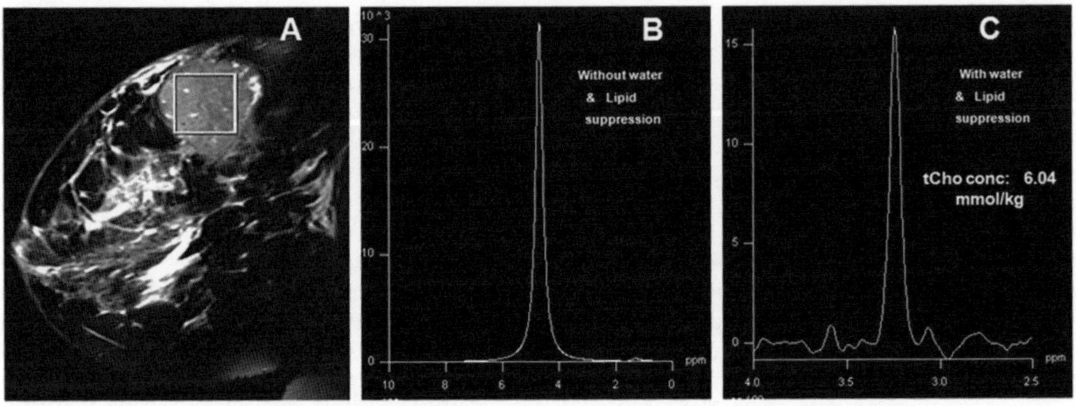

Fig. 4 (**A**) MRI of a patient with locally advanced breast cancer showing the voxel position from which the single-voxel (SV) ^{1}H MR in-vivo spectrum was obtained without water and fat (lipid) suppression (**B**) and with the suppression of water and fat (lipid) resonances (**C**) (Reproduced with permission from John Wiley and Sons, Inc. from ref. 19)

positioning of the voxel (Fig. 3). Figure 4 shows the representative ^{1}H MR spectrum (without and with water and fat suppression) acquired from a patient with infiltrating ductal carcinoma along with MR images used for positioning of the voxel.

6. Parameters used for short inversion recovery images are TR/TE = 6940/58 ms; matrix size = 320 × 256; and slice thickness = 3 mm.

7. Parameters used for fat-suppressed MR images are: TR/TE = 6270/102 ms; slice thickness = 3 mm and no gap; and matrix size = 512 × 440.

8. Dynamic contrast-enhanced (DCE) MR imaging is used for voxel placement, especially in small-sized tumors for clear visualization of tumor and demarcation of the tumor margins. Typical parameters used for axial dynamic contrast-enhanced MRI DCEMRI are: fat-saturated 3D fast low angle shot sequence; TR/TE = 5.46/2.53 ms; matrix size = 305 × 448; flip angle 12°; and slice thickness = 1.4 mm with no gap.

9. Use either ionic or nonionic contrast agents for clear visualization of tumor margins to distinguish between benign and malignant breast lesions. Typical dose of the contrast gadodiamide (GdDTPA-BMA) is 0.1 mmol/kg. It is given at the rate of 2 mL/s using an automatic injector followed by saline flush.

10. In DCE imaging, a series of MR images are acquired as a function of time. Typically, one pregadolinium series of images followed by 5 postgadolinium image series are acquired in a total acquisition time of 5.5 min (6 × 55 s).

11. Pulse sequences: Two pulse sequences, the point-resolved spectroscopy (PRESS) and stimulated echo acquisition mode

(STEAM) are used for localized MR spectroscopy. However, the PRESS sequence is more commonly used.

12. Typical parameters used for SVS in-vivo ^{1}H MRS are: PRESS pulse sequence with TR = 1500 ms; TE = 100 ms; data size = 1024; spectral width = 1000 Hz; averages = 128; and a total acquisition time of 3.18 min [12].

13. Use long echo times (135 ms) for better detection of tCho (*see* **Note 11**).

14. Positioning of voxel: MR images are used as a reference for positioning the voxel within the lesion. Voxel size can be varied depending on the tumor volume (Fig. 3). Fat and necrotic regions should be avoided while positioning the voxel. Sensitivity of detection of the tCho peak depends on the voxel size and its position (*see* **Note 12**).

15. Shimming: Shimming is essential to increase the homogeneity of the magnetic field in the selected voxel. In SVS, global shimming followed by localized shimming is performed to achieve this goal. For the water peak, a line width of 5–15 Hz at 1.5 T is acceptable for high signal-to-noise ratio (SNR) and better spectral resolution. At 3 T, the water peak line width is in the range of 25–30 Hz.

16. Water and fat suppression: Breast tissue constitutes high concentrations of water and fat (in the range of M); therefore, it is essential to suppress their signals to detect tCho signal, which is in the range of mM.

17. Use chemical shift-selective RF pulses for suppression of the water signal. In this method, the water signal is excited with a narrow frequency band (~60 Hz) and simultaneously, "crusher" gradients are applied to dephase the excited water signal.

18. Pulse sequences such as MEGA provide simultaneous suppression of both water and fat. A pre-saturation frequency-selective pulse with a bandwidth of 50 Hz is applied to the water signal while fat is suppressed by a bandwidth of 1.8 ppm applied at the frequency range 0.4–1.8 ppm. An echo-time averaging method was reported by Bolan et al. to reduce the sideband of the fat peak [22]. Alternatively, water suppression sequences like VAPOR can be used.

19. Suppression of fats from outside the voxel region of interest is carried out using 4–6 saturation bands.

20. For quantification of tCho, an additional ^{1}H MR spectrum without water and fat suppression is acquired from the same voxel with the same TR and TE parameters.

21. Referencing: The water peak at 4.7 ppm is used as a chemical shift reference [12, 14, 15].

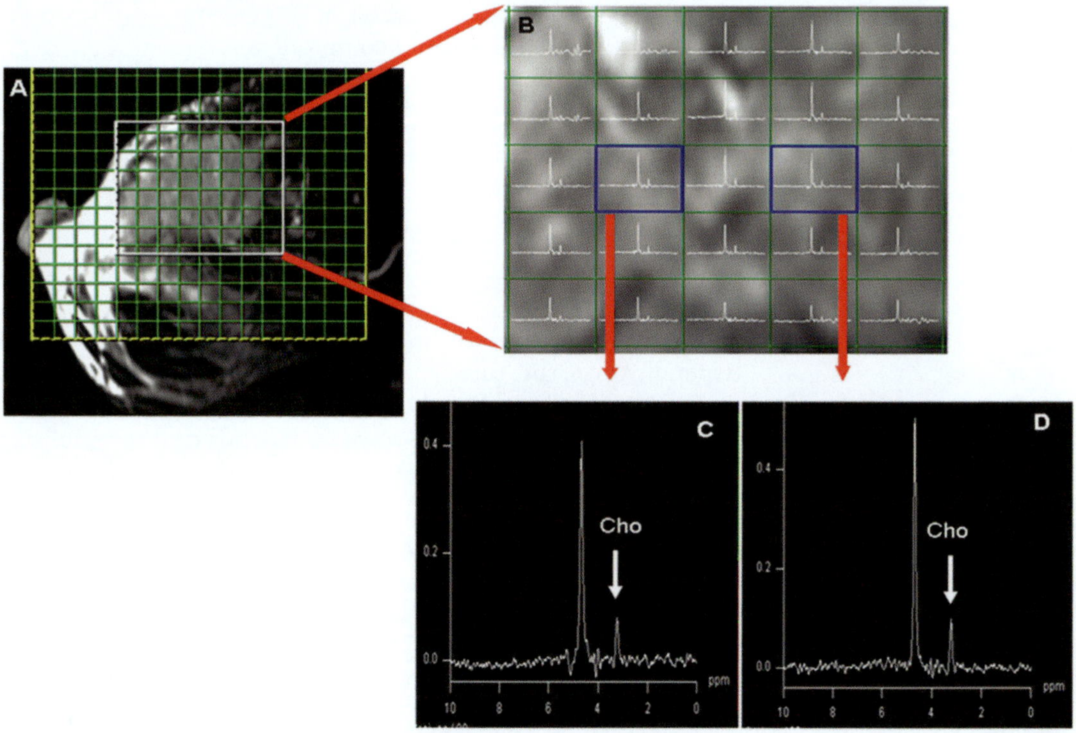

Fig. 5 (**A**) Pre-therapy sagittal T_2-weighted fat-suppressed breast image of an LABC patient with the MRSI grid. (**B**) The spectral map obtained from the tumor region. (**C**, **D**) MR spectral pattern obtained from two individual voxels that are highlighted in (**B**) (Reproduced with permission from John Wiley and Sons, Inc. from ref. 17)

3.7.2 Multi-Voxel Spectroscopic Imaging

1. ^{1}H MR spectra from multi-voxels are obtained using chemical shift imaging or spectroscopic imaging technique (MRSI) (Fig. 5). For MRSI, the PRESS sequence is used as for single voxel spectroscopy (SVS) (*see* **Note 13**).
2. Instead of a single voxel, a 2D MRSI slab which is divided into multiple small voxels is positioned on the tumor region using MR images. The spectrum is acquired from these multiple small voxels. The MRSI grid is carefully positioned so that maximum coverage of the lesion is achieved [17].
3. The steps for the acquisition of the spectrum are similar to SVS; both global and voxel level shimming are carried out. The full-width at half maximum of the water peak should be $\leq$20 Hz.
4. Typical parameters used for 2D MRSI are: TR =2000 ms and TE = 135 ms [17]. For good SNR, four averages are acquired in 9 min of acquisition time.

3.7.3 SVS Spectral Post-processing and Analyses

1. Spectral data are usually processed with the manufacturer-supplied post-processing package. Post-processing of the time domain data includes multiplication by a line broadening factor

of 2.0 Hz followed by Fourier transformation, phase correction, and baseline correction using polynomial order 5 fitting. The integrated area of tCho is calculated using a curve fitting program (i.e., integral value).

2. In MRS, the concentration of a metabolite is proportional to the amplitude of the signal. It is calculated by fitting either the time domain (TD) or the frequency domain (FD) data to an appropriate model and the goodness of fit is evaluated.
3. The ^{1}H MR spectrum with suppression of both water and fat from a patient with infiltrating ductal carcinoma shows a distinct peak at 3.22 ppm due to tCho (*see* Fig. 4). Assessment of malignancy using tCho is carried out using two approaches namely, qualitative and quantitative.
4. Qualitative assessment of tCho signal: Observation of the tCho peak is taken as the indication of malignancy in this approach [10, 11].
5. Semiquantitative assessment: In the semiquantitative assessment, the signal-to-noise ratio of the tCho signal (ChoSNR) is calculated with the formula given below [17].

$$\text{ChoSNR} = \text{tCho amplitude}/\text{RMS amplitude of noise}$$

   tCho amplitude is calculated from the peak at 3.2 ppm, noise amplitude is calculated from the spectral region that has no signal (>9 or <0 ppm) (*see* **Note 14**).

6. Quantitative assessment: Similar to that used in-vitro quantification, the concentration of tCho is obtained by comparison of the integrated area of tCho with the area of the reference signal. Corrections are applied to take into account the effects of relaxation, *J* coupling, as well as instrumental and pulse sequence imperfections. For referencing, two types of approaches are available, namely external [16] and internal referencing [12–15].
   (a) The MR spectrum is acquired from a known concentration of phosphocholine contained in a phantom placed along with the patient and is used as an external reference.
   (b) The signal intensity of phosphocholine from the phantom is compared to the signal intensity of tCho from the lesion and the absolute concentration of tCho is calculated.
   (c) In the internal referencing method, two ^{1}H MR spectra are obtained from the same VOI, one water suppressed and other unsuppressed.
   (d) The concentration of tCho is calculated by comparing the tCho integral to the integral of unsuppressed water. Additionally, corrections for receiver gain and relaxation rates

of both the water and tCho signals are carried out for accurate calculation of tCho concentration.

(e) Automated normalization with an internal water reference signal was carried out using post-processing software and the relative normalized integral of tCho is determined. The concentration of tCho is calculated using the formula given by Bolan et al. at 4.0 T [14] which was modified by Baik et al. for 1.5 T [15] (*see* **Note 15**).

$$[\text{Cho}] = \left(I_{\text{Cho}} \times 10^{-5}\right) \times 8792.78 \text{ mmol/kg}$$

where (I_{Cho}) represents the tCho peak integral.

3.7.4 MRSI Data Analyses

All MR spectra are analyzed using the software provided by the manufacturer. The SNR of the tCho peak at 3.2 ppm is calculated, while signal a free region (>9 ppm or <0 ppm) is used for calculating the noise in the spectrum [17]. All voxels that showed a choline signal ChoSNR ≥2 are considered positive with malignancy in the spectrum [17].

4 Notes

1. Cryo-probes provide an improvement in signal-to-noise ratio (SNR) by a factor of up to 4 due to increased sensitivity of the RF coil. Nano-probes are available for smaller quantities of sample.
2. For more detailed analysis, conventional NMR spectroscopy using a 5 mm diameter NMR tube is preferred. Smaller tubes with 1–3 mm can be used with micro-coil probes with sample volume as low as 10 μL.
3. Large-scale, high-throughput metabolomics studies can be performed with the robotics handling automated systems.
4. Tissue, blood, and urine samples can be stored at −70 °C for about 9 months as reported by studies and seen in our laboratory. It is important to freeze tissue biopsies as quickly as possible in liquid nitrogen while collecting and also it is important to keep thawing time of tissue as small as possible to minimize metabolic alterations due to enzymatic activity and degradation processes. Pulverized samples should not be thawed prior to the addition of perchloric acid or other solvents used for the extraction.
5. TSP serves as an internal reference for the chemical shift calibration and quantification of metabolites in tissue and urine. TSP generally binds to proteins or macromolecules present in samples and hence using its signal can deleteriously affect the

accuracy of metabolite quantitation. In our laboratory, we have optimized the concentration of 0.5 mM TSP for the quantification of metabolites in tissue and urine samples using ^{1}H NMR spectroscopy, which is consistent with the previously reported protocols [20, 23].

6. The pH of tissue extract samples and urine samples must be kept constant as it can lead to changes in chemical shifts.
7. Ensure salt concentrations in buffers used in urine samples do not affect tuning and matching of the probe. Shimming is performed to obtain a homogeneous magnetic field around the sample.
8. Generally, ^{1}H NMR spectroscopy is used in metabolomics studies. Acquisition of the 2D spectrum requires longer experiment time; 2D experiments are needed for the assignment of unknown resonances. Use of other nuclei like ^{13}C, ^{15}N or ^{31}P may be useful in the assignment.
9. In-vivo breast MRS should be carried out after about a week of core biopsy when acute inflammatory signs like edema have settled.
10. Prone position of the patient prevents the signal contamination from the chest wall and also artifacts due to breathing and chest motion.
11. A long TE reduces the fat signal due to its short T_2 and provides better signal contrast [10, 11].
12. If DCE MRI is carried out then position the voxel in the area that has maximum uptake of contrast MR images by avoiding any necrotic areas. A decrease in the tCho signal is seen due to the paramagnetic effect of contrast media (both ionic and neutral). Diffusion-weighted imaging may also be used to visualize tumor and voxel positioning.
13. SVS is less sensitive to patient motion and spectrum quality is better. Also, the quantitative analysis of the spectrum is relatively simple compared to multi-voxel spectroscopy. MRSI provides the spatial variation of tCho in normal breast parenchyma and lesions. It is important that experiments are performed on a lactate containing phantom at regular intervals and the SNR is compared to ensure the stability of the MRSI experiments. Contamination of the signal from outside the VOI occurs in MRSI due to discrete sampling. Further, shimming is difficult at larger volumes and the experimental times are longer for MRSI compared to SVS.
14. A value of ChoSNR ≥ 2 is considered indicative of malignancy using both the SV and MRSI methods [17]. In another semi-quantitative approach, the integral value of the tCho peak is used. Color-coded metabolic maps obtained through MRSI

are useful in providing a visual assessment of pathology and represent the relative levels of the metabolite.

15. When tissue water is used as an internal reference, no separate calibration experiment is required as factors like B_0 shimming effects, partial volume effects, radiofrequency transmission efficiency, and receive sensitivity are taken into account. Quantitative estimates of the tCho signal provide an objective assessment of results in sequential studies, where the therapeutic response is monitored periodically using MRS. There are many fitting programs like jMURI, LC Model, etc., that may be used for determination of the concentration. Estimation of error can be done using Cramer–Rao bounds but these are not suitable at low SNR. Alternatively, parametric or confidence interval methods can be used.

Acknowledgments

The Science and Engineering Research Board (SERB), Government of India is acknowledged for financial support (SP/S0/B27/95; SP/S0/B21/2001; SP/SO/HS-80/2006 and SR/SO/HS/213/2012) and a J.C. Bose Fellowship to NRJ.

References

1. Jagannathan NR, Sharma U (2017) Breast tissue metabolism by magnetic resonance spectroscopy. Meta 7:25
2. Bezabeh T, Ijare OB, Nikulin AE et al (2014) MRS-based metabolomics in cancer research. Magn Reson Insights 7:1–14
3. Tenori L, Oakman C, Morris PG et al (2015) Serum metabolomic profiles evaluated after surgery may identify patients with oestrogen receptor negative early breast cancer at increased risk of disease recurrence. Results from a retrospective study. Mol Oncol 9:128–139
4. GD T, Miller JA, Chow HH et al (2014) Metabolomic characterization of nipple aspirate fluid by (1) H NMR spectroscopy and GC-MS. J Proteome Res 13:883–889
5. Nagana Gowda GA, Barding GA Jr, Dai J et al (2018) A metabolomics study of BPTES altered metabolism in human breast cancer cell lines. Front Mol Biosci 5:49
6. Giskeødegård GF, Grinde MT, Sitter B et al (2010) Multivariate modeling and prediction of breast cancer prognostic factors using MR metabolomics. J Proteome Res 9:972–979
7. Choi JS, Baek HM, Kim S et al (2013) Magnetic resonance metabolic profiling of breast cancer tissue obtained with core needle biopsy for predicting pathologic response to neoadjuvant chemotherapy. PLoS One 8:e83866
8. Maria RM, Altei WF, Selistre-de-Araujo HS et al (2017) Impact of chemotherapy on metabolic reprogramming: characterization of the metabolic profile of breast cancer MDA-MB-231 cells using 1H HR-MAS NMR spectroscopy. J Pharm Biomed Anal 146:324–328
9. Fuss TL, Cheng LL (2016) Evaluation of cancer metabolomics using ex vivo high resolution magic angle spinning (HRMAS) magnetic resonance spectroscopy (MRS). Meta 6:11
10. Roebuck JR, Cecil KM, Schnall MD et al (1998) Human breast lesions: characterization with proton MR spectroscopy. Radiology 209:269–275
11. Jagannathan NR, Kumar M, Seenu V et al (2001) Evaluation of total choline from *in-vivo* volume localized proton MR spectroscopy and its response to neoadjuvant chemotherapy in locally advanced breast cancer. Br J Cancer 84:1016–1022
12. Sah RG, Sharma U, Parshad R et al (2012) Association of estrogen receptor, progesterone receptor, and human epidermal growth factor receptor 2 status with total choline concentration and tumor volume in breast cancer

patients: an MRI and in vivo proton MRS study. Magn Reson Med 68:1039–1047

13. Bolan PJ (2013) Magnetic resonance spectroscopy of the breast: current status. Magn Reson Imaging Clin N Am 21:625–639
14. Bolan PJ, Meisamy S, Baker EH et al (2003) In vivo quantification of choline compounds in the breast with ^{1}H MR spectroscopy. Magn Reson Med 50:1134–1143
15. Baik HM, Su MY, Yu HJ et al (2006) Quantification of choline containing compounds in malignant breast tumors by ^{1}H MR spectroscopy using water as an internal reference at 1.5 T. MAGMA 19:96–104
16. Bakken IJ, Gribbestad IS, Singstad TE et al (2001) External standard method for the in vivo quantification of choline-containing compounds in breast tumors by proton MR spectroscopy at 1.5 tesla. Magn Reson Med 46:189–192
17. Danishad KK, Sharma U, Sah RG et al (2010) Assessment of therapeutic response of locally advanced breast cancer (LABC) patients undergoing neoadjuvant chemotherapy (NACT) monitored using sequential magnetic resonance spectroscopic imaging (MRSI). NMR Biomed 23:233–241
18. Tozaki M, Sakamoto M, Oyama Y et al (2010) Predicting pathological response to neoadjuvant chemotherapy in breast cancer with quantitative ^{1}H MR spectroscopy using the external standard method. J Magn Reson Imaging 31:895–902
19. Sharma U, Baek HM, Su MY et al (2011) In vivo 1H MRS in the assessment of the therapeutic response of breast cancer patients. NMR Biomed 24:700–711
20. Beckonert O, Keun HC, Ebbels TM et al (2007) Metabolic profiling, metabolomic and metabonomic procedures for NMR spectroscopy of urine, plasma, serum and tissue extracts. Nat Protoc 2:2692–2703
21. Blekherman G, Laubenbacher R, Cortes DF et al (2011) Bioinformatics tools for cancer metabolomics. Metabolomics 7:329–343
22. Bolan PJ, DelaBarre L, Baker EH et al (2002) Eliminating spurious lipid sidebands in ^{1}H MRS of breast lesions. Magn Reson Med 48:215–222
23. Emwas AH, Roy R, McKay RT et al (2016) Recommendations and Standardization of Biomarker Quantification Using NMR-Based Metabolomics with Particular Focus on Urinary Analysis. J Proteome Res 15:360–73. https://doi.org/10.1021/acs.jproteome.5b00885[PNJ1][PNJ1]

Chapter 13

NMR-Based Urinary Metabolomics Applications

Tedros Bezabeh, Ana Capati, and Omkar B. Ijare

Abstract

The field of metabolomics has been growing tremendously over the recent years and, consistent with that growth, a number of investigators have been looking at the potential of NMR-based urinary metabolomics for several applications. While such applications have shown promising results, there still remains an enormous amount of work to be done before this approach becomes accepted and widely used in clinical diagnostics and other biomedical applications. To achieve such goals, optimization of parameters and standardization of protocols are of paramount importance. In view of this, in this chapter, we present some recommended methods and procedures that can help researchers in the field. Furthermore, we have highlighted some of the challenges encountered in such applications and suggested some possible ways to overcome those challenges.

Key words Metabolic profile, Metabolomics, Multivariate analysis, NMR spectroscopy, Spectral profile, Urinalysis, Urine

1 Introduction

Metabolomics, a relatively newer member of the "omics" family, deals with the systematic analysis of the complete set of metabolites found in biological samples. Metabolomics represents a comprehensive and systematic profiling of metabolite concentrations and their response to different environmental and/or physiological stimuli [1]. With the aid of analytical tools, a metabolic profile of a body fluid that is reflective of the health/disease status of an individual can be captured. Furthermore, such a profile can be monitored over time to assess the response to treatment or disease progression. At times, these metabolic changes can precede clinical symptoms, and thus make early successful intervention possible [2–4]. Metabolomics is the closest to and likely the most predictive of the phenotypic properties of a biological system [5, 6]. Metabolites are the downstream products of numerous biochemical interactions and can be very sensitive measures of an organism's phenotype, making metabolomics very useful in the analysis of perturbations and interactions involving genetic and environmental

G. A. Nagana Gowda and Daniel Raftery (eds.), *NMR-Based Metabolomics: Methods and Protocols*, Methods in Molecular Biology, vol. 2037, https://doi.org/10.1007/978-1-4939-9690-2_13, © Springer Science+Business Media, LLC, part of Springer Nature 2019

factors [2, 7, 8]. The relatively small number of metabolites analyzed also makes it easier to handle the data compared to those generated from genomics and proteomics [9].

Two essential components of a biomarker identification strategy are the analytical technique and the associated statistical analysis [10]. Nuclear magnetic resonance (NMR) spectroscopy and mass spectrometry (MS) are the two analytical techniques commonly employed in metabolomics. This chapter, however, will only focus on NMR-based metabolomics. Although NMR and MS are complimentary techniques, NMR has certain advantages over MS. Among them, NMR does not require preselection of analysis conditions and the sample preparation is rather straightforward [11]. Furthermore, it is nondestructive and hence it allows further analysis of the same sample, if needed [12]. A single NMR experiment can provide information on multiple metabolites simultaneously unlike other techniques, which require separate assays for the analyses of specific metabolites. A human metabolome database (HMDB), with detailed information on small-molecule metabolites, has already been established and is available to the public at no cost; it contains over 114,000 metabolite entries, including both water and lipid soluble metabolites [13]. The availability of such a database facilitates the interpretation of the metabolomics data.

Although several body fluids have been the subject of metabolomics studies, urine is perhaps the most studied/analyzed biofluid to date. This is because, urine is readily available, it can be obtained noninvasively, and requires minimal sample preparation. Furthermore, it is rich in small molecule metabolites and has a very low concentration of macromolecules. The most common metabolites identified in urine are those associated with major endogenous pathways along with their intermediates, such as citrate succinate, oxaloacetate, and α-ketoglutarate [12]. Figure 1 shows a high-resolution ^{1}H NMR spectrum of human urine with the assignments of various metabolites that have been identified and quantified to date [8]; this study reports identification and quantification of 209 metabolites in urine using ^{1}H NMR spectroscopy.

2 Materials

2.1 Sample Collection and Storage

1. Samples: Collect midstream urine specimens.
2. Sample containers: Samples should be collected in leak-proof, break-resistant sterile collection cups, and sample aliquots must be stored in cryovials that can withstand very low temperatures (liquid Nitrogen temperatures).
3. Refrigeration: A regular refrigerator for short-term storage (−20 °C) and a cryofreezer (−80 °C) for long-term storage. For very long-term storage, a liquid nitrogen Dewar may be used.

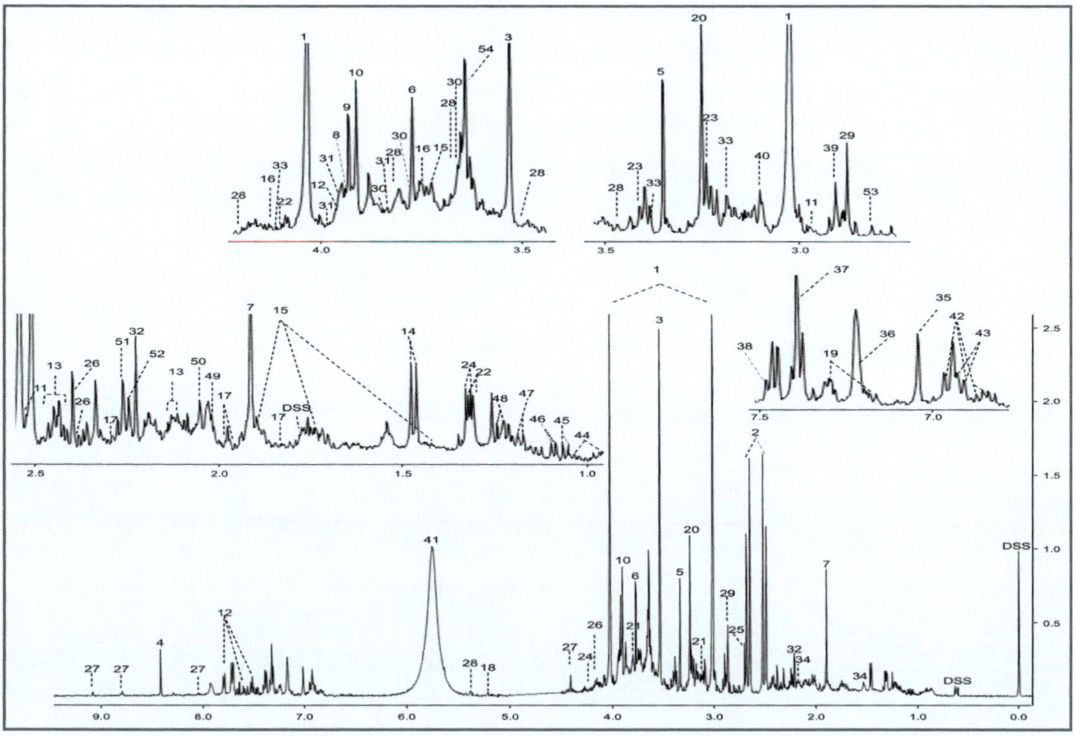

Fig. 1 Typical 500 MHz ^{1}H-NMR spectra of urine from human urine (Reproduced with permission from Bouatra et al., 2013, PLOS One) [8]. Assignment of metabolite peaks: (1) creatinine; (2) citric acid; (3) glycine; (4) formic acid; (5) methanol; (6) guanidoacetic acid; (7) acetic acid; (8) L-cysteine; (9) glycolic acid; (10) creatine; (11) isocitric acid; (12) hippuric acid; (13) L-glutamine; (14) L-alanine; (15) L-lysine; (16) gluconic acid; (17) 2-hydroxyglutaric acid; (18) D-glucose; (19) indoxyl sulfate; (20) trimethyl-N-oxide; (21) ethanolamine; (22) L-lactic acid; (23) taurine; (24) L-threonine; (25) dimethylamine; (26) pyroglutamic acid; (27) trigonelline; (28) sucrose; (29) trimethylamine; (30) mannitol; (31) L-serine; (32) acetone; (33) L-cystine; (34) adipic acid; (35) L-histidine; (36) L-tyrosine; (37) imidazole; (38) mandelic acid; (39) dimethylglycine; (40) Cis-aconitic acid; (41) urea; (42) 3-(3-hydroxyphenyl)-3-hydroxypropanoic acid (HPHPA); (43) phenol; (44) unknown; (45) isobutyric acid; (46) methylsuccinic acid; (47) 3-aminoisobutyric acid; (48) L-fucose; (49) *N*-acetylaspartic acid; (50) *N*-acetylneuraminic acid; (51) acetoacetic acid; (52) Alpha-aminoadipic acid; (53) methylguanidine; (54) phenylacetylglutamine

4. Preservatives: Addition of 3 mM sodium azide (NaN_3) to urine is preferred to prevent bacterial growth; NaN_3 does not affect the chemical shifts of metabolites in urine.

2.2 Preanalytics for Sample Collection/Storage

The quality of urine samples collected will undoubtedly affect the quality of the data generated and its subsequent interpretation. Hence, uniform and strict guidelines have to be followed in the collection and storage of the samples.

1. First morning specimens are generally recommended (*see* **Note 1**).

2. It is preferable that midstream urine specimens are collected to avoid any contamination with bacteria residing at the opening of the urinary tract (*see* **Note 2**).
3. Samples should be collected in sterile containers.
4. Samples should be collected over different times during a single day or over a few consecutive days if possible (*see* **Note 3**).
5. If there are plans to run multiple tests on the same samples at a later date, the samples should be divided into small aliquots using cryovials at this point. The cryovials should be appropriately labeled prior to freezing (*see* **Note 4**).
6. While analysis of freshly collected urine specimens may be the ideal choice, logistical considerations generally preclude such an approach. Hence, frozen specimens are typically used for analysis. Freeze the samples at −20 °C within 8 h of sample collection and then transfer the samples to lower-temperature freezers (−40 °C or − 80 °C) within a week for long-term storage. For very long-term storage, samples may be stored in liquid nitrogen vapors in a Dewar (*see* **Note 5**).
7. If access to a freezer is limited, the addition of a preservative (3 mM NaN_3) right after the collection of the specimen will help hinder microbial growth and can be a viable option (*see* **Note 6**).

2.3 Sample Preparation

1. Buffer Solution: Prepare a 0.33 M phosphate buffer solution in D_2O (pH = 7.4). For highly concentrated urine samples, a 1.0 M phosphate buffer solution is recommended.
2. Chemical Shift Reference: Use 0.1 or 0.5 mM solution of trimethylsilyl propanoic acid (TSP) or 4,4-dimethyl-4-silapentane-1-sulfonic acid (DSS). Since TSP/DSS bind to urinary protein, their use may introduce errors in the quantitative analysis.

2.4 NMR Instrumentation, Software, and Consumables

1. NMR Spectrometer: ≥400 MHz (Bruker, JEOL, Agilent). Typically: 400–600 MHz spectrometer is required (*see* **Note 7**).
2. Automation: Autosample changers (capacity can be 24, 30, 60, 100, or 120 samples). 96-well plates can be used while using flow injection (FI) probes (*see* **Note 8**).
3. Computer Hardware: NMR console, SGI, Sun, or similar workstations for off-line NMR processing.
4. Software: SIMCA-P, Chenomx NMR Suite, MetaboAnalyst, MNova, or similar.
5. NMR tubes: Standard 5 mm NMR tubes (Wilmad-LabGlass, Norell, or similar).

3 Methods

3.1 Sample Preparation

1. Thaw frozen urine samples on ice and vortex for 30 s before use.
2. Transfer aliquots of 500 μL of the samples into Eppendorf tubes and add 250 μL of 0.33 M phosphate buffer prepared in D_2O (pH = 7.4) and a chemical shift reference (TSP or DSS). A buffer concentration of 0.33 M is sufficient to minimize pH-related variations in the NMR chemical shifts for most urine samples [14] (*see* **Note 9**).
3. Centrifuge at 1200 × *g* and transfer 600 μL supernatant liquid into a 5 mm NMR tube for analysis [14–16] (*see* **Note 8**).

3.2 NMR Data Acquisition

1. Run replicate samples from the same assay for quality control purposes (*see* **Note 10**).
2. Use the 1D NOESY pulse sequence with water presaturation, which is robust, readily available on modern spectrometers and typically used in urinalysis in metabolomics applications (*see* **Note 11**). *See* Table 1 for additional details of data collection and processing parameters using the 1D NOESY pulse sequence (*noesypr1d* on Bruker spectrometer) [17].

3.3 NMR Data Processing and Analysis

1. Normalization: Given the varying dilution issues encountered in urine specimens due to the variation in the subjects' water intake, the first step of the data analysis is to ensure that the data are normalized [18]. One of the following approaches is used for the normalization.
   (a) Normalize each metabolite peak integral to the creatinine level based on the assumption that the creatinine levels are constant between different individuals or within an individual over time (*see* **Note 12**).
   (b) Perform normalization with reference to the total spectrum area/integral. This, however, assumes that the total area/integrals of all metabolites is constant for all samples (*see* **Note 13**).
   (c) Perform probabilistic quotient normalization (PQN) [19]. This method calculates a most probable dilution factor (e.g., the median of the quotients) among all the metabolite variables, which is then utilized to perform normalization of all the variables in the sample (*see* **Note 14**).
2. Peak alignment: Use the internal chemical shift reference to perform peak alignment (*see* **Note 15**).
3. Peak integration: Given the complex nature of urine samples, its ^1H NMR spectrum shows well-resolved peaks for only a few

Table 1
Acquisition and processing parameters used in 1D NOESY pulse sequence for the collection of ^1H NMR spectra of urine samples with water suppression in metabolomics applications

Parameters	Recommended value
Acquisition	
Time domain (TD) points	64k
Dummy scans (DS)	4–8
Number of scans (NS)	64–128
Spectral width (SW)	12–16 ppm
Relaxation delay	≥5 s
Acquisition time (AQ)	2–2.5 s
Excitation pulse	90°
Receiver gain	Optimize (depends on concentration)
Mixing time	100 ms
Sample temperature	300 K
Processing	
Window function	Exponential
Line broadening	0.3–0.5 Hz
Zero filling	2 × TD
Phase correction	Manual
Baseline correction	Automatic/manual

metabolites. Such metabolites can be quantified by manual integration. However, metabolites, which are found in low concentration generally overlap with other metabolites present in high concentration. To overcome such challenges, more robust quantification methodologies such as the targeted profiling [20] and the Bayesian deconvolution [21] have been developed. The targeted profiling relies on a database of compounds modeled to behave like the pure spectra of the individual compounds under comparable experimental conditions of pH, and ionic strength. A Lorentzian peak shape model of each reference compound is generated from the database information and superimposed with the actual spectrum. The linear combination of all modeled metabolites gives rise to the total spectral fit, which gives an estimate of peak areas of the experimental spectral peak. Using this information, concentrations of various metabolites can be estimated. Analysis of samples using targeted profiling allows both identification and quantification of individual compounds. Chenomx NMR Suite is a commercial software which provides a comprehensive database of metabolites (>350 compounds) for targeted profiling [22]; it has been used in many metabolomics studies for the analysis of biofluids including urine [23] (*see* **Note 16**).

4. Pattern recognition: Data analysis in metabolomics is usually performed via unsupervised or supervised analysis. Unsupervised analysis involves the application of statistical models without prior knowledge of the sample identity or classification assignment. This is often times done to see if the data separate into certain patterns or clusters and are usually the first step in data pattern exploration. Principal component analysis (PCA) is a representative of the unsupervised method and is commonly applied to reduce the dimensionality and examine the structure of the data set [24–26]. A score plot is generated to assess the clustering of different samples and the corresponding loading plot identifies metabolites that account for the most variation in the specified principal component. In the supervised analysis, information of sample class labels (e.g., disease and control) is utilized for building the statistical models. One commonly used supervised analysis is partial least squares discriminant analysis (PLS-DA), which maximizes the covariance between predictor variables (e.g., metabolite intensities from NMR measurements) and the response variables (e.g., the classes of each sample) [24]. SIMCA, Soft Independent Modelling of Class Analogies, a part of the readily available software SIMCA-P, is commonly used in metabolomics analysis [25]. SIMCA-P relies almost exclusively on Principal Component Analysis (PCA); for classification, it relies on the supervised versions, Partial Least Squares (PLS) or Principal Component Regression (PCR) (for 2-class problems, PLS and PCR are equivalent) [25, 26] (*see* **Note 17**).
5. Peak identification: While the primary focus of metabolomics is finding spectral profiles and patterns that distinguish between different classes, it can also be beneficial to identify those metabolites that have discriminatory power. Some of the common NMR detectable metabolites present in urine are branched chain amino acids, lactate, alanine, acetate, citrate, dimethylamine, creatine, creatinine, TMA-O, hippurate, and formate which can be identified based on their chemical shifts. However, identification of other metabolites, which are in low concentration and overlap with high concentration metabolites is challenging. The identification process mainly involves comparison of the 1D and 2D spectral data of the metabolites with that of NMR metabolites repositories such as HMDB [13, 27], the Biological Magnetic Resonance Data Bank (BMRB) [28], and/or the Birmingham Metabolite Library (BML) [29]. 2D

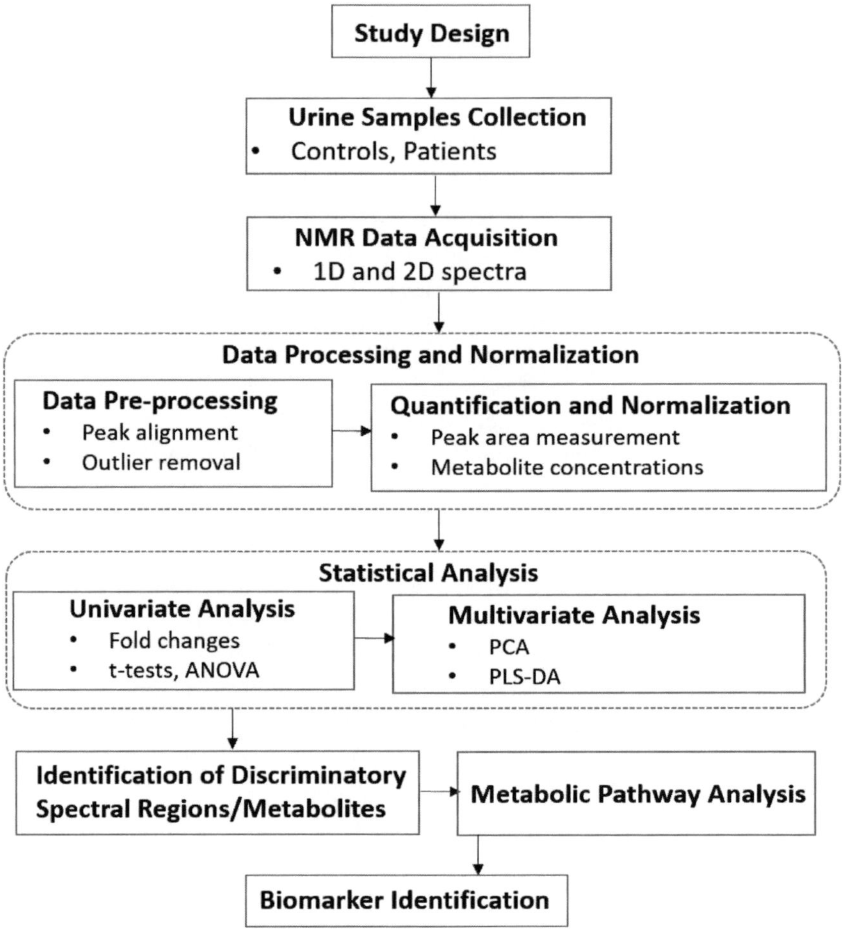

Fig. 2 A schematic diagram of the metabolomics workflow for urinalysis

JRES, COSY, TOCSY, and HMBC data of reference compounds from these databases are valuable for the identification of metabolites (*see* **Note 18**).

6. A schematic diagram of the complete urinary metabolomics workflow is depicted in Fig. 2 [30].

4 Notes

1. Urine is generally more concentrated in the morning (due to the length of time it is allowed to stay in the bladder).
2. This becomes a more serious consideration with female subjects compared to male subjects due to the anatomical differences.
3. Significant metabolic changes can be expected due to intra-subject and inter-subject variabilities. The following factors can

explain some of the expected variations and should be considered in designing a study protocol (or clinical trial) and analyzing/interpreting the data. Diet may be the most significant source for urine spectral variation. High protein diets such as fish have proven to cause increases in creatinine concentrations, which can cause a distortion in some ^{1}H NMR spectra [31]. A study by Lenz et al. has suggested that endogenous urinary profiles are affected by cultural and severe dietary influences, emphasizing that variation in profiles can occur between different populations [32]. The study compared the urinary metabolic profile of the Swedish and the British population, in which they found that high levels of trimethylamine N-oxide (TMAO) were observed in Swedish population, due to their fish-diet. This effect was also consistent in a study by Dumas et al., in which a similar trend was observed in the Japanese population, where a fish diet is also dominant [33]. This variation is an important factor that needs to be considered when interpreting NMR spectral profiles for diagnostic purposes. Similarly, Zuppi et al. tested subjects living in different parts of Europe and found that a diet rich in carbohydrates resulted in increased excretion of citrate, lactate, and glycine [34]. In a recent study by Jakobsen et al., urinary excretions of acetate and anserine were found to be higher after chicken intake, while carnosine, fumarate, and TMAO excretion were found to be higher after beef intake [35]. Rasmussen et al. suggest that it is important to perform diet standardization before subject recruitment and sample collection for a metabolomics study. This will help reduce intra- and inter-subject variability [31]. Slupsky et al. also investigated the effects of diurnal variation, gender, and age in urinary metabolomic profiles [36]. They showed that gender and age affected metabolites related to energy metabolism, while diurnal variation affected metabolites and dietary components associated with circadian rhythms. In a different study by Psihogios et al., gender-associated metabolites such as citrate, creatinine, and TMAO were identified [37].

4. Repeated thaw–freeze cycles should be avoided since it can alter the spectral profiles and interfere with the interpretation of the data.

5. According to a study by Rist et al., the variation in the freezing procedures can cause significant variations in the spectral profile [38], and hence the same protocol needs to be applied throughout the study. Samples kept at 4 °C for an extended period of time were found to show elevated levels of acetate possibly due to microbial contamination [14]. Similar observations were also made by Saude et al. with urine specimens kept at room temperature for 4 weeks [39]. Samples kept at

temperatures below −40 °C should not exhibit any significant spectral changes for at least 26 weeks [14]. In a related study, Rotter et al. determined the stability of metabolite profiles of urine specimens under different conditions. Their conclusion was that most metabolites were stable at −20 °C or 4 °C for a period of 24 h. At higher temperatures, however, some of the metabolite levels showed significant changes [40]. Their recommendation was to avoid shipping or storing urine samples on cool packs (~9 °C) or room temperatures for more than 8 h.

6. The addition of NaN_3 (3 mM) does not affect the spectral profile, whereas the addition of NaF for the same purpose was found to shift the citrate resonance [14].
7. Comparison of urinary metabolomics data obtained using 500 MHz and 600 MHz NMR spectrometers has shown very similar results [41].
8. Given the availability of automatic sample preparation and robotic liquid handling technologies, large-scale studies can now be carried out for high-throughput applications [42]. Moreover, with the combined use of flow injection (FI) probes and liquid handling systems, it is possible to perform "tubeless" NMR using 96-well plates. Samples can be identified using barcode, prepared for analysis using a robot handling 96-well plates, and finally transferred to the FI probe for NMR analysis. Such large-scale metabolomics studies have already been reported in the analysis of urine samples, which allow ^{1}H NMR data collection for approximately 100 samples per day using an acquisition period of about 5 min per sample [1].
9. The pH of human urine samples generally lies in the range 5.5–6.5. Under physiological stress, however, the pH range can be much wider, 4.6–8.0. Such a pH variation causes spectral variability between samples, particularly for metabolites such as citrate [43]. Lauridsen et al. suggested using a 1.0 M phosphate buffer solution for highly concentrated urine samples [14]. DSS is less sensitive than TSP to variations in the pH of the sample. However, given its more hydrophobic nature, its resonance can broaden due to binding with macromolecules such as proteins in the sample [44].
10. Given the homogenous nature of urine specimens, not much variation is expected between the replicates. However, this ensures the stability of the spectrometer during analysis.
11. A high concentration of water in urine makes it difficult to detect very low concentration metabolites. Water presaturation is often applied to samples in order to mitigate the effects of excess water. In this technique, a transmitter is set to the water frequency and a selective pulse is employed to saturate the

water resonance. However, solvent suppression can sometimes cause unintended damping effect on the adjacent peaks which may result in artificially lowered concentrations [15]. There are several water suppression techniques available for metabolomics applications [45]. The experimental details of these techniques have been discussed in detail in a recent review article [46]. One drawback of using presaturation would be in the quantification of urea. The pulse train in a typical hard pulse-based WATERGATE water suppression sequence may negatively impact the urea signal due to lack of sufficient selectivity. In a study by Liu et al., the authors used WET for water suppression, and chose its pulses to be long enough to achieve sufficient selectivity and exert minimal impact on the urea signal [17]. The pulse duration was also kept short to make sure that no significant proton exchange takes place between water and urea [17]. Recently, Le Guennec et al. tested another water suppression sequence "PURGE" and concluded that it can be on par with the 1D NOESY for metabolomics applications [47]. However, its adaptability on modern spectrometers and robustness for metabolomics applications needs to be evaluated. Other techniques such as excitation sculpting entirely eliminate the water signal from the spectrum, but the intensity of the signals adjacent to the water signal is greatly affected [45].

12. One should keep in mind that creatinine normalization is not a robust approach since factors unconnected with the dilution such as muscle mass, age, gender, diurnal creatinine production, physical activity, diet, emotional stress, and/or disease state affect the urine creatinine levels [18].
13. This method, however, is not suitable if drug metabolite signals appear in NMR spectra, differentially.
14. A recent study by Hertel et al. questioned the PQN approach and proposed a new analyte-specific normalization method, which takes into consideration the time-span since the subject's last urination [48].
15. Despite the efforts to minimize the pH variation using buffer solution, peak positions still vary among the samples. One reason for such a variation is the differences in metal ion concentrations [49]. Peak variations induced by common metal ions in urine such as Na^+, K^+, Ca^{2+}, and Mg^{2+} were measured for 33 metabolites in urine recently by Tredwell et al. and such data could be useful for peak alignment and peak identification using automation.
16. An alternative approach based on the Bayesian model makes extensive use of prior information on the characteristic spectral pattern of each metabolite. It also accounts for shifts in peak

positions in NMR spectra. It is an "R"-based public domain software package "Bayesian automated metabolite analyzer for NMR spectra (BATMAN)", which deconvolutes peaks from one-dimensional NMR spectra, assigns them to specific metabolites from a target list, and provides an estimate of metabolite concentrations. It applies a Markov chain Monte Carlo algorithm to sample from a joint posterior distribution of the model parameters and obtains concentration estimates with reduced errors compared to the conventional numerical integration and comparable to that of manual deconvolution [21–23].

17. One of the important steps in metabolomics data analysis is identifying which of the original features in the spectra are relevant and meaningful in the final outcome. A Genetic-Algorithm-based Optimal Region Selection algorithm (GA_ORS) has been developed specifically for such feature extraction [23]. An important advantage of this algorithm is that it retains spectral identity; the new features, being functions (typically the averages) of ***adjacent*** spectral data points, are readily interpretable. Such an algorithm has been part of a statistical classification strategy employed in several studies involving NMR data [50–54].

18. A detailed methodology formetabolite identification using 1D (^{1}H, ^{13}C) and 2D (JRES, COSY, TOCSY, and HMBC) is discussed in recent review articles [29, 55–57].

References

1. Beckonert O, Keun H, Ebbels T, Bundy J, Holmes E, Lindon J, Nicholson J (2007) Metabolic profiling, metabolomics and metabonomic procedures for NMR spectroscopy of urine, plasma, serum and tissue extracts. Nat Protoc 2(11):2692–2703. https://doi.org/10.1038/nprot.2007.376
2. Roessner U, Bowne J (2009) What is metabolomics all about? BioTechniques 46:363–365. https://doi.org/10.2144/000113133
3. Van QN, Veenstra TD, Issaq HJ (2011) Metabolic profiling for the Detection of Bladder Cancer. Curr Urol Rep 12(1):34–40
4. Duarte LF, Gil AM (2012) Metabolic signatures of cancer unveiled by NMR spectroscopy of human biofluids. Prog Nucl Magn Reson Spectrosc 62:51–74
5. Fiehn O (2002) Metabolomics—the link between genotypes and phenotypes. Plant Mol Biol 48(1-2):155–171
6. Weckweth W (2003) Metabolomics in Systems Biology. Annu Rev Plant Biol 54:669–689
7. Duarte L, Diaz S, Gil A (2014) NMR metabolomics of human blood and urine in disease research. J Pharm Biomed Anal 93:17–26. https://doi.org/10.1016/j.jpba.2013.09.025
8. Bouatra S, Aziat F, Guo AC et al (2013) The Human Urine Metabolome. PLoS One 8(9): e73076. https://doi.org/10.1371/journal.pone.0073076
9. Kim K, Aronov P, Zakharkin S (2008) Urine metabolomics analysis for kidney cancer detection and biomarker discovery. Mol Cell Proteomics 8:558–570. https://doi.org/10.1074/mcp.M800165-MCP2000
10. Serkova NJ, Niemann CU (2006) Pattern recognition and biomarker validation using quantitative ^{1}H-NMR-based metabolomics. Expert Rev Mol Diagn 6(5):717–731. https://doi.org/10.1586/14727159.6.5.717

11. Dunn W, Bailey N, Johnson H (2005) Measuring the metabolome: current analytical technologies. Analyst 130:606–625
12. Lehman-McKeeman LD, Car B (2004) Metabonomics: application in predictive and mechanistic toxicology. Toxicol Pathol 32:94–96. https://doi.org/10.1080/01926230490462084
13. Wishart DS, Feunang YD, Marcu A et al (2018) (2013) HMDB 4.0: the human metabolome database for 2018. Nucleic Acids Res 46(D1):D608–D617. https://doi.org/10.1093/nar/gkx1089
14. Lauridsen M, Hansen SH, Jaroszewski JW, Cornnet C (2007) Human urine as test material in ^{1}H NMR-based metabonomics: recommendations for sample preparation and storage. Anal Chem 79(3):1181–1186. https://doi.org/10.1021/ac061354x
15. Vitols C, Fu H (2006) Targeted profiling of common metabolites in urine. Chenomx, Edmonton, Alberta
16. Rosewell R, Vitols C (2006) Identifying metabolites in biofluids. Chenomx Inc, Edmonton, Alberta
17. Liu L, Mo H, Wei S, Raftery D (2012) Quantitative analysis of urea in human urine and serum by ^{1}H nuclear magnetic resonance. Analyst 137(3):595–600. https://doi.org/10.1039/c2an15780b
18. Tang KW, Toh QC, Teo BW (2015) Normalisation of urinary biomarkers to creatinine for clinical practice and research--when and why. Singapore Med J 56(1):7–10. https://doi.org/10.11622/smedj.2015003
19. Weljie AM, Newton J, Mercier P, Carlson E, Slupsky CM (2006) Targeted profiling: quantitative analysis of 1H NMR metabolomics data. Anal Chem 78(13):4430–4442. https://doi.org/10.1021/ac060209g
20. Chenomx NMR suite. http://www.chenomx.com/software/software.php
21. Hao J, Liebeke M, Astle W, De lorio M, Bundy JG, Ebbels TM (2015) Bayesian deconvolution and quantification of metabolites in complex 1D NMR spectra using BATMAN. Nat Protoc 9(6):1416–1427. https://doi.org/10.1038/nprot.2014.090
22. Friedrich N, Budde K, Suhre K, Volker U, John U, Felix SB et al (2015) Sex differences in urine metabolites related with risk of diabetes using NMR spectroscopy: results of the study of health in Pomerania. Metabolomics 11:1405–1415
23. Barker M, Ravens W (2003) Partial least squares for discrimination. J Chemom 3:166–173
24. Eriksson L, Johansson E, Kettaneh-Wold N, World S (2001) Multi- and megavariate data analysis-principles and applications. Umetrics Academy, Dublin OH
25. Jackson JE (1991) A user's guide to principal components. Wiley, New York
26. Nikulin AE, Dolenko B, Bezabeh T, Somoriaj RL (1998) Near-optimal region selection for feature space reduction: novel-preprocessing methods for classifying NMR spectra. NMR Biomed 11:209–216
27. Xiao J, Wishart DS (2011) Web-based inference of biological patterns, functions, and pathways from metabolomics data using metaboanlayst. Nat Protoc 6:743–760. https://doi.org/10.1038/nprot.2011.319
28. Ulrich EL, Akutsu H, Doreleijers JF, Harano Y, Ioannidis YE, Lin J et al (2008) BioMagResBank. Nucleic Acids Res 36: D402–D408. https://doi.org/10.1093/nar/gkm957
29. Ludwig C, Easton JM, Lodi A, Tiziani S, Manzoor SE, Southam AD et al (2012) Birmingham metabolite library. Metabolomics 8:8–18. https://doi.org/10.1007/s11306-011-0347-7
30. Capati A, Ijare OB, Bezabeh T (2017) Diagnostic applications of nuclear magnetic resonance-based urinary metabolomics. Magn Reson Insights 10:1178623X17694346. https://doi.org/10.1177/1178623X17694346
31. Rasmussen LG, Winning H, Savorani F et al (2012) Assessment of the effect of high or low protein diet on the human urine metabolome as measured by NMR. Nutrients 4 (2):112–131. https://doi.org/10.3390/nu4020112
32. Lenz EM, Bright J, Wilson I et al (2004) Metabonomics, dietary influences and cultural differences: NMR-based study of urine samples obtained from healthy British and Swedish subjects. J Pharm Biomed Anal 36(4):841–849. https://doi.org/10.1016/j.pba.2004.08.002
33. Dumas ME, Maibaum EC, Teague C et al (2006) Assessment of analytical reproducibility of 1H NMR spectroscopy based metabonomics for large-scale epidemiological research: the INTERMAP study. Anal Chem 78(7):2199–2208
34. Zuppi C, Messana I, Forni F, Ferrari F, Rossi C, Giardina B (1998) Influence of feeding on metabolite excretion evidenced by urine ^{1}H NMR spectral profiles: a comparison between subjects living in Rome and subjects living at arctic latitudes (Svaldbard). Clin Chim Acta 278:75–79

35. Jakobsen LM, Yde CC, Van Hecke T et al (2017) Impact of red meat consumption on the metabolome of rats. Mol Nutr Food Res 61(3). https://doi.org/10.1002/mnfr.201600387
36. Slupsky CM, Rankin KN, Wagner J et al (2007) Investigations of the effects of gender, diurnal variation, and age in human urinary metabolomics. Anal Chem 79(18):6995–7004
37. Psihogios N, Gazi I, Elisaf M, Seferiadis K, Bairaktari E (2008) Gender-related and age-related urinalysis of healthy subjects by NMR-based metabonomics. NMR Biomed 21:195–207. https://doi.org/10.1002/nbm.1176
38. Rist MJ, Muhle-Goll C, Gorling B, Bub A, Heissler S, Watzl B, Luy B (2012) Influence of freezing and storage procedure on human urine samples in NMR-based metabolomics. Metabolites 3:243–258. https://doi.org/10.3390/metabo3020243
39. Saude EJ, Sykes BD (2007) Urine stability for metabolomics studies: effects of preparation and storage. Metabolomics 3:19–27. https://doi.org/10.1007/s11306-006-0042-2
40. Rotter M, Brandmaier S, Prehn C et al (2017) Stability of targeted metabolite profiles of urine samples under different storage conditions. Metabolomics 13(1):4. https://doi.org/10.1007/s11306-016-1137-z
41. Keun HC, Ebbels TM, Antti H, Bollard ME, Beckonert O, Scholtterbeck G (2002) Analytical reproducibility in (1) H NMR-based metabonomic urinalysis. Chem Res Toxicol 15(11):1380–1386
42. Da Silva L, Godejohann M, Martin FP et al (2013) High –resolution quantitative metabolome analysis of urine by automated flow injection NMR. Anal Chem 84(12):5801–5809. https://doi.org/10.1021/ac4004776
43. Miyataka H, Ozaki T, Himeno S (2007) Effect of pH on 1H-NMR spectroscopy of mouse urine. Biol Pharm Bull 30(4):667–670
44. Emwas AH, Luchinat C, Turano P, Tenori P, Roy R, Salek RM et al (2015) Standardizing the experimental conditions for using urine in NMR-based metabolomics studies with a particular focus on diagnostic studies: a review. Metabolomics 11(4):872–894. https://doi.org/10.1007/s11306-014-0746-7
45. Giraudeau P, Silvestre V, Akoka S (2015) Optimizing water suppression for quantitative NMR-based metabolomics: a tutorial review. Metabolomics 11:1041–1055. https://doi.org/10.1007/s11306-015-0794-7
46. Emwas AH, Roy R, McKay RT et al (2016) Recommendations and Standardization of Biomarker Quantification Using NMR-Based Metabolomics with Particular Focus on Urinary Analysis. J Proteome Res 15(2):360–373. https://doi.org/10.1021/acs.jproteome.5b00885
47. Le Guennec A, Tayyari F, Edison AS (2017) Alternatives to Nuclear Overhauser Enhancement Spectroscopy Presat and Carr—Purcell—Meiboom—Gill Presat for NMR-Based Metabolomics. Anal Chem 89(17):8582–8588. https://doi.org/10.1021/acs.analchem.7b02354
48. Hertl J, Rotter M, Frenzel S et al (2018) Dilution correction for dynamically influenced urinary analyte data. Anal Chim Acta 1032:18–31. https://doi.org/10.1016/j.aca.2018.07.068
49. Tredwell GD, Bundy JG, De lorio M, Ebbels TMD (2016) Modelling the acid/base ^{1}H NMR chemical shift limits of metabolites in human urine. Metabolomics 12(10):152. https://doi.org/10.1007/s11306-016-1101-y
50. Cl L, Somorjai RL, Smith ICP, Russell P, Mountford CE (2002) Accurate diagnosis and prognosis of human cancers by proton MRS and a three-stage classification strategy. Annu Rep NMR Spectro 48:71–111
51. Bezabeh T, Somorjai RL, Dolenko B et al (2009) Detecting colorectal cancer by ^{1}H magnetic resonance spectroscopy of fecal extracts. NMR Biomed 22:593–600
52. Bezabeh T, Ijare OB, NIkulin AE, Somoriai R, Smith ICP (2014) MRS-based metabolomics in cancer research. Magn Reson Insights 7:1–14
53. Somorjai RL, Nikulin AE, Pizzi N et al (1995) Computerized consensus diagnosis: a classification strategy for the robust analysis of MR spectra. Application to ^{1}H spectra of thyroid neoplasms. Magn Reson Med 33:257–263
54. Somorjai RL, Dolenko B, Nikulin AK et al (1996) Classification of ^{1}H MR spectra of human brain neoplasms: the influence of preprocessing computerized consensus diagnosis on classification accuracy. J Magn Reson Imaging 6:437–433
55. Dona AC, Kyriakides M, Scott F, Shephard EA, Varshavi D, Veselkov K, Everett JR (2016) A guide to the identification of metabolites in NMR-based metabonomics/metabolomics experiments. Comput Struct Biotechnol 12:135–153. https://doi.org/10.1016/j.csbj.2016.02.005

56. Everett JR (2015) A new paradigm for known metabolite identification in metabonomics/metabolomics: metabolite identification efficiency. Comput Struct Biotechnol 13:131–144. https://doi.org/10.1016/j.csbj.2015.01.002

57. Oman T, Tessem MB, Bathen TF, Bertilsson H, Angelsen A, Hedenstrom M, Andreassen T (2014) Identification of metabolites from 2D (1) H-(13)C HSQC NMR using peak correlation plots. Bioinformatics 15:413. https://doi.org/10.1186/s12859-014-0413-z

Chapter 14

NMR-Based Metabolomics in Gallbladder Cancer Research

Renuka Ranjan and Neeraj Sinha

Abstract

Gallbladder cancer (GBC) is a common form of malignancy, which has high incidence rates in the northern parts of India, South America, Japan, and in Native American populations. A few metabolomic studies have revealed changes in the composition of biofluids, tissues, and gallstones by nuclear magnetic resonance (NMR), which may prove useful in understanding GBC. In this chapter, we focus on the use of NMR in unveiling the metabolomics of GBC.

Key words Gallbladder cancer, Nuclear magnetic resonance, Statistical analysis

1 Introduction

Cancer is a global health issue of major concern as cancer death tolls continue to rise. It is one among the top ten causes of deaths worldwide according to world health statistics, a report by the WHO. It is important to gather detailed knowledge about this disease and understand factors that might play a causal role in the progression of cancer in order to devise effective measures to combat the malignancy. Gallbladder cancer (GBC) is one such common form of biliary tract malignancy which is more prevalent in women in northern India and hence called an Indian disease, according to Indian council of medical research (ICMR) reports [1]. The prognosis of this disease is still poor in its early stages; as only the advanced progression of malignancy can be detected, which has a 5-year survival of less than 5–10% [2].

Metabolomics is the study of small molecule metabolites that can be observed in the metabolic pathways active in a biological microenvironment. Metabolomics analysis is easier and less labor intensive compared to genomics, transcriptomics, and proteomics. Metabolites could be detected through spectroscopic techniques and their distribution in disease and control groups can be differentiated through various quantitative as well as statistical methods. Common methods used in metabolomics are mass spectrometry

G. A. Nagana Gowda and Daniel Raftery (eds.), *NMR-Based Metabolomics: Methods and Protocols*, Methods in Molecular Biology, vol. 2037, https://doi.org/10.1007/978-1-4939-9690-2_14, © Springer Science+Business Media, LLC, part of Springer Nature 2019

and nuclear magnetic resonance (NMR) spectroscopy. Sample preparation for mass spectrometry can be relatively tedious and cumbersome. In addition, the samples cannot be reused after data acquisition. On the other hand, NMR spectroscopy involves minimal sample preparation. There have been tremendous developments in NMR technologies in the past few years, which have opened doors for metabolomics applications focused on better understanding of biological systems.

Gallstones are a major risk for gallbladder disease. Analysis of gallbladder stones using solid-state NMR from different geographical regions shows correlations between environmental factors and the occurrence of gallbladder stones. A comparison between gallstones from patients from different geographical regions such as North India, South India, and UAE show different constituents and levels of cholesterol, bilirubin, and calcium carbonate [3]. Chronic cholecystitis (CC), an inflammation of the gallbladder may be closely associated with GBC. Therefore, serum profiling of CC may provide avenues for the early detection, prevention, and treatment of GBC. Serum profiling of GBC and other types of inflammation (CC and xanthogranulomatous cholecystitis (XGC)) in the gallbladder by one-dimensional ^{1}H NMR (NOESY and CPMG pulse sequences) as well as two-dimensional ^{1}H–^{1}H COSY, ^{1}H–^{1}H TOCSY, and ^{1}H–^{13}C HSQC experiments reveal important alterations in certain metabolites [4, 5]. Concentrations of low-density lipids (LDL), very-low-density lipids (VLDL), branched chain amino acids (BCAA), alanine, glutamine, creatinine, and tyrosine are reduced in GBC patients in comparison to healthy individuals whereas 1,2-propanediol, pyruvate, glutamate, and formate show reduction in concentration levels when compared to healthy individuals [5]. Such changes have also been detected in gallbladder tissues based on HR-MAS NMR. In particular, differences in the levels of triacylglycerol (TAG) and other small metabolites were detected among CC, XGC, and GBC patients [6]. Lipid profiling of gallbladder disease patients using NMR of tissue extracts showed alterations in lipid metabolism between benign as well as malignant gallbladder diseases [2]. Solid-state NMR studies of gallbladder stones by ^{13}C CPMAS experiments reveal that different gallbladder diseases show different composition and microstructure of gallstones [7]. Using ^{1}H NMR, minerals like calcium and magnesium were also found to be in higher proportion in the extracts of gallstones with GBC when compared to non-cancerous gallbladder inflammation [8]. 1D ^{1}H and 2D ^{1}H–^{13}C HSQC NMR and metabolite profiling of bile fluid reveal that levels of cholic acid and phospholipids could be used to differentiate between different types of gallbladder diseases [9]. ^{31}P NMR of bile gives information about the altered levels of phosphorus-containing metabolites such as inorganic phosphate, lysophosphatidylcholine, and phosphatidylcholine in cholelithiasis patients, liver transplant recipients, and liver donors [10]. This chapter provides protocols for GBC metabolomics based on the NMR methods important for its study.

2 Materials

2.1 Samples

Powdered gallstone samples or gallstone powder dissolved in a deuterated solvent; serum; bile collected through percutaneous trans-hepatic biliary drainage (PTBD), endoscopic nasobiliary drainage (ENBD), or surgical procedures [11] (*see* Fig. 1). Tissue samples, either intact or extracts.

2.2 Chemicals and Solvents

Methanol, chloroform, D_2O, CD_3OD, $CDCl_3$. Internal NMR standards: TSP (trimethylsilyl propionate), DSS (4,4-dimethyl-4-silapentane-1-sulfonic acid) or TMS (trimethylsilane).

3 Methods

NMR spectroscopy involves minimal sample preparation. Intact samples are often used for analysis with no need for sample processing. These characteristics make NMR spectroscopy a preferable method for clinical application. Numerous biological specimens including malignant tissue, tissue extract, biofluids, and gallstones are used to identify biomarkers of GBC using NMR spectroscopy-based metabolomics.

Solid as well as solution-state NMR have been successfully carried out for these types of samples. To acquire a good-quality NMR spectrum of nonhomogeneous samples like gallstones, some minor sample processing is required. The following methods have been used for processing of each type of sample.

3.1 Sample Preparation

3.1.1 Gallstones

1. Powdered gallstone samples are used for solid-state NMR.
2. Gallstones can be powdered by grinding the stones using a mortar and pestle.
3. For solid-state NMR, the powder is filled into the rotor. The rotor size depends on the type of NMR probe used.
4. For solution state NMR, gallstone powder is dissolved in a deuterated solvent.
5. The type of solvent depends on the nature of metabolites under study [3, 7, 8]. To study hydrophilic metabolites, deuterium oxide is used. To study hydrophobic metabolites, deuterated chloroform is used.
6. Gallstones can also be dissolved in deuterated methanol for analysis of lipids.

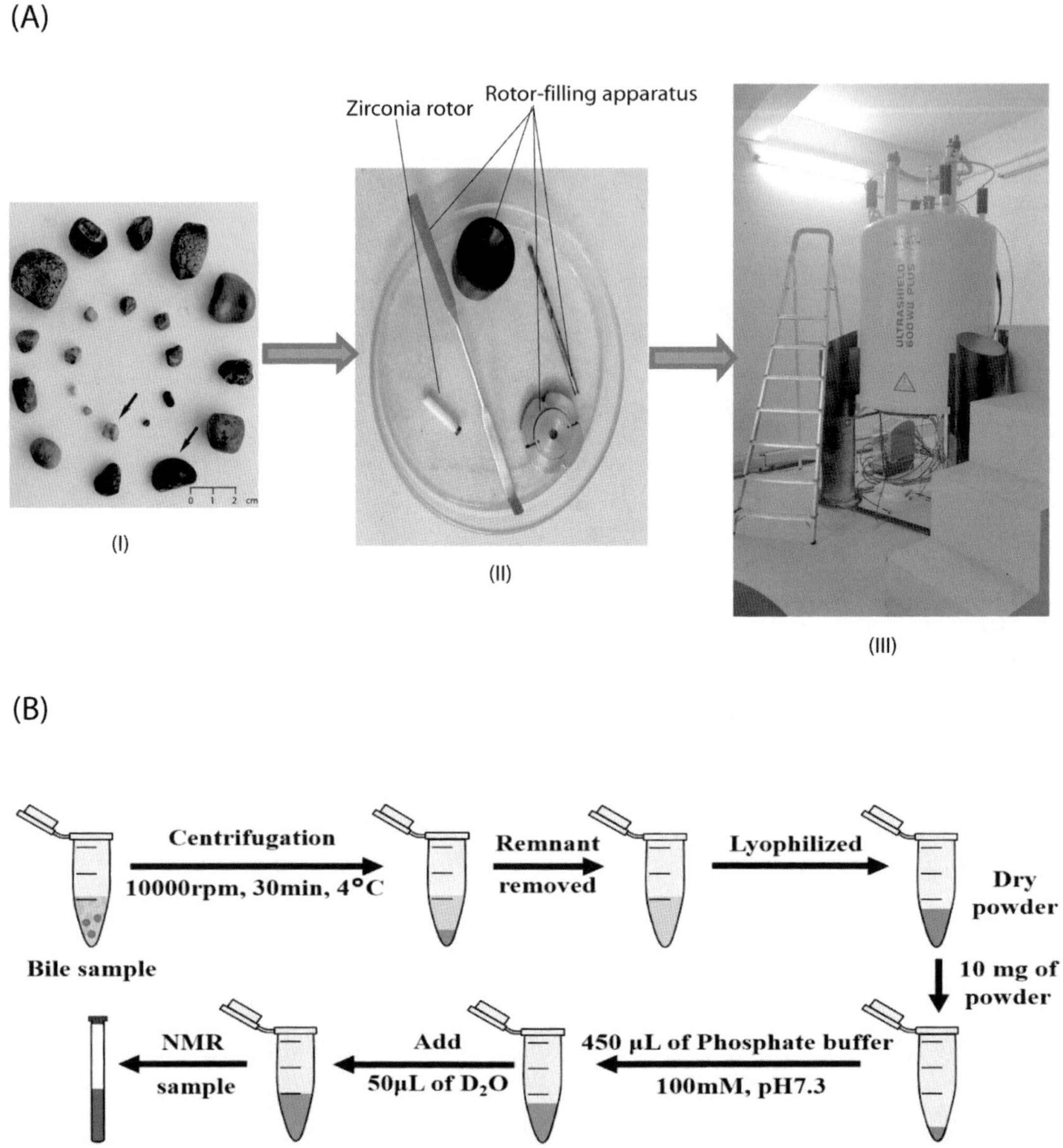

Fig. 1 (**a**) Representation of (I) gallbladder stones which are powdered (Adapted from [7]); (II) zirconia rotor (3.2 mm) filled with gallstone powder using rotor-filling apparatus; (III) rotor transferred to the probe in the NMR instrument. (**b**) Steps used for processing bile samples for NMR analysis (Reprinted by permission from Applied Spectroscopic Review 51:7-9, 706-717, Copyright 2016 Taylor and Francis Online)

3.1.2 Serum

1. Blood collected from patients is centrifuged at 1500 × *g* for 10 min at 4 °C.
2. The supernatant portion, serum, is collected and stored at −80 °C until used for analysis.
3. For NMR data acquisition, serum samples are thawed and transferred to 5 mm NMR tubes.

4. A solution of the reference standard (TSP, TMS, or DSS in deuterated solvent) in a coaxial tube is inserted into the NMR tube [4, 5].
5. Serum can also be mixed with deuterated buffer to maintain physiological pH.

3.1.3 Bile

1. Bile samples are immediately stored at −80 °C.
2. For NMR analysis, bile samples are thawed and centrifuged at 1500 × *g* for 30 min at 4 °C.
3. The supernatant is then freeze dried.
4. The dried residue is dissolved in a buffer containing D_2O or CD_3OD depending on the type of metabolites to be studied (aqueous or lipophilic metabolites) and transferred to 5 mm NMR tubes for analysis using NMR spectroscopy [9].
5. Lipid extraction from bile is carried out as follows: chloroform is added to bile diluted with water (0.8:1, v/v) and then methanol is added such that chloroform, methanol, and water is in the ratio 1:2:0.8 (v/v) before dilution of sample and 2:2:1.8 after dilution.
6. A phase separation in the solution occurs and the bottom chloroform layer contains lipids.
7. The methanol/water layer is removed by aspiration.
8. Chloroform containing the lipids is vacuum dried and dissolved in $CDCl_3$ for NMR acquisition of the sample [10, 12].

3.1.4 Tissue Extract

1. For lipids profiling using tissue, tissue is first pulverized under liquid nitrogen and ground into powder using mortar and pestle.
2. Lipid extraction is then carried out by sequential treatment of the tissue powder with methanol and chloroform in the ratio of 1:3.
3. Extracted lipid solution is vacuum dried and the residue is dissolved in deuterated chloroform and transferred to 5 mm NMR tube for NMR analysis [2].

3.1.5 Intact Tissues

1. Tissue samples are thawed at room temperature after taking out from freezer (−80 °C).
2. The samples are washed with deuterated saline to remove blood contents from tissues.
3. The tissue samples are weighed and then filled in zirconium oxide rotors.
4. 20 μl of D_2O is added to each rotor.
5. The filled rotor is then transferred to the spectrometer for analysis [6, 13].

3.2 NMR Experiments

NMR experiments for metabolite analysis in gallbladder cancer samples mostly involve detection through ^{1}H, ^{13}C, or ^{31}P nuclei. Acquisition of NMR spectra involves pre-acquisition steps such as locking of the magnetic field by deuterated solvent, tuning and matching of radiofrequency irradiation of nuclei in observation, shimming of the magnet (*see* **Note 1**) and water signal suppression (*see* **Notes 2** and **3**).

3.2.1 ^{1}H NMR

1. ^{1}H solution NMR can be carried out using a simple one pulse experiment to get a complete profile of the metabolite content of the sample.
2. For the one pulse experiment, a recycle delay of 5 s with 32,000 data points and spectral width of 12 ppm can be used [2]. The CPMG (Carr–Purcell–Meiboom–Gill) pulse sequence enables suppression of signals from macromolecules such as proteins and fatty acids. For CPMG, a relaxation delay of 5 s and 400 ms echo time can be used with a spectral width of 20 ppm and 64,000 data points [4] (*see* Fig. 2).
3. ^{1}H HR-MAS for tissue can also utilize the above-mentioned pulse sequences [6]. HR-MAS of tissue samples, which are semi-solids, involves the use of magic angle spinning (MAS) to suppress signals from anisotropic interactions. The spinning speed can be about 4 kHz.

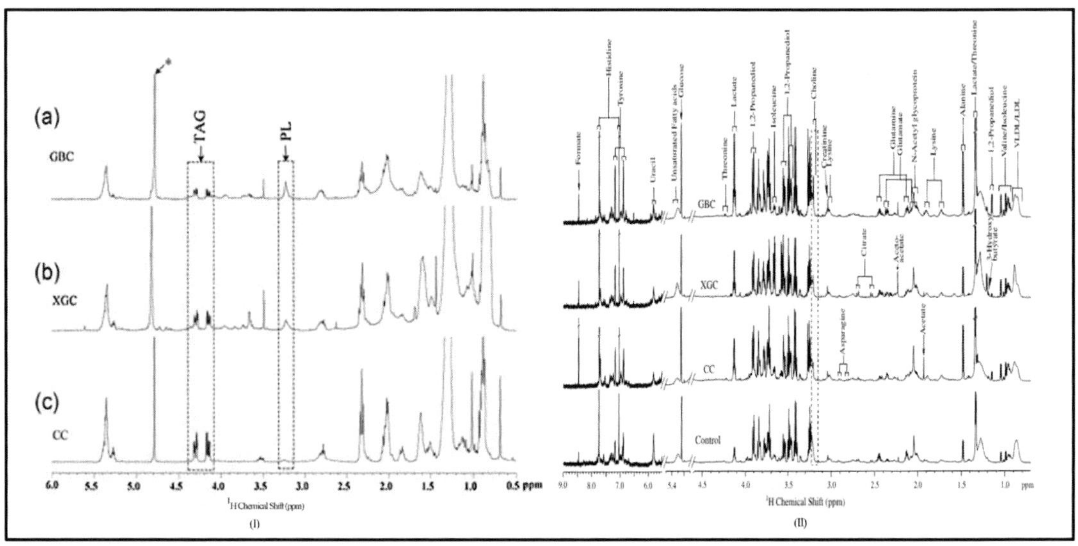

Fig. 2 (I) ^{1}H NMR spectra of lipid extracts of gallbladder tissues from the disease types: (**a**) gallbladder cancer (GBC); (**b**) xanthogranulomatous cholecystitis (XGC); and (**c**) chronic cholecystitis (CC), showing differences in some of the lipid components. PL, choline-containing phospholipids; TAG, triacylglyceride. (II) Stack plot of representative ^{1}H NMR CPMG spectra of control serum and serum from patients suffering from CC, XGC and GBC (from bottom to top) (Reprinted by permission from NMR Biomed., 24: 335-342, Copyright 2010 John Wiley & Sons, Ltd)

4. For samples containing water, water signal suppression can be achieved using weak radiofrequency irradiation methods. TSP for aqueous samples as well as TMS for hydrophobic samples is used for external referencing to 0 ppm.

3.2.2 ^{13}C NMR

1. Solid-state NMR experiments using cross polarization (CP) as well as direct polarization (^{13}C one pulse) of gallstones have been utilized to look at ^{13}C nuclei in the metabolites (*see* Fig. 3).
2. A CP contact time of 1 ms, recycle delay of 10 s, and MAS speed of 5 kHz with 2000 data points and spectral width of 315 ppm can be used.
3. Spectral referencing is carried out externally using carbonyl carbon peak of glycine at 174.14 ppm.

3.2.3 ^{31}P NMR

1. High-resolution proton-decoupled ^{31}P NMR reveals information about metabolites containing phosphorus (*see* Fig. 4).
2. A relaxation delay of 5 s with 16,000 data points and spectral width of 4000 Hz can be used for acquiring the spectra.
3. 85% H_3PO_4 solution in H_2O is used for external referencing of the chemical shift to 0 ppm [10].

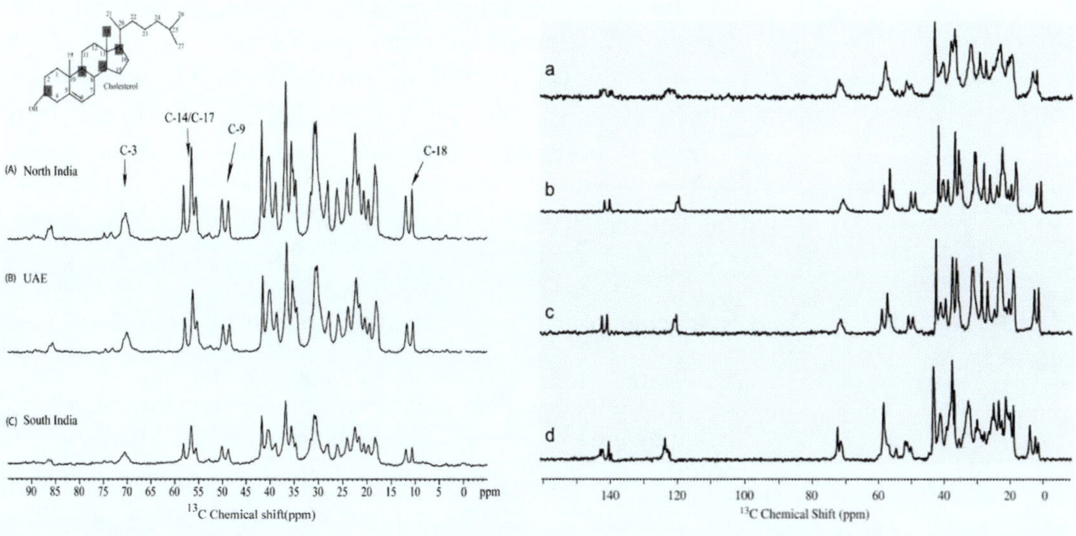

Fig. 3 (I) ^{13}C CPMAS NMR spectra of gall stones (GS) from different geographical regions showing: (**a**) high prevalence rates of both GS disease and GBC (North India), the inset shows the structure of cholesterol, which is an abundant constituent of GS; (**b**) a high prevalence rate of GS disease but low prevalence rate of GBC (UAE); and (**c**) low prevalence rate of both GS disease and GBC (South India). (II) Natural abundance ^{13}C CP-MAS NMR spectra of different types of human gallstones with cholesterol (**a**) in anhydrous form; (**b**) in monohydrate crystalline and amorphous form; and (**c**) in monohydrate crystalline form. The spectrum in (**d**) corresponds to pure cholesterol (Reprinted with permission from PLoS ONE (2016) 11(11): e0166351)

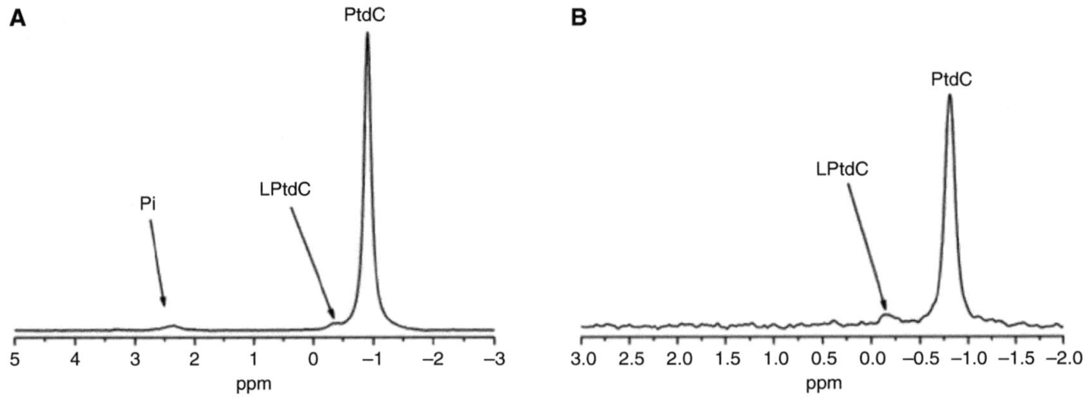

Fig. 4 (**a**) A portion of the ^{31}P NMR spectrum of intact bile obtained from a living donor with the labeling of phosphatidylcholine (PtdC) at −0.87 ppm, lysophosphatidylcholne (LPtdC) at −0.3 ppm, and inorganic phosphate (P_i) at 2.55 ppm. The chemical shifts are referenced to the external standard (H_3PO_4) at 0 ppm. (**b**) Same as in (**a**) obtained after lipid extraction using chloroform (Reprinted with permission from Biocybernetics and Biomedical Engineering, 31, 63-71, Copyright 2011 Elsevier Urban & Partner)

3.2.4 2D Methods

Apart from 1D NMR experiments, 2D methods such as TOCSY, COSY, and HSQC have also been utilized for metabolite identification.

1. For ^{1}H–^{1}H COSY, 2000 data points are collected in the t_2 domain with a spectral width of 12,820 Hz, a relaxation delay of 1.5 s, and an acquisition time of 95 ms.
2. For ^{1}H–^{13}C HSQC, 2k data points are collected in the t_2 domain with a spectral width of 12,820 Hz, a relaxation delay of 2.0 s, an acquisition time of 80 ms, and a 90° pulse of 9 μs [5, 6].
3. For ^{1}H–^{1}H TOCSY, 2k data points are collected in the t_2 domain with a spectral width of 10,683 Hz [5].

3.3 NMR Spectral Processing

The following points are taken into consideration for the data processing.

1. After zero filling and apodization, Fourier transformation followed by phase correction as well as baseline correction is applied.
2. Alignment of the spectra is done to facilitate binning of the spectra for the regions where changes are observed.
3. Integration of the spectral area provides relative quantitation of metabolite peaks.
4. Normalization and scaling methods are used for correction of various factors that arise due to dilution of the sample as well as due to changes in the intensity of certain peaks which could affect the contribution of other significant metabolites.

5. There are several software packages available online such as Bruker Topspin, PERCH, MVAPACK, MestReNOva, NMRProcFlow, Automics, and Chenomx NMR, which can easily process the spectra according to the user's requirements [14].

3.4 NMR Spectral Peak Assignment

The following steps are followed for NMR peak assignments.

1. Metabolite peak identification can be achieved by comparison of chemical shifts of peaks with online NMR chemical shift libraries that enable metabolite searching based on the chemical shift range for each peak.
2. Biological Magnetic Resonance databank (BMRB) and Human Metabolome Database (HMDB) online libraries are widely used for this purpose [14].
3. There are many other online software packages such as the Chenomx NMR suite profiler for automatic assignment of peaks.

3.5 Statistical Analysis

1. NMR data are subjected to statistical analyses to distinguish between control and disease groups. Statistical analysis also facilitates identification of genetic, environmental as well as other factors which could be responsible for the disease. It is, however, very important to validate the results obtained from such analysis using large and independent sample cohorts.
2. Univariate analyses such as Mann–Whitney *U*-test, Kruskal-- Wallis test, Analysis of variance (ANOVA), Bonferroni correction, Student's *t*-test, and Pearson's correlation coefficient are usually used for statistical analysis of NMR data.
3. Principal component analysis (PCA), partial least squares (PLS), discriminant analysis (DA), and a combination of these methods such as PLS-DA, orthogonal partial least squares (OPLS), and OPLS-DA are some of the multivariate statistical methods which are frequently utilized in analyzing NMR data for GBC. Apart from these methods, pathway enrichment analysis and heat maps are also useful in comparing different experimental groups.
4. It is now easy to use online software programs such as STATISTICA, MATLAB, MestReNova, Graphpad Prism, Unscrambler X, Metaboanalyst, Amix toolkit, etc. to apply statistical methods to NMR data, and to automatically analyze data (*see* Fig. 5).

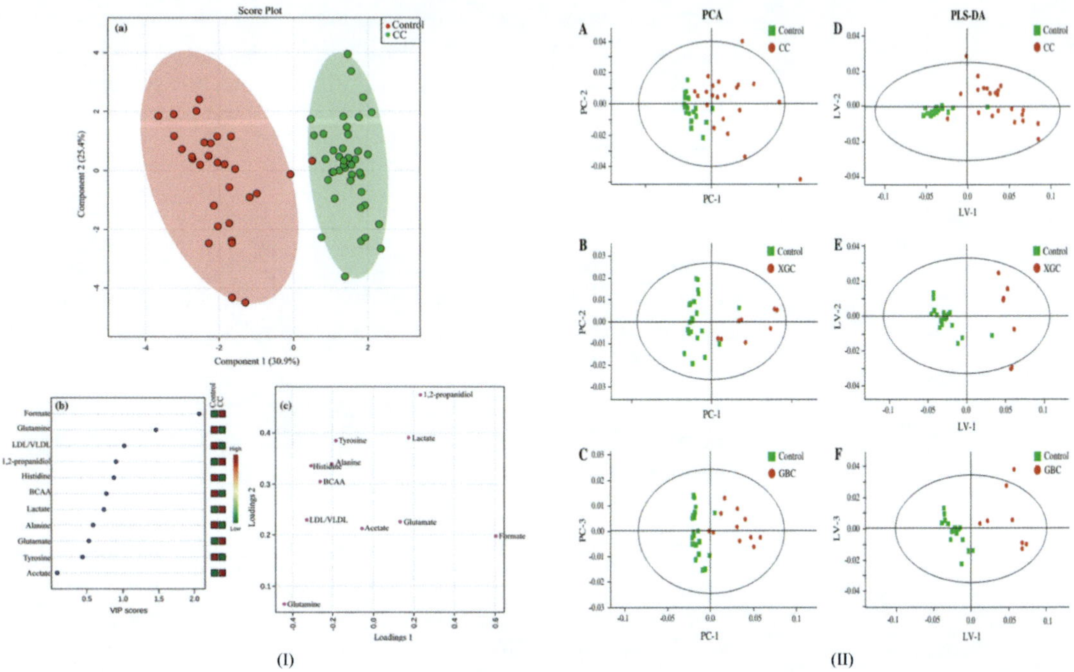

Fig. 5 (I) Statistical plots for data obtained from metabolomic analysis of CPMG NMR spectra of serum samples of patients with chronic cholecystitis (CC) and control group: (**a**) partial least squares discriminant analysis score plot; (**b**) variable importance in projection (VIP) plot showing the effect of the most important metabolites in the model: elevated levels for formate, lactate, glutamate and acetate, and reduced levels for glutamine, BCAA, glutamine, histidine, alanine, and tyrosine in CC; and (**c**) loadings plot obtained from PLS-DA analysis. (II) Results from the statistical analysis of low molecular weight metabolites in serum detected by CPMG NMR of CC, xanthogranulomatous cholecystitis (XGC), and gallbladder cancer (GBC): PCA score plot for (**a**) CC and control; (**b**) XGC and control; (**c**) GBC and control. PLS-DA score plot for (**d**) CC and control; (**e**) XGC and control; and (**f**) GBC and control (Reprinted with permission from Inflamm. Res. (2017) 66: 97. Copyright 2016 Springer International Publishing)

4 Notes

1. Shimming of the NMR magnet after loading the sample in the instrument is an important step to ensure good signal to noise as well as minimize the width at half height of the NMR peaks. This also ensures correct chemical shifts for the peaks.
2. Solvent suppression in solution NMR of aqueous samples is important in order to improve receiver response and avoid the overlapping of peaks from metabolites with the strong solvent peak.
3. Solvent suppression can be achieved using pulse sequences designed for water suppression. Numerous sequences exist including gradient-based methods with pulse sequences using shaped pulses, excitation sculpting, MEGA and SOGGY, pre-saturation, and combined methods such as WET are effective.

In addition, suppression of "far away" solvent, NMR relaxation differentiation, such as WEFT, radiation damping-related methods enhanced presaturation methods, enhanced WATERGATE methods, multi-resonance suppression, such as SLP, and PGSE diffusion methods, such as CONVEX, are used for solvent suppression in 1D as well as multidimensional NMR experiments, which have been listed in many review articles [15].

References

1. Kapoor V, McMichael A (2003) Gallbladder cancer: an "Indian" disease. Natl Med J India 16(4):209–213
2. Jayalakshmi K, Sonkar K, Behari A, Kapoor VK, Sinha N (2011) Lipid profiling of cancerous and benign gallbladder tissues by ^{1}H NMR spectroscopy. NMR Biomed 24(4):335–342
3. Sharma RK, Sonkar K, Sinha N, Rebala P, Albani AE, Behari A, Reddy DN, Farooqui A, Kapoor VK (2016) Gallstones: a worldwide multifaceted disease and its correlations with gallbladder carcinoma. PLoS One 11(11): e0166351
4. Sharma RK, Mishra K, Farooqui A, Behari A, Kapoor VK, Sinha N (2017) ^{1}H nuclear magnetic resonance (NMR)-based serum metabolomics of human gallbladder inflammation. Inflamm Res 66(1):97–105
5. Sonkar K, Behari A, Kapoor V, Sinha N (2013) ^{1}H NMR metabolic profiling of human serum associated with benign and malignant gallstone diseases. Metabolomics 9(2):515–528
6. Bharti SK, Behari A, Kapoor VK, Kumari N, Krishnani N, Roy R (2013) Magic angle spinning NMR spectroscopic metabolic profiling of gall bladder tissues for differentiating malignant from benign disease. Metabolomics 9(1):101–118
7. Jayalakshmi K, Sonkar K, Behari A, Kapoor V, Sinha N (2009) Solid state ^{13}C NMR analysis of human gallstones from cancer and benign gall bladder diseases. Solid State Nucl Magn Reson 36(1):60–65
8. Srivastava M, Sharma A, Kapoor VK, Nagana Gowda GA (2008) Stones from cancerous and benign gallbladders are different: a proton nuclear magnetic resonance spectroscopy study. Hepatol Res 38(10):997–1005
9. Ko H, Choi I, Chang K, Jeong G, Gong G, Seo H, Ryu D, Lee KG, Choi D, Chung H (2016) Amphiphilic metabolites in gallbladder bile: potential biomarkers for gallbladder diseases. Appl Spectrosc Rev 51(7–9):706–717
10. Toczyłowska B, Piotrowski M, Chalimoniuk M (2011) ^{31}P high resolution NMR spectroscopy in analysis of phosphate-containing compounds of bile. Biull Eksp Biol Med 31 (1):63–71
11. Wen H, Yoo SS, Kang J, Kim HG, Park J-S, Jeong S, Lee JI, Kwon HN, Kang S, Lee D-H (2010) A new NMR-based metabolomics approach for the diagnosis of biliary tract cancer. J Hepatol 52(2):228–233
12. Bligh EG, Dyer WJ (1959) A rapid method of total lipid extraction and purification. Can J Biochem Physiol 37(8):911–917
13. Srivastava S, Roy R (2010) 1H HRMAS NMR based metabolomics of benign and malignant neuro-endocrine tumors and its comparision with oral cancer and benign gall bladder tissues. Proc. Intl. Soc. Mag. Reson. Med. 18,4796
14. Ellinger JJ, Chylla RA, Ulrich EL, Markley JL (2013) Databases and software for NMR-based metabolomics. Curr Metabolomics 1 (1):28–40
15. Ranjan R, Sinha N (2018) Nuclear magnetic resonance (NMR)-based metabolomics for cancer research. NMR Biomed:e3916

Chapter 15

Biomarker Discovery Using NMR-Based Metabolomics of Tissue

Maria T. Grinde, Guro F. Giskeødegård, Trygve Andreassen, May-Britt Tessem, Tone F. Bathen, and Siver A. Moestue

Abstract

NMR-based metabolomics has shown promise in the diagnosis of diseases as it enables identification and quantification of metabolic biomarkers. Using high-resolution magic-angle-spinning (HR-MAS) NMR spectroscopy, metabolic profiles from intact tissue specimens can be obtained with high spectral resolution. In addition, HR-MAS NMR requires minimal sample preparation and the sample is kept intact for subsequent analyses. In this chapter, we describe a typical protocol for NMR-based metabolomics of tissue samples. We cover all major steps ranging from tissue sample collection to determination of biomarkers, including experimental precautions taken to ensure reproducible and reliable reporting of data in the area of clinical application.

Key words High-resolution magic-angle-spinning MR spectroscopy, Tissue analysis, Biopsies, Metabolism, Metabolomics, Biomarkers, Protocols

1 Introduction

Endogenous metabolites serve as diagnostic biomarkers for a variety of medical conditions, since disease frequently is associated with altered concentrations of one or more metabolites. A universally known example is phenylketonuria, an inherited genetic disorder where dysfunctional metabolic degradation causes buildup of phenylalanine to potentially toxic levels [1]. An elevated phenylalanine level in blood is the diagnostic biomarker used when newborns are screened for this disease. Thus, blood and other biofluids serve as matrices for metabolic phenotyping and biomarker detection in many diseases. However, in the search for novel biomarkers, direct analysis of the affected tissue has been suggested as a potential method for identification of metabolites or metabolic patterns that hold prognostic or predictive information. This approach has been widely applied to cancer, but analysis of biopsy material is also

G. A. Nagana Gowda and Daniel Raftery (eds.), *NMR-Based Metabolomics: Methods and Protocols*, Methods in Molecular Biology, vol. 2037, https://doi.org/10.1007/978-1-4939-9690-2_15,

reported in various inflammatory diseases, pre-eclampsia, neurological disorders, and more [2–5].

High-resolution magic-angle-spinning (HR-MAS) NMR spectroscopy has emerged as a technique that is highly suitable for metabolic analyses of intact tissue specimens ex vivo. The technique, based on the principles of NMR, was first described by Andrew in 1958 [6], and differs from the solution-state NMR spectroscopy in that it involves spinning the solid or semisolid specimen at the magic angle, 54.7°, with respect to the main magnetic field. This causes averaging of orientation-dependent interactions between the nuclear spins and their surroundings, including chemical shift anisotropy, dipole–dipole, and quadrupolar couplings. This averaging causes narrowing of the signals in the NMR spectra and results in high spectral resolution similar to the solution-state spectra. Generally, the rotor containing the tissue specimen is spun at high speed (typically 3–5 kHz) activated by a flow of pressurized air through the NMR probe, which also leaves the sample intact for further analyses. From the resulting NMR spectra, individual metabolites can be identified and quantified. Alternatively, metabolic patterns or "fingerprints" representing various disease states can be identified.

For discovery and validation of novel biomarkers for potential use in clinical practice, it is of outmost importance that the analytical methods can be trusted to provide accurate and reproducible data. Thus, attention must be paid to rigorous quality assurance and validation of the methods, and standardized protocols optimized for generation of reliable data must be used.

Obtaining metabolic information using HR-MAS NMR for clinical use depends on built-in systems for quality assurance and validation to ensure that the laboratory meets relevant criteria for robustness and analytical accuracy. The performance of the NMR spectrometer must be monitored regularly. Sample handling protocols must ensure that the metabolic state of the sample is kept stable from collection to analysis. Importantly, the NMR spectra reflect the composition of a complex biological sample and care must be taken to process and analyze the raw NMR data in an objective and standardized way, without introducing systematic bias or error. Therefore, a reliable and robust pipeline for data management must be defined. In this chapter, we provide detailed guidance on how to perform HR-MAS NMR analysis of tissue specimens for the purpose of identification and reporting of diagnostic metabolic biomarkers. Important steps from harvesting of tissue material to analysis of the NMR spectra are covered.

2 Materials

2.1 Sample Collection

The protocol includes materials/equipment needed to handle tissue samples following collection during open surgery, needle biopsy, or similar procedures. Equipment needed for surgery/needle biopsy varies between procedures and is not described.

1. Cryo-bucket filled with liquid nitrogen.
2. Safety equipment: Lab coat, safety glasses, and disposable gloves (*see* **Note 1**).
3. Tweezers, scalpel, blades, and cryo-tubes.

2.2 Sample Preparation

1. Safety equipment: Lab coat, safety glasses, and disposable gloves (*see* **Note 1**).
2. Cryo-bucket with liquid nitrogen for sample storage and cap removal.
3. Workstation with liquid nitrogen (*see* **Note 2**) or ice block.
4. Scalpel and blades, tweezers, scale, and weighting boats.
5. D_2O (99.8%) with HCOONa (~25 mM). Pipette and tips (pipette volume >3 μL).
6. HR-MAS accessories: (*see* Fig. 1):

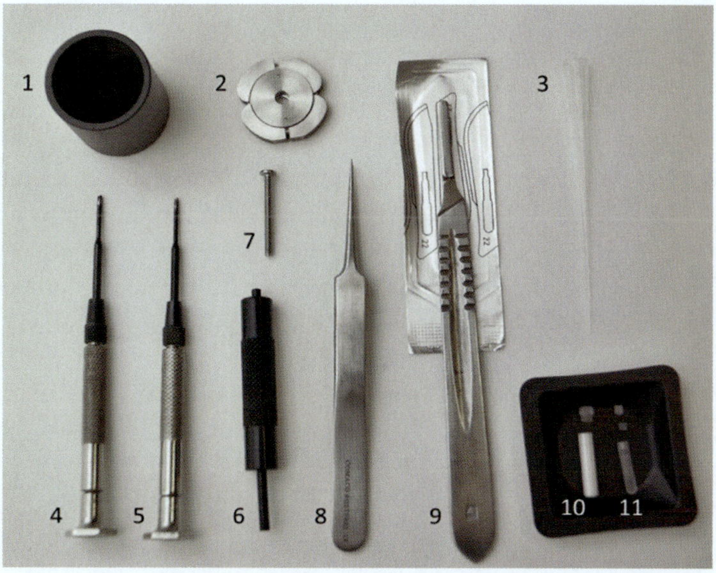

Fig. 1 Typical accessories used for HR-MAS NMR analysis of tissue samples. (1) Filling funnel; (2) rotor cap remover; (3) narrow pipette tip for adding D_2O with HCOONa; (4) blunt-ended screwdriver; (5) forked screwdriver; (6) rotor packer; (7) extraction screw; (8) tweezers; (9) scalpel; (10) rotor and spinning cap; (11) insert with taper and screw cap

(a) 4 mm ZrO_2 rotors (either volume 50 or 80 μL) with spinning caps (*see* **Note 3**).
(b) Spacers (for 50 μL rotors only).
(c) Disposable inserts, plugs, and screws (80 μL rotors only).
(d) Filling funnel, blunt-ended screwdriver, and forked screwdriver. Permanent marker.

7. Rotor cap remover and extraction screw for recovery of the sample after analysis.
8. Chemicals for temperature calibration, magic angle adjustments, and stability test: KBr salt, 4% methanol dissolved in methanol-d_4.

2.3 HR-MAS NMR Acquisition

1. NMR spectrometer with an HR-MAS probe (*see* **Note 4**).
2. Compatible user interface software, such as TopSpin for Bruker instruments (*see* **Note 5**).

2.4 Data Analysis

1. Database and software for identification of metabolites, for example, Human Metabolome Database (HMDB), Chenomx, or MNova.
2. Software for quantification of metabolites, for example, TopSpin, LCModel, MATLAB, or R.
3. Software for data analysis, for example, MATLAB, R, The Unscrambler, SIMCA, or MetaboAnalyst.

3 Methods

3.1 Sample Collection and Storage

Tissue samples must be collected under conditions which ensure that the integrity of the tissue metabolite profiles is preserved. However, patient safety is of paramount importance. The sample collection procedure must therefore be adapted to the biopsy procedure/surgical procedure.

1. Collect the biopsy (needle biopsy, surgical biopsy, etc.) from the organ or tissue of interest. The procedure must be performed by a health care professional. The time from sample collection until freezing should be minimized to avoid tissue degradation. We recommend <5 min from collection until snap freezing. For surgical biopsies, the time from stopping blood circulation to the tissue to snap freezing should be as short as possible.
2. Using a pair of tweezers, put the tissue sample into the cryotube and snap freeze in liquid nitrogen.
3. Store the sample either in liquid nitrogen or at −80 °C until analysis.

3.2 Sample Preparation for HR-MAS NMR

The sample needs to be cut with a scalpel to fit into a disposable insert or the rotor itself. To avoid tissue degradation, it is important to keep samples cold or, preferably, frozen during sample preparation. We recommend preparing all samples on an ice block or on a workstation filled with liquid nitrogen. We also recommend thorough cleaning of rotors and inserts prior to analysis, and inspection of rotors and inserts to ensure the absence of any contaminants that can interfere with tissue analyses (*see* **Note 6**).

1. Place the insert or the rotor (50 μL) on the scale and tare.
2. Add at least 3.0 μL D_2O with HCOONa (25 mM) to the insert (or rotor) and weigh. Tare the scale afterwards.
3. Put the weighing boat either on an ice block or on a workstation with liquid nitrogen.
4. Put the frozen tissue sample on the weighing boat on the ice block/cold workstation and use a scalpel to carefully cut the sample into an appropriate size (typically between 5 and 45 mg) so that it fits in the sample insert or rotor (*see* **Note 7**).
5. Transfer the sample carefully into the bottom of the insert or into the rotor itself directly using tweezers. Make sure the sample is evenly distributed (*see* **Note 8**) and avoid air bubbles (*see* **Note 9**).
6. Weigh the sample.
7. If sample insert is used, put the plug into the insert using a blunt-ended screwdriver. Place the screw into the plug and tighten (*see* **Note 10**), then put the insert inside the rotor. If a 50 μL rotor is used, put the spacer into the rotor using the screw, and adjust the depth of the spacer using a rotor packer.
8. Put the cap on the rotor. This can be done using a filling funnel.
9. Label the bottom of the rotor using a permanent marker. The sample is now ready for HR-MAS NMR acquisition.

3.3 HR-MAS NMR Acquisition

The following section describes how 1D and 2D ^{1}H and ^{13}C HR-MAS NMR spectra from tissue samples can be acquired. Before acquiring MR spectra, it is important to calibrate the temperature (*see* **Note 11**) and to ensure that the magic angle is adjusted correctly (*see* **Note 12**). We recommend performing temperature calibration and magic angle adjustment at least once per week during the analysis period. We also recommend analysis of a standard sample weekly and logging the results (*see* **Note 13**), to detect any changes in NMR system performance.

1. Transfer the sample into the magnet.
2. Spin the sample at 5000 Hz.

3. Set the temperature to a desired value based on the last temperature calibration (*see* **Note 14**). Wait until temperature is stable.
4. Tune and match the probe for the nuclei of interest (^1H, ^{13}C, etc.). Use the rods that extend from the bottom of the HR-MAS probe labelled "T" (for tuning) and "M" (for matching) (*see* **Note 15**).
5. Lock the B_0 field using D_2O (*see* **Note 16**).
6. Acquire a spectrum with a fewer number of scans (NS $\approx$ 4) without applying line broadening. Measure full width at half maximum (FWHM) from one peak (*see* **Note 17**).
7. If FWHM > ~1.5 Hz, perform shimming by adjusting the current in the shim coils: X, XZ, XZ^2 and Z^4 (or Y, YZ, YZ^2, and Z^4, depending on the probe orientation) (*see* **Note 18**).
8. Determine the 90° ^1H pulse (P1) by acquiring an NMR spectrum using the one-pulse sequence (*see* **Note 19**).
9. Determine the optimal frequency for water suppression (O1) using an NMR sequence with water suppression (*see* **Note 20**).
10. Set up NMR experiment using an appropriate pulse sequence and insert optimized parameters (pulse angles, transmitter frequency offset (O1), etc.). Some commonly used pulse sequences are described in the next section.

3.4 NMR Pulse Sequences and Acquisition Parameters

The aim of this section is to briefly describe some of the most commonly used NMR pulse sequences used during tissue analysis. Each NMR pulse sequence is a series of RF pulses, delays, and gradient pulses designed to excite nuclei and create desired signals from samples. The resulting NMR signal is detected by the receiver coil, digitized, and stored as a free induction decay (FID) on a computer. After Fourier transformation, this FID is transformed into a one-dimensional (1D) NMR spectrum. By introducing a second time variable in the pulse sequence (i.e., an incremented delay) and double Fourier transformation of the data collected as a function of two time variables, a two-dimensional (2D) NMR spectrum is generated. 2D NMR spectra are useful for identifying unknown metabolites. Generally, the pulse sequences used for HR-MAS NMR are the same as used for liquid samples, but some sequences require special consideration due to sample spinning. This is especially important for TOCSY experiments, where rotor synchronization is recommended.

1. One-pulse sequence: The one-pulse sequence is the simplest of the NMR pulse sequences. It consists of a single 90° pulse that flips the z-magnetization to the xy-plane. The freely evolving signal is then detected as FID and processed to get 1D spectrum.

2. One-pulse sequence with presaturation: The is the same as the one-pulse sequence, but has an additional low power pulse to saturate the signal from water abundantly present in all tissue samples. Effectively suppressing the water signal is critical for detecting low concentration metabolite signals from tissue.
3. 1D NOESY with presaturation: 1D nuclear Overhauser effect spectroscopy (NOESY) is probably the most frequently used experiment in NMR-based metabolomics studies. It offers better water signal suppression; it consists of three 90° RF pulses apart from a long presaturation pulse for water suppression. Gradient pulses are used in the sequence to purge the residual water signal.
4. Carr–Purcell–Meiboom–Gill pulse sequence with presaturation: Macromolecules (mainly lipids in tissue) will cause broad peaks in the NMR spectra. The Carr–Purcell–Meiboom–Gill (CPMG) sequence is designed to suppress NMR signals from lipids. This sequence uses a 90° pulse and a series of 180° pulses to create spin echos. Signals from molecules with short T_2 relaxation (large molecules such as lipids) decay faster and consequently are suppressed in the NMR spectra. The pulse sequence is generally used with presaturation to suppress signal from H_2O.
5. COrrelated SpectroscopY (COSY): COSY is a homonuclear two-dimensional experiment used to identify scalar coupled ^1H nuclei in the sample. There are many variants of COSY pulse sequences available, all having their pros and cons. In its most basic form, the COSY experiment consists of a 90° pulse, a t_1 delay which is increased stepwise, and a second 90° pulse. The FID is acquired after the second 90° pulse. A problem with the basic COSY is that diagonal and cross peaks have a 90° phase difference, meaning that large dispersive tails from diagonal peaks can mask nearby cross peaks. This can be remedied by using double-quantum filtered (DQF) COSY, where both diagonal and cross peaks have the same phase. DQF-COSY also suppresses signals from singlets (including water) and provides information of *J*-couplings from the cross peak fine structure. The downside of DQF-COSY is a reduction of sensitivity (theoretically 50%) compared to conventional COSY and it requires high resolution to obtain the multiplet fine structure.
6. TOtal Correlated SpectroscopY (TOCSY): TOCSY is similar to COSY spectra; however, it shows cross peaks between all sequentially connected spins. The pulse sequence for TOCSY is similar to that of COSY; in place of the second 90° pulse, TOCSY uses spin-lock pulses to transfer magnetization between all sequentially connected spins. Selecting the appropriate spin-lock duration is crucial for determining how far the

magnetization will be transferred through the spin system. Because of faster relaxation, however, it is recommended to use a shorter spin-lock time in HR-MAS NMR compared to liquid-state NMR. The optimal value will depend on the sample type, but a compromise value of 80 ms should be a reasonable starting point for tissue samples [7]. The choice of the spin-lock sequence is also of importance, and it is recommended to use an adiabatic pulse scheme to better handle RF field inhomogeneity.

7. ^{1}H *J*-resolved: The 2D ^{1}H homonuclear *J*-resolved pulse sequence produces an NMR spectrum with chemical shift values along the first axis and ^{1}H–^{1}H couplings along the second axis. Briefly, the projection along the first axis is a decoupled proton spectrum, while the second axis shows the splitting patterns from ^{1}H–^{1}H couplings.
8. 1D ^{13}C pulse sequence: Despite the low sensitivity, ^{13}C NMR has several advantages over ^{1}H. First, a large chemical shift dispersion in 1D ^{13}C NMR allows easy identification of metabolites with less peak overlap (~200 ppm for ^{13}C, compared to ~15 ppm for ^{1}H). In addition, ^{13}C NMR is most suitable for isotopic labeling studies to track the passage of labeled ^{13}C through metabolic pathways in cell cultures, animal models, and even humans. ^{13}C NMR spectra can be acquired using a single pulse sequence with no need for water suppression. With off-resonance decoupling, ^{13}C peaks show multiplicity due to *J* coupling with ^{1}H. Proton decoupling removes ^{13}C–^{1}H *J* couplings. Power-gated decoupling, where decoupling is on during the entire experiment, is commonly used. This leads to nuclear Overhauser enhancement (NOE) to carbon signals disproportionately, and hence, such spectra are not quantitative. On the other hand, inverse-gated decoupling, in which decoupling is on only during acquisition, removes the NOE effect and provides quantitative spectrum. The use of a shorter (30°) excitation pulse allows the use of shorter relaxation delays and more rapid pulsing compared to a 90° pulse.
9. Heteronuclear single-quantum coherence (HSQC): 2D ^{1}H–^{13}C HSQC NMR spectroscopy provides single-bond correlations between ^{1}H and ^{13}C nuclei. The experiment is based on proton detection, and hence, it offers higher sensitivity compared to carbon-detected experiments. The spectral resolution in the indirect dimension (^{13}C) is usually kept low to shorten experiment time, and thus, closely spaced peaks may be difficult to distinguish if they are not resolved well in the direct (^{1}H) dimension. Carbon atoms not bound to hydrogen (i.e., carbonyls) are not detected by the HSQC experiment. The sensitivity-enhanced HSQC sequence with a second refocusing period doubles the theoretical signal-to-noise ratio. For

HR-MAS NMR, this prolonged pulse sequence could actually provide lower signal-to-noise ratio, due to faster relaxation. In general, we suggest using the sensitivity-enhanced version, but for samples with broader peaks, a sequence without the additional refocusing is recommended for better results (*see* **Note 21**).

10. Heteronuclear multiple-bond correlation (HMBC): 2D ^{1}H–^{13}C HMBC NMR is similar to 2D ^{1}H–^{13}C HSQC NMR. However, in HMBC cross peaks for two, three, and sometimes four bonds, ^{1}H–^{13}C couplings can be detected, unlike HSQC. HMBC is useful for the assignment of quaternary and carbonyl carbons, and protons bound to other nuclei than ^{13}C. Above all, it is very useful for identifying chemical structure for unknown metabolites. There are many variants of HMBC. It has been shown that constant-time (CT) HMBC performs well under HR-MAS conditions [8]. The intensity of cross peaks depends on the ^{1}H–^{13}C coupling constants and the delay (Δ) in the pulse sequence. Although longer delays are useful for smaller ^{1}H–^{13}C J couplings, the prolonged pulse sequence due to longer delays may cause signal loss due to extended relaxation. As a compromise, a delay of 70 ms is often used, which is optimal for couplings of 7 Hz, even though most coupling constants are smaller than 5 Hz. In HR-MAS, relaxation effects can be more pronounced than for liquid samples, and it might be necessary to reduce delays for better sensitivity. The low resolution in the ^{13}C dimension can sometimes make it difficult to distinguish closely spaced peaks; as an alternative, the selective HMBC experiment, which excites a part of the ^{13}C region (e.g., carbonyls) and provides higher resolution for the selected spectral window, can be used.

3.5 Choice of Acquisition Parameters

This section provides a practical guide for various acquisition protocols to get high-quality HR-MAS NMR spectra with examples for trade-offs between experimental duration and signal-to-noise ratio.

1. Non-uniform sampling (NUS): The data acquisition time for multidimensional NMR can be reduced using NUS, which only acquires a subset of the data points in the indirect dimension, and using these points, the remaining points can be predicted. Sampling only a fraction of the data points reduces the experiment time by the same fraction. The time saved using NUS can be reinvested in more scans, allowing better signal-to-noise ratio for weak samples.
2. Number of scans (NS): NS used depends on the amount of sample and the pulse sequence. For ^{1}H NMR spectra using NOESY and CPMG pulse sequences (5–40 mg samples),

128 or 256 scans are commonly used. For 1D ^{13}C NMR acquisition, the typical NS is between 4k and 32k.

3. Transmitter frequency offset (O1): It is important to set the correct O1 value in the ^{1}H dimension for optimal water suppression.
4. Pulse angle: For ^{1}H experiments, determine the 90° pulse length to set correct pulse angles. This should also be done for other nuclei utilizing the ^{1}H channel. The ^{13}C pulse angle, however, is much less effected by the sample, and hence, it is not calibrated each time.
5. Parameters for quantitative NMR spectra: These spectra are typically acquired using 1D pulse sequences such as NOESY for ^{1}H and zgig for ^{13}C. For quantitation, a long repetition time (TR) needs to be used to avoid T_1 relaxation bias. The T_1 relaxation effect is negligible when TR is more than five times T_1 (*see* **Note 22**). If a short TR is used, it is also possible to compensate to achieve absolute quantification, as described in [9]. For absolute quantification, spectra from calibration standards (e.g., creatine or formate) need to be obtained as described in Subheading 3.8.
6. Receiver gain: The receiver gain (RG) should be the same for all samples if quantitative NMR spectra shall be acquired (*see* **Note 23**).
7. Repetition time: For 1D NMR spectra, the repetition time (TR) can be estimated by summing the relaxation delay and the acquisition time (D1 + AQ).

3.6 Processing of NMR Spectra

Preprocessing of spectra can be performed using TopSpin or other software such as SpinWorks and MNova. Post-processing is typically performed in more general software platforms like MATLAB or R. We describe below pre-processing using the TopSpin user interface and post-processing using a customized MATLAB pipeline.

1. Pre-processing of 1D spectra.
   (a) Zero filling, apodization, and Fourier transformation:
      - Zero filling refers to adding additional zero data points to the FID before Fourier transformation. In TopSpin, this is done by increasing the size of real spectrum (SI) in the ProcPars tab (*see* **Note 24**). Zero filling is used to create spectra with more defined multiplets, but does not add any additional information to the dataset.
      - Apodization involves applying a window function to the FID before Fourier transformation. The goal is to enhance either the SNR or resolution in the NMR

spectra. The most commonly used window function for enhancing SNR is the exponential multiplication (EM) function using a line broadening (LB). The magnitude of line broadening (LB) can be specified (*see* **Note 25**). The second group of window functions aim to enhance resolution at the expense on SNR. Most commonly used window functions are Gaussian (GM) and sine bell (SINE) multiplication functions.

- The resulting FID must then be Fourier transformed to generate the NMR spectrum.

(b) Phasing: Automatic phase correction can be done in TopSpin. However, NMR spectra from tissue samples may in some cases benefit from manual adjustment. Manual phasing needs to be done in two steps. First, phase the reference peak (typically the tallest peak in the spectrum, but not the residual water peak) using the zero-order phase correction. Then, phase the remaining peaks of the spectrum using the first-order phase correction.

(c) Baseline correction: The spectra can be baseline corrected automatically in TopSpin. Manual baseline correction is also possible using different baseline functions (polynomial, sine, and exponential).

(d) Chemical shift calibration: Proper chemical shift calibration is important. Select a reference peak, zoom into the peak, use the cursor to define the middle of the peak, and set its chemical shift value (in ppm) (*see* **Note 26**).

2. Processing of 2D NMR spectra.

(a) Fourier transformation and apodization.

- The FID must be Fourier transformed in both dimensions to generate the NMR spectrum.
- Window functions can be applied in 2D NMR spectra. Commonly used window functions for 2D are sine-bell squared (QSINE) and sine bell (SINE).

(b) Phasing: We recommend manual phasing for 2D NMR spectra. Define two or three rows/columns that contain peaks (right-click using the cursor and click add). Then, select phase correction for columns/rows. Perform zero-order phase correction on the first peak and first-order phase correction on the other peaks. HMBC spectra are not phase sensitive and phasing can therefore be omitted.

(c) Chemical shift calibration: Use a peak in the 2D spectrum with known ppm values for both dimensions. Select calibrate axis and right-click in the middle of the peak and set the correct ppm values.

3.7 Post-processing of 1D NMR Spectra for Metabolomics Analyses

To evaluate multiple NMR spectra, peaks in the spectra must be aligned and the spectra normalized.

1. Peak alignment:
   (a) Choose a method for peak alignment that is suitable for your data, for instance, icoshift [10] or COW [11] (*see* **Note 27**). Alternatively, for smaller peak shifts, binning procedures may be applied. In binning, small spectral regions are summed into one bin (*see* **Note 28**).
   (b) Most methods require alignment to a reference spectrum. The spectrum with the highest correlation to the remaining spectra might be a good choice. The mean or median spectrum may also be a good candidate [12].
   (c) After peak alignment, visually inspect the results to ensure that the correct peaks have been aligned to each other.
2. Spectral normalization: In order to correct for differences in tissue amount and tissue content, and to make all spectra comparable, normalization of each spectrum should be performed. Methods used for normalization are scaling to sample weight, mean area normalization, scaling all spectra to an equal total peak area, and PQN normalization [13].
   (a) Remove spectral regions with no peaks, the residual water peak and potential contaminants (*see* **Note 29**). In addition, remove residual lipid peaks from tissue with high lipid content such as breast tissue.
   (b) Normalize the spectra using the chosen method.

3.8 Metabolite Peak Assignment

Identification of most of the metabolites detected in tissue by NMR has been described previously. A number of metabolite databases are now freely available including the Human Metabolome Database (HMDB, www.hmdb.ca) [14], which is currently the largest repository of spectroscopic data from human samples. HMDB includes a search function that allows the user to match peaks from 1D or 2D NMR data with the database; it provides a list of potential candidates. The problem, however, is that most spectral data in the database are from solution state and at different temperatures than normally used for HR-MAS NMR. Slight deviations in chemical shifts should, therefore, be considered in the metabolite search.

2D NMR spectra help confirm the assignment of 1D NMR spectra and/or elucidate the structure of novel metabolites. In the interest of time, a common practice is to acquire 2D spectra on selected samples after 1D NMR analysis of all samples. Thus, selected samples must have high levels of the metabolites that need confirmation or structure elucidation. Between the two analytic runs, the samples must be stored at −80 °C or lower to avoid degradation.

Although it is possible to elucidate chemical structures from HR-MAS spectra, usually the data are insufficient for unknown metabolite identification. Tissue extraction and analysis using other spectroscopic methods (usually MS with chromatographic separation) will therefore be required. Alternatively, if the standard compound for the suspected metabolite is available, perform a spike-in experiment to confirm the peak identity; the chemical shifts of the spiked-in compound must match perfectly with the peaks under investigation.

Commercial software platforms such as Chenomx, MNova, and Amix are also useful tools for assigning peaks using reference spectra.

3.9 Quantification of Peaks in NMR Spectra

Quantification of individual metabolites is based on accurate measurement of area under the peak. This can be challenging since peak overlap and broad signals from lipids and macromolecules are frequently observed in HR-MAS. Metabolites can be quantified based on either manual peak fitting or automatic fitting [15–18]. For absolute quantification, one possibility is to use the calibrated ERETIC (Electronic REference To access In vivo Concentrations) signal as a reference [16]. The ERETIC signal is calibrated using a solution of a standard compound. The inherent stability of modern NMR spectrometers also allows quantification using pulse length based concentration determination (PULCON) [17]. This enables direct quantification based on the spectrum of a reference sample. Other options for quantification are using the BATMAN algorithm in R [19] or commercial software such as Chenomx. It should, however, be noted that absolute quantification may not always be feasible or necessary; depending on the experimental design and research objective, relative concentrations or metabolite ratios may be used.

3.10 Multivariate Analysis of ^1H NMR Spectra

Spectral data are generally complex and hence require specialized analysis methods that can handle such data. The unsupervised, principal component analysis (PCA), and the supervised, partial least squares (PLS), methods are most commonly used multivariate methods for spectral data or quantified metabolites. The analyses can be performed using software such as R, MATLAB (with or without PLS_Toolbox), Unscrambler, and SIMCA.

In PCA, principal component 1 (PC1) explains the direction of largest variation in the data; the second component, PC2, which is orthogonal to PC1, explains the next largest variation, and so on. In PLS analysis, latent variables are defined to maximize the covariance between the input data (spectra or quantified metabolites) and a response variable. PLS regression (PLSR) is used for a continuous response variable, while categorical variables can be modeled by PLS discriminant analysis (PLS-DA). Usually, the first few principal components or latent variables are sufficient to describe important aspects of the data.

The results of PCA and PLS can be visualized in scores and loadings plots. Scores represent the coordinates of the samples in the new coordinate system defined by the model, while loadings represent the weights of each original variable that represent the model.

PCA is particularly useful for an initial analysis of the data. It can detect possible outliers and natural groupings of the samples. Spectral data should be mean-centered before analysis, while quantified data should be either mean-centered or autoscaled (*see* **Note 30**). On the other hand, PLS analysis is used to model the data according to a clinical variable of interest. PLS-R is used for a continuous variable and PLS-DA for categorical variables. Validation of the PLS model is important as multivariate methods are prone to overfitting. Choose a method for validation suitable for your sample size [20, 21].

4 Notes

1. When handling liquid nitrogen, you should use cryo-gloves, lab coat with long sleeves, and pants worn outside the shoes.
2. The cooling workstation can be a homemade system consisting of a styrofoam block containing liquid nitrogen, a cooling block made of metal, and plexiglass for minimizing nitrogen evaporation, as described in [22]. The cooling block must first be cooled by placing it into liquid nitrogen. After 5–10 min, transfer the cooling block to the top of the workstation. Pour liquid nitrogen into the workstation to keep the cooling block cold for several hours. Place the plexiglass on top to reduce nitrogen evaporation.
3. The choice of rotor depends on the design of the HR-MAS probe and the type of application. We recommend a 50 μL rotor when the availability of tissue is high and if sensitivity is an issue (i.e., 1D ^{13}C spectra). However, 50 μL rotors require thorough cleaning compared to disposable inserts placed in 80 μL rotors. For small samples (<20 mg), we recommend disposable inserts placed in 80 μL rotors.
4. The following description pertains mainly to Bruker spectrometers and the TopSpin user interface. However, many of the general features presented also apply to instruments from other manufacturers.
5. Table of relevant commands in the Bruker TopSpin user interface:

Command	Action
wobb	Wobbling for tuning and matching magnet
lock	Lock the magnetic field
gs	Repeated execution of the current pulse program
zg	Start acquisition command
zg	Single pulse NMR sequence
zgpr	Single pulse NMR sequence with presaturation
ft	Fourier transform
efp	Fourier transform with window function and phase correction
lb	Line broadening
apk	Automatic phasing of spectra
abs	Automatic baseline correction
cal	Calibration of spectrum chemical shift
xfb	Fourier transform in both directions (2D data)
rsh	Read existing shim file
wsh	Write new shim file

6. We have observed various contaminants in batches of disposable inserts (traces of production oil) and in new rotors. We recommend the following cleaning procedures for the zirconium rotors: wash rotor with dichloromethane (DCM) by pipetting up and down at least three times, and then place rotor in a glass beaker with DCM for 15 min. Make sure that the rotor is filled with DCM. Repeat the washing procedure by pipetting DCM up and down three times, and then place rotor in DCM for 15 more min. Afterwards, rinse rotor with DCM and let the rotor dry in fume hood for 30 min.

   For quality-control purposes, we additionally recommend periodically acquiring the NMR spectrum of a disposable insert from newly procured batches as well as the rotor containing only D_2O to make sure inserts and rotor are devoid of contaminants. Before use, we recommend that inserts be soaked overnight in 100% ethanol and then dried at room temperature for a couple of days.

7. If the tissue sample is too cold, it is very difficult to cut. The trick is to use a warm scalpel (~25 °C) and allow enough time (~1–2 min) when cutting the sample.

8. We have experienced problems with spinning the rotor if the sample is not evenly distributed in the insert or rotor.

9. Shimming will be poor if air bubbles are present.
10. Keeping the screw too loose may cause the insert to rotate inside the rotor during spinning. Tightening too much may cause the insert to expand, which can make it difficult to place the insert into the rotor. The trick is to place the insert halfway in the rotor while tightening and then screw/adjust until the insert fits perfectly in the rotor without rotating.
11. There is usually a discrepancy between the set temperature and the actual temperature inside the rotor. Since temperature will affect the NMR spectra, it is important to know the actual temperature of the sample. This can be done by acquiring NMR spectra from a solution with methanol (typically 4% dissolved in methanol-d_3). The distance between the two methanol peaks in NMR spectrum depends on the temperature. By measuring the distance, one can accordingly determine and adjust the temperature inside the sample [23].
12. The sensitivity and spectral resolution are dependent on how accurate the magic angle (54.7°) is set. It is important to adjust the magic angle every time the HR-MAS probe has been changed and otherwise regularly (weekly) during the experiments to achieve high-quality NMR spectra. The magic angle can be measured using a rotor filled with solid KBr. The spinning sidebands of KBr are sensitive to magic angle misalignments, and the number and intensities of the sidebands are highest when the magic angle is correctly aligned. Adjustment of the magic angle (a rod extends from the bottom of the NMR probe to adjust the angle) should be performed if the intensities of KBr sidebands are low. As a general rule, the intensity of the left sideband peak should be approximately 10– 12% (or more) compared to the intensity of the main peak when the KBr sample is spun at 5000 Hz.
13. We recommend using an insert filled with 2 mM sucrose in 90% H_2O and 10% D_2O. Sucrose can also be used to set up baseline shim values if shim values are far off.
14. We recommend a sample temperature of ~278 K.
15. Tune the probe so that its frequency corresponds to the transmitter frequency and matching maximizes the power transfer between the probe and the transmitter and receiver. Tuning and matching can be carried out by observing either the wobble curve in TopSpin window (*see* Fig. 2) or the green and red LEDs on the high-performance preamplifier (HPPR) cover display.
16. When there is a small B_0 drift, the ^2H field lock system is able to detect and automatically compensate for the drift. The B_0 field will, thus, be kept at a constant value during acquisition.

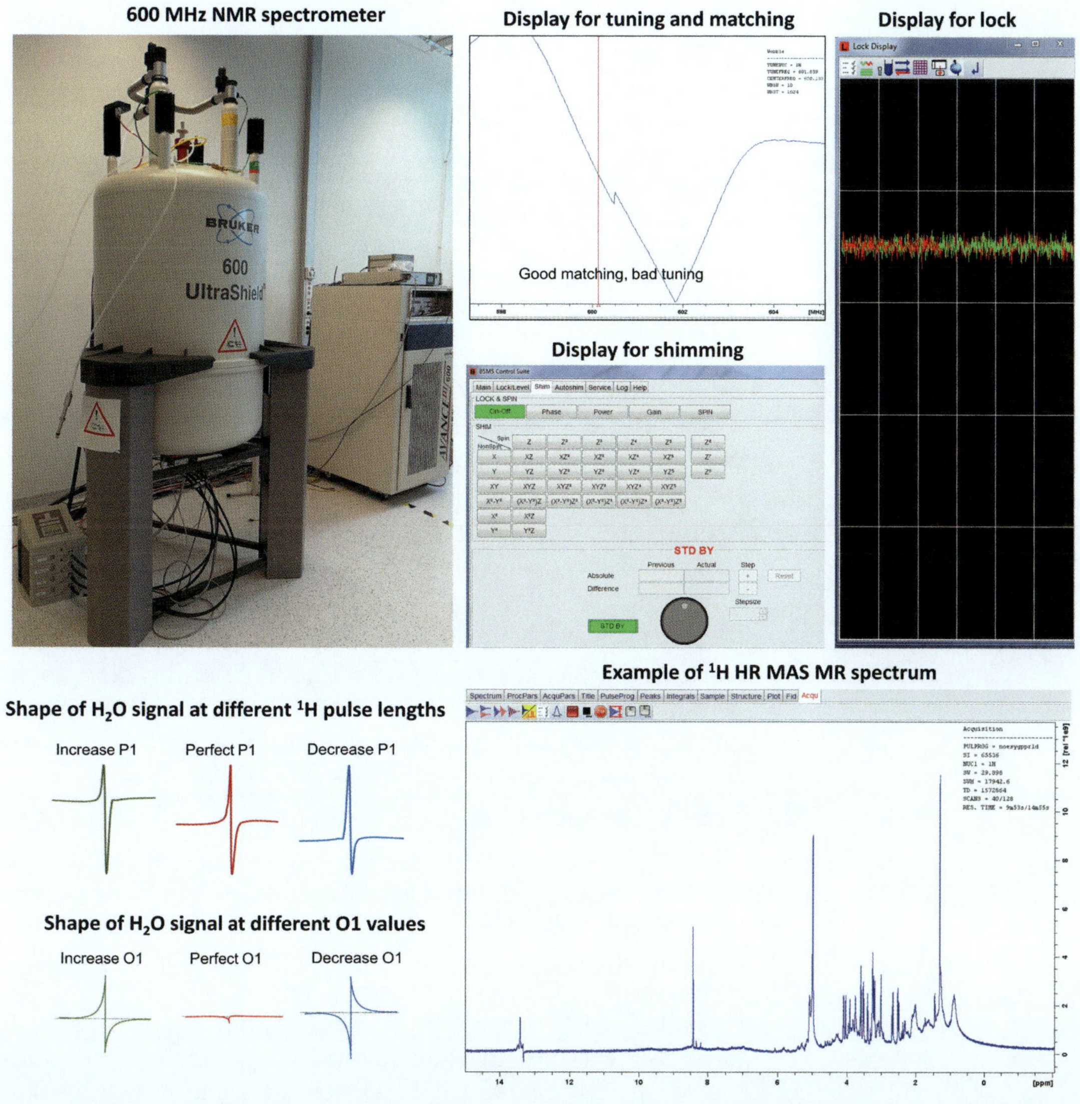

Fig. 2 Some important steps in the HR-MAS NMR experiment: Top left: Typical NMR spectrometer equipped with an HR-MAS NMR probe. Top middle: The wobble curve that indicates whether the probe is properly tuned (the dip of the curve is positioned at the red vertical line) and matched (the dip extends all the way down to the *x*-axis). Top right: Lock display shows the amplitude of the D_2O signal. When the sample is not locked, the amplitude of the lock signal will either be very low or a ringing signal is observed. The sample is considered locked when the amplitude of the lock signal (red and green) is high and flat as shown in the figure. Center: The shim interface in the BSMS Control Suite; the shim is optimized by adjusting the current in specific shim coils. Bottom left: Shape of the H_2O signal used to optimize the ^{1}H 90° pulse length (P1) and the frequency for water suppression (O1). Bottom right: A typical HR-MAS NMR spectrum

Locking is critical because even small B_0 field drifts will affect the spectral resolution.

17. Sodium formate (HCOONa) at 8.44 ppm may be a suitable option to measure the line width.

18. Shimming is important in order to achieve a homogenous magnetic field across the sample. Poor shim values will cause low spectral resolution. This step may be tricky, particularly if the sample contains high amounts of fat. We recommend shimming in "gs-mode" (continuous acquisition of NMR spectra). Adjust the current in the shim coils: X, XZ, XZ^2, and Z^4 until the FWHM of HCOONa is less than 1.5 Hz. For a practical guide on how to perform shimming of HR-MAS probes, *see* [24].
19. Acquire an NMR spectrum using P1 ≈ 90° pulse (pulse program: zg). Phase the spectrum. Then, acquire another spectrum using P1 ≈ 360°. Observe if the water peak in the spectrum is below (too short P1) or above (too long P1) the baseline. Repeat until you see 50% of the signal above the baseline and 50% below the baseline as shown in Fig. 2.
20. Acquire an NMR spectrum with water presaturation (pulse program zgpr) using different values for O1. Repeat until suppression of water signal is optimal as shown in Fig. 2.
21. In 2D ^{1}H–^{13}C NMR, signals are detected through the highly sensitive ^{1}H nuclei; hence, it is beneficial to use this experiment considering the low sensitivity of ^{13}C. Such 2D NMR spectra can be used for semiquantitative analysis of metabolite levels and multivariate analysis, similar to 1D NMR.
22. The T_1 relaxation times for ^{13}C carbonyls are very long, often up to 20 s.
23. The observed NMR signal size depends on the receiver gain (RG). Too low RG leads to low SNR. If the receiver gain is too high, the signal will exceed the limits of the analogue-to-digital converter (ADC), and the amplitude of the FID will be clipped. The RG may be automatically optimized by the instrument. We recommend using a fixed RG, which is optimized for the sample type/size used. Tissue with high lipid content might require lower RG setting due to large lipid signals. RG is usually set to the maximum value in ^{13}C NMR.
24. On Bruker NMR instruments, when the spectrum size (SI) = time domain points (TD), it means you have zero filled once. If SI = 2TD, you have zero filled twice, etc.
25. Typical LB for 1D ^{1}H NMR spectra varies, but is typically 0.3 or 1 Hz, and for ^{13}C NMR spectra, 3 Hz. The optimal LB value to achieve the highest possible SNR value is called the matched filter and is defined as the (non-broadened) linewidth at half-height.
26. One can use the alanine doublet at 1.48 ppm (middle of the doublet) for ^{1}H spectra, and [3-^{13}C] alanine at 18.9 ppm for ^{13}C NMR spectra for chemical shift scale calibration.

27. Icoshift has been shown to perform well on HR-MAS data.
28. Bins may be of fixed width (typically 0.04 ppm) or varying in width such as AI binning [25].
29. Spectra from tissue samples can contain signals from exogenous compounds such as local anesthetics or disinfectants used during sample collection, or from solvents used to clean laboratory equipment. Care must be taken to identify and remove such signals from the spectra before further analyses.
30. Mean centering is done by subtracting the mean of each variable from each variable. In autoscaling, the mean of the variables is subtracted before each variable is divided by its standard deviation. After autoscaling, all variables have the same mean and standard deviation of 0 and 1, respectively, and both small and large variables influence the model equally.

References

1. Følling A (1934) Utskillelse av fenylpyrodruesyre i urinen som stoffskifteanomali i forbindelse med imbecillitet. Nord Med 8:1054–1059
2. Austdal M, Thomsen LC, Tangeras LH, Skei B, Mathew S, Bjorge L, Austgulen R, Bathen TF, Iversen AC (2015) Metabolic profiles of placenta in preeclampsia using HR-MAS MRS metabolomics. Placenta 36 (12):1455–1462
3. Hu JZ, Rommereim DN, Minard KR, Woodstock A, Harrer BJ, Wind RA et al (2008) Metabolomics in lung inflammation:a high-resolution (1)h NMR study of mice exposedto silica dust. Toxicol Mech Methods 18 (5):385–398
4. Prabakaran S, Swatton JE, Ryan MM, Huffaker SJ, Huang JT, Griffin JL et al (2004) Mitochondrial dysfunction in schizophrenia: evidence for compromised brain metabolism and oxidative stress. Mol Psychiatry 9(7):684–697, 643
5. Sun G, Wang J, Zhang J, Ma C, Shao C, Hao J et al (2014) High-resolution magic angle spinning (1)H magnetic resonance spectroscopy detects choline as a biomarker in a swine obstructive chronic pancreatitis model at an early stage. Mol BioSyst 10(3):467–474
6. Andrew ER, Bradbury A, Eades RG (1958) Nuclear magnetic resonance spectra from a crystal rotated at high speed. Nature 182:1659
7. Hashim F, Denis C-M, Ronald S, Wolfgang B, William MK, André JS (2013) HR-MAS NMR spectroscopy: a practical guide for natural samples. Curr Org Chem 17(24):3013–3031
8. Ramadhar TR, Amador F, Ditty MJT, Power WP (2008) Inverse H-C ex situ HRMAS NMR experiments for solid-phase peptide synthesis. Magn Reson Chem 46(1):30–35
9. Bharti S, Sinha N, Shankar Joshi B, Kumar Mandal S, Roy R, Khetrapal CL (2008) Improved quantification from ^{1}H-NMR spectra using reduced repetition times. Metabolomics 4:367–376
10. Savorani F, Tomasi G, Engelsen SB (2010) Icoshift: a versatile tool for the rapid alignment of 1D NMR spectra. J Magn Reson 202 (2):190–202
11. Nielsen N-PV, Carstensen JM, Smedsgaard J (1998) Aligning of single and multiple wavelength chromatographic profiles for chemometric data analysis using correlation optimised warping. J Chromatogr A 805 (1):17–35
12. Giskeødegård GF, Bloemberg TG, Postma G, Sitter B, Tessem M-B, Gribbestad IS et al (2010) Alignment of high resolution magic angle spinning magnetic resonance spectra using warping methods. Anal Chim Acta 683 (1):1–11
13. Dieterle F, Ross A, Schlotterbeck G, Senn H (2006) Probabilistic quotient normalization as robust method to account for dilution of complex biological mixtures. Application in 1H NMR metabonomics. Anal Chem 78 (13):4281–4290
14. Wishart DS, Jewison T, Guo AC, Wilson M, Knox C, Liu Y et al (2013) HMDB 3.0--the human Metabolome database in 2013. Nucleic Acids Res 41(Database issue):D801–D807

15. Giskeodegard GF, Bertilsson H, Selnaes KM, Wright AJ, Bathen TF, Viset T et al (2013) Spermine and citrate as metabolic biomarkers for assessing prostate cancer aggressiveness. PLoS One 8(4):e62375
16. Heinzer-Schweizer S, De Zanche N, Pavan M, Mens G, Sturzenegger U, Henning A et al (2010) In-vivo assessment of tissue metabolite levels using 1H MRS and the electric REference to access in vivo concentrations (ERETIC) method. NMR Biomed 23(4):406–413
17. Wider G, Dreier L (2006) Measuring protein concentrations by NMR spectroscopy. J Am Chem Soc 128(8):2571–2576
18. Wright AJ, Fellows GA, Griffiths JR, Wilson M, Bell BA, Howe FA (2010) Ex-vivo HRMAS of adult brain tumours: metabolite quantification and assignment of tumour biomarkers. Mol Cancer 9:66
19. Hao J, Astle W, De Iorio M, Ebbels TMD (2012) BATMAN—an R package for the automated quantification of metabolites from nuclear magnetic resonance spectra using a Bayesian model. Bioinformatics 28 (15):2088–2090
20. Euceda LR, Haukaas TH, Bathen TF, Giskeodegard GF (2018) Prediction of clinical endpoints in breast cancer using NMR metabolic profiles. Methods Mol Biol 1711:167–189
21. Westerhuis JA, Hoefsloot HCJ, Smit S, Vis DJ, Smilde AK, van Velzen EJJ et al (2008) Assessment of PLSDA cross validation. Metabolomics 4(1):81–89
22. Giskeodegard GF, Cao MD, Bathen TF (2015) High-resolution magic-angle-spinning NMR spectroscopy of intact tissue. Methods Mol Biol 1277:37–50
23. Findeisen M, Brand T, Berger S (2006) A 1H-NMR thermometer suitable for cryoprobes. Magn Reson Chem 45(2):175–178
24. Piotto M, Elbayed K, Wieruszeski JM, Lippens G (2005) Practical aspects of shimming a high resolution magic angle spinning probe. J Magn Reson 173(1):84–89
25. De Meyer T, Sinnaeve D, Van Gasse B, Tsiporkova E, Rietzschel ER, De Buyzere ML et al (2008) NMR-based characterization of metabolic alterations in hypertension using an adaptive, intelligent binning algorithm. Anal Chem 80(10):3783–3790

Part VI

Drug Discovery

Chapter 16

NMR Metabolomics Protocols for Drug Discovery

Fatema Bhinderwala and Robert Powers

Abstract

Drug discovery is an extremely difficult and challenging endeavor with a very high failure rate. The task of identifying a drug that is safe, selective, and effective is a daunting proposition because disease biology is complex and highly variable across patients. Metabolomics enables the discovery of disease biomarkers, which provides insights into the molecular and metabolic basis of disease and may be used to assess treatment prognosis and outcome. In this regard, metabolomics has evolved to become an important component of the drug discovery process to resolve efficacy and toxicity issues and as a tool for precision medicine. A detailed description of an experimental protocol is presented that outlines the application of NMR metabolomics to the drug discovery pipeline. This includes (1) target identification by understanding the metabolic dysregulation in diseases, (2) predicting the mechanism of action of newly discovered or existing drug therapies, (3) and using metabolomics to screen a chemical lead to assess biological activity. Unlike other OMICS approaches, the metabolome is "fragile" and may be negatively impacted by improper sample collection, storage, and extraction procedures. Similarly, biologically irrelevant conclusions may result from incorrect data collection, preprocessing or processing procedures, or the erroneous use of univariate and multivariate statistical methods. These critical concerns are also addressed in the protocol.

Key words Metabolomics, Chemometrics, NMR, Drug discovery

1 Introduction

Nuclear magnetic resonance (NMR) spectroscopy is an instrument of choice for drug discovery and design (*see* Fig. 1) [1, 2]. For decades, NMR has provided structural details for druggable protein targets and is routinely used for investigating protein-drug interactions and conducting high-throughput ligand-affinity screens. In this regard, NMR is able to obtain atomic-level resolution of protein-ligand structures in near physiological conditions that results in better quality chemical leads [3, 4]. Accordingly, NMR has enhanced the success rate of the drug discovery process and has made numerous contributions to the development of new therapeutics [5–7]. In addition to its role in structural biology, NMR has also made significant contributions to the field of metabolomics, especially in the application of metabolomics to drug discovery [8–12].

G. A. Nagana Gowda and Daniel Raftery (eds.), *NMR-Based Metabolomics: Methods and Protocols*, Methods in Molecular Biology, vol. 2037, https://doi.org/10.1007/978-1-4939-9690-2_16, © Springer Science+Business Media, LLC, part of Springer Nature 2019

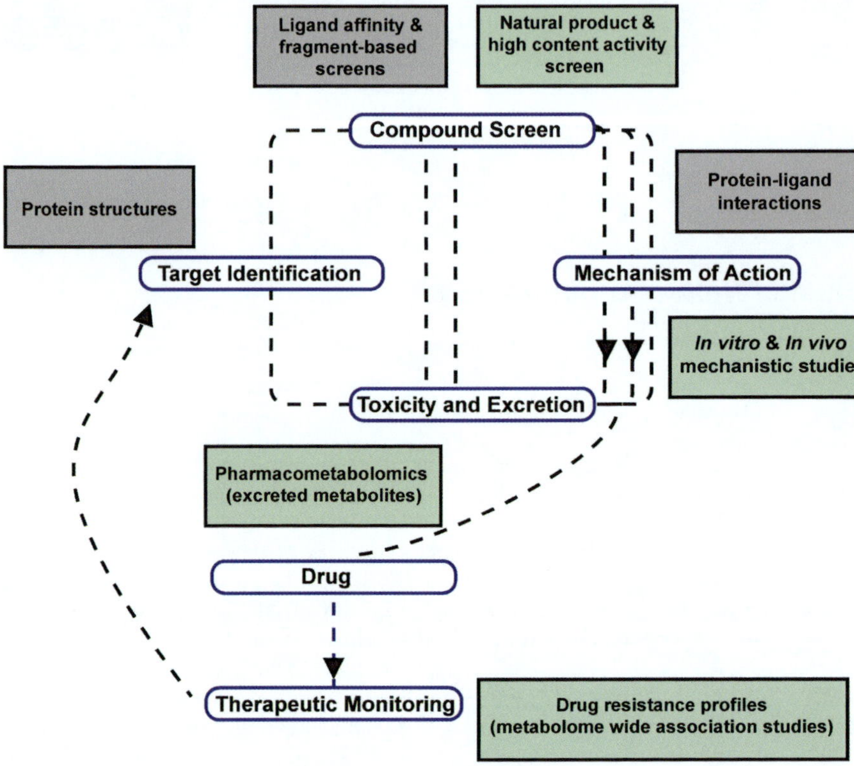

Fig. 1 A schematic diagram outlines the overall drug discovery process and highlights the role of NMR- and NMR-based metabolomics at every step in the process. Traditional drug discovery roles for NMR are boxed in grey, while NMR metabolomics applications are boxed in green

A traditional view of a disease is as a genome or proteome level dysfunction, but there is a growing appreciation for the role of metabolomic perturbations in a variety of human diseases [13]. For example, an alteration in glucose metabolism is a well-known hallmark of cancer (i.e., Warburg effect), but other metabolic changes are also associated with cancer, such as glutaminolysis, redox imbalance, and flux changes in the tricarboxylic acid (TCA) cycle or the pentose phosphate pathway (PPP) [14]. Similarly, antibiotic resistance in *Staphylococcus aureus* has been shown to result from a re-direction of carbon flow into metabolic processes that counteract antibiotic activity [15, 16]. Another example is mitochondrial metabolism, especially changes in NAD levels, which is an indicator of aging- and age-related disorders [17]. The NAD-sirtuin pathway regulates neuronal protection and is an important process that prevents axon degradation, which is a primary cause of Parkinson's disease and Alzheimer's. Thus, lowered NAD^+ levels is a crucial metabolic observation in neurodegeneration that is being manipulated for disease prevention [15, 17]. Thus, the characterization of metabolite levels in any biological system coupled with the ability to trace the metabolic fate of individual atoms has the potential to beneficially impact a variety of biomedical needs [12]. Specifically, metabolomics is an invaluable tool of drug discovery, since metabolic changes may be used to identify therapeutic targets, verify in vivo drug efficacy, decipher mechanisms of action, and identify potential toxic side effects.

The drug discovery pipeline is an extremely resource intensive endeavor. The process is time-consuming and may require upward of 12 years before a new drug obtains FDA approval [18] and costly with an average expense of $2.6 billion dollars per approved drug [19]. The process is also prone to a very high failure rate where only a few lead compounds that emerge from traditional medicinal chemistry efforts can be transferred to clinical trials [20]. Furthermore, very few drug leads make it past phase I clinical trials, and most compounds eventually fail in phase III [21]. On average, less than one in ten compounds receive FDA approval and are launched as new drug entities [22]. Of course, FDA-approved drugs are still routinely pulled from the market or encounter safety concerns when broadly prescribed [23]. This low overall success rate is a result of numerous problems routinely encountered throughout the drug discovery process [24, 25]. These problems are multifactorial, complex in nature, disease specific, and, in some cases, poorly understood; they have also been extensively discussed and reviewed in the literature [26–31]. Accordingly, it is well appreciated that a single technological advancement is not going to resolve the entirety of the challenges facing drug discovery research, but a new approach may lead to better chemical leads, to a reduction in toxicity issues, or to an improvement in selectivity and efficacy. Thus, incorporating NMR metabolomics into a standard drug discovery pipeline may lead to an improved success rate, but it won't solve all of the field's problems.

Metabolomics has the potential to identify therapeutic targets in the drug discovery process. First, metabolomics may improve our fundamental understanding of cellular physiology, disease development, or disease progression, which may lead to the discovery of new therapeutic targets. In addition, metabolic profiles resulting from a disease state or drug resistance may identify cellular or metabolic processes that are critical to the system. In this manner, specific proteins associated with these upregulated or downregulated pathways may be targeted in either a single drug therapy or as part of a multidrug treatment. The underlying assumption is that the perturbed metabolic processes are required to maintain the drug resistance or disease state. Accordingly, inhibiting these proteins may reverse the drug resistance or eliminate the disease. Of course, the metabolic profile of a disease state may also be used to monitor a response to treatment [32]. The return of the metabolic profile to a healthy state would be indicative of a positive therapeutic response. Similarly, a lack of a response would suggest an ineffective treatment, and, conversely, a drastic change in the metabolic profile may be indicative of a toxic response. Thus, in total, metabolomics may beneficially impact all the steps of the drug discovery process, from target identification to therapeutic monitoring.

Cancer biology has pioneered the idea of metabolic dysregulation as the source of disease pathology. Warburg aerobic glycolysis and glutaminolysis are essential features of cancer pathogenesis and are known to occur in most tumors [33]. These metabolic processes are potentially linked to oncogene and tumor suppressor regulation [34]. Thus, new therapeutic targets to improve cancer management may be based on modulating these altered metabolic processes. For example, a new experimental treatment for cancer is 4-phenylbutyrate [35], which is a sacrificial metabolite that induces an alternative metabolic pathway to overcome a defective urea cycle and repair a nitrogen imbalance. Presumably, 4-phenylbutyrate may indirectly address a dysregulated glutaminolysis by facilitating the excretion of glutamine in the urine [36]. Specifically, 4-phenylbutyrate is converted to 4-phenylacetate, which then reacts with glutamine to form the excretion product phenylacetylglutamine.

In a similar manner, ongoing efforts to develop novel cancer treatments by targeting metabolic dysregulation (e.g., glycolysis, glutaminolysis, PPP) has highlighted the therapeutic importance of metabolic enzymes such as pyruvate dehydrogenase kinase, lactate dehydrogenase A, glutaminase 1, thioredoxin reductase, and gamma-glutamylcysteine synthetase [37, 38]. In fact, inhibitors developed against all of these metabolic targets are currently at various stages of preclinical and clinical trials. Indeed, anti-metabolite compounds and enzyme inhibitors are among the most commonly prescribed drugs. For example, gemcitabine, a first-line treatment for pancreatic cancer, is a nucleotide (metabolite) analog that interferes with DNA synthesis [34]. Notably, gemcitabine resistance results from a metabolic shift that increases the cellular production of deoxycytidine triphosphate to function as a competitive inhibitor of gemcitabine [34]. Similarly, azathioprine is a purine analog that disrupts RNA and DNA synthesis and is used to treat autoimmune diseases like Crohn's disease and rheumatoid arthritis [39]. Sulfanilamide antibiotics inhibit bacterial enzymes that require para-aminobenzoic acid (PABA) for synthesizing folic acid, a coenzyme in DNA synthesis. Treatments for inborn errors of metabolism are other successful examples of employing metabolite analogs as a therapy [40]. Clearly, a metabolic view of disease pathology has beneficially contributed to existing drugs and the creation of next-generation therapies and has the potential of significantly improving patient outcomes [41].

As illustrated above, NMR-based metabolomics is an asset for drug discovery since it helps characterize metabolite level perturbations and altered metabolic pathways that may be used to identify protein targets for designing new drugs. NMR is also routinely used as a high-throughput screen (HTS) to identify and evolve chemical leads against these protein targets. NMR ligand-affinity screens such as SAR-by-NMR, MS/NMR, multi-step NMR, SHAPES, and FAST-NMR that make use of saturation transfer difference (STD), WaterLOGSY, ^{19}F NMR, HSQC chemical shift perturbations (CSP), transfer NOEs, and fragment-based libraries are established means of identifying and validating chemical leads through experimentally observed protein-ligand interactions [42–47]. Consequently, combining NMR HTS techniques with metabolomics to implement a high content drug screen is a logical next step. One such example is the NMR-based metabolomics screen of a library of 56 kinase inhibitors targeting eEF-2 kinase and NF-kB [48]. Tiziani et al. monitored global metabolic changes resulting from compound treatment to a panel of cancer cell lines (e.g., HeLa, CCRF-CEM, SKOV-3, and acute myeloid leukemia (AML) primary cells) in a 96-well format. SDS-induced cell lysates were then used to collect one-dimensional (1D) ^{1}H CPMG and two-dimensional (2D) ^{1}H JRES NMR spectra followed by multivariate statistical analysis of the entire spectral dataset. The resulting scores and backscaled loadings plot calculated relative to controls were used to identify active lead compounds based on the magnitude of the metabolic response. Notably, approximately 100 samples were screened in less than 24 h.

NMR metabolomics combined with traditional medicinal chemistry can, thus, be used to validate in vitro efficacy, selectivity, and specificity while avoiding off-target effects (*see* Fig. 2). Metabolic perturbations resulting from a drug treatment may be correlated to the inhibition of the desired protein target and/or as a result of unexpected off-target effects [49]. If the lead compound inhibits additional protein targets beside the specifically designed drug target, then the lack of selectivity may help predict and anticipate toxicity issues in future clinical trials. Thus, medicinal chemistry techniques can then be used to evolve the chemical lead in an iterative fashion using insights and feedback from the metabolomics experiments. The goal is to maintain the metabolic profile associated with inhibiting the desired protein target while diminishing or eliminating metabolic changes resulting from unwanted off-target effects. This global approach of monitoring all possible changes to the system in response to a drug's activity would be difficult, if not impossible, to replicate in an in vitro setting. Even if the specific off-target proteins could be identified, it is not feasible to reproduce the synergistic and simultaneous impact on multiple cellular processes.

In a similar manner, an in vivo mechanism of action for a new chemical lead may be inferred by comparing its induced metabolic profile against metabolic profiles for drugs with known mechanisms

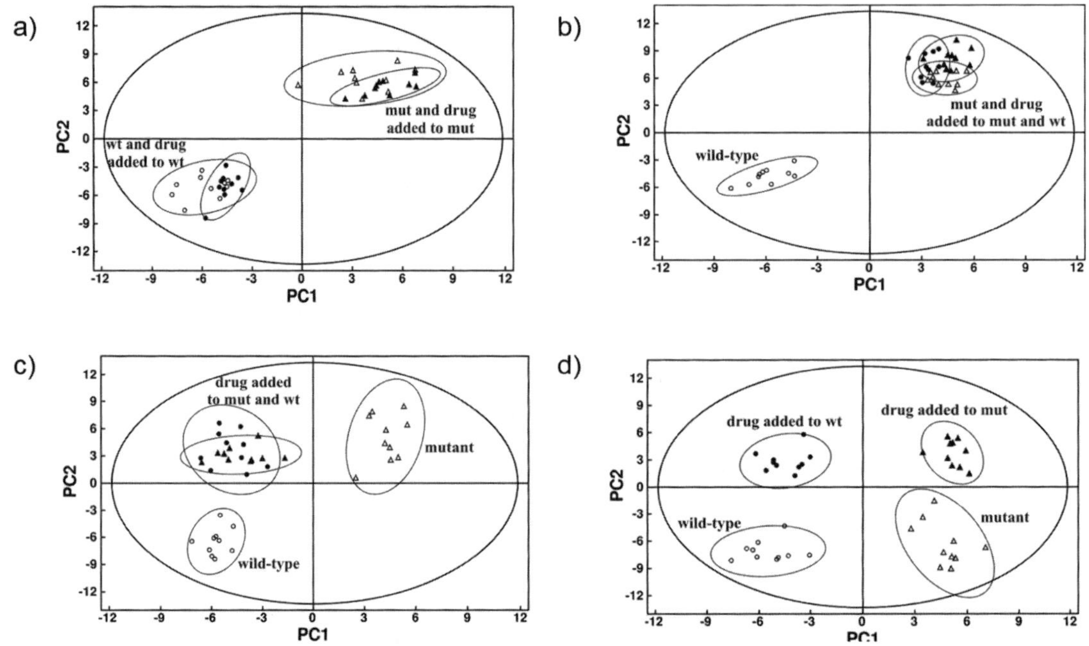

Fig. 2 Cartoon illustrations of hypothetical PCA scores plot for the following scenarios: (**a**) inactive compound; (**b**) active and selective inhibitor; (**c**) active, nonselective inhibition of target and secondary protein; and (**d**) active, nonselective preferential inhibition of secondary protein. Labels correspond to wild-type cells (wt) and mutant cells (mut) (Reproduced with permission from ref. 51. Copyright © 2006 American Chemical Society)

of action [50]. Compounds that share the same mechanisms of action would be expected to have similar impacts on the metabolome. Technological advances have reduced NMR analyses times to less than a few minutes per sample [51]. Thus, like other secondary assays routinely used in a drug discovery pipeline, metabolomics can be applied in a high-throughput fashion to interrogate an entire library of chemical leads. Accordingly, metabolomics is a cost-effective means of filtering out problematic lead compounds prior to expensive clinical trials.

Mechanistic understanding of in vivo drug activity is an important and, typically, challenging stage of the drug discovery process. Knowledge of in vitro and in vivo mechanisms of action improves the overall success rate and will beneficially enhance the drug design effort [50]. Unfortunately, elucidating a mechanism of action for a specific chemical lead may be an arduous, time-intensive, and resource-intensive task. Multiplying this effort across a large screening library is impractical. Instead, predicting the in vivo mechanism of action for new chemical leads by leveraging known entities is a valuable alternative to resolving this problem. Furthermore, the availability of a set of drugs with a diverse range of biological activities for such a comparison will increase the likelihood of successfully assigning a mechanism of action. In addition, identifying compounds that target various biological processes and have a range of mechanisms of action may also assist in disease management and in combating the emergence of resistance. This may be accomplished through a multidrug therapy that combines compounds with distinct modes of action. Halouska et al. [50] demonstrated the general approach by comparing the metabolic profile of ***Mycobacterium smegmatis*** following treatment with a range of antibiotics with known mechanisms of action and some promising chemical leads from high-throughput cell-based screens. 1D ^{1}H NMR spectra combined with multivariate analysis were used to characterize the relative global metabolomes. As expected, antibiotics with similar modes of action clustered together in the orthogonal projections to latent structures (OPLS) scores plot. The HTS chemical leads were clustered together with known cell wall inhibitors implying a similar smechanism of action, which was supported by literature reports. It is also possible to invert the protocol and use a drug with a known mechanism of action to modulate a metabolic pathway predicted to be important to disease progression of drug resistance. In this manner, if the targeted metabolic pathway is reversed following drug treatment and the desired therapeutic outcome is achieved (i.e., cell death), then an in vivo mechanism of action is likely indicated. A recent example of this process involved a vancomycin-intermediate resistant strain (VISA) of ***Staphylococcus aureus*** that exhibited an increase in purine biosynthesis relative to a vancomycin-susceptible strain (VSSA) [15]. Treating VISA strains with a combination drug therapy consisting of a sub-lethal dose of vancomycin and 6-mercaptopurine (i.e., purine substrate analog) led to a statistically significant increase in staphylococcal killing. In addition to reversing the acquired vancomycin resistance, the metabolomics results also indicate (as expected) that the in vivo mechanism of action for 6-mercaptopurine is an inhibitor of purine biosynthesis.

Metabolomics can also provide valuable information during a clinical trial. Metabolomics can assist with therapeutic monitoring and drug dosage adjustments and by differentiating responders from non-responders. Simply, specific metabolic profiles may be attributed to each group, which then can be used to classify individual patients by monitoring metabolite changes in common biofluids such as urine, serum, or fecal material. Another aspect of a clinical trial that has benefited from metabolomics is monitoring patient compliance. In this regards, patient non-compliance is readily identified by the presence of distinct and/or strong signals in the NMR spectrum that are not normally present, such as metabolites from drug metabolism or from consuming caffeine, ethanol, or nicotine. As a result of these clear benefits to drug discovery, metabolomics continues to be adopted by the pharmaceutical and biotechnology industry. A prime example is COMET, the Consortium for Metabonomic Toxicity [52], which brought together pharmaceutical companies to utilize large datasets of NMR metabolomics analysis of urine and serum samples to predict drug toxicity.

To address the expanding interest in NMR metabolomics, we present a detailed, step-by-step description of a metabolomics workflow with a specific attention to details important to its application to a drug discovery and design process. The protocol includes methods relevant to understanding the mechanism of action of novel drug leads in comparison with existing therapeutics. Furthermore, we describe the application of metabolomics to a wide variety of sample sources commonly encountered in drug discovery programs. This includes bacterial and mammalian cell cultures, human and animal tissue samples, and different biofluids (e.g., urine, serum, and fecal material). The entire process from cell culturing and sample collection to biological interpretation is described. Importantly, the protocol addresses critical concerns regarding the proper (1) preparation and handling of metabolomics samples; (2) preprocessing, processing, and acquisition of spectral data; (3) metabolite identification; and (4) statistical analysis.

Unlike other OMICS techniques, the composition of the metabolome can easily change from the processing, handling, and storage of samples [53]. Metabolites are chemically labile or unstable, are subject to thermal degradation or oxidation, and are readily transformed by residual enzymatic activity. For example, the oxidation of cysteine and glutathione or the conversion of ATP to AMP (because of a labile phosphate group) is a common occurrence in metabolomics samples [54]. These biologically irrelevant sample variations may distort our understanding of a system and lead to erroneous conclusions. Accordingly, metabolomics requires a robust, efficient, and fast workflow to yield reproducible and reliable data. Univariate and multivariate statistical analyses are essential components of any metabolomics study and provide the means to interpret the dataset and identify the key group variances (i.e., metabolite changes) [55]. Nevertheless, issues such as the incorrect application of statistical techniques, the subsequent lack of proper model validation, or the over-interpretation of models are common occurrences in the field of metabolomics [56]. Accordingly, the protocol presented herein addresses these issues by providing instructions to avoid common mistakes.

2 Materials

All samples should be prepared using nanopure water (H_2O) and/or deuterated water (D_2O). All safety regulations should be followed for the proper handling and disposal of chemicals and biological samples used in the study.

2.1 Laboratory Equipment

1. Bruker AVANCE III HD 700 MHz NMR spectrometer equipped with a 5 mm quadruple resonance QCI-P cryoprobe (^{1}H, ^{13}C, ^{15}N, and ^{31}P) with *z*-axis gradients, an automatic tune and match system (ATM), and a SampleJet automated sample changer system with Bruker ICON-NMR software (Bruker Biospin, Billerica, MA).
2. BSL-2 biosafety level grade hood (e.g., Biological Safety Cabinet, LF BSC class 2 type A, Thermo Fischer Scientific, Waltham, MA).
3. Nanopure ultra water system (Barnstead Inc., Dubuque, IA).
4. Lab Armor bead bath (Chemglass Life Sciences, Vineland NJ).
5. Incubator capable of maintaining physiological temperature and proper carbon dioxide levels (e.g., HERA CELL VIOS 250i CO_2 Incubator, Thermo Fischer Scientific, Waltham, MA).
6. pH meter and probe.
7. Refrigerated centrifuge capable of speeds up to 21,100 × *g* (e.g., SORVALL micro 21R centrifuge, Thermo Fischer Scientific, Waltham, MA).
8. Speed Vac for solvent removal (e.g., SAVANT SC210A SpeedVac concentrator, Thermo Fischer Scientific, Waltham, MA).
9. Freeze dryer to remove water (e.g., FreeZone 4.5, LABCONCO, Kansas City, MO).
10. 1 μL to 1000 μL pipettes
11. FastPrep-96 homogenizer (MP Biomedicals, Santa Ana, CA).
12. ACCU-SCOPE 3030ph microscope (Commac, NY).
13. Cryogenic storage container (Taylor Wharton, Theodor, Al).
14. −80 °C freezer.
15. Flask shaker incubator (up to 250 rpm).
16. Sonic Dismembrator (Sonicator).

2.2 Disposable Supplies

1. 1 μL to 1 mL pipette tips.
2. 10 mL aspirating pipettes.
3. 15 mL Falcon tubes.
4. 2 mL Eppendorf tubes.
5. 1 mL screw-cap microcentrifuge tubes.

2.3 Isotopically Labeled Solvents and Reagents (See Notes 1 and 2)

1. Deuterium oxide (D_2O, 99.8 atom% D).
2. 3-(trimethylsilyl) propionic-2,2,3,3-d_4 acid sodium salt (TMSP-d_4, 99.8 atom% D).
3. Dimethyl sulfoxide-d_6 (DMSO-d_6, 99.8 atom% D).

4. $^{13}C_6$-glucose (99% ^{13}C).
5. $^{13}C_2$-acetate (99% ^{13}C).
6. Other potential ^{13}C-carbon labeled or ^{15}N-nitrogen labeled reagents.

2.4 Buffers

1. Wash buffer, phosphate-buffered saline (PBS) at pH 7.4: 137 mM NaCl, 2.7 mM KCl, 10 mM Na_2HPO_4, and 2 mM KH_2PO_4. To prepare 1 L PBS buffer at pH 7.4, add 8.0 g of NaCl, 0.2 g of KCl, 2.68 g of $Na_2HPO_4{\cdot}7H_2O$, and 0.24 g of KH_2PO_4 to a final volume of 1 L of Nanopure water.
2. NMR buffer: 50 mM phosphate buffer at pH 7.2 (uncorrected, *see* **Note 3**) in 600 μL of 99.8% D_2O. Add 50 μM (1D NMR experiment) or 500 μM (2D NMR experiment) TMSP-d_4 as an internal chemical shift reference.

2.5 Cell Media (See Note 4)

1. Dulbecco's modified eagle medium (DMEM) and Roswell Park Memorial Institute (RPMI) medium with the addition of an appropriate amount of fetal bovine serum (5–20% v/v) and/or supplemental glucose (1–25 mM) are commonly used to culture mammalian cells.
2. Bacterial cells are typically cultured in complex media, such as Luria-Bertani (LB) media for *E. coli*, Middlebrook's albumin dextrose catalase media (MADC) for *Mycobacterium smegmatis*, or M9 minimal media.

2.6 Software and Databases

1. Bruker ICON-NMR software for automated NMR data acquisition (Bruker Biospin, Billerica, MA).
2. MVAPACK metabolomics toolkit for processing and analyzing chemometric data (http://bionmr.unl.edu/mvapack.php) [57].
3. PCA/PLS-DA utilities for quantifying separation in PCA, PLS-DA, and OPLS-DA scores plots (http://bionmr.unl.edu/pca-utils.php) [58].
4. NMRPipe software for processing and visualizing NMR data (https://www.ibbr.umd.edu/nmrpipe/install.html) [59].
5. NMRViewJ software for processing and visualizing NMR data (One Moon Scientific, Inc. Westfield, NJ; https://nmrfx.org/) [60].
6. R statistical package (https://www.r-project.org/) [44].
7. Chenomx (Chenomx, Inc., Edmonton, AB, Canada) software for automated metabolite assignment and quantification from 1D ^1H NMR spectra (https://www.chenomx.com/).
8. MetaboAnalyst software for the statistical, functional, and integrative analysis of metabolomics data (http://www.metaboanalyst.ca/) [61].

9. Human Metabolomics Database (HMDB) of reference NMR data for known metabolites (http://www.hmdb.ca/) [62].
10. Biological Magnetic Resonance Data Bank (BMRB) of reference NMR data for known metabolites http://www.bmrb.wisc.edu/metabolomics/) [63].
11. Non-uniform schedule (NUS) generator (http://bionmr.unl.edu/dgs-gensched.php) for NUS NMR data acquisition [64].
12. Small Molecule Pathway Database (SMPDB) (http://www.smpdb.ca) for modeling metabolic pathway changes from list of metabolite changes [65].

3 Methods

Extreme consistency is key to the success of any metabolomics experiment. Accordingly, it is essential to apply *all* experimental protocols and parameters uniformly to *all* groups in the study. Any variation in the preparation, handling, or storage of the sample may perturb the metabolome and lead to erroneous results. For example, different sample processing times, samples handled by different investigators, and cell cultures grown in different incubators or shakers; different sources or lot numbers of chemicals, media, or disposables; and different buffer/media preparations (even if the same recipe was prepared by the same individual) are all likely to induce a biologically irrelevant bias in the samples ***and needs to be avoided***. In effect, the *only* difference between groups should be the intended variable (e.g., wild type versus mutant; treated versus untreated, healthy versus disease, etc.). In general, the defined variable(s) is expected to induce a measurable change in the metabolome that is detectable by NMR. For example, a proper dosage for a drug used in a metabolomics project needs to be experimentally determined. The drug dosage needs to be high enough to induce a detectable metabolic response, but not too high to induce a generic cell death phenotype (*see* **Note 5**).

3.1 Experimental Design

A scientific hypothesis needs to be clearly defined in order to properly design a metabolomics experiment (*see* Fig. 3). In other words, simply characterizing the metabolome of a cell, tissue, or biofluid is likely to fail to provide any scientific insight or value without a logically designed comparative analysis. In this regard, a metabolomics experiment is fundamentally a comparison between two or more groups (e.g., control vesus drug treated), where a difference or similarity in a metabolome is expected to answer a scientific question. The next critical step is to identify an appropriate source for the metabolome or the type of biological samples to be compared (i.e., established or primary cell lines, tissue samples from an animal model, human biofluids, etc.). The choice of system

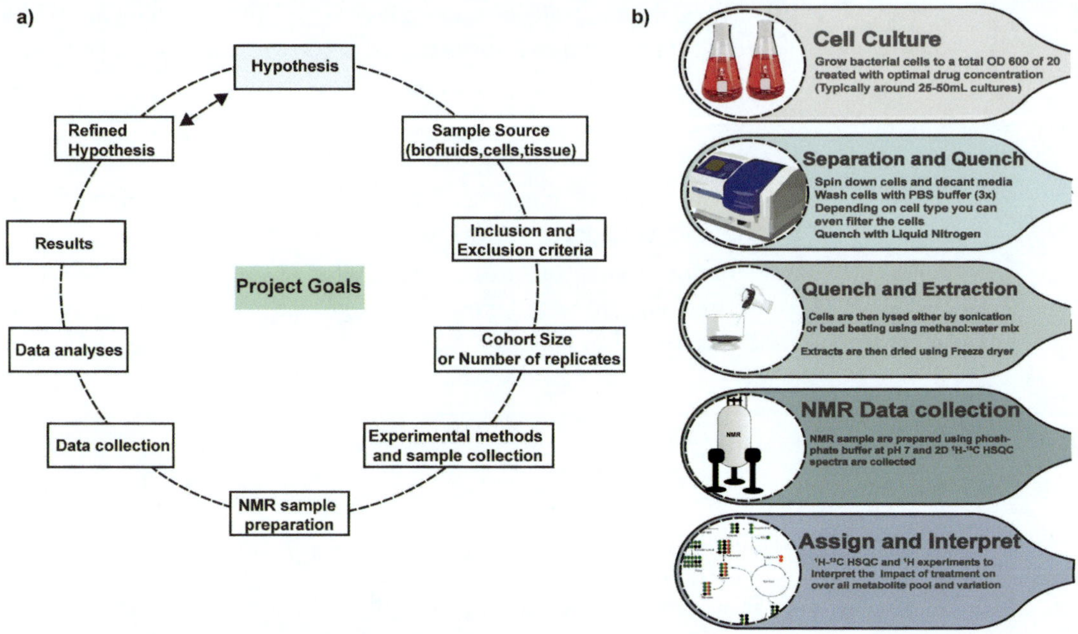

Fig. 3 (**a**) A summary of the design and planning steps in a typical metabolomics-based drug discovery project. (**b**) A workflow for a typical cell-based NMR metabolomics experiment focused on understanding the mechanism of action for a chemical lead

will then determine the number of biological replicates that can be practically obtained, which, in turn, will define the statistical significance of any observed results. Given a testable hypothesis and the type of biological samples, there are essentially an endless number of experimental variables that can be manipulated, where each permutation defines a new group. Practically speaking, it would be impossible to account for every possible permutation in a single experimental design. Instead, the identification of which groups to use is fundamentally driven by the scientific question the study is designed to answer (*see* Fig. 3) Thus, a general set of protocols is presented for a generic metabolomics experiment that can be applied to a drug discovery project. The protocol provides a means to determine a drug's mechanism of action and to monitor a patient's response to treatment.

The simplest experimental design for a metabolomics drug discovery project will consist of two groups: untreated samples and drug-treated samples. Nevertheless, it is typical to significantly expand the experimental design beyond just two groups (*see* Figs. 2 and 3). Common groups include:

1. Controls (e.g., wild-type cells or healthy subjects)
2. Control with treatment (positive control, e.g., wild-type cells or healthy animals treated with a drug lead)

3. Experimental model (e.g., animal disease model or treatment naïve patients)
4. Experimental model with a drug treatment (e.g., animal disease model or patients treated with a drug lead)
5. Negative control (e.g., mutant cells with no treatment, patients with alternative disease)
6. Negative control with a drug treatment (e.g., mutant cell lines treated with a drug lead)

The metabolomics study may consist of any combination of biological samples originating from cell cultures, tissue samples, and/or biofluids. Human or animal biofluids may consist of urine, serum, plasma, blood, fecal material, saliva, sweat, condensed breath, cerebrospinal fluid (CSF), or any other body fluid or excretion. Similarly, tissue samples usually consist of organs, such as the heart, lungs, brain, liver, or kidneys from sacrificed animals. Tissue samples from human subjects typically originate from biopsies (i.e., muscle, adipose, etc.), surgeries (i.e., tumors), or tissue banks. Each biofluid, tissue type, or time point would constitute a different group. Of course, there are numerous experimental protocols that could also be varied, such as different media, nutrients, temperature, diets, etc.

Besides sample type and experimental protocols, there are other variations that may be applied to form additional groups. For example, different groups may be formed by varying the drug dosage or by using a placebo. Of course, multiple drugs could be tested in a single study, and group membership would be defined by what drug a cell culture or an individual receives. In addition, the number of drug treatments received by a cohort could vary. Time could be another variable in the experimental design (e.g., when samples are collected and when treatments are administered). Similarly, different groups may consist of different mutant or wild-type cell lines or strains. For example, a mutant cell line may correspond to the drug protein target being inactivated (i.e., negative control) through a variety of genetic mutations or gene knockout methods. Alternatively, the mutant cell lines could be various human isolates with variable levels of resistance or susceptibility to the tested drug. The cell lines could just be different types of bacteria (e.g., *E. coli*, *S. aureus*, etc.) or different types of cancers (e.g., pancreatic, breast, etc.). Similarly, animal models or human cohorts may have different stages or severity of the disease or even different diseases. Furthermore, multiple biofluid samples (i.e., both urine and serum) may be collected from the same animal or human participant at multiple time points during the study. Similarly, more than one tissue or organ sample may be harvested from each sacrificed animal.

Another important decision regarding the experimental design is the number of biological replicates required per group. In general, more biological replicates led to a better statistical significance in differentiating between the various groups. But of course, there are practical limits to the number of cell cultures, animals, or human cohorts that can be prepared or recruited for any study. Facility and equipment capacities, the availability of personnel, and cost are all major factors that greatly limit the number of samples or cohorts that are possible. There is also an obvious multiplication factor based on the number of desired groups and the number of replicates per group. Accordingly, there may be a tendency to maximize the number of groups at the expense of the number of replicates per group. This is inadvisable since it would likely lead to statistically insignificant conclusions. Instead, the number of groups should be reduced to achieve an acceptable minimal number of biological replicates per group. In fact, sample size calculation for metabolomics studies is complex and not well defined, where recent efforts suggest sample sizes may need to be >20 per group [66, 67]. In the case of metabolomics studies involving human cohorts, the need for replicates is even greater and more challenging to estimate, but sample sizes >60–100 cohorts per group are typical targets.

One final factor to consider for the design of a metabolomics study involving human cohorts is the inclusion and exclusion criteria. Simply, what individual characteristics, traits, or behavior are needed for the goals of the study? Conversely, what are potential confounding factors that may confuse or obscure the desired outcome of the study and should be avoided? The specific inclusion and exclusion criteria are likely unique to each study; and there are no general guidelines to apply. Criteria commonly considered include age, body mass index, diet, comorbidities, ethnicity, gender, health test panel (e.g., blood pressure, lipid panel, metabolic panel, urinalysis, etc.), physical activity, race, and the use of alcohol, recreational drugs, or tobacco (*see* Table 1).

Table 1
A list of typical exclusion criteria for a metabolomics clinical study

Exclusion criteria
• Acute illness within 2 weeks • Comorbidity (e.g., chronic lung disease, depression, etc.) • Excessive intake of alcohol • Obesity • Pregnant women or menstruating women • Recreational drug use • Risk of transmitting disease through urine (e.g., hepatitis B or C) • Significant abnormalities in clinical chemistry of hematology • Smoker

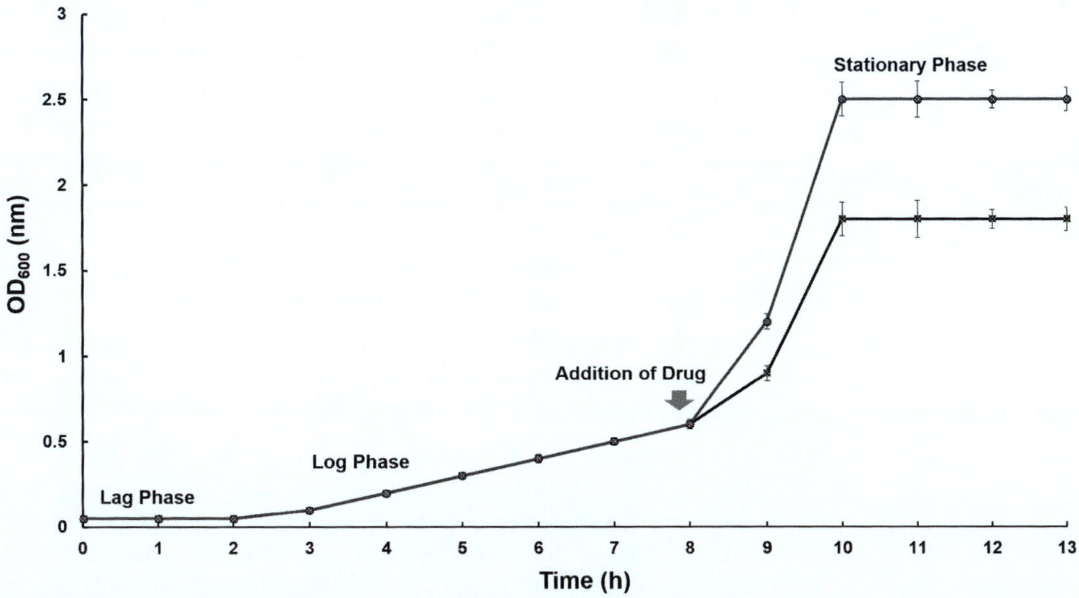

Fig. 4 An example of bacterial growth curve for a cell line treated with a drug at an OD_{600} of 0.6. The cellular metabolome should be extracted during the log phase, which captures the majority of metabolic variations

3.2 Determining Drug Dosage for Cultured Cells

The reported minimum inhibitory concentration (MIC) for a drug against a specific cell strain is a good starting point for determining the optimal drug dosage for a metabolomics study. Typically, drug concentrations that range between 1 and 24 times the reported MIC or MIC50 are tested to identify an optimal drug dosage. The goal is to identify a drug dosage that stresses the cells, but does not induce cell death. This is accomplished by identifying a drug dosage that results in a decrease in cell growth by approximately 50% (MIC50, *see* Fig. 4).

Along with dosage, the time point for administering the drug needs to be experimentally determined. First, enough biomass needs to be harvested in order to extract a detectable metabolome. Second, the timing of the drug dosage should induce the desired perturbation in the metabolome. For example, a particular metabolic process may only be active during a specific phase of the growth curve. Typically, a drug treatment is administered during the exponential or log phase for a bacterial culture. The cells are allowed to grow for at least one generation or one doubling time before harvesting.

1. Approximately 1–20 mM of a stock solution is prepared for each compound to be tested by dissolving the required amount of the compound into 0.5 mL of ethanol or DMSO-d_6 (*see* **Note 6**).
2. Bacterial cells are typically grown at 37 °C with shaking at 200 rpm in 50 mL of complex or minimal media in a 125 mL Erlenmeyer flask. Mammalian cells are typically grown in a

6-well plate with 3 mL of media at 37 °C and 5% percent CO_2 in an incubator. A total of 21 cell cultures are prepared.

3. The bacterial cells are allowed to grow until an average optical density at 600 nm (OD_{600}) of 0.6 is achieved (*see* Fig. 4). Mammalian cells are grown to approximately 40% confluency.
4. Each cell culture receives a different aliquot of the stock drug solution such that the final drug concentration ranges from 0 to 24 times the MIC value. Each drug dosage is made in triplicate for a total of 21 cell cultures (*see* **Note 7**).
5. Following the addition of the drug leads, the cells are allowed to grow for an additional amount of time. This is typically the cell doubling time.
6. For bacterial cells, the optical density (OD_{600}) is recorded at the end of the additional cell growth time. The growth rate inhibition is calculated by comparing the optical density of the treated cells to the untreated cells. Typically, untreated cells will reach an OD_{600} of 1.2 (i.e., a doubling from an initial OD_{600} of 0.6), while the OD_{600} of drug treated cells will be lower. For mammalian cells, the cells are visualized under a microscope to measure the percent confluency and assess the amount of detached cells [53, 68, 69]. Typically, untreated cells should reach a confluency of approximately 80% (i.e., doubling from an initial 40% confluency).
7. The desired dosage corresponds to the drug concentration that inhibits cell growth by approximately 50% relative to untreated cells (*see* **Note 5**).
8. For mammalian cells, an additional consideration is the length of time following the drug treatment that is required before the cells can be harvested. It is important to provide a significant amount of time to allow a perturbation in the metabolome to occur, but it should be short enough to prevent a significant accumulation of dead cells. No dead cells are a preferred outcome. After the proper drug dosage is identified, repeat experimental **steps 2–5** with the following modifications:
   (a) A constant drug dosage is used for each replicate cell culture. Each cell culture replicate is a different time point instead of a different drug dosage.
   (b) The time after drug treatment that the cells are harvested are varied from zero (i.e., control) to an end point of approximately 12–24 h.
9. For each time point, the percentage of live and dead cells is determined with a cell counter. The optimal time point for harvesting the cells would be the latest time point following the drug treatment where the vast majority of the cells are still alive.

3.3 Determining Drug Dosage for Cohorts

How a drug is administered (e.g., oral, intravenous, intraperitoneal, etc.), how often the drug is administered, how the drug is formulated, and the drug's dosage will all impact how the drug is adsorbed, distributed, metabolized, and excreted (ADME) in an animal or human [70, 71]. This, in turn, will determine both the peak and steady-state concentration of the drug and the resulting efficacy and toxicity of the treatment. Determining the proper drug dosage and the route of administering a drug lead for an animal model or a clinical trial is, thus, a complex question that is not easily answered. In fact, the entire field of pharmacokinetics and pharmacodynamics is dedicated to addressing these very challenging questions [72]. Available information, such as a lethal dose (LD50) from cell-based assays or reported dosages for similar compounds from the scientific literature, may provide initial guidance. Unfortunately, there is not a simple algorithm to convert an LD50 to an animal dosage. Instead, a detailed efficacy and toxicity study needs to be conducted to determine the proper drug dosage prior to embarking on a metabolomics study.

Drug dosages are typically administered according to a species' body-mass (e.g., mg/kg). Accordingly, it may appear trivial to translate an animal drug dosage to a human clinical trial by simply scaling the dosage based on relative body masses. Unfortunately, this approach is incorrect since pharmacokinetics and toxicity issues vary widely across species. Instead, a variety of factors, such as differences in allometry, metabolic rates, and physiological, need to be considered [73, 74]. Thus, the protocol for determining the proper dose and means of administering a drug candidate is beyond the intended scope of this chapter.

3.4 Collecting and Storage of Biofluids from Human Cohorts (See Table 1)

1. Human volunteers in a metabolomics study must sign an informed consent form before participating.
2. Participants in a metabolomics study are typically required to complete a questionnaire that confidentially collects information regarding the individual's age, sex, race/ethnicity, residing state, diet (e.g., vegetarian, vegan, etc.), dietary restrictions (e.g., no peanuts, no gluten, low carb, low calorie, no lactose, etc.), and a list of daily medications or supplements.
3. Biofluid collection should be done after 12 h of fasting with no fish consumed for 24 h [75].
4. Daily supplements and over-the-counter medications should be stopped at least 24 h prior to collecting the biofluid(s).
5. Each type of biofluid requires specific and unique collection procedures. For example, urine specimens are best collected midstream and in the morning and can be collected by the participant. Conversely, blood needs to be collected by a medical professional in a clinical or medical facility. A detailed

discussion of common protocols employed for collecting biofluids from human and animal cohorts is beyond the scope of this chapter, but has been previously described in detail [19, 53, 68, 69, 76–78].

6. After collecting a biofluid, the samples should be processed immediately, kept on ice at all times, and stored at −80 °C.

3.5 Preparation of Mammalian Cells

1. Cell culture media and supplements are obtained from commercial vendors (*see* Subheading 2.5). The cell culture media may be supplemented with isotopically labeled nutrients (e.g., ^{13}C-glucose, ^{13}C-acetate, ^{15}N-ammonium chloride, etc.) as dictated by the goals of the experiment (*see* Subheading 3.7).
2. Cell culture procedures must follow published guidelines for the cell type to avoid misidentification and contamination [79].
3. A minimum target of ten replicate cultures is prepared per group (*see* **Note 8**).
4. Mammalian cells are typically grown in a 100 mm^2 dish with 10 mL of media at 37 °C and 5% percent CO_2 in an incubator. Equally distribute replicates from each group between incubators if more than one incubator is required (*see* **Note 9**).
5. The cells are grown until reaching approximately 80–90% confluency.
6. The drug is added to the cell culture for the treated groups using the dosage determined from Subheading 3.3. A blank solvent bolus is added to all non-treatment groups.
7. The drug is typically added to the cell culture after 80–90% confluency. But, the drug may also be added at the start of the cell culture or at other time points as required.
8. The cells are harvested after a pre-determined incubation time with the drug.

3.6 Preparation of Bacterial Cell Cultures

1. Perform **steps 1–3** from Subheading 3.5.
2. The drug is added to the cell culture for the treated groups using the dosage determined from Subheading 3.3. A blank solvent bolus is added to all non-treatment groups.
3. Typically, the drug is added at the start of the cell culture, but it may be added at various time points along the growth curve depending on the specific goals of the study or the proposed mechanism of action of the drug.
4. Start with a 25 mL culture of cells with an OD_{600} of 0.05 from glycerol frozen stocks.
5. For any type of bacterial cell, it is important to obtain a final OD_{600} of 20 for the entire culture volume. For example, if the total volume is 20 mL, then the minimum OD_{600} should be

1.0 per mL. This will ensure that a high-quality NMR spectrum will be obtained from each cell lysate in a reasonable amount of time.

6. Cell cultures are harvested at a specific time point of the growth curve (e.g., exponential growth, stationary phase, etc.). Importantly, all cell cultures must be harvested at the identical phase of the growth curve, not necessarily the same OD_{600} value.
7. All replicate cell cultures are normalized based on the final OD_{600} value. The total volume for each individual cell culture is adjusted so that each cell culture has the same effective OD_{600} value.

3.7 Isotopically Labeled Metabolomics Samples

1. Identify a ^{13}C-, ^{15}N-, or other isotopically labeled compound or nutrient to label the metabolome and act as a tracer. The tracer should be associated with the metabolic pathway of interest and expected to be affected by the experimental treatment. $^{13}C_6$-glucose is a common choice for a tracer since it highlights central carbon metabolism (e.g., glycolysis and TCA cycle).
2. The culture media should not contain the selected nutrient or metabolite.
3. The culture media is supplemented with an appropriate molar concentration of the tracer metabolite (*see* **Notes 4** and **10**). The concentration of the ^{15}N- or ^{13}C-labeled metabolite needs to be high enough (≥1–5 μM) to be detected by NMR. Accordingly, the culture media should be supplemented with approximately 2.5–4 g/L (0.5–25 mM) of a major nutrient source like $^{13}C_6$-glucose or ~10–100 mg/L of a targeted metabolite like ^{13}C-D-alanine (~1 mM). Multiple labeled isotopes may be used together if the type of label (^{13}C or ^{15}N) differs between the compounds.
4. In general, the isotopically labeled metabolite is added to the culture medium at the start of the cell culture. Of course, it may be added at different time points to emphasize a specific cellular process or timed according to the drug treatment.

3.8 Extracting Water-Soluble Metabolites from Mammalian Cells

Samples should be stored at −80 °C, but, ideally, samples should be immediately analyzed. All samples should be kept on ice or at 4 °C during the extraction process and/or handling. There are a few critical aspects to keep note of throughout the process that can affect the quality of the resultant data (1) speed, (2) consistency, (3) random processing of samples, and (4) the efficient removal of all biomolecules and cell debris [80]. Quick and prompt sample processing and preparation are ideal to avoid extended temperature variations and chances of environmental contamination and degradation of metabolites. For example, protein degradation due to improper shipping processes may lead to alterations in amino acid levels in biofluids. Furthermore, metabolites can chemically degrade or transform within milliseconds due to enzymatic activity, oxidation, chemical instability, or any number of other chemical processes [81]. Thus, rapid inactivation and removal of all macromolecular cell debris (usually through methanol/ethanol precipitation) that may bind or transform a metabolite are necessary steps of the protocol (*see* **Note 11**).

1. The extraction process for intracellular metabolites begins after the cells have grown to the desired time point. In addition to the cell extract, the media should also be analyzed for metabolomics changes as many metabolites get exchanged or effluxed outside of the cell.
2. Pre-chill the extraction solvents methanol and methanol:water mixture on dry ice.
3. Collect 1 mL of the media and immediately freeze the samples with liquid nitrogen and store at −80 °C until ready for NMR sample preparation.
4. For the cell extracts, wash the cells still attached to the plates twice with 5 mL of PBS to remove debris. Discard the wash.
5. Lyse and quench cells with 1 mL of pre-chilled methanol at −20 °C. Incubate cells at −80 °C for 15 min. This process should detach the cells from the plate. Confirm cell detachment using a microscope and repeat lyse and quenching step if necessary.
6. Using a cell scraper, detach and transfer the cell suspension in methanol to a 2 mL microcentrifuge tube.
7. Centrifuge the 2 mL microcentrifuge tube for 5 min at 15,000 × g and 4 °C to pellet the cell debris.
8. Collect the supernatant and transfer to a new 2 mL microcentrifuge tube. Again, keep all samples on ice throughout the process.
9. Perform a second extraction to improve the overall efficiency. Add 0.5 mL of an 80%/20% mixture of methanol/water pre-chilled at −20 °C to the cell pellet.
10. Centrifuge the suspension of cell debris with 80%/20% methanol/water for 5 min at 15,000 × g at 4 °C to pellet the cell debris.
11. Collect the supernatant and transfer to the 2 mL microcentrifuge tube containing the original methanol extract, thereby combining the two extractions.
12. For a third time, extract the pellet by adding 0.5 mL of ice cold water.

13. Centrifuge the cell suspension for 5 min at 15,000 × *g* at 4 °C to pellet the remaining cell debris.
14. Transfer the supernatant to the original microcentrifuge tube. Combine all extractions together in a 2 mL microcentrifuge tube.
15. Repeat **steps 1–14** for each replicate and for each group either in a randomized fashion or at the same time depending on the total sample number and your centrifuge capacity (*see* **Note 9**).
16. Use a SpeedVac or a rotary evaporator to remove the methanol. Chill the condenser with liquid nitrogen if using a rotary evaporator.
17. Following evaporation of the methanol solvent, poke a pin hole using a thumbtack and flash-freeze the samples in liquid nitrogen.
18. Freeze dry the samples using a lyophilizer; the pin hole helps establish low pressure in each sample tube.
19. Store samples in a −80 °C freezer or proceed to preparing the NMR samples (*see* Subheading 3.11).

3.9 Extracting Water-Soluble Metabolites from Gram-Negative Bacteria

1. Spin down the suspension culture at 3200 × *g* at 4 °C for 20 min to obtain a pellet.
2. Aspirate the media, keeping 1 mL of the media for NMR analysis.
3. Wash the pellet with ice cold PBS before extraction.
4. Add 1.5 mL of a 1:1 mixture of methanol and water to the pellet and transfer to a 2 mL microcentrifuge tube.
5. Pipet to mix the pellet with a 1:1 mixture of methanol and water pre-chilled to −20 °C.
6. Lyse the cells using a handheld sonicator by pulsing for six 10 s bursts. Take precautions to avoid any sample heating during the sonication process.
7. Incubate the microcentrifuge tubes on ice for 2 min to extract the metabolome.
8. Centrifuge at 1000 × *g* for 10 min at 4 °C to remove cell debris.
9. Collect the supernatant (0.75 mL) and transfer to a new 2 mL microcentrifuge tube.
10. Repeat the metabolome extraction by adding 0.75 mL of 1:1 mixture of methanol and water pre-chilled to −20 °C to the cell pellet.
11. Repeat **steps 6–8** and combine the supernatant with the previous extract.
12. Repeat **steps 1–11** for all replicate samples and groups (*see* **Note 9**).

13. Use a SpeedVac or a rotary evaporator to remove the methanol.
14. Flash-freeze the samples in liquid nitrogen.
15. Remove the water and bring to dryness using a lyophilizer.
16. Store samples in a −80 °C freezer or proceed to preparing the NMR samples (*see* Subheading 3.11).

3.10 Extracting Water-Soluble Metabolites from Gram-Positive Bacteria

1. Perform **steps 1–5** from Subheading 3.9.
2. Lyse the cells using Matrix B lysing tubes using a FastPrep system. Each sample is bead beaten for 45 s burst at 1200 rpm for three rounds Take precautions to avoid any sample heating during the process.
3. Perform **steps 7–16** from Subheading 3.9.

3.11 Preparation of NMR Samples from Cell Lysates or Tissue Extracts

1. For 1D NMR experiments, lyophilized cell-free lysates or tissue extracts are suspended in 550 μL of 100% 50 mM D_2O phosphate buffer at pH 7.2 (uncorrected, *see* **Note 3**) with 50 μM TMSP-d_4 as a chemical shift and concentration reference.
2. For 2D NMR experiments, lyophilized cell-free lysates or tissue extracts are suspended in 550 μL of 100% 50 mM D_2O phosphate buffer at pH 7.2 (uncorrected, *see* **Note 3**) with 500 μM TMSP-d_4 as a chemical shift and concentration reference.
3. Centrifuge the sample at 14,000 × *g* for 10 min to remove any particulates.
4. The sample is transferred to a 4″ 5 mM SampleJet NMR tube with a RAININ XLS pipette (*see* **Note 12**).
5. Repeat **steps 1–4** for each replicate and for each group (*see* **Note 9**).
6. Each sample is added to a 96-well plate SampleJet configuration equilibrated to 4 °C to prevent metabolite degradation (*see* Fig. 5).

3.12 Preparation of NMR Samples from Urine

1. Thaw the urine samples stored at −80 °C on ice.
2. Samples are centrifuged at 15,000 × *g* for 10 min at 4 °C to pellet any particulates in the sample.
3. 500 μL of urine is transferred into a new Eppendorf tube and mixed with 50 μL of 50 mM phosphate buffer in 99.8% D_2O at pH 7.2 (uncorrected, *see* **Note 3**). 50 μM (for 1D ^{1}H NMR) or 500 μM (for 2D NMR) TMSP-d_4 are added to each sample as a chemical shift and concentration reference.
4. The sample is transferred to a 4″ 5 mM SampleJet NMR tube with a RAININ XLS pipette (*see* **Note 12**).
5. Repeat **steps 1–7** for each replicate and for each group (*see* **Note 9**).

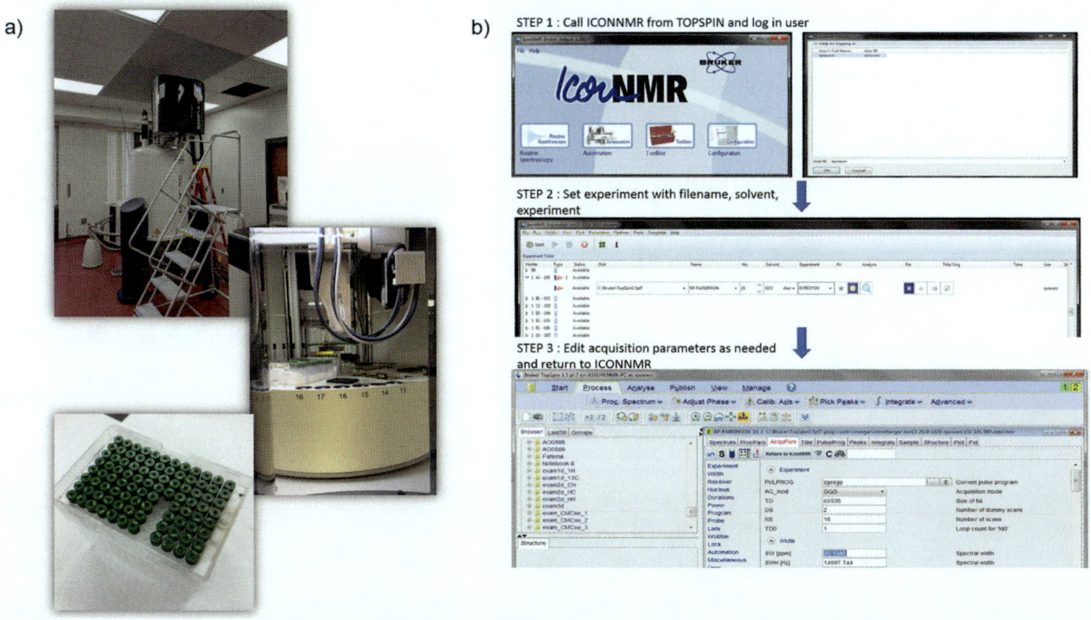

Fig. 5 High-throughput sample preparation and data collection. (**a**) Images illustrating the loading of replicate metabolomics samples into the 96-well plate SampleJet configuration. (**b**) ICONNMR screenshots illustrating the stepwise workflow for setting up a high-throughput 1D ^{1}H NMR metabolomics screen

6. Each sample is added to a 96-well plate SampleJet configuration equilibrated to 4 °C to prevent metabolite degradation (*see* Fig. 5).

3.13 Preparation of NMR Samples from Serum

1. Thaw the serum samples stored at −80 °C on ice.
2. 500 μL of serum sample is transferred into a new Eppendorf tube and mixed with 1 mL of methanol to precipitate proteins.
3. Samples are centrifuged at 15,000 × *g* for 10 min at 4 °C.
4. Transfer the supernatants into new microcentrifuge tubes.
5. Remove the methanol by rotary evaporation using a SpeedVac.
6. The samples are flash-frozen in liquid nitrogen and then dried by lyophilization.
7. The dried samples are reconstituted into 550 μL of 50 mM phosphate buffer in 99.8% D_2O at pH 7.2 (uncorrected, *see* **Note 3**). 50 μM (for 1D ^{1}H NMR) or 500 μM (for 2D NMR) TMSP-d_4 are added to each sample as a chemical shift and concentration reference.
8. The sample is transferred to a 4″ 5 mM SampleJet NMR tube with a RAININ XLS pipette (*see* **Note 12**).
9. Repeat **steps 1–8** for each replicate and for each group (*see* **Note 9**).

10. Each sample is added to a 96-well plate SampleJet configuration equilibrated to 4 °C to prevent metabolite degradation (*see* Fig. 5).

3.14 Preparation of NMR Samples from Fecal Matter

1. Thaw the fecal samples stored at −80 °C on ice.
2. Weigh 3–50 mg of the fecal material in a microcentrifuge tube containing matrix D type lysing beads.
3. Add 1 mL of 1:1 water:methanol mixture to extract the metabolome and precipitate proteins.
4. Using a FastPrep, bead beat the samples for 30 s at 1200 rpm thrice.
5. Centrifuge at 15,000 × *g* for 10 min at 4 °C.
6. Collect the supernatants in a separate microcentrifuge tube.
7. Repeat **steps 3–6** to re-extract the pellet and combine both supernatants.
8. Remove the methanol by rotary evaporation using a SpeedVac and water by lyophilization.
9. Reconstitute in 550 μL of 50 mM phosphate buffer in 99.8% D_2O at pH 7.2 (uncorrected, *see* **Note 3**). 50 μM (for 1D NMR) or 500 μM (for 2D NMR) of TMSP-d_4 will be added to seach sample as a chemical shift and concentration reference.
10. The sample is transferred to a 4″ 5 mM SampleJet NMR tube with a RAININ XLS pipette (*see* **Note 12**).
11. Repeat **steps 1–10** for each replicate and for each group (*see* **Note 9**).
12. Each sample is added to a 96-well plate SampleJet configuration equilibrated to 4 °C to prevent metabolite degradation (*see* Fig. 5).

3.15 NMR Data Collection

All NMR experiments are conducted at 298 K using a Bruker AVANCE III HD 700 MHz spectrometer equipped with a 5 mm quadruple resonance QCI-P cryoprobe (^1H, ^{13}C, ^{15}N, and ^{31}P) with *z*-axis gradients. An automatic tune and match system (ATM) and a SampleJet automated sample changer system with Bruker ICON-NMR software were used to automate the NMR data collection (*see* Fig. 5).

3.15.1 1D ^1H NMR

1. Load the NMR metabolomics samples into the SampleJet automated sample changer system (*see* Fig. 5 and **Note 9**). Check that the SampleJet is in the correct mode (i.e., 5 mm tubes).
2. Log into an account on the spectrometer workstation and start the Topspin software.

3. The first NMR sample is lowered into the magnet using the Bruker command, *sx 101*, where 101 corresponds to sample one in rack one.
4. The spectrometer is locked onto the D_2O solvent frequency using the Bruker command, *lock* D_2O (*see* **Note 13**).
5. The NMR sample is shimmed for optimal signal and suppression of the water signal by typing the Bruker command *topshim*. This will initiate an automated gradient shimming procedure, which may take a few min to complete (*see* **Note 13**).
6. The sample is automatically tuned and matched using the ATM system by typing the Bruker command, *atma*.
7. The 90° pulse length (μs) is determined by measuring a null spectrum with an approximate 360° pulse using the Bruker **zg** pulse sequence (*see* **Note 14**).
8. A 1D ^{1}H NMR spectrum is obtained for each sample using a standard excitation sculpting water suppression pulse program (Bruker **zgesgp** pulse sequence) that provides optimal suppression of the residual water signal while maintaining a flat baseline (*see* **Note 15**).
9. Typical experimental parameters for a 1D ^{1}H NMR spectrum obtained on a Bruker 700 MHz spectrometer with a cryoprobe correspond to 128 scans, 16 dummy scans, 32,768 data points, a spectral width of 11,160.7 Hz, and a relaxation delay of 1.5 (*see* **Note 16**).
10. Automated data collection of the entire set of metabolomics samples is accomplished using Bruker ICONNMR 5 (*see* Fig. 5).
11. The sample filename, solvent, pulse program, and temperature parameters are all defined in Bruker ICONNMR 5 (*see* **Notes 15–19**).
12. Collect the 1D ^{1}H NMR spectrum for each replicate and each group (*see* **Note 9**).
13. The data is processed initially with Topspin to verify spectral quality but exported for further analysis.

3.15.2 2D ^{1}H–^{13}C-HSQC NMR (See Note 20)

1. Follow **steps 1–7** from Subheading 3.15.1.
2. Using ICONNMR 5, the sample filename, solvent, pulse program, and temperature parameters are adjusted (*see* **Notes 16–19**).
3. The ICONNMR setup is similar to a 1D ^{1}H NMR data collection as shown in Fig. 5.
4. A standard 2D ^{1}H–^{13}C-HSQC experiment (Bruker *hsqcetgpsisp2* pulse program) is used to determine the ^{1}H–^{13}C chemical shift correlations for all ^{13}C-labeled metabolites in the metabolomics sample (*see* **Note 20**).

5. Typical experimental parameters for a 2D ^{1}H–^{13}C-HSQC NMR spectrum obtained on a Bruker 700 MHz spectrometer with a cryoprobe correspond to 128 scans, 32 dummy scans, and a 1.0 s relaxation delay. The spectrum is collected with 2k data points and a spectrum width of 4734 Hz in the direct dimension and 64 data points and a spectrum width of 18,864 Hz in the indirect dimension (*see* **Note 21**).
6. Implementation of fast NMR methods that includes non-uniform sampling significantly decreases data acquisition time and/or increases spectral resolution but may introduce artifacts (*see* **Note 22**).
7. Collect the 2D ^{1}H–^{13}C-HSQC NMR spectrum for each replicate and each group (*see* **Note 9**).
8. Process the data initially with Topspin to verify spectral quality but exported for further analysis (*see* Figs. 6 and 7).

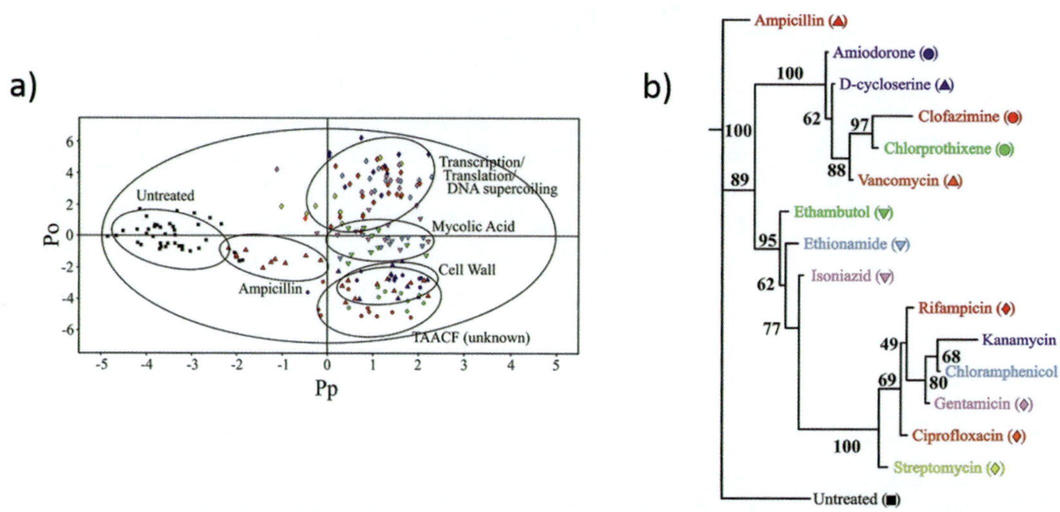

Fig. 6 A representative example of bacterial cells treated with existing drug therapies to decipher the in vivo mechanism of action of unknown drugs using 1D ^{1}H NMR-based metabolomics. (**a**) 2D OPLS-DA scores plot demonstrating the clustering pattern for 12 antibiotics with known biological targets and 3 compounds of unknown in vivo activity: untreated cells (■ black), chloramphenicol (◆ teal), ciprofloxacin (◆ orange), gentamicin (◆ pink), kanamycin (◆ purple), rifampicin (◆ red), streptomycin (◆ yellow), ethambutol (▼ green), ethionamide (▼ blue), isoniazid (▼ pink), ampicillin (▲ red), D-cycloserine (▲ purple), vancomycin (▲ orange), amiodorone (● purple), chlorprothixene (● green), and clofazimine (● red)-treated *M. smegmatis* cells. The ellipses correspond to the 95% confidence limits from a normal distribution for each cluster. The untreated *M. smegmatis* cells (■ black) were designated the control class, and the remainder of the cells was designated as treated. The OPLS-DA used one predictive component and six orthogonal components to yield a R^2_X of 0.715, R^2_Y of 0.803, and Q^2 of 0.671. (**b**) Metabolomics tree diagram was determined from the OPLS-DA scores plot. The coloring scheme and associated symbol for each compound in the tree diagram correlate with colored symbols in the OPLS-DA scores plot. The bootstrap numbers for each node are indicated on the tree diagram (Reproduced with permission from ref. 52. Copyright © 2012 American Chemical Society)

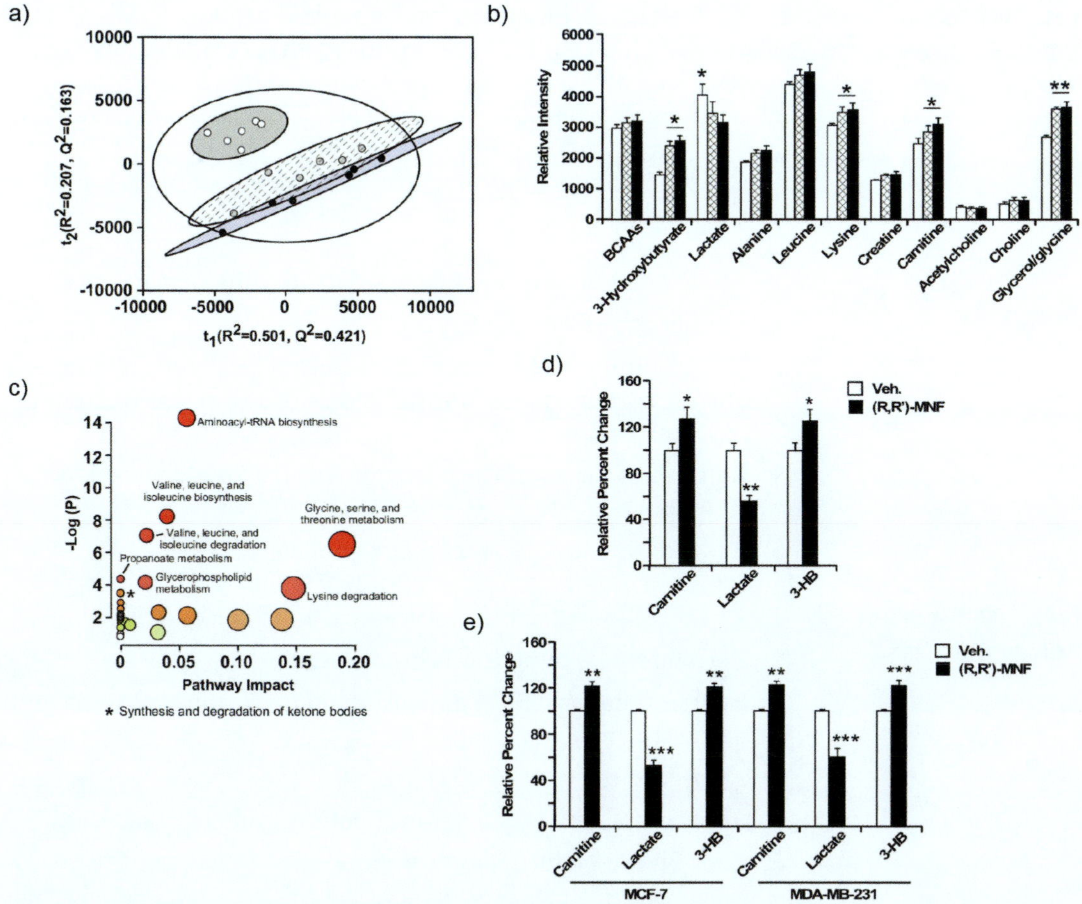

Fig. 7 Metabolomics analysis on the effect of (R,R′)-MNF in PANC-1 cells. (**a**) PCA scores resulting from modeling of the 1D ^{1}H NMR data matrix from untreated PANC-1 cells (open symbols) and PANC-1 cells treated with 0.5 μM (R,R′)-MNF (gray symbols) or 1 μM (R,R′)-MNF (black symbols). A statistically significant degree of separation is observed between treated and untreated groups. The ellipses correspond to 95% confidence intervals for a normal distribution. Each principal component is labeled with the corresponding R^2 and Q^2 values. (**b**) Bar graph of 1D ^{1}H NMR peak intensities (relative metabolite concentrations) resulting from the analysis of PANC-1 cellular extracts after a 1-h incubation with 0.5 μM (hatched bars) or 1 μM (filled bars) of (R,R′)-MNF versus untreated cells (open bars). Metabolites were identified from the backscaled loadings plots as the major contributors to the group separations in the OPLS models. Benjamini-Hochberg corrected Student's *t*-test *p*-values from pairwise comparisons are indicated. An asterisk indicates significance between the control and 1 μM of (R,R′)-MNF, while an asterisk with underlying bar indicates significance between the control and both treated groups. (**c**) Metabolic pathway analysis of the identified metabolites found in the 1D ^{1}H NMR spectra of the PANC-1 extracts after 1-h incubation with 0.5 μM or 1 μM of (R,R′)-MNF. Each circle represents a matched pathway and is colored according to its *p* values from the pathway enrichment analysis. Statistically significant pathways ($p < 0.05$) are labeled with their common name. (**d** and **e**) Targeted metabolomics on the effect of 1 μM (R,R′)-MNF on intracellular concentrations of carnitine, L-lactate, and 3-hydroxybutyrate in PANC-1 cells (**d**) as well as MCF-7 and MDA-MD-321 breast tumor cells (**e**). $*p \leq 0.05$; $**p \leq 0.01$; $***p \leq 0.001$ (Reproduced with permission from M. Bernier, J. Catazaro, N. S. Singh, Q. Shi. M. Wang, Q. Yang, A. Wnorowski, K. Jozwiak, R. Powers, and I. W. Wainer∗ (2017) "GPR55 Receptor Antagonists Decrease Glycolytic Activity in PANC-1 Pancreatic Cancer Cell Lines and Tumor Xenografts", International Journal of Cancer, 141(10):2131–2142 Copyright © 2017 John Wiley and Sons)

3.16 NMR Data Processing (See Note 23)

All NMR data is processed and analyzed with our MVAPACK software [57], PCA/PLS-DA utilities [58], NMRPipe [59], and NMRViewJ [60]. See example processing scripts at http://bionmr.unl.edu/wiki/Scripts. Please see the NMRPipe and nmrDraw tutorial (https://spin.niddk.nih.gov/NMRPipe/doc1/) and NMRViewJ documentation (http://docs.nmrfx.org/) for more details.

3.16.1 1D ^{1}H NMR (See Figs. 6 and 8)

1. Apply a 1.0 Hz exponential apodization function to the FID.
2. Fourier transform the FID.
3. The resulting NMR spectrum is automatically simultaneously phased corrected and normalized with the phase-scatter correction algorithm [82].
4. Reference the NMR spectrum to the peak of TMSP-d_4 (0.0 ppm).
5. Remove the noise and solvent regions manually.

3.16.2 2D ^{1}H–^{13}C-HSQC NMR (See Figs. 7 and 8)

1. Apply a sine-bell apodization function to the t_2 dimension.
2. Zero fill the t_2 dimension three times.
3. Fourier transform he t_2 dimension, manually phase correct, and delete the imaginary data.
4. Transpose the data matrix.
5. Apply a sine-bell apodization function to the t_1 dimension.
6. Zero fill the t_1 dimension three times.
7. Fourier transform the t_1 dimension and phase correct manually.
8. Reference the NMR spectrum in both dimensions to the peak of TMSP-d_4 (0.0 ppm).

3.17 NMR Data Preprocessing for Multivariate Modeling

To obtain an accurate and reliable multivariate statistical model, it is essential that the dataset is properly preprocessed to remove normal systematic variations resulting from biological variability, instrument instability, and inconsistency in sample handling and preparation. Key preprocessing steps include (1) alignment, (2) normalization, (3) binning, and (4) scaling, which is illustrated in Fig. 9. All NMR datasets are processed with our MVAPACK software [82] and our PCA/PLS-DA utilities [58]. See example MVAPACK scripts at http://bionmr.unl.edu/wiki/Scripts.

3.17.1 1D ^{1}H NMR

1. Spectra may first be normalized based on either the total cell count or the total protein concentration using the BCA (bicinchoninic acid) protein estimation assay. The BCA assay uses parallel dishes that are similarly treated.
2. Normalize the spectra with the PSC algorithm [82].

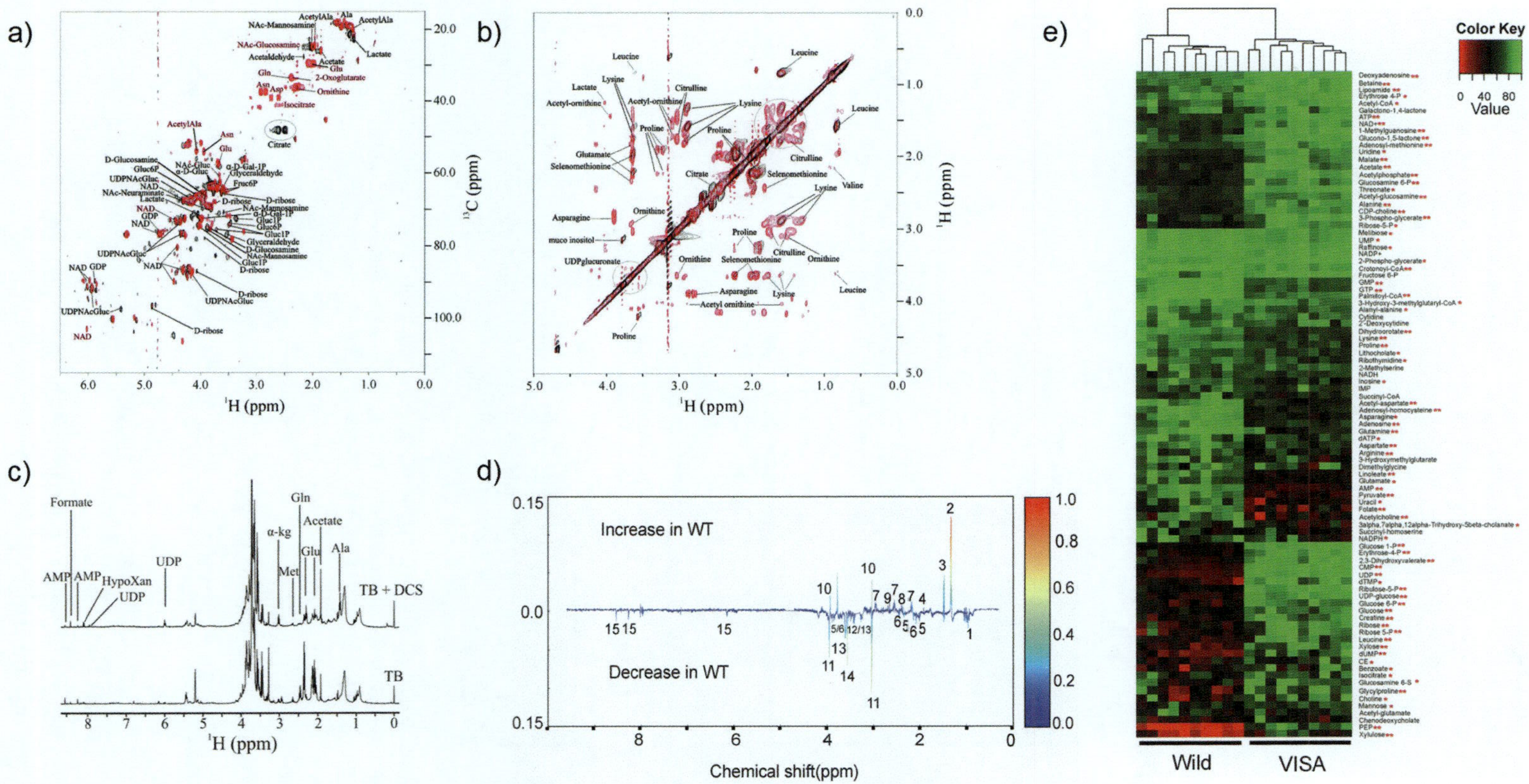

Fig. 8 Examples of a typical (**a**) 2D ^{1}H–^{13}C HSQC spectrum and (**b**) 2D ^{1}H–^{1}H TOCSY spectrum acquired from bacterial cells (Reproduced with permission from Zhang, B.; Halouska, S.; Schiaffo, C. E.; Sadykov, M. R.; Somerville, G. A.; Powers, R. NMR analysis of a stress response metabolic signaling network. J. Proteome Res., 2011, 10, 3743–3754. Copyright © 2011 American Chemical Society). (**c**) 1D ^{1}H NMR spectrum acquired from bacterial cells (Reproduced with permission from S. Halouska, R. J. Fenton, D. K. Zinniel, D. D. Marshall, R. G. Barletta∗ and R. Powers∗ (2014) "Metabolomics Analysis Identifies D-Alanine-D-alanine Ligase as the Primary Lethal Target of D-cycloserine in Mycobacteria", J. Proteome Res., 13(2):1065-1076. Copyright © 2014, American Chemical Society). (**d**) Back-scaled loadings plot for 1D ^{1}H NMR spectra showing 15 different metabolites that vary between groups (Reproduced with permission from ref. 15. Copyright © 2017 American Society for Microbiology). (**e**) Heatmap showing differences between the vancomycin-susceptible *Staphylococcus aureus* (VSSA) strains and *S. aureus* with intermediate-level resistance to vancomycin (VISA) metabolome. The heatmap was generated by using 2D ^{1}H–^{13}C HSQC NMR spectrum-normalized peak intensities of the VSSA or VISA strain, and hierarchical clustering is depicted by the dendrogram. The color scale ranges from 0 (red; less intense) to 1 (green; intense). Statistical significance is denoted by a single asterisk if determined using Student's t test at the 95% confidence level or by two asterisks if determined by Student's t test with a Bonferroni correction for multiple hypothesis testing (Reproduced with permission from ref. 15. Copyright © 2017 American Society for Microbiology)

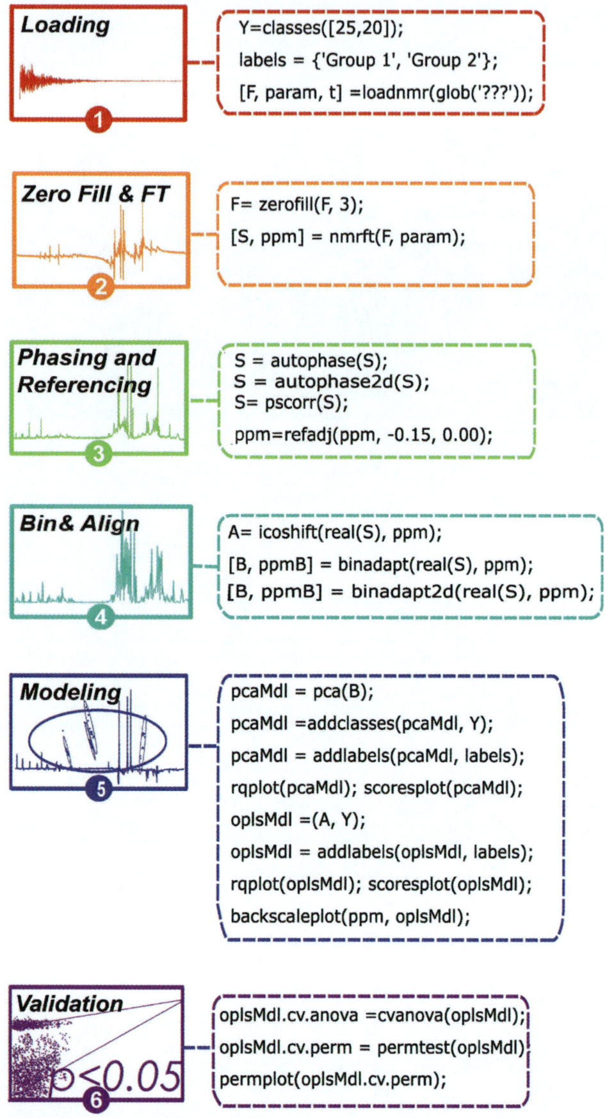

Fig. 9 Example of an MVAPACK processing script for a 1D ^{1}H NMR dataset. The numbered steps in the flow diagram correspond to the numbered lines in the processing script

3. Align and/or bin the spectra. For principal component analysis (PCA), use the following parameters:
   (a) Globally align the spectral data to the peak of TMSP-d_4 at 0.0 ppm.
   (b) Regionally align the spectral data using the icoshift algorithm [83].
   (c) Bin the spectral data using the adaptive, intelligent binning algorithm [84].

4. For orthogonal projection to latent structures (OPLS), use the following parameters:
   (a) Globally align the spectral data to the peak of TMSP-d_4 at 0.0 ppm.
   (b) Regionally align the spectral data using the icoshift algorithm [83].
   (c) The spectral data are not binned. Instead, use the full-resolution spectral data to build the model.
5. Manually remove the solvent peaks and noise regions.
6. Scale the dataset using Pareto scaling.
7. Generate the PCA or OPLS model from the data matrix.

3.17.2 2D ^{1}H–^{13}C-HSQC NMR

1. Spectra may first be normalized based on either the total cell count or the total protein concentration as explained above.
2. Normalize the spectral data using standard normal variate normalization.
3. Bin the spectral data using a generalized adaptive, intelligent binning algorithm [84].
4. Pareto scale the data.
5. Generate a PCA or OPLS model from the data matrix.

3.18 Statistical Analysis

Datasets are analyzed with our MVAPACK software [57], PCA/PLS-DA utilities [58], and R [85]. See example MVAPACK and R scripts at http://bionmr.unl.edu/wiki/Scripts.

Improper application of statistical methods is a serious issue in the field of metabolomics, which results from some common misconceptions. For example, validation of supervised multivariate statistical methods are routinely lacking. Instead, supervised OPLS and PLS models need to be validated using a cross-validation score like CV-ANOVA [86] and/or response permutation testing [87]. R^2 and Q^2 values from these supervised models provide a measure of model fitness from the original data and a measure of consistency between the original and predicted data, respectively. R^2 and Q^2 values do not provide for model validation without a proper standard of comparison.

Metabolomics datasets are often described using PCA or PLS/OPLS models. It is important to note that PCA finds the largest source of variance in the dataset irrespective of the intent of the study. Separation of clusters observed in a PCA model may not have anything to do with a drug treatment if there are other major sources of variance in the data. Thus, a loadings plot generated from a PCA model represents all of the variation in the dataset, not just the expected changes induced by the drug treatment. Supervised methods, like PLS and OPLS, address this issue by aggressively forcing group separation based on the defined group membership. So, by design, group separation is almost always achieved in a supervised model and should not be interpreted as a novel discovery in itself. Again, group separation may occur irrespective of the validity of the model. This happens because PLS and OPLS models are easily over-fitted, especially for metabolomics datasets since the number of variables (e.g., metabolites) are typically larger than the number of replicates. This forced separation of groups also contributes to a common misconception in the metabolomics community, the incorrect belief that OPLS or PLS outperforms PCA in identifying group separation and the underlying dataset variances. Remember, PCA finds the largest source of variance in the dataset, so if there is no variance in a PCA model, smaller group-specific variations are also not present. Thus, if PCA fails to identify group separation, it is unlikely that PLS or OPLS will yield a valid model despite the appearance of group separation [87]. Simply, PCA, PLS, and OPLS generate different models by extracting different information to achieve different goals.

PLS and OPLS appear to provide a similar model. In fact, a comparison of PLS and OPLS scores plots generated from the same dataset may suggest the only difference is a relative rotation of the group-defined ellipses. Nevertheless, this apparently subtle change highlights a critical difference. PLS intermingles both group-independent and group-dependent variance, whereas OPLS places group-independent variance (e.g., confounding factors such as differences in diet, age, race, etc.) orthogonal to group-dependent variance. Therefore, a metabolite identified as a major contributor to an OPLS model is strictly the result of the defined group difference. Accordingly, we strongly recommend using OPLS instead of PLS.

3.18.1 Univariate Analysis

1. Determine relative metabolite abundances from NMR spectral peak heights and/or peak volumes (*see* Fig. 8).
2. Normalize relative metabolite abundances on a per spectrum basis. One common approach (standard normal variate) is to convert the absolute peak intensities (arbitrary units) to a *Z*-score:

$$Z = \frac{I_i - \bar{I}}{\sigma} \tag{1}$$

where $\bar{I}$ is the average peak intensity for the spectrum, I_i is the intensity of peak i, and σ is the standard deviation of peak intensities. Peak intensities may also be normalized to the total number of cells, to the total protein concentration (*see* Subheading 3.17.1), to the average spectral noise, or to an internal standard (*see* **Note 24**). Relative metabolite abundances may also be converted to fold changes:

$$F = \frac{I_i}{I_o} \quad (2)$$

where I_i is the normalized peak intensity of metabolite *i* from a treated spectrum and I_o is the normalized peak intensity of metabolite *i* from the control or untreated spectrum.

3. A standard Student's *t*-test is commonly used to determine statistical significance only for a pairwise comparison of metabolite changes based on either fold changes or normalized peak intensities (*see* **Note 25**). A statistically significant difference is typically identified by a *p*-value <0.05 (*see* Fig. 8).
4. A Student's *t*-test is insufficient for the multiple comparisons that are common to a metabolomics study [87, 88]. In order to identify the set of metabolites that exhibit a statistically significant change, a multiple hypothesis test correction method such as a Benjamini-Hochberg [89] or a Bonferroni [90] correction must be applied (*see* **Notes 25** and **26**).
5. A heatmap with hierarchical clustering is commonly generated from the fold changes or normalized peak intensities using R (*see* example R script at http://bionmr.unl.edu/wiki/Scripts). The heatmap may contain relative metabolite abundances for each individual replicate in the study or simply the replicate-averages for each group (*see* **Note 27**).

3.18.2 Multivariate Analysis

1. Generate a PCA and or OPLS model from the data matrix.
2. Fractions of explained variation ($R^2{}_X$ and $R^2{}_Y$) are computed during PCA or OPLS model training.
3. Internally cross-validate the PCA or OPLS model using sevenfold Monte Carlo cross-validation [91] to compute Q^2 values (*see* **Note 28**).
4. For an OPLS model, compare the Q^2 value against a distribution of null model Q^2 values in 1000 rounds of response permutation testing [87]. Group membership is randomly reassigned to generate the set of null models. A *p*-value is calculated from a comparison of the true Q^2 value to the set of null model Q^2 values (*see* **Note 28**).
5. Further validate the model using CV-ANOVA significance testing, which is used to calculate another model *p*-value [86] (*see* **Note 28**).
6. Scores plots (*see* Fig. 6), backscaled loadings plots (*see* Fig. 8), S-plots, and/or SUS-plots are often generated from OPLS models.
7. PCA/PLS-DA utilities [58] are used to define group membership by drawing an ellipse per group onto the scores plots (*see* Figs. 6 and 7). Each ellipse corresponds to 95% confidence interval for a normal distribution. The PCA/PLS-DA utilities

also generate a metabolomics tree diagram that identifies the statistical significance (*p*-value and/or bootstrap value) and the relative similarity of each group in the scores plot (*see* Fig. 6). The *p*-value or bootstrap number from the pairwise comparison is labeled at each node in the tree.

3.19 Data Analysis: Metabolite Assignment from 1D ^{1}H NMR Data

NMR data are analyzed with NMRPipe [59], NMRViewJ [60], and Chenomx. See example scripts at http://bionmr.unl.edu/wiki/Scripts. Please see the NMRPipe and nmrDraw tutorial (https://spin.niddk.nih.gov/NMRPipe/doc1/) and NMRViewJ documentation (http://docs.nmrfx.org/) for more details.

1. Perform the identification of metabolites in a 1D ^{1}H NMR spectrum with software programs such as Chenomx. Chenomx matches the experimental 1D ^{1}H NMR spectrum to a database of 1D ^{1}H NMR spectra of known metabolites. Chenomx attempts to explain or describe the experimental NMR spectrum by combining or summing as many of the individual reference metabolite NMR spectra as needed. In addition to metabolite identification, Chenomx also provides an estimate of the metabolite concentration (*see* **Note 29**).
2. Upload the 1D ^{1}H NMR spectrum for processing. The NMR spectra can be batch processed or processed one at a time.
3. Phase the 1D ^{1}H NMR spectrum.
4. Calibrate and reference the 1D ^{1}H NMR spectrum to TMSP-d_4, using the known concentration of TMSP-d_4.
5. Send the properly phased and calibrated 1D ^{1}H NMR spectrum to the Chenomx profiler where the spectrum is compared against the metabolite library.
6. Chenomx will overlay a 1D ^{1}H NMR reference spectrum for each metabolite identified in the experimental 1D ^{1}H NMR spectrum. The spectral overlay needs to be manually adjusted to optimize the alignment of the experimental 1D ^{1}H NMR spectrum with the reference spectrum. Figure 8 shows an example of a labeled 1D ^{1}H NMR spectrum.

3.20 Data Analysis: Metabolite Assignment from 2D ^{1}H–^{13}C-HSQC NMR Data

NMR data are analyzed with NMRPipe [59] and NMRViewJ [60]. See example scripts at http://bionmr.unl.edu/wiki/Scripts. Please see the NMRPipe and nmrDraw tutorial (https://spin.niddk.nih.gov/NMRPipe/doc1/) and NMRViewJ documentation (http://docs.nmrfx.org/) for more details.

3.20.1 NMRPipe Processing to Obtain .ft2 and .nv Files

1. The data files from ICONNMR can be used directly by NMRPipe to process the 2D ^{1}H–^{13}C-HSQC spectra.

2. On a Linux workstation, open a terminal and go to the directory that contains the NMR data. Type *bruker* to start the NMRpipe software.
3. Read in the experimental parameters file by clicking *Read Parameters* and verify that all the parameters have been correctly updated. Confirm that the mode of data collection has been set to *echo-antiecho* if the NMR spectrum was collected with the *hsqcetgpsisp2* pulse program.
4. Click *Update Script* to save an NMRPipe processing script *fid.com* file in the working directory.
5. Type *./fid.com* to start the NMRPipe processing script.
6. When the NMRPipe processing has finished, type *nmrDraw* to view the processed NMR spectrum. Please see the NMRPipe and nmrDraw tutorial (https://spin.niddk.nih.gov/NMRPipe/doc1/) for detailed instructions.
7. Phase the NMR spectrum in NMRpipe and note the p_0 and p_1 values for both the ^{1}H and ^{13}C dimensions.
8. Edit the NMRPipe processing script *hsqcproc.com* and replace the parameters associated with the NMRPipe phase correction command, *ps*, with the p_0 and p_1 values obtained from **step 7**.
9. Type *./hsqcproc.com* to start the NMRPipe processing script.
10. Repeat **steps 3–9** for each 2D ^{1}H–^{13}C-HSQC NMR spectrum in the dataset. This produces a set *.ft2* files. One *.ft2* file is created for each 2D ^{1}H–^{13}C-HSQC NMR spectrum collected for each replicate from each group.
11. Copy all of the *.ft2* files into a new folder and use the NMRPipe script *addnmr.com* to generate NMRviewJ files from the *.ft2* files. A *.nv* file will be generated for each individual spectrum (*.ft2* file) with a numerically incremented root name of *Final_*. In addition, the script will combine all of the NMR spectra together into a single file called *results.nv*. The script will also generate the text file, *rate.txt*, which lists all of the individual *.nv* files (*Final_*).

3.20.2 Peak Picking and Peak Integration of 2D ^{1}H–^{13}C-HSQC Spectra in NMRviewJ

1. Type *nmrviewj* to start NMRviewJ. Refer to NMRViewJ documentation (http://docs.nmrfx.org/) for more details.
2. From the *Dataset* toolbar in the *main* window, use the *Open and Draw Datasets* function to select the *result.nv* file.
3. Right click and select *attributes* to open the *attributes* window.
4. In the *attributes* window, select the *PeakPick* tab.
5. In the blank *Lists* field in the *attribute* window, type a filename (i.e., *lists*) for the new peak pick list. Click the *Pick* button. The software will automatically peak pick the displayed spectrum and populate *lists* with the peak ID number, chemical shifts, and intensity.

6. Choose *Show Peak Table* from the *Peak* toolbar on the ***main*** window. A ***peak table*** window will open that lists the peak ID, peak intensity, and the peak chemical shifts.
7. Manually edit the peak list and remove solvent peaks, noise peaks, or other spectral artifacts. Peaks are deleted from the peak table by using the delete function in the ***PeakPick*** tab in the ***attributes*** window along with the ***spectrum display*** window. In the ***spectrum display*** window, use the mouse to position the two cursors around any peak or spectral region to form a box. Then, click the *Delete* button under the ***PeakPick*** tab in the ***attributes*** window to remove the peak(s).
8. After the peak table has been completely edited, on the ***peak table*** window, choose the ***Edit*** tab and select *Compress and Degap*. Answer *yes* to the pop-up question. This will finalize changes to the peak list and prevent any further edits.
9. On the ***peak table*** window, choose the ***Edit*** tab and then select *Save Table*. A file browser window will open in order to choose a name and location to save the new peak list file. The saved peak pick file can be viewed and edited in Excel.
10. In order to obtain peak intensities across the entire set of NMR spectra in the dataset, click on the *Analysis* tab on the main window and select *Rate Analysis*. A setup window for the *Rate Analysis* will open.
11. In the *Rate Analysis* setup window:
    (a) Set the Prefix for matrix numbers field to Final_.
    (b) Set the *Peaklist* field to *lists* (defined in Subheading 3.20.2, **step 5**).
    (c) Make sure *Auto fit* is checked.
    (d) Use all other default settings.
    (e) Click *Load time file*.
    (f) In the file browser window, select *rate.txt* (created in Subheading 3.20.1, **step 11**).
    (g) Click *Measure All*. The software automatically populates the table in the *Rate Analysis* setup window with all of the peak intensities across the entire NMR dataset.
    (h) Click *Save Table*. In the file browser window, save the peak intensities table to a new filename (i.e., ***intensities***).
12. The peak list (i.e., *list*) and the peak intensities (i.e., ***intensities***) files are merged in Microsoft Excel using the common peak ID column. The ***ppm1*** (^{1}H ppm) and ***ppm2*** (^{13}C ppm) columns are added to the peak intensities columns to generate a complete matrix of NMR peaks and intensities across the entire dataset.
13. The merged Excel file is saved to a new filename.

3.20.3 Metabolite Assignments from 2D ^{1}H–^{13}C-HSQC Peak Lists

1. The complete list of peaks obtained from the NMRviewJ analyses is searched using NMR metabolomics databases such as HMDB [62], BMRB [63], or other databases (*see* **Note 30**).
2. On the HMDB homepage, choose the *Search* tab and select *2D NMR Search*.
3. From the *Spectra Library* pull-down menu, choose *^{13}C HSQC*.
4. Cut and paste the 2D ^{1}H–^{13}C-HSQC peak lists into the *Peak List* field. One set of ^{1}H and ^{13}C chemical shifts, respectively, per line. Chemical shift values should only be separated by white space.
5. Set the ^{1}H chemical shift error tolerance to 0.05 ppm (*X-axis Peak Tolerance* ± field) and the ^{13}C chemical shift error tolerance to 0.10 ppm (*Y-axis Peak Tolerance* ± field).
6. Click the *Search* button. Depending on the size of the peak list, the software will return a ranked-order list of possible metabolites based on the number of chemical shift matches to reference spectrum.
7. Manually curate the list of potential metabolite assignments based on the number of chemical shift assignments, the quality of the spectral overlap (i.e., chemical shift match), the number of other metabolites in the same metabolic pathway, and the biological system (i.e., is it a reasonable or possible metabolite for the organism).
8. Obtain additional NMR (e.g., HMBC, HSQC-TOCSY) and/or MS spectral data to confirm or refute the assignment.
9. An assigned 2D ^{1}H–^{13}C-HSQC spectrum is shown in Fig. 8.

4 Notes

1. NMR isotope labeled compounds are non-radioactive and do not require any special handling. It is imperative you wear typical personal protective gear like gloves and eye protection and follow standard safety protocols.
2. Always keep deuterated solvents in a desiccator or dry box and keep them sealed or wrapped in parafilm while in storage.
3. Typically, metabolomics sample from cell lysates is prepared in 100% D_2O, and using a standard pH probe may not report pH accurately. Standard corrections are pD = pH + 0.4 although Rubinson et al. [92] suggest that phosphate buffer pH is not significantly altered to warrant corrections.
4. The choice of culture medium is dependent on the specific needs of the cell type and the specific goals of the experiment. There are a variety of cell culture media available either through

commercial sources or from published recipes. Also, it may be possible to use different media for the same cell type. Again, the choice of medium will be dictated by the specific needs of the experiment. Accordingly, only a few representative culture media are listed as illustrative examples, and these are not intended to be an exhaustive list.

5. For the success of a metabolomics experiment, cell survival is an essential requirement. This presents a challenge when investigating the efficacy, side effects, mechanism of action, or resistance of a drug that impacts cell survival or is designed to kill cells. Similarly, cell mutations or experimental conditions (e.g., nutrient depletion, high temperature, anaerobic growth, etc.) that may lead to an increase in cell death are also potential problems. The goal is to identify a drug dosage and/or experimental condition that only stress the cells, but does not result in *any* cell death (*see* Subheading 3.3). In this regard, the observed metabolomic changes will only be a result of the cell's immediate response to the biological activity of the drug or the experimental condition. Conversely, if the metabolome is harvested from a mixture of live and dead cells, then it will be difficult, if not impossible, to deconvolute metabolome changes resulting from the stressor or cell death. A correct drug dosage or experimental condition is usually determined by acquiring a series of growth curves in which the drug or stressor is titrated over a wide range of values.
6. The choice of solvent and concentration is determined by the compound's solubility. The goal is to maximize the concentration of the stock solution so as to minimize the amount of solvent that needs to be added to each cell culture. The maximal amount of solvent added to a cell culture is 5–10 μL.
7. The following range of drug concentrations is typically used: 0×, 1×, 2×, 4×, 8×, 16×, and 24× the MIC value. The amount of solvent (i.e., DMSO-d_6 or ethanol) added to any cell culture should be 5–10 μL or less.
8. The number of replicates per group will have a significant impact on the quality of the study and the statistical validity of the outcomes. In general, it is best to maximize the number of replicates per group, with a typical target of ten replicates per group. A variety of experimental considerations may impact the number of replicates that are practical for a given study. For example, a large number of groups may require a reduction in the number of replicates per group. Another consideration is the impact of the number of replicates on the quality of the metabolomics samples. Sacrificing quality for a greater number of replicates will not likely lead to a successful outcome. Conversely, a limited number of replicates <4 per group will likely provide meaningless results.

9. In order to avoid the introduction of bias, always randomize sample handling steps. If the samples are always processed in the order of the sample number, then a time bias will be induced across the entire dataset. Instead, if the order is constantly changed at each step, the processing time and any impact on the metabolome will be randomized, which in turn will minimize or eliminate any bias.
10. The experimental conditions for optimal labeling should be experimentally verified for every cell line or cell type used for a metabolomics experiment. Factors to consider include the composition of the culture media and the ^{13}C-carbon source (e.g., glucose, pyruvate, or glutamine) that are used for the cell culture. For mammalian cells, a time course experiment should be conducted where cells are cultured from 1 to 48 h to assess the rate of carbon consumption. For bacterial cells, a time course experiment should be conducted where cells are cultured for one to three doubling times to assess the rate of carbon consumption. Examples of media used for ^{13}C-carbon labeling of mammalian cell metabolome are Dulbecco's modified Eagle medium (DMEM) and Roswell Park Memorial Institute (RPMI) media. Examples of media used for ^{13}C-carbon labeling of bacterial cell metabolome are Middlebrook's albumin dextrose catalase (MADC) media or M9 minimal media.
11. Removal of proteins and other biomolecules by methanol or ethanol precipitation is preferred over mechanical filtration methods or the application of Carr-Purcell-Meiboom-Gill (CPMG) NMR T_2 filtering techniques. Filtering techniques are known to impart biologically irrelevant group differences, due to metabolite-macromolecule binding [87].
12. Smaller diameter NMR tubes of 3 mm (160 μL) or 1.7 mm (45 μL) may be needed if the available metabolomics sample is limited. Filling of these smaller diameter NMR tubes may be improved by a liquid handling robot, such as a Gilson 215 Liquid Handler. In addition, the NMR acquisition parameters will likely need to be adjusted to account for the lower sensitivity due to the lower number of nuclei in the samples.
13. Topshim requires the sample to contain either a D_2O or H_2O solvent. It is advisable to create a shim file with a parameter set that produces an optimal set of shims for your sample type. Read in a shim file using the Bruker command *rsh* and select the appropriate Topshim shim file. If you are doing this for the first time, complete the command *topshim*; if you are not satisfied with the shim performance, use command *topshim tuneb tunea* to obtain an improved set of shims. Write the shim set parameters with the Bruker command *wsh* and save it to a new file name for future reference.

14. The 90° pulse length is commonly measured by incrementing the P1 pulse in the *zg* pulse program by 1 μs or smaller increments and by plotting the relative peak heights or intensities. A maximum peak height should be observed at the pulse length corresponding to the 90° pulse. Conversely, a minimum or null spectrum should be observed at the pulse length corresponding to the 360° pulse length. In practice, a more accurate measure of the 90° pulse is obtained by measuring the 360° pulse length and dividing by four to obtain the 90° pulse length. A typical 90° pulse length for a metabolomics sample ranges from approximately 8–13 μs or longer. Among other factors, the relative salt concentration of the metabolomics sample affects the 90° pulse, in which a higher salt concentration results in a longer 90° pulse. Other factors also contribute to the observed 90° pulse, so it is always necessary to experimentally determine the 90° pulse for each sample or set of samples.

15. Excitation sculpting parameters (*zgesgp*)—32,768 data points (TD), SW = 12.02 ppm, O1P (transmitter offset) = 4.70 ppm, D1 = 1 s, NS (number of scans) = 128, DS (dummy/steady state scans) = 16, P1 = 9.5–13.5 us, SPNAM (shaped pulse for water suppression) = SINC1.1000 at 26.39 dB or 0.00228 W.

16. The NMR data acquisition parameters need to be adjusted to compensate for differences in the field strength and sensitivity of the NMR spectrometer actually used for the data collection. Specifically, the number of scans, the number of data points, the sweep width (13.79 ppm, ^{1}H frequency range), and the frequency offset (centered on water peak at 4.70 ppm) need to be adjusted according to the type and configuration of the NMR spectrometer used for the study.

17. For high-throughput NMR data collection, please refer to the Bruker ICONNMR manual to explore various configuration options. For example, composite experiments allow for the collection of multiple 1D and 2D experiments for the same metabolomics sample. An experimental set consisting of a 1D ^{1}H, a 2D ^{1}H–^{13}C HSQC, and a 2D ^{1}H–^{13}C HMBC experiment may be subsequently collected for the same sample before moving to the next sample in the queue.

18. It is imperative that NMR data are collected at the same temperature for a queue of metabolomics samples. ICONNMR assists this by allowing for a temperature delay when a large number of samples are in the SampleJet queue. For example, a 15–60-s delay may be inserted prior to data acquisition to allow each sample to equilibrate to the probe temperature. We recommend a 60-s delay for both pre- and post-sample insertion to prevent any temperature variation.

19. Parameters to check before you queue experiments in ICONNMR for 1D ^{1}H NMR are number of scans ***ns***, number of dummy scans ***ds***, 90° pulse ***p1***, delay ***d1***, sweep width ***sw***, receiver gain ***rg***, experiment temperature ***te***, and automation setup ***aunm***. We recommend using ***au_zgonly*** as the automation setup. This will collect all samples at the same receiver gain, which will avoid peak intensity variation across the dataset.
20. NMR metabolomics studies are beginning to make use of HMBC, TOCSY, HSQC-TOCSY, 2D J-Resolved spectra, or other experiments along with the traditional 2D ^{1}H–^{13}C HSQC approach. Heteronuclear isotope labels like ^{15}N, ^{31}P, and other isotope-labeled metabolites may be detected in addition to ^{1}H- and ^{13}C-labeled metabolites [93]. Accordingly, experimental parameters, data processing and preprocessing methods, and data analysis techniques all need to be adjusted to accommodate the specifics of each NMR experiment.
21. 2D ^{1}H–^{13}C-HSQC parameters (*hsqcetgpsisp2*)—1024 data points in F2 and F1, non-uniform sampling at 25%, O1P = 4.7 ppm, O2P (offset for ^{13}C) = 75 ppm, NS = 64, DS = 16, d1 = 2, P1 = 10–13 μs depending on salinity, CPDPRG2 = garp (decoupling program), PCPD2 = 55 μs at PLW12 = 4.09 W.
22. Non-uniform sampling of 2D ^{1}H–^{13}C HSQC data can be performed on metabolomics samples albeit with some caution. From our experience we recommend no lower than 20% sparsity using the burst augmented scheduler available at http://bionmr.unl.edu/dgs-gensched.php. Download the sampling schedules as a text file for Topspin. Similarly, processing scripts need to be adapted for NUS datasets and a quick guide to that can be found online at http://bionmr.unl.edu/wiki/NUS.
23. A minimalistic approach to the processing of NMR data is optimal for a metabolomics analysis based on multivariate statistics models such as PCA and OPLS. The resulting multivariate statistical model is dependent on the choice of processing and preprocessing protocols. Baseline corrections and alignment, type of weighting (apodization) function, the type of spectral alignment or referencing, the resulting phase correction or phase correction algorithm, the number of zero-fills or the application of linear prediction, or any other data manipulation method will affect the outcome of the statistical model. Accordingly, it is best to avoid data processing steps since it is difficult to ascertain if the data processing induced a biologically irrelevant bias to the data or improved the model.
24. For NMR, relative peak intensities are averaged across all replicates per group and for each NMR peak assigned to the

metabolite. Most metabolites will have more than one peak in an NMR spectrum, and commonly all NMR peaks are incorporated into an average relative peak intensity. If a metabolite is differentially labeled due to contributions from different pathways, it is then appropriate to show single resonances for the metabolite. NMR peaks may need to be scaled by the number of attached hydrogens, since peak intensity is proportional to the number of nuclei.

25. There are a wide variety of statistical tests available to analyze the data. Some examples are the Mann-Whitney *U* test [94], Welch's *t*-test [95], Hotelling's *t*-squared statistic [96], and one-way analysis of variance [87], among others. The proper choice of a statistical test depends on several factors. A good introduction to what dictates the choice of statistical tests can be found in *A Biologist's Guide To Statistical Thinking And Analysis* [95].

26. In effect, the uncertainty in each pairwise comparison (as determined by the Student's *t*-test) is compounded with the addition of each metabolite to a set. The actual p value for a set of metabolites is defined as:

$$p = 1 - (1 - \alpha)^m \tag{3}$$

where m is the number of hypotheses (metabolites) and α is typically defined as 0.05. Accordingly, a set of ten metabolites becomes an insignificant $p = 0.401$ even though each individual metabolite is statistically significant based on a pairwise Student's *t*-test with a $p < 0.05$.

27. It is more informative to display all replicates from each group in the heatmap. Hierarchical clustering of each replicate is indicative of the relative group separation and provides further confirmation of an observed group separation from a PCA, PLS, or OPLS scores plot.

28. Typical R^2 values $>Q^2$ values, and Q^2 values >0.4. While *p*-values <0.05 are typically acceptable for a supervised PLS-DA or OPLS model, usually, high-quality PLS/OPLS models from metabolomics datasets yield *p*-values $\ll 0.001$.

29. NMR field strength and sample pH are important aspects of any chemical shift database matching. Thus, we strongly recommend performing a chemical shift match with the correct database (e.g., same NMR field strength and pH) while using Chenomx or similar software.

30. BMRB, HMDB, and other such databases operate in a similar fashion. Simply upload a peak list with a set of chemical shift tolerances and obtain a list of potential matches.

Acknowledgments

This material is based upon work supported by the National Science Foundation under Grant Number (1660921). This work was supported in part by funding from the Redox Biology Center (P30 GM103335, NIGMS) and the Nebraska Center for Integrated Biomolecular Communication (P20 GM113126, NIGMS). The research was performed in facilities renovated with the support from the National Institutes of Health (RR015468-01). Any opinions, findings, and conclusions or recommendations expressed in this material are those of the author(s) and do not necessarily reflect the views of the National Science Foundation.

References

1. Hajduk PJ, Greer J (2007) A decade of fragment-based drug design: strategic advances and lessons learned. Nat Rev Drug Discov 6:211. https://doi.org/10.1038/nrd2220
2. Pellecchia M, Sem DS, Wüthrich K (2002) Nmr in drug discovery. Nat Rev Drug Discov 1:211. https://doi.org/10.1038/nrd748
3. Pellecchia M, Bertini I, Cowburn D, Dalvit C, Giralt E, Jahnke W, James TL, Homans SW, Kessler H, Luchinat C, Meyer B, Oschkinat H, Peng J, Schwalbe H, Siegal G (2008) Perspectives on NMR in drug discovery: a technique comes of age. Nat Rev Drug Discov 7 (9):738–745
4. Hughes JP, Rees S, Kalindjian SB, Philpott KL (2011) Principles of early drug discovery. Br J Pharmacol 162(6):1239–1249. https://doi.org/10.1111/j.1476-5381.2010.01127.x
5. Chessari G, Woodhead AJ (2009) From fragment to clinical candidate—a historical perspective. Drug Discov Today 14 (13):668–675. https://doi.org/10.1016/j.drudis.2009.04.007
6. Harner MJ, Frank AO, Fesik SW (2013) Fragment-based drug discovery using NMR spectroscopy. J Biomol NMR 56(2):65–75. https://doi.org/10.1007/s10858-013-9740-z
7. Huth JR, Mendoza R, Olejniczak ET, Johnson RW, Cothron DA, Liu Y, Lerner CG, Chen J, Hajduk PJ (2005) ALARM NMR: a rapid and robust experimental method to detect reactive false positives in biochemical screens. J Am Chem Soc 127(1):217–224. https://doi.org/10.1021/ja0455547
8. Powers R (2014) The current state of drug discovery and a potential role for NMR metabolomics. J Med Chem 57(14):5860–5870. https://doi.org/10.1021/jm401803b
9. Powers R (2009) NMR metabolomics and drug discovery. Magn Reson Chem 47(S1): S2–S11. https://doi.org/10.1002/mrc.2461
10. Robertson DG, Reily MD, Baker JD (2007) Metabonomics in pharmaceutical discovery and development. J Proteome Res 6 (2):526–539. https://doi.org/10.1021/pr060535c
11. Wishart DS (2008) Applications of metabolomics in drug discovery and development. Drugs R&D 9(5):307–322. https://doi.org/10.2165/00126839-200809050-00002
12. Wishart DS (2016) Emerging applications of metabolomics in drug discovery and precision medicine. Nat Rev Drug Discov 15:473. https://doi.org/10.1038/nrd.2016.32
13. Zhang L, Hatzakis E, Patterson AD (2016) NMR-based metabolomics and its application in drug metabolism and cancer research. Curr Pharmacol Rep 2(5):231–240. https://doi.org/10.1007/s40495-016-0067-9
14. Wu W, Zhao S (2013) Metabolic changes in cancer: beyond the Warburg effect. Acta Biochim Biophys Sin 45(1):18–26. https://doi.org/10.1093/abbs/gms104
15. Gardner SG, Marshall DD, Daum RS, Powers R, Somerville GA (2018) Metabolic mitigation of *Staphylococcus aureus* Vancomycin intermediate-level susceptibility. Antimicrob Agents Chemother 62(1). https://doi.org/10.1128/aac.01608-17
16. Gaupp R, Lei S, Reed JM, Peisker H, Boyle-Vavra S, Bayer AS, Bischoff M, Herrmann M, Daum RS, Powers R, Somerville GA (2015) *Staphylococcus aureus* metabolic adaptations during the transition from a daptomycin susceptibility phenotype to a daptomycin nonsusceptibility phenotype. Antimicrob Agents Chemother 59(7):4226–4238. https://doi.org/10.1128/AAC.00160-15

17. Verdin E (2015) NAD+ in aging, metabolism, and neurodegeneration. Science 350 (6265):1208
18. Van Norman GA (2016) Drugs, devices, and the FDA: part 1: an overview of approval processes for drugs. JACC Basic Transl Sci 1 (3):170–179. https://doi.org/10.1016/j.jacbts.2016.03.002
19. DiMasi JA, Grabowski HG, Hansen RW (2016) Innovation in the pharmaceutical industry: new estimates of R&D costs. J Health Econ 47:20–33
20. Mohs RC, Greig NH (2017) Drug discovery and development: role of basic biological research. Alzheimers Dement (N Y) 3 (4):651–657
21. Harrison RK (2016) Phase II and phase III failures: 2013–2015. Nat Rev Drug Discov 15:817. https://doi.org/10.1038/nrd.2016.184
22. Smietana K, Siatkowski M, Moeller M (2016) Trends in clinical success rates. Nat Rev Drug Discov 15(6):379–380. https://doi.org/10.1038/nrd.2016.85
23. Downing NS, Shah ND, Aminawung JA et al (2017) Postmarket safety events among novel therapeutics approved by the us food and drug administration between 2001 and 2010. JAMA 317(18):1854–1863. https://doi.org/10.1001/jama.2017.5150
24. Settleman J, Cohen RL (2016) Communication in drug development: "translating" scientific discovery. Cell 164(6):1101–1104. https://doi.org/10.1016/j.cell.2016.02.050
25. Scannell JW, Bosley J (2016) When quality beats quantity: decision theory, drug discovery, and the reproducibility crisis. PLoS One 11(2): e0147215/0147211-e0147215/0147221. https://doi.org/10.1371/journal.pone.0147215
26. Kamb A, Wee S, Lengauer C (2007) Why is cancer drug discovery so difficult? Nat Rev Drug Discov 6(2):115–120. https://doi.org/10.1038/nrd2155
27. Swinney DC (2004) Opinion: biochemical mechanisms of drug action: what does it take for success? Nat Rev Drug Discov 3 (9):801–808. https://doi.org/10.1038/nrd1500
28. Pink R, Hudson A, Mouries M-A, Bendig M (2005) Opportunities and challenges in Antiparasitic drug discovery. Nat Rev Drug Discov 4(9):727–740. https://doi.org/10.1038/nrd1824
29. Koul A, Arnoult E, Lounis N, Guillemont J, Andries K (2011) The challenge of new drug discovery for tuberculosis. Nature 469 (7331):483–490. https://doi.org/10.1038/nature09657
30. Paul SM, Mytelka DS, Dunwiddie CT, Persinger CC, Munos BH, Lindborg SR, Schacht AL (2010) How to improve R&D productivity: the pharmaceutical industry's grand challenge. Nat Rev Drug Discov 9(3):203–214. https://doi.org/10.1038/nrd3078
31. Payne DJ, Gwynn MN, Holmes DJ, Pompliano DL (2007) Drugs for bad bugs: confronting the challenges of antibacterial discovery. Nat Rev Drug Discov 6(1):29–40. https://doi.org/10.1038/nrd2201
32. Gebregiworgis T, Powers R (2012) Application of NMR metabolomics to search for human disease biomarkers. Comb Chem High Throughput Screen 15(8):595–610
33. Tennant DA, Durán RV, Gottlieb E (2010) Targeting metabolic transformation for cancer therapy. Nat Rev Cancer 10:267. https://doi.org/10.1038/nrc2817
34. Shukla SK, Purohit V, Mehla K, Gunda V, Chaika NV, Vernucci E, King RJ, Abrego J, Goode GD, Dasgupta A, Illies AL, Gebregiworgis T, Dai B, Augustine JJ, Murthy D, Attri KS, Mashadova O, Grandgenett PM, Powers R, Ly QP, Lazenby AJ, Grem JL, Yu F, Matés JM, Asara JM, J-w K, Hankins JH, Weekes C, Hollingsworth MA, Serkova NJ, Sasson AR, Fleming JB, Oliveto JM, Lyssiotis CA, Cantley LC, Berim L, Singh PK (2017) MUC1 and HIF-1alpha signaling crosstalk induces anabolic glucose metabolism to impart gemcitabine resistance to pancreatic cancer. Cancer Cell 32(1):71–87. e77. https://doi.org/10.1016/j.ccell.2017.06.004
35. Al-Keilani MS, Al-Sawalha NA (2017) Potential of Phenylbutyrate as adjuvant chemotherapy: an overview of cellular and molecular anticancer mechanisms. Chem Res Toxicol 30 (10):1767–1777. https://doi.org/10.1021/acs.chemrestox.7b00149
36. Choi Y-K, Park K-G (2018) Targeting glutamine metabolism for cancer treatment. Biomol Ther (Seoul) 26(1):19–28
37. Seo M, Crochet RB, Lee Y-H (2014) Chapter 14—targeting altered metabolism—emerging cancer therapeutic strategies. In: Neidle S (ed) Cancer drug design and discovery, 2nd edn. Academic Press, San Diego, pp 427–448. https://doi.org/10.1016/B978-0-12-396521-9.00014-0
38. Jones NP, Schulze A (2012) Targeting cancer metabolism—aiming at a tumour's sweet-spot. Drug Discov Today 17(5):232–241. https://doi.org/10.1016/j.drudis.2011.12.017
39. Copeland RA, Harpel MR, Tummino PJ (2007) Targeting enzyme inhibitors in drug

discovery. Expert Opin Ther Targets 11 (7):967–978. https://doi.org/10.1517/14728222.11.7.967

40. Kinch MS, Umlauf S, Plummer M (2015) An analysis of FDA-approved drugs for metabolic diseases. Drug Discov Today 20(6):648–651

41. Phan LM, Yeung S-CJ, Lee M-H (2014) Cancer metabolic reprogramming: importance, main features, and potentials for precise targeted anti-cancer therapies. Cancer Biol Med 11(1):1–19. https://doi.org/10.7497/j.issn.2095-3941.2014.01.001

42. Mercier KA, Baran M, Ramanathan V, Revesz P, Xiao R, Montelione GT, Powers R (2006) FAST-NMR: functional annotation screening technology using NMR spectroscopy. J Am Chem Soc 128 (47):15292–15299. https://doi.org/10.1021/ja0651759

43. Powers R, Mercier KA, Copeland JC (2008) The application of FAST-NMR for the identification of novel drug discovery targets. Drug Discov Today 13(3):172–179. https://doi.org/10.1016/j.drudis.2007.11.001

44. Venkitakrishnan RP, Benard O, Max M, Markley JL (2012) Use of NMR saturation transfer difference spectroscopy to study ligand binding to membrane proteins. Methods Mol Biol 914:47–63. https://doi.org/10.1007/978-1-62703-023-6_4

45. Mercier KA, Shortridge MD, Powers R (2009) A multi-step NMR screen for the identification and evaluation of chemical leads for drug discovery. Comb Chem High Throughput Screen 12(3):285–295

46. Moy FJ, Haraki K, Mobilio D, Walker G, Tabei K, Tong H, Siegel MM, Powers R (2001) MS/NMR: a structure-based approach for discovering protein ligands and for drug design by coupling size exclusion chromatography, mass spectrometry, and nuclear magnetic resonance spectroscopy. Anal Chem 73 (3):571–581

47. Dalvit C, Fagerness PE, Hadden DTA, Sarver RW, Stockman BJ (2003) Fluorine-NMR experiments for high-throughput screening: theoretical aspects, practical considerations, and range of applicability. J Am Chem Soc 125(25):7696–7703. https://doi.org/10.1021/ja034646d

48. Tiziani S, Kang Y, Choi JS, Roberts W, Paternostro G (2011) Metabolomic high-content nuclear magnetic resonance-based drug screening of a kinase inhibitor library. Nat Commun 2:545. https://doi.org/10.1038/ncomms1562. https://www.nature.com/articles/ncomms1562#supplementary-information

49. Forgue P, Halouska S, Werth M, Xu K, Harris S, Powers R (2006) NMR metabolic profiling of Aspergillus nidulans to monitor drug and protein activity. J Proteome Res 5 (8):1916–1923

50. Halouska S, Fenton RJ, Barletta RG, Powers R (2012) Predicting the in vivo mechanism of action for drug leads using NMR metabolomics. ACS Chem Biol 7(1):166–171. https://doi.org/10.1021/cb200348m

51. Weljie AM, Newton J, Mercier P, Carlson E, Slupsky CM (2006) Targeted profiling: quantitative analysis of 1H NMR metabolomics data. Anal Chem 78(13):4430–4442. https://doi.org/10.1021/ac060209g

52. Lindon JC, Keun HC, Ebbels TMD, Pearce JMT, Holmes E, Nicholson JK (2005) The consortium for Metabonomic toxicology (COMET): aims, activities and achievements. Pharmacogenomics 6(7):691–699. https://doi.org/10.2217/14622416.6.7.691

53. Torell F, Bennett K, Rännar S, Lundstedt-Enkel K, Lundstedt T, Trygg J (2017) The effects of thawing on the plasma metabolome: evaluating differences between thawed plasma and multi-organ samples. Metabolomics 13 (6):66. https://doi.org/10.1007/s11306-017-1196-9

54. Catala A, Culp-Hill R, Nemkov T, D'Alessandro A (2018) Quantitative metabolomics comparison of traditional blood draws and TAP capillary blood collection. Metabolomics 14 (7):100. https://doi.org/10.1007/s11306-018-1395-z

55. Worley B, Powers R (2013) Multivariate analysis in metabolomics. Curr Metabolomics 1 (1):92–107. https://doi.org/10.2174/2213235X11301010092

56. Worley B, Powers R (2016) PCA as a practical indicator of OPLS-DA model reliability. Curr Metabolomics 4(2):97–103. https://doi.org/10.2174/2213235X04666160613122429

57. Worley B, Powers R (2014) MVAPACK: a complete data handling package for NMR metabolomics. ACS Chem Biol 9 (5):1138–1144. https://doi.org/10.1021/cb4008937

58. Worley B, Halouska S, Powers R (2013) Utilities for Quantifying Separation in PCA/PLS-DA scores plots. Anal Biochem 433 (2):102–104. https://doi.org/10.1016/j.ab.2012.10.011

59. Delaglio F, Grzesiek S, Vuister GW, Zhu G, Pfeifer J, Bax A (1995) NMRPipe: a multidimensional spectral processing system based on UNIX pipes. J Biomol NMR 6(3):277–293

60. Johnson BA (2004) Using NMRView to visualize and analyze the NMR spectra of

macromolecules. In: Downing AK (ed) Protein NMR Techniques. Humana Press, Totowa, NJ, pp 313–352. https://doi.org/10.1385/1-59259-809-9:313

61. Xia J, Sinelnikov IV, Han B, Wishart DS (2015) MetaboAnalyst 3.0-making metabolomics more meaningful. Nucleic Acids Res 43(W1): W251–W257. https://doi.org/10.1093/nar/gkv380
62. Wishart DS, Jewison T, Guo AC, Wilson M, Knox C, Liu Y, Djoumbou Y, Mandal R, Aziat F, Dong E, Bouatra S, Sinelnikov I, Arndt D, Xia J, Liu P, Yallou F, Bjorndahl T, Perez-Pineiro R, Eisner R, Allen F, Neveu V, Greiner R, Scalbert A (2013) HMDB 3.0-the human Metabolome database in 2013. Nucleic Acids Res 41(D1):D801–D807. https://doi.org/10.1093/nar/gks1065
63. Markley JL, Anderson ME, Cui Q, Eghbalnia HR, Lewis IA, Hegeman AD, Li J, Schulte CF, Sussman MR, Westler WM, Ulrich EL, Zolnai Z (2007) New bioinformatics resources for metabolomics. World Scientific Publishing Co. Pte. Ltd, Singapore, pp 157–168. https://doi.org/10.1142/9789812772435_0016
64. Worley B, Powers R (2015) Deterministic multidimensional nonuniform gap sampling. J Magn Reson 261:19–26. https://doi.org/10.1016/j.jmr.2015.09.016
65. Jewison T, Su Y, Disfany FM, Liang Y, Knox C, Maciejewski A, Poelzer J, Huynh J, Zhou Y, Arndt D, Djoumbou Y, Liu Y, Deng L, Guo AC, Han B, Pon A, Wilson M, Rafatnia S, Liu P, Wishart DS (2014) SMPDB 2.0: big improvements to the small molecule pathway database. Nucleic Acids Res 42(D1): D478–D484. https://doi.org/10.1093/nar/gkt1067
66. Blaise BJ, Correia G, Tin A, Young JH, Vergnaud A-C, Lewis M, Pearce JTM, Elliott P, Nicholson JK, Holmes E, Ebbels TMD (2016) Power analysis and sample size determination in metabolic Phenotyping. Anal Chem 88(10):5179–5188. https://doi.org/10.1021/acs.analchem.6b00188
67. Nyamundanda G, Gormley IC, Fan Y, Gallagher WM, Brennan L (2013) MetSizeR: selecting the optimal sample size for metabolomic studies using an analysis based approach. BMC Bioinformatics 14:338
68. Bando K, Kawahara R, Kunimatsu T, Sakai J, Kimura J, Funabashi H, Seki T, Bamba T, Fukusaki E (2010) Influences of biofluid sample collection and handling procedures on GC–MS based metabolomic studies. J Biosci Bioeng 110(4):491–499. https://doi.org/10.1016/j.jbiosc.2010.04.010
69. Emwas A-HM, Salek RM, Griffin JL, Merzaban J (2013) NMR-based metabolomics in human disease diagnosis: applications, limitations, and recommendations. Metabolomics 9 (5):1048–1072. https://doi.org/10.1007/s11306-013-0524-y
70. Lipinski CA (2001) Drug-like properties and the causes of poor solubility and poor permeability. J Pharmacol Toxicol Methods 44 (1):235–249. https://doi.org/10.1016/S1056-8719(00)00107-6
71. Li AP (2001) Screening for human ADME/Tox drug properties in drug discovery. Drug Discov Today 6(7):357–366. https://doi.org/10.1016/S1359-6446(01)01712-3
72. Heller AA, Lockwood SY, Janes TM, Spence DM (2018) Technologies for measuring pharmacokinetic profiles. Annu Rev Anal Chem 11:79–100. https://doi.org/10.1146/annurev-anchem-061417-125611
73. Reigner BG, Blesch KS (2002) Estimating the starting dose for entry into humans: principles and practice. Eur J Clin Pharmacol 57 (12):835–845
74. Nair AB, Jacob S (2016) A simple practice guide for dose conversion between animals and human. J Basic Clin Pharm 7(2):27–31
75. O'Gorman A, Gibbons H, Brennan L (2013) Metabolomics in the identification of biomarkers of dietary intake. Comput Struct Biotechnol J 4:e201301004
76. Bouatra S, Aziat F, Mandal R, Guo AC, Wilson MR, Knox C, Bjorndahl TC, Krishnamurthy R, Saleem F, Liu P, Dame ZT, Poelzer J, Huynh J, Yallou FS, Psychogios N, Dong E, Bogumil R, Roehring C, Wishart DS (2013) The human urine Metabolome. PLoS One 8(9):e73076. https://doi.org/10.1371/journal.pone.0073076
77. Emwas A-H, Luchinat C, Turano P, Tenori L, Roy R, Salek RM, Ryan D, Merzaban JS, Kaddurah-Daouk R, Zeri AC, Nagana Gowda GA, Raftery D, Wang Y, Brennan L, Wishart DS (2015) Standardizing the experimental conditions for using urine in NMR-based metabolomic studies with a particular focus on diagnostic studies: a review. Metabolomics 11(4):872–894. https://doi.org/10.1007/s11306-014-0746-7
78. Psychogios N, Hau DD, Peng J, Guo AC, Mandal R, Bouatra S, Sinelnikov I, Krishnamurthy R, Eisner R, Gautam B, Young N, Xia J, Knox C, Dong E, Huang P, Hollander Z, Pedersen TL, Smith SR, Bamforth F, Greiner R, McManus B, Newman JW, Goodfriend T, Wishart DS (2011) The human serum metabolome. PLoS One 6(2): e16957. https://doi.org/10.1371/journal.pone.0016957

79. Geraghty RJ, Capes-Davis A, Davis JM, Downward J, Freshney RI, Knezevic I, Lovell-Badge R, Masters JRW, Meredith J, Stacey GN, Thraves P, Vias M (2014) Guidelines for the use of cell lines in biomedical research. Br J Cancer 111:1021. https://doi.org/10.1038/bjc.2014.166
80. Halouska S, Zhang B, Gaupp R, Lei S, Snell E, Fenton RJ, Barletta RG, Somerville GA, Powers R (2013) Revisiting protocols for the NMR analysis of bacterial metabolomes. J Integr OMICS 3(2):120–137. https://doi.org/10.5584/jiomics.v3i2.139
81. Westerhoff HV, Chen Y-D (1984) How do enzyme activities control metabolite concentrations? Eur J Biochem 142(2):425–430. https://doi.org/10.1111/j.1432-1033.1984.tb08304.x
82. Worley B, Powers R (2014) Simultaneous phase and scatter correction for NMR datasets. Chemom Intell Lab Syst 131:1–6. https://doi.org/10.1016/j.chemolab.2013.11.005
83. Savorani F, Tomasi G, Engelsen SB (2010) Icoshift: a versatile tool for the rapid alignment of 1D NMR spectra. J Magn Reson 202(2):190–202. https://doi.org/10.1016/j.jmr.2009.11.012
84. De Meyer T, Sinnaeve D, Van Gasse B, Tsiporkova E, Rietzschel ER, De Buyzere ML, Gillebert TC, Bekaert S, Martins JC, Van Criekinge W (2008) NMR-based characterization of metabolic alterations in hypertension using an adaptive, intelligent binning algorithm. Anal Chem 80(10):3783–3790. https://doi.org/10.1021/ac7025964
85. Development Core Team R (2011) R: a language and environment for statistical computing, vol 1
86. Eriksson L, Trygg J, Wold S (2008) CV-ANOVA for significance testing of PLS and OPLS® models. J Chemom 22(11–12):594–600. https://doi.org/10.1002/cem.1187
87. Triba MN, Le Moyec L, Amathieu R, Goossens C, Bouchemal N, Nahon P, Rutledge DN, Savarin P (2015) PLS/OPLS models in metabolomics: the impact of permutation of dataset rows on the K-fold cross-validation quality parameters. Mol BioSyst 11(1):13–19. https://doi.org/10.1039/C4MB00414K
88. Goodacre R, Broadhurst D, Smilde AK, Kristal BS, Baker JD, Beger R, Bessant C, Connor S, Capuani G, Craig A, Ebbels T, Kell DB, Manetti C, Newton J, Paternostro G, Somorjai R, Sjöström M, Trygg J, Wulfert F (2007) Proposed minimum reporting standards for data analysis in metabolomics. Metabolomics 3(3):231–241. https://doi.org/10.1007/s11306-007-0081-3
89. Benjamini Y, Hochberg Y (1995) Controlling the false discovery rate: a practical and powerful approach to multiple testing. J R Stat Soc B Methodol 57(1):289–300
90. Bland JM, Altman DG (1995) Multiple significance tests: the Bonferroni method. BMJ 310(6973):170. https://doi.org/10.1136/bmj.310.6973.170
91. Xu Q-S, Liang Y-Z (2001) Monte Carlo cross validation. Chemom Intell Lab Syst 56(1):1–11. https://doi.org/10.1016/S0169-7439(00)00122-2
92. Rubinson KA (2017) Practical corrections for p(H,D) measurements in mixed H_2O/D_2O biological buffers. Anal Methods 9(18):2744–2750. https://doi.org/10.1039/C7AY00669A
93. Bhinderwala F, Lonergan S, Woods J, Zhou C, Fey PD, Powers R (2018) Expanding the coverage of the Metabolome with nitrogen-based NMR. Anal Chem 90(7):4521–4528. https://doi.org/10.1021/acs.analchem.7b04922
94. Ruxton GD (2006) The unequal variance t-test is an underused alternative to Student's t-test and the Mann–Whitney U test. Behav Ecol 17(4):688–690. https://doi.org/10.1093/beheco/ark016
95. Fay MP, Proschan MA (2010) Wilcoxon-Mann-Whitney or t-test? On assumptions for hypothesis tests and multiple interpretations of decision rules. Statist Surv 4:1–39. https://doi.org/10.1214/09-SS051
96. Hotelling H (1931) The economics of exhaustible resources. J Polit Econ 39(2):137–175

Part VII

Environmental Studies

Chapter 17

Practical Aspects of the Analysis of Low- and High-Field NMR Data from Environmental Samples

Jun Kikuchi

Abstract

NMR has considerable advantages for investigation of environmental samples including the ability to measure solid-, gel-, and solution-state samples using both high-field superconducting magnets and low-field permanent magnets. Samples from diverse geochemical environments and microbial, plant, and animal systems can be analyzed using common protocols and the NMR analytical platform. To handle broad range of spectra from solid-state NMR or low-field benchtop NMR, two preprocessing tools have been developed and evaluated in my laboratory: one, SENSI, which enhances the signal-to-noise ratio by spectral integration, and the other, PKSP, which separates overlapping peaks using algorithms, such as non-negative sparse coding. In addition, the SpinLIMS database stores numerous standard spectra for molecular species ranging from small molecules to macromolecules in solid and solution states (dissolved in polar/nonpolar solvents). The database can be searched under various conditions using the InterAnalysis molecular assignment tool, which improves the accuracy of molecular assignments by integrated analysis of 2D ^{1}H–^{13}C correlation peaks, and ^{1}H–J resolved peaks of small molecules dissolved in solvents such as D_2O, MeOD, and DMSO-d_6. With emphasis on the above, the current chapter provides practical aspects of the analysis of data from diverse environmental samples obtained using low- and high-field NMR instruments.

Key words Environmental NMR, Pattern recognition, Spectral deconvolution, Web tool, Metabolite annotation, Data science, Machine learning

1 Introduction

Among various analytical methods, NMR has several advantages for analysis of complex environmental samples, including ease of sample treatment and applicability to various states of samples (e.g., solution, gel, and solid) as well as in vivo biological samples [1–7]. The high interlaboratory reproducibility of NMR spectra worldwide is also a distinct merit for cross-site analytical validity studies [8, 14]. In this regard, the solid-state NMR approach for macromolecular complexes and low-solubility samples has a unique benefit for various environmental samples [9–11] because physical

G. A. Nagana Gowda and Daniel Raftery (eds.), *NMR-Based Metabolomics: Methods and Protocols*, Methods in Molecular Biology, vol. 2037, https://doi.org/10.1007/978-1-4939-9690-2_17, © Springer Science+Business Media, LLC, part of Springer Nature 2019

separation of the sample (e.g., by chromatography) is not needed for measurements of environmental molecular complexity [12].

Excellent reproducibility and quantification are particular strengths for NMR; in the case of environmental complexity, however, relative quantification within a set of NMR spectra is a major challenge. Multiple datasets can be obtained using NMR with high reproducibility and interinstitution compatibility; analysis of such data using multivariate analysis and machine-learning (ML) approaches offers numerous benefits [13]. The "target analysis" methods often used in the field focus on a single compound or specific type of compounds in environmental samples. Such methods generally require sample preprocessing including pretreatment and extraction of the target molecule of interest, before analysis. In contrast, NMR, as a global analysis method, stands out among various other techniques for analyzing environmental samples because of its unique characteristics including high accuracy and throughput. The NMR-based approach has potential for investigating agricultural and fishery products (e.g., evaluating geographical origins and controlling product quality). It can also be applied to environmental metabolomics (i.e., the interaction between an organism and its environment) [14] and for laboratory-bred organisms, for example, where high-field NMR instruments and laboratory-specific experiments such as stable isotope labeling offer numerous benefits.

Coupled with the ease of sample preparation relative to other methods such as mass spectrometry, the recent development of cost-effective (e.g., benchtop NMR) instruments promise immense practical applications to assess environmental homeostasis and on-site evaluation of ecosystem features [15, 16]. In modern food production, for example, the primary production or distribution sites (i.e., farms or fishing grounds) are located far from laboratories or analytical centers (i.e., food companies or facilities). In such cases, benchtop NMR may potentially revolutionize the quality control processes that identify metabolic changes in food resulting from storage and fermentation [17]. Innovations in NMR hardware that facilitated high magnetic field and high sensitivity instruments are also useful for homeostatic assessments of health [18–20], and quality control in the fields of agriculture [21–23], forestry [11, 24, 25], fishery [26–29], foods [16, 28], natural materials, and biomass utilization [30, 31].

To aid the practical use of the broad spectra obtained from solid-state and low-field NMR, specific tools for the pretreatment and processing of the data are needed. For example, we previously developed FoodPro, a database and web tool for predicting the taste and longevity of foods based on the similarity of desktop NMR spectra of food substances [16]. More recently, we developed two preprocessing tools for sensitivity enhancement (SENSI) and peak separation (PKSP), which are expected to lead to improved

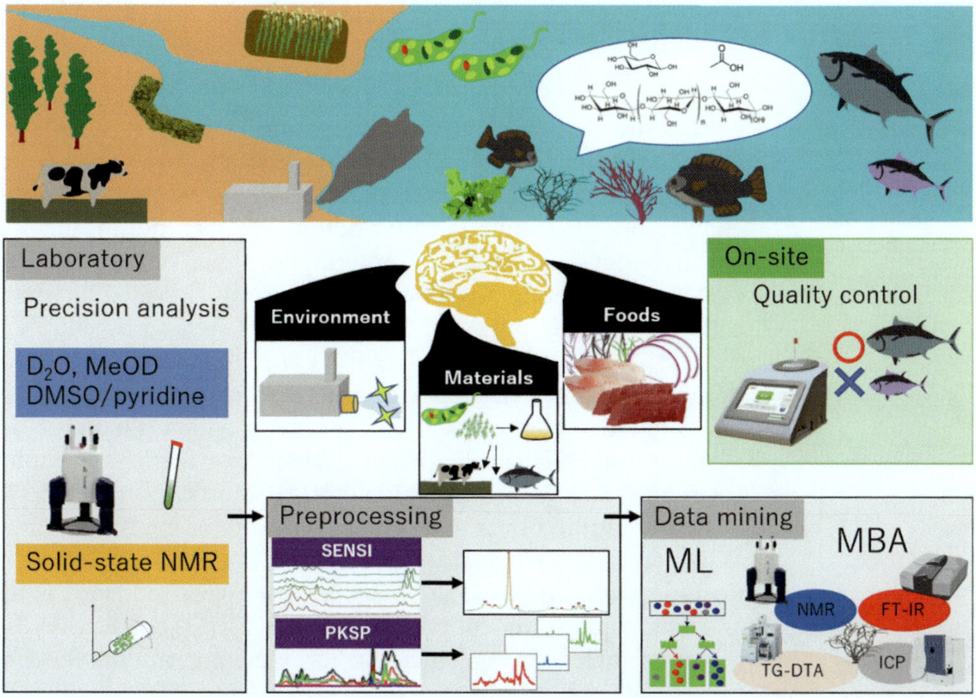

Fig. 1 Conceptual diagram showing analysis of environmental complexity by low- and high-field NMR. Both geochemical and biological samples can be analyzed by solution- and solid-state NMR. For broad signals arising from solid-state NMR and low-field benchtop NMR, it is better to perform preprocessing by using web tools for sensitivity enhancement (SENSI) and peak separation (PKSP). In addition to high-resolution NMR spectra and other analytical data, feature extraction can be computed by data science approaches, such as machine learning (ML) and market basket analysis (MBA)

cost-effectiveness of benchtop NMR by facilitating annotation of the broad spectra obtained from in situ low-field NMR [32, 33].

In this chapter, we summarize approaches to processing and analyzing low- and high-field NMR spectra of environmental complexities (*see* Fig. 1). We describe pulse sequences used for solid- and solution-state NMR, data pretreatment for analysis of molecular complexities, pattern recognition approaches, and databases and tools, focusing on the newly developed SENSI and PKSP methods.

2 Materials

Sample preparation is relatively easy for NMR because complex samples can be measured without column purification. In the case of solution-state NMR, deuterated solvents such as water, methanol-d_4, dimethyl sulfoxide (DMSO)-d_6, and chloroform-d_3 can be used [34].

3 NMR Experiments and Data Analysis

3.1 Pulse Sequences for Solution-State NMR

Traditional one-dimensional (1D) NMR-based approaches are commonly used for metabolomics studies; however, signal overlaps due to splitting by spin–spin coupling constitutes a serious obstacle to the analysis of complex biological mixtures. To overcome this challenge, "pure shift" techniques are gaining significant interest because they remove spin–spin couplings and hence offer significant improvement in spectral resolution. Pure shift methods include real-time pure shift and pure shift yielded by chirp excitation (PSYCHE). However, a problem with pure shift methods is the low sensitivity compared to the traditional 1D NMR.

Alternatively, the 2D-*J*res experiment, which is mainly used for structure analysis, facilitates an evaluation of spin–spin coupling constants [35]. Multidimensional NMR enables correlation signals to be determined by homo- and heteronuclear measurements such as heteronuclear single quantum coherence (HSQC) [36]. Correlation spectroscopy (COSY), total correlated spectroscopy (TOCSY), and single quantum-double quantum (SQ-DQ) correlation experiments are helpful to enhance the accuracy of metabolite identification [37]. As an alternative or supplementary to the HSQC experiment, an extended network of correlations can be determined using HSQC-TOCSY, 3D-HCCH-COSY/TOCSY, or (H)CCH-COSY/TOCSY experiment. Diffusion-ordered spectroscopy (DOSY) provides the capability to separate signals from mixtures on the basis of differences in mobility and to evaluate the physicochemical properties of molecules.

3.2 Pulse Sequences for Solid-State NMR

Solid-state NMR can be used to analyze samples of low solubility. Although many pulse sequences for solid-state NMR have been developed, the application of multidimensional NMR pulse sequences has been delayed because the anisotropic effect of dipole interactions is strong, and experiments that employ ^{1}H detection are met with challenges to obtain good spectra. Two approaches have been developed to overcome these problems: 1D cross-polarization (CP) magic-angle-spinning (MAS) combined with a greater number of scans to detect low-sensitivity ^{13}C; and 1D nonquaternary suppression (NQS)-MAS, which emphasizes quaternary carbons [9, 38].

In solid-state NMR, several relaxation parameters including longitudinal relaxation (T_1), relaxation in the rotating frame ($T_{1\rho}$), and transverse relaxation greatly influence the spectra and constitute important sample characteristics. Notably, ^{1}H relaxation time constants represent the measure of molecular motion within domains, whereas ^{13}C relaxation time constants represent local molecular motions. The use of uniform ^{13}C labeling allows NMR measurements at sensitivities that are 100-fold greater for ^{1}H–^{13}C moieties and 10,000-fold greater for ^{13}C–^{13}C moieties compared to natural abundance ^{13}C [10]. ^{1}H–^{13}C couplings can be used to evaluate crystalline polysaccharides, lipids, and low-molecular-weight molecules using double CP, solid-state HSQC, and high-resolution (HR)-MAS.

3.3 Data Pretreatment

Valuable information from the experimentally obtained NMR spectra can be extracted via multivariate analysis and ML algorithms. Before such analyses, however, appropriate pretreatment and preprocessing of datasets are essential, as described below.

We recently proposed a novel analytical strategy, SENSI (Signal enhancement by Spectral Integration), to resolve the low signal-to-noise (S/N) problem in ^{13}C NMR and 2D-*J*res NMR by the integration of multiple spectra without additional measurements for analysis and annotation. In addition to integration, SENSI also measures the distribution of each data point in large datasets of multiple NMR spectra by calculating coefficient of variation (CV). Notably, both the improvement in S/N ratio and calculation of CV in SENSI are achieved via a single data-processing stage [39]. Furthermore, in order to preprocess a broad spectrum from either solid-state NMR or benchtop NMR, we have developed the web tool, InterSpin, which incorporates PKSP (PeaKs SeParation) and SENSI.

3.3.1 Instructions for Using SENSI

Figure 2 shows the opening window of SENSI, which prompts the user to perform the following steps:

1. Select either demo-data (benchtop/solid-state NMR) or user file.
2. In the case of "Use your file" in **step 1**, use a tab-separated spectral data file (chemical shift versus signal intensity). Figure 3 shows the required data format for SENSI.
3. Press and select the "Import" button.
4. Turn on "Peak pick mode" on the SENSI Plot tab (*see* Fig. 4).
5. Select the peak to pick (blue) on the spectrum.
6. Press the "Pick" button.
7. To enlarge the spectrum, specify the display range.
8. If necessary, set the allowable range by using the cursor.

Figures 4 and 5 show visualization of SENSI results with CV and the option to download the data file on the "Pick peak" tab, respectively.

3.3.2 Data Normalization

Normalization, including scaling, is a key process in data pretreatment. Considerable differences in scales between variables can result in inappropriate interpretation of the data. A number of normalization methods have been developed. The use of

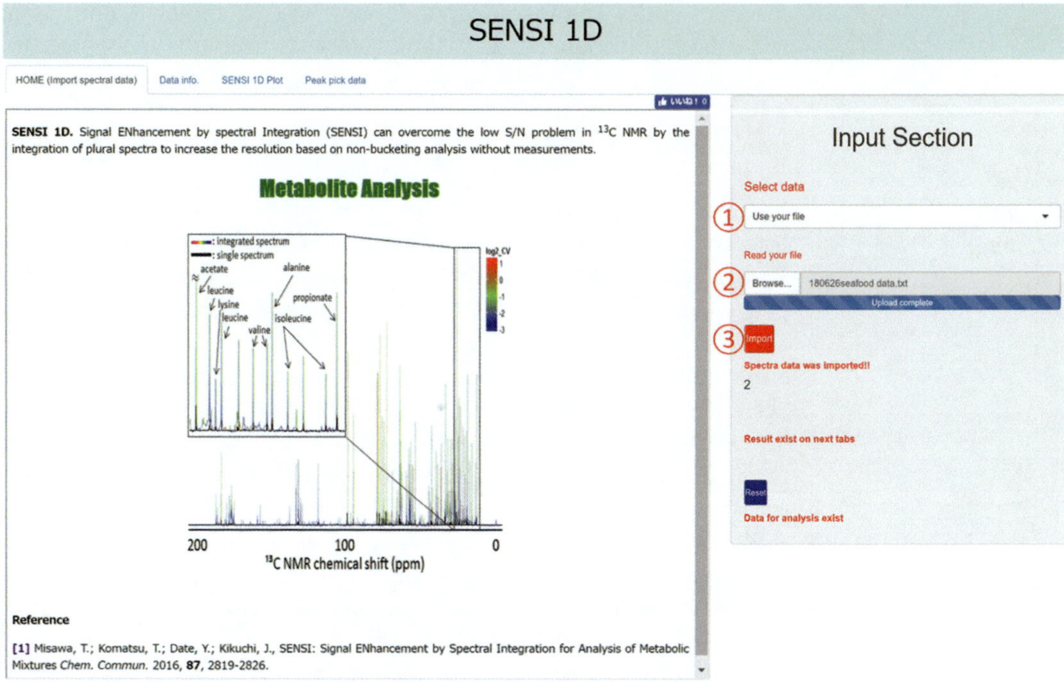

Fig. 2 Overview of the initial window of the SENSI (sensitivity enhancement by spectral integration) web tool

Unit (NMR: Chemical shift)

	1	2	3	4	5	· · ·	X
Sample 1	0.36	0.36	0.36	0.36	0.36	· · ·	0.00
Sample 2	0.39	0.39	0.39	0.39	0.39	· · ·	0.00
Sample 3	0.30	0.30	0.30	0.30	0.30	· · ·	0.00
·	·	·	·	·	·	· · ·	·
Sample Y	0.32	0.32	0.32	0.32	0.32	· · ·	0.00

Delete the sample name column.

Input format (txt or csv)

Fig. 3 Required data format for the SENSI web tool

4,4-dimethyl-4-silapentane-1-sulfonic acid (DSS) or tetramethylsilane (TMS) as an internal standard of known concentration is more commonly used for correcting chemical shift errors between spectra and spectral normalization. Alternatively, *Z*-scoring is often used to preprocess datasets; the *Z*-score is calculated by setting the average and standard deviation values of each metabolite signal to 0 and 1, respectively. However, the problem with applying this method to analyze an entire spectrum is that the normalization is influenced by noise. Constant sum normalization (CSN) normalizes data by dividing the signal intensity for each bin by total spectrum area; however; this method occasionally leads to inaccurate data interpretation. Another method, probabilistic quotient normalization (PQN), uses a median of multiple estimated values and it is not susceptible to outliers; however, it is inadequate for datasets

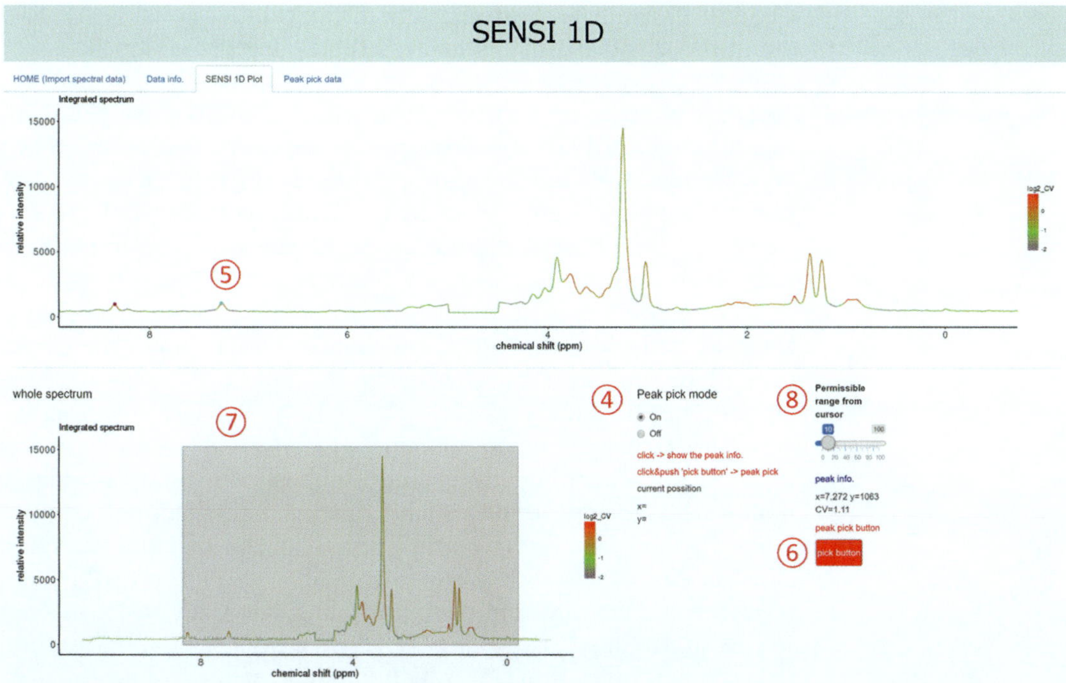

Fig. 4 Visualization of SENSI results by coefficient of variation

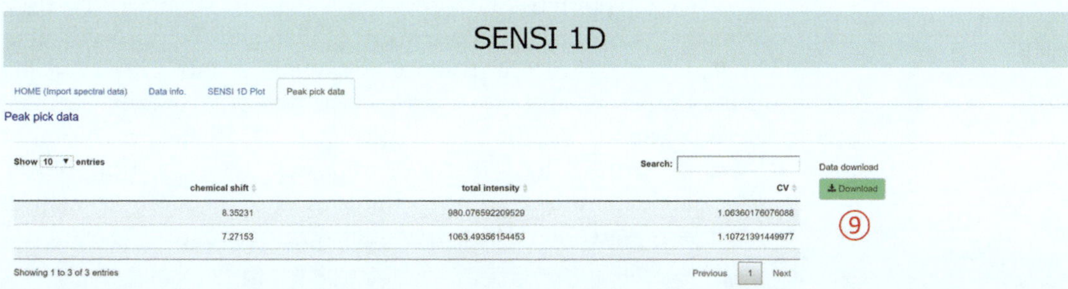

Fig. 5 Downloaded data file on the "Peak pick" data tab of SENSI

involving simultaneous changes of multiple metabolite concentrations. Group aggregating normalization (GAN) collects the data into a group that is centered in the subspace of principal component analysis (PCA); this method is known to perform better than CSN or PQN in multivariate analysis.

3.3.3 Peak Alignment

Spectral bucketing and binning methods are often used for peak alignment. Binning is the splitting of spectral regions into buckets; peaks in each bucket region are replaced by one representative peak. The major drawback of binning is the considerable decline in resolution. To address this problem, improved binning techniques have been developed, including kernel-based binning, the optimized bucketing algorithm (OBA), and the *i*coshift algorithm. In a different approach, Ito et al. reported chemical shift alignments of

^{1}H and ^{13}C signals from a variety of multidimensional NMR data using the kernel density function and 2D-*J*res projection data.

3.3.4 Peak Separation

Overlap of peaks in NMR spectra is one of the biggest problems encountered in the analysis of complex mixtures. Solid-state NMR typically has very low resolution; thus, a single peak is likely to contain multiple signals derived from different chemical components. Different peak separation techniques can be applied to address this problem.

The secondary differentiation method is a traditional technique for peak detection in overlapping signals, and facilitates the detection of buried peaks by identifying the inflection point of overlapping peaks; however, the calculated peaks lack quantitative accuracy. Independent component analysis (ICA) is a calculation method for separating multivariate signals into multiple additive components using latent variables called "independent components," which are linear combinations of the original variables. This method is a type of multivariate curve resolution (MCR) and has been successfully applied to analyzing CP-MAS NMR spectra. MCR–alternating least squares (ALS) is a peak separation method that is based on optimization by iterative least squares for calculating components [31]. This method has been applied in several NMR-based metabolomic studies.

Nonnegative matrix factorization (NMF) is an algorithm used for the analysis of matrices consisting of zero and/or positive values [30]. It decomposes multivariate and spectral data into a smaller number of factors, and extracts local features. NMF can be used for peak separation in solution- and solid-state NMR spectra. Notably, the use of MCR-based methods requires determination of the appropriate separation number for a peak via a method such as the Durbin–Watson (DW) or residual sum of squares (RSS) plot.

3.3.5 Instructions for Using PKSP

The recently developed web tool PKSP, in our laboratory, uses a number of different algorithms. Figure 6 shows an overview of the opening window of PKSP and the input column, which prompts the user to perform the following steps:

1. Set estimate (Semi-Automatic, Manual), components (Min and Max), method (MCR–ALS, Fast ICA, NMF, NNSC), checkbox (Show the data), and data (user tab-separated chemical shift data file or demo data). The data format required for PKSP is shown in Fig. 7.
2. Press the "Calculation" button.
3. Show tabs (Data, Scores Plot, Loadings Plot, Separation, Residue, Estimation). For semiautomatic estimates, the components can be adjusted.
4. When the analysis has completed, the analysis time is displayed.
5. Click the "Set DL Comp." button to download the results.

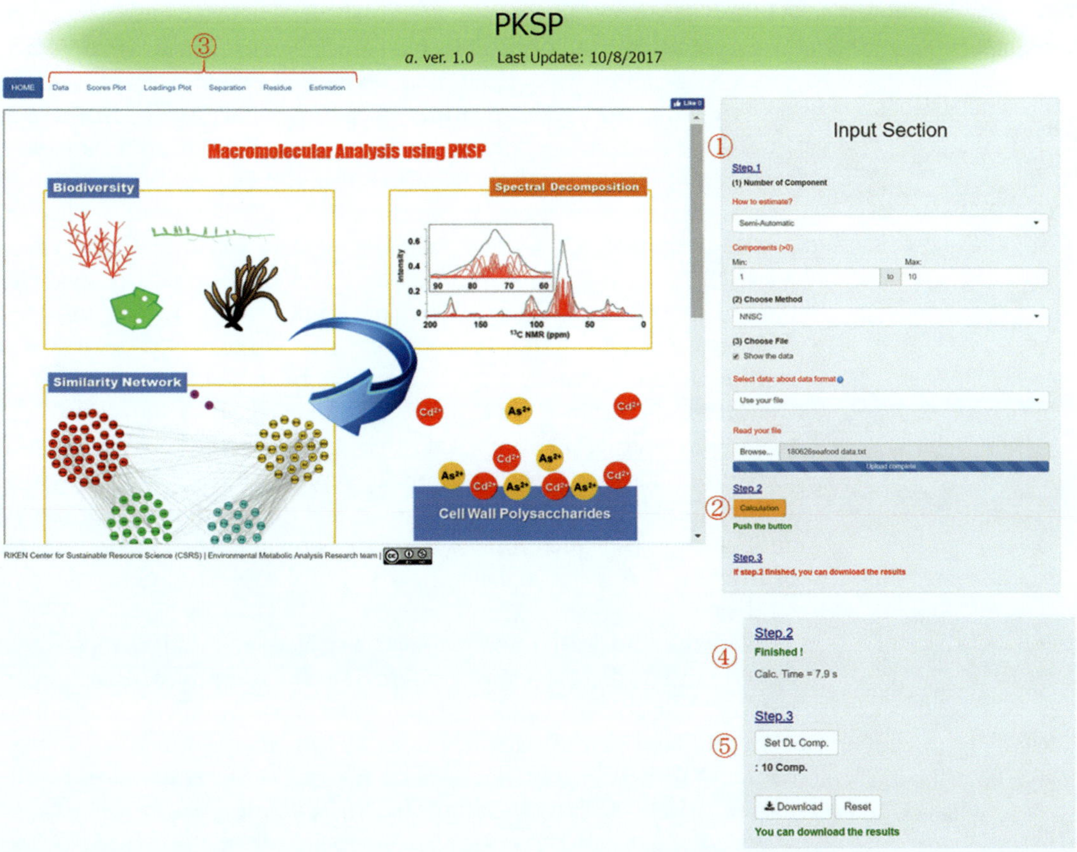

Fig. 6 Overview of the top window of PSKP (left) and the input column for analytical data matrices (right)

NMR: Chemical shift, (Relaxation: Time)
TG-DTA: Temperature
FT-IR: Wave number

Unit

	1	2	3	4	5	• • •	X
Sample 1	0.36	0.36	0.36	0.36	0.36	• • •	0.00
Sample 2	0.39	0.39	0.39	0.39	0.39	• • •	0.00
Sample 3	0.30	0.30	0.30	0.30	0.30	• • •	0.00
•	•	•	•	•	•	• • •	•
Sample Y	0.32	0.32	0.32	0.32	0.32	• • •	0.00

Delete the sample name column.

Input format (txt or csv)

Fig. 7 Required data format for the PKSP web tool

3.4 Pattern Recognition Tools

3.4.1 Multivariate Analysis

Multivariate analysis plays an essential role in NMR-based metabolomics studies. One of the most commonly used multivariate methods is PCA, an unsupervised method used for dimension reduction, overviewing, and identification of trends in datasets. Numerous PCA-based approaches have been applied to NMR data analysis, including SIMCA, and principal component regression.

Alternative unsupervised methods include hierarchical clustering analysis (HCA), *k*-means clustering, and self-organizing maps (SOM). HCA has been used in a data-driven approach to evaluate altered metabolism, microbial communities, and minerals induced by daily dietary intakes in humans [18]. *k*-means clustering is a nonhierarchical cluster analysis and has been further improved in the method of *x*-means clustering. Similar to HCA, *k*-means clustering has been applied in a data-driven approach to environmental metabolomics for biogeochemical typing of rice paddy fields. Lastly, SOM is a neural-network-based method used for trend and cluster identification. A method of variable selection using SOM has also been reported to be applicable to metabolomics studies.

Correlation-based analysis uses correlation coefficients such as Spearman's rank correlation coefficient to identify a relationship between variables. Visualization of correlation coefficients as a heat map or network diagram has been shown to be effective in environmental metabolomics studies for evaluation of estuarine sediments [40], seasonal variations in chemical components of seaweeds [31], and chemical diversity in various seaweed species [30]. An industrial process using anaerobic microbial ecosystems has also been investigated using correlation-based analysis and visualized by network diagrams [41].

Market basket analysis (MBA) is also a useful statistical method to explore variable co-occurrences and was recently applied to metabolomics studies. Overall, the supervised methods of partial least squares (PLS) and related analyses are the most common classification and regression tools used to analyze metabolomics data generated by NMR. PLS-related analyses are powerful data mining methods because they facilitate the analysis of large, highly complex datasets that contain collinearity and noise.

3.4.2 Machine Learning (ML)

ML, which is based on various algorithms, is used to identify patterns and capture trends in datasets. The ML algorithms construct models from the data using iterative improvements and can predict values and trends by applying new data to the model. Development of ML is a key research challenge in artificial intelligence and is increasingly used in devices such as search engines, e-mail programs, and mobile phones. Deep learning is an AI technology that has led to tremendous improvements in image recognition, text mining, and speech and language recognition. Various algorithms applicable to metabolomics data analysis include deep learning, a neural-network-based method [42, 43], random forest (RF), and support vector machine (SVM).

In metabolomics studies that use ML algorithms, the analytical procedure is the same as that in studies using supervised multivariate analysis [44]. Spectroscopic data obtained by NMR are usually preprocessed, and the resulting matrices of digitized spectroscopic data are applied to and computed by ML algorithms. During the ML computation, parameter optimization is an important step for improving performance. The parameter optimization step is usually performed using some of the original data in an independent test to avoid "overlearning" of the constructed model, whereby the predictive model overfits the dataset at training, resulting in poor prediction accuracy for an unknown dataset. Cross-validation is used to verify and confirm the validity of the constructed model. This analytical procedure is used not only for model optimization and optimal parameter determination but also to evaluate parameter-determined and constructed models. Therefore, original datasets are commonly used in two steps in so-called double cross-validation: first, cross-validation of optimal parameters (using training data and test data) and second, evaluation of model performance (using modeling data and evaluation data). Owing to this complex procedure, ML approaches are liable to be computationally expensive, and a great deal of time is needed to complete calculations and obtain final results.

RF (also called randomized trees or random decision forests) is an ensemble learning algorithm used for classification, regression, and clustering. It enables extraction of important parameters (e.g., important metabolites contributing to a classification or regression model). An improved RF method (cforest) has also been reported [45].

SVM uses the kernel method to perform linear and nonlinear classification and regression, and is applicable to numerous fields including metabolomics, as well as to studies that use data from other spectroscopic techniques. In metabolomics studies, the performance of SVM has been shown to be superior to that of PLS-DA in terms of classification accuracy and feature selection. In addition, several studies analyzing metabolomics data by SVM have been reported. An advantage of SVM over other ML approaches is its suitability for analyses based on a small sample size [46]. Although "overlearning" is a big problem with ML approaches, especially in cases with a limited number of samples for training data, one study has reported that SVM is applicable to metabolomics data even if the sample size is small.

3.5 Databases and Tools for Annotating NMR Signals

3.5.1 Databases

Identification of compounds is one of the main challenges in the analysis of NMR spectra of environmental complexity. Multidimensional NMR can be used to identify and determine the structures of isolated and purified compounds, but it is difficult to apply to complex samples, such as metabolic mixtures. In such cases, standard spectra and chemical shift databases are typically used to assign and annotate spectra. Many chemical shift databases are available online, which are simple to use and convenient. However, if the database includes candidate compounds that are not present in the sample but have similar partial structures to sample components, the result may be a false-positive identification. Prior knowledge of the types of chemicals present in the sample is needed to avoid such false-positive identification.

The Biological Magnetic Resonance Data Bank (BMRB), Human Metabolome Database (HMDB) [47], and Madison Metabolomics Consortium Database (MMCD) are commonly used chemical shift databases for metabolomics studies [48]. The BMRB contains NMR data for proteins, peptides, nucleic acids, and other biomolecules. Candidate compounds are identified from ^{1}H and ^{1}H–^{13}C HSQC peak data entered by the user. This database also includes spectral data for metabolite standards. HMDB is a database of the human metabolome and is also useful for NMR data from environmental sources. The MMCD was developed for the identification and quantification of metabolites in biological samples and includes both experimental and theoretical NMR data.

The SpinAssign program developed in our laboratory can annotate metabolites in D_2O and MeOD buffer using ^{1}H, ^{13}C, and ^{1}H–^{13}C HSQC data [36]. The probability or "*P* value" has been adopted as a mathematical indicator to evaluate annotation. A metabolomics database for NMR analysis (TOCCATA) was also developed recently. Highly reliable annotation is possible with database searching using TOCSY peaks of metabolites in a mixture. The Birmingham Metabolite Library contains ^{1}H and 2D-*J*res data for metabolites, which can be used to annotate 2D-*J*res spectra of metabolic mixtures. Our laboratory has also developed SpinCouple, an annotation tool for 2D-*J*res spectra of metabolic mixtures [35]. Users of SpinCouple can find candidate compounds by entering ^{1}H chemical shifts, spin–spin coupling constants, and intensity values. High-precision annotation is expected if the HSQC, TOCSY, and 2D-*J*res databases are used in combination.

The above annotation tools cannot be applied to metabolites, known or unknown, that are not in the database. For this purpose, chemical shift prediction tools that use experimental data have been developed. These tools can be used for identifying metabolites. One such tool, Nmrshiftdb2, predicts chemical shifts from 2D or 3D hierarchical organization of spherical environments (HOSE) codes of planar structures [49]. ECOMICS is a web tool for trans-omics research on environmental and metabolic systems

[50]. Bm-Char in ECOMICS characterizes chemical structures from chemical shifts of biomass samples. HeatMap constructs a correlation matrix between NMR data and other datasets and enables visualization of the relationships. FoodPro is a database and tool for evaluating the taste of food and predicting food taste based on similarities in food spectra [16].

3.5.2 InterSpin, An Integrated Supportive Web Tool

To gather together various web tools for the preprocessing, analysis, and annotation of NMR spectra of environmental complexity, we developed the web tool InterSpin, which comprises the following three elements [33]:

1. Spectrum preprocessing tools: In the case of a broad spectrum obtained from low-field benchtop ^{1}H-NMR or solid-state ^{13}C-CP-MAS, SENSI overcomes the problem of low S/N ratio by increasing resolution through the integration of multiple spectra, while PKSP performs effective peak separation by a multivariate spectral decomposition method, as described above.
2. Molecular assignment tools: SpinMacro simplifies the macromolecular assignment of a solid CP-MAS spectrum or a DMSO-solubilized ^{1}H–^{13}C HSQC spectrum. SpinAssign searches the SpinLIMS database for a compound corresponding to the HSQC NMR peaks. SpinCouple can assign ^{1}H–*J* 2D-*J*res NMR peaks. InterAnalysis is a Venn diagram–type highly accurate annotation tool that narrows down candidate molecules by using correlation peaks from both the HSQC spectrum and the 2D-*J*res spectrum (*see* Fig. 8).
3. SpinLIMS (InterSpin Laboratory Information Management System) database: This database includes reference solid-state CP-MAS spectra and solution-state HSQC spectra (in DMSO) for macromolecules, and reference solution-state HSQC and 2D-*J*res spectra (D_2O and MeOD) for small molecules.

4 Conclusions

NMR spectroscopy plays an important role among the various techniques used to analyze complex mixtures owing to its high-throughput nature and high analytical precision. As demonstrated in this chapter, many applications demonstrate the potential of NMR-based approaches to investigate geographic origins, quality control, and processing of agricultural and fishery products. Innovations in NMR instrumentation are anticipated to make greater impacts on the assessment of the environment as well as health and quality control in agriculture, forestry, and fishery. Because the physical and chemical properties of compounds can be extremely diverse, various sample preparation methods and pulse sequences

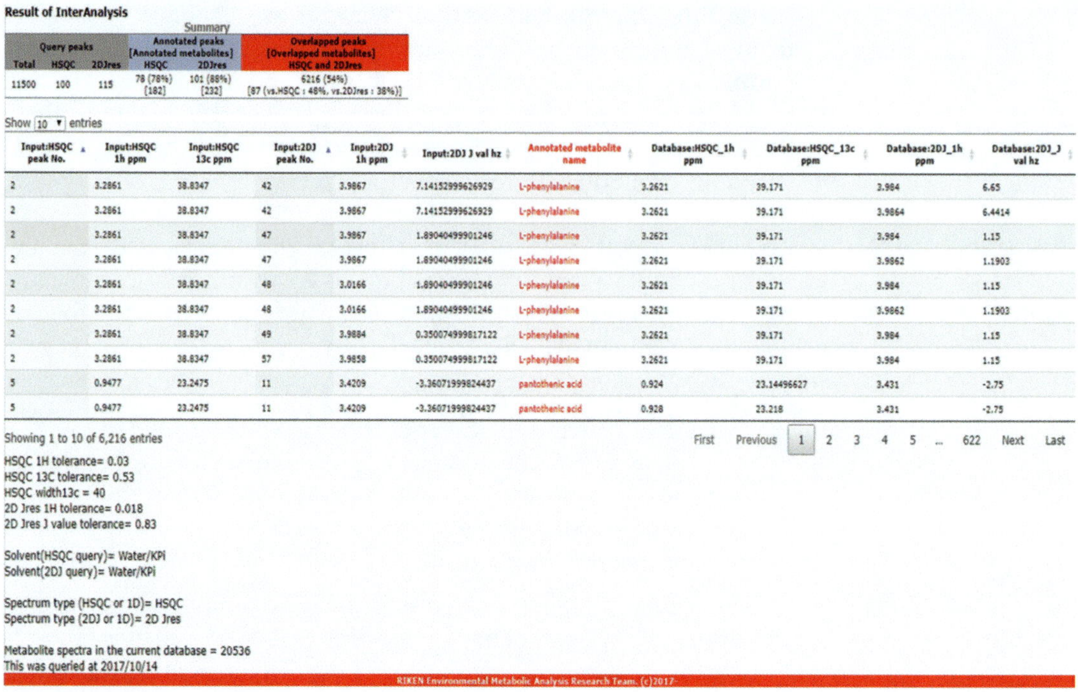

Result of InterAnalysis

Summary

Query peaks			Annotated peaks [Annotated metabolites]		Overlapped peaks [Overlapped metabolites]
Total	HSQC	2DJres	HSQC	2DJres	HSQC and 2DJres
11500	100	115	78 (78%) [182]	101 (88%) [232]	6216 (54%) [87 (vs.HSQC : 48%, vs.2DJres : 38%)]

Show 10 entries

Input:HSQC peak No.	Input:HSQC 1h ppm	Input:HSQC 13c ppm	Input:2DJ peak No.	Input:2DJ 1h ppm	Input:2DJ J val hz	Annotated metabolite name	Database:HSQC_1h ppm	Database:HSQC_13c ppm	Database:2DJ_1h ppm	Database:2DJ_J val hz
2	3.2861	38.8347	42	3.9867	7.14152999626929	L-phenylalanine	3.2621	39.171	3.984	6.65
2	3.2861	38.8347	42	3.9867	7.14152999626929	L-phenylalanine	3.2621	39.171	3.9864	6.4414
2	3.2861	38.8347	47	3.9867	1.89040499901246	L-phenylalanine	3.2621	39.171	3.984	1.15
2	3.2861	38.8347	47	3.9867	1.89040499901246	L-phenylalanine	3.2621	39.171	3.9862	1.1903
2	3.2861	38.8347	48	3.0166	1.89040499901246	L-phenylalanine	3.2621	39.171	3.984	1.15
2	3.2861	38.8347	48	3.0166	1.89040499901246	L-phenylalanine	3.2621	39.171	3.9862	1.1903
2	3.2861	38.8347	49	3.9884	0.350074999817122	L-phenylalanine	3.2621	39.171	3.984	1.15
2	3.2861	38.8347	57	3.9858	0.350074999817122	L-phenylalanine	3.2621	39.171	3.984	1.15
5	0.9477	23.2475	11	3.4209	-3.36071999824437	pantothenic acid	0.924	23.14496627	3.431	-2.75
5	0.9477	23.2475	11	3.4209	-3.36071999824437	pantothenic acid	0.928	23.218	3.431	-2.75

Showing 1 to 10 of 6,216 entries

First Previous 1 2 3 4 5 ... 622 Next Last

HSQC 1H tolerance= 0.03
HSQC 13C tolerance= 0.53
HSQC width13c = 40
2D Jres 1H tolerance= 0.018
2D Jres J value tolerance= 0.83

Solvent(HSQC query)= Water/KPi
Solvent(2DJ query)= Water/KPi

Spectrum type (HSQC or 1D)= HSQC
Spectrum type (2DJ or 1D)= 2D Jres

Metabolite spectra in the current database = 20536
This was queried at 2017/10/14

RIKEN Environmental Metabolic Analysis Research Team. (c)2017-

Fig. 8 Screenshot of interanalysis, a venn diagram–type web tool for analysis of 2D-*J*res and ^{1}H–^{13}C HSQC data. In the bottom right, the blue, yellow, and red circles represent a set of search results; the green star represents the narrowed down set

have been used for mixture analysis. Making spectral assignments based on 2D-NMR experiments, such as HSQC and JRES, represents a powerful and accurate strategy. Our newly developed Venn diagram–type web tool, InterAnalysis, streamlines the annotation of environmentally and biologically derived small-molecule mixtures. The InterSpin comprises an integrated supportive web tool that is effective not only for analysis in conventional laboratories but also for on-site analysis using benchtop NMR. As a platform linking the laboratory and the real world, it will promote sustainable development based on NMR data.

Acknowledgments

I wish to thank all members of Environmental Metabolic Analysis Research Team (RIKEN) for their support with NMR data acquisition and handling. Namely, Shunji Yamada (Nagoya Univ.) was kindly involved development of InterSpin tool and initial drawing of figures in this article. This work was partially supported by the Agriculture, Forestry and Fisheries Research Council.

References

1. Lens PN, Hemminga MA (1998) Nuclear magnetic resonance in environmental engineering: principles and applications. Biodegradation 9(6):393–409
2. Nestle N, Baumann T, Niessner R (2002) Magnetic resonance imaging in environmental science. Environ Sci Technol 36 (7):154A–160A
3. Simpson AJ, McNally DJ, Simpson MJ (2011) NMR spectroscopy in environmental research: from molecular interactions to global processes. Prog Nucl Magn Reson Spectrosc 58 (3–4):97–175
4. Telgmann L, Sperling M, Karst U (2013) Determination of gadolinium-based MRI contrast agents in biological and environmental samples: a review. Anal Chim Acta 764:1–16
5. Gunduz O, Yetmez M, Sonmez M, Georgescu M, Alexandrescu L, Ficai A et al (2015) Mesoporous materials used in medicine and environmental applications. Curr Top Med Chem 15(15):1501–1515
6. de Prado Bert P, Mercader EMH, Pujol J, Sunyer J, Mortamais M (2018) The effects of air pollution on the brain: a review of studies interfacing environmental epidemiology and neuroimaging. Curr Environ Health Rep. https://doi.org/10.1007/s40572-018-0209-9
7. Singh P, Singh MK, Beg YR, Nishad GR (2019) A review on spectroscopic methods for determination of nitrite and nitrate in environmental samples. Talanta 191:364–381
8. Soininen P, Kangas AJ, Würtz P, Suna T, Ala-Korpela M (2015) Quantitative serum nuclear magnetic resonance metabolomics in cardiovascular epidemiology and genetics. Circ Cardiovasc Genet 8(1):192
9. Ogura T, Date Y, Kikuchi J (2013) Differences in cellulosic supramolecular structure of compositionally similar rice straw affect biomass metabolism by Paddy soil microbiota. PLoS One 8(6):e66919
10. Komatsu T, Kobayashi T, Hatanaka M, Kikuchi J (2015) Profiling planktonic biomass using element-specific, multicomponent nuclear magnetic resonance spectroscopy. Environ Sci Technol 49(11):7056–7062
11. Komatsu T, Kikuchi J (2013) Selective signal detection in solid-state NMR using rotor-synchronized dipolar dephasing for the analysis of hemicellulose in lignocellulosic biomass. J Phys Chem Lett 4(14):2279–2283
12. Kikuchi J, Yamada S (2017) NMR window of molecular complexity showing homeostasis in superorganisms. Analyst 142(22):4161–4172
13. Kikuchi J, Ito K, Date Y (2018) Environmental metabolomics with data science for investigating ecosystem homeostasis. Prog Nucl Magn Reson Spectrosc 104:56–88
14. Viant MR (2008) Recent developments in environmental metabolomics. Mol BioSyst 4 (10):980–986
15. Mitchell J, Gladden LF, Chandrasekera TC, Fordham EJ (2014) Low-field permanent magnets for industrial process and quality control. Prog Nucl Magn Reson Spectrosc 76:1–60
16. Chikayama E, Yamashina R, Komatsu K, Tsuboi Y, Sakata K, Kikuchi J et al (2016) FoodPro: a web-based tool for evaluating covariance and correlation NMR spectra associated with food processes. Meta 6(4):E36
17. Parker T, Limer E, Watson AD, Defernez M, Williamson D, Kemsley EK (2014) 60 MHz 1H NMR spectroscopy for the analysis of edible oils. Trends Analyt Chem 57 (100):147–158
18. Misawa T, Date Y, Kikuchi J (2015) Human metabolic, mineral, and microbiota fluctuations across daily nutritional intake visualized by a data-driven approach. J Proteome Res 14 (3):1526–1534
19. Motegi H, Tsuboi Y, Saga A, Kagami T, Inoue M, Toki H et al (2015) Identification of reliable components in multivariate curve resolution-alternating least squares (MCR-ALS): a data-driven approach across metabolic processes. Sci Rep 5:15710
20. Osaka T, Moriyama E, Arai S, Date Y, Yagi Jm Kikuchi J et al (2017) Meta-analysis of fecal microbiota and metabolites in experimental colitic mice during the inflammatory and healing phases. Nutrients 9(12):E1329. https://doi.org/10.3390/nu9121329
21. Ogura T, Date Y, Masukujane M, Coetzee T, Akashi K, Kikuchi J (2016) Improvement of physical, chemical, and biological properties of aridisol from Botswana by the incorporation of torrefied biomass. Sci Rep 6:28011
22. Ogura T, Date Y, Tsuboi Y, Kikuchi J (2015) Metabolic dynamics analysis by massive data

integration: application to tsunami-affected field soils in Japan. ACS Chem Biol 10 (8):1908–1915

23. Mega R, Abe F, Kim JS, Tsuboi Y, Tanaka K, Kobayashi H et al (2019) Water-saving wheat: tuning water use efficiency and drought tolerance using ABA receptors . Nature Plants in press
24. Komatsu T, Ohishi R, Shino A, Kikuchi J (2016) Structure and metabolic-flow analysis of molecular complexity in a (13) C-labeled tree by 2D and 3D NMR. Angew Chem Int Ed Engl 55(20):6000–6003
25. Komatsu T, Kikuchi J (2013) Comprehensive signal assignment of 13C-labeled lignocellulose using multidimensional solution NMR and 13C chemical shift comparison with solid-state NMR. Anal Chem 85 (18):8857–8865
26. Yoshida S, Date Y, Akama M, Kikuchi J (2014) Comparative metabolomic and ionomic approach for abundant fishes in estuarine environments of Japan. Sci Rep 4:7005
27. Asakura T, Sakata K, Yoshida S, Date Y, Kikuchi J (2014) Noninvasive analysis of metabolic changes following nutrient input into diverse fish species, as investigated by metabolic and microbial profiling approaches. PeerJ 2:e550
28. Mekuchi M, Sakata K, Yamaguchi T, Koiso M, Kikuchi J (2017) Trans-omics approaches used to characterise fish nutritional biorhythms in leopard coral grouper (Plectropomus leopardus). Sci Rep 7(1):9372
29. Wei F, Sakata K, Asakura T, Date Y, Kikuchi J (2018) Systemic homeostasis in Metabolome, Ionome, and microbiome of wild Yellowfin goby in estuarine ecosystem. Sci Rep 8 (1):3478
30. Wei F, Ito K, Sakata K, Date Y, Kikuchi J (2015) Pretreatment and integrated analysis of spectral data reveal seaweed similarities based on chemical diversity. Anal Chem 87 (5):2819–2826
31. Ito K, Sakata K, Date Y, Kikuchi J (2014) Integrated analysis of seaweed components during seasonal fluctuation by data mining across heterogeneous chemical measurements with network visualization. Anal Chem 86 (2):1098–1105
32. Misawa T, Komatsu T, Date Y, Kikuchi J (2016) SENSI: signal enhancement by spectral integration for the analysis of metabolic mixtures. Chem Commun (Camb) 52 (14):2964–2967
33. Yamada S, Ito K, Kurotani A, Yamada Y, Chikayama E, Kikuchi J (2019) InterSpin: integrated supportive webtools for low- and high-field NMR analysis of small- to macro-toward molecular mixture complexity. ACS Omega (Submitted)
34. Sekiyama Y, Chikayama E, Kikuchi J (2011) Evaluation of a semipolar solvent system as a step toward heteronuclear multidimensional NMR-based metabolomics for 13C-labeled bacteria, plants, and animals. ACS Omega 4:3361–3369
35. Kikuchi J, Tsuboi Y, Komatsu K, Gomi M, Chikayama E, Date Y (2016) SpinCouple: development of a web tool for analyzing metabolite mixtures via two-dimensional J-resolved NMR database. Anal Chem 88 (1):659–665
36. Chikayama E, Sekiyama Y, Okamoto M, Nakanishi Y, Tsuboi Y, Akiyama K et al (2010) Statistical indices for simultaneous large-scale metabolite detections for a single NMR spectrum. Anal Chem 82(5):1653–1658
37. Shima H, Masuda S, Date Y, Shino A, Tsuboi Y, Kajikawa M et al (2017) Exploring the impact of food on the gut ecosystem based on the combination of machine learning and network visualization. Nutrients 9(12):E1307. https://doi.org/10.3390/nu9121307
38. Yamazawa A, Iikura T, Shino A, Date Y, Kikuchi J (2013) Solid-, solution-, and gas-state NMR monitoring of (1)(3)C-cellulose degradation in an anaerobic microbial ecosystem. Molecules 18(8):9021–9033
39. Misawa T, Wei F, Kikuchi J (2016) Application of two-dimensional nuclear magnetic resonance for signal enhancement by spectral integration using a large data set of metabolic mixtures. Anal Chem 88(12):6130–6134
40. Asakura T, Date Y, Kikuchi J (2014) Comparative analysis of chemical and microbial profiles in estuarine sediments sampled from Kanto and Tohoku regions in Japan. Anal Chem 86 (11):5425–5432
41. Yamazawa A, Date Y, Ito K, Kikuchi J (2014) Visualizing microbial dechlorination processes in underground ecosystem by statistical correlation and network analysis approach. J Biosci Bioeng 117(3):305–309
42. Date Y, Kikuchi J (2018) Application of a deep neural network to metabolomics studies and its performance in determining important variables. Anal Chem 90(3):1805–1810

43. Asakura T, Date Y, Kikuchi J (2018) Application of ensemble deep neural network to metabolomics studies. Anal Chim Acta 1037:230–236
44. Oita A, Tsuboi Y, Date Y, Oshima T, Sakata K, Yokoyama A et al (2018) Profiling physicochemical and planktonic features from discretely/continuously sampled surface water. Sci Total Environ 636:12–19
45. Shiokawa Y, Date Y, Kikuchi J (2018) Application of kernel principal component analysis and computational machine learning to exploration of metabolites strongly associated with diet. Sci Rep 8(1):3426
46. Asakura T, Sakata K, Date Y, Kikuchi J (2018) Regional feature extraction of various fishes based on chemical and microbial variable selection using machine learning. Anal Methods 17:16–26
47. Wishart DS, Jewison T, Guo AC, Wilson M, Knox C, Liu Y et al (2013) HMDB 3.0--the human Metabolome database in 2013. Nucleic Acids Res 41(Database issue):D801–D807
48. Cui Q, Lewis IA, Hegeman AD, Anderson ME, Li J, Schulte CF et al (2008) Metabolite identification via the madison metabolomics consortium database. Nat Biotech 26 (2):162–164
49. Steinbeck C, Kuhn S (2004) NMRShiftDB -- compound identification and structure elucidation support through a free community-built web database. Phytochemistry 65 (19):2711–2717
50. Ogata Y, Chikayama E, Morioka Y, Everroad RC, Shino A, Matsushima A et al (2012) ECOMICS: a web-based toolkit for investigating the biomolecular web in ecosystems using a trans-omics approach. PLoS One 7(2):e30263

Part VIII

Food, Nutrition, and Plants

Chapter 18

NMR-Based Metabolomics of Food

Ulrik K. Sundekilde, Nina Eggers, and Hanne Christine Bertram

Abstract

NMR spectroscopy is one of the major analytical techniques used in the metabolomics studies of food. There are many applications of metabolomics on food-related topics and on the food itself. Here, we describe protocols for performing NMR-based metabolomics of foods ranging from simple beverages to solid foods and semisolid foods. Beverages can be analyzed either directly or after sample preprocessing to remove interfering macromolecules, muscle-based foods can be analyzed after extraction, and semisolid foods can be analyzed directly using high-resolution magic-angle spinning (HR-MAS) NMR. Finally, we discuss metabolomic data analysis as well as different procedures and strategies for targeted and untargeted approaches.

Key words Food constituents, Food composition, Beverages, Milk, Muscle, Food metabolites, Micronutrients, Solid phase, Liquid phase, Nuclear magnetic resonance spectroscopy

1 Introduction

Proton nuclear magnetic resonance spectroscopy (^{1}H NMR) is an appealing analytical technique for analyzing food to obtain information on the content of micronutrients and metabolites. Liquid-phase ^{1}H NMR is optimal for liquid-based food items such as milk, fruit and vegetable juices, and alcoholic beverages, but also offers the opportunity to analyze solid-based foods following appropriate extractions to solubilize compounds. High-resolution magic-angle spinning (HR-MAS) ^{1}H NMR spectroscopy offers the ability to analyze and profile semisolid foods and tissues as it enables measurement of intact samples [1].

NMR-based metabolomics typically enables detection of 30–100 metabolites from food items depending on the extraction procedure [2–5]. The metabolites measured by NMR spectroscopy are diverse and comprise carbohydrates, organic acids, amino acids, small peptides, vitamins, flavor-active nucleotides, and more. In food science, metabolomics has been used for food component analysis [5, 6], quality elucidation [7], authenticity [8], and

G. A. Nagana Gowda and Daniel Raftery (eds.), *NMR-Based Metabolomics: Methods and Protocols*, Methods in Molecular Biology, vol. 2037, https://doi.org/10.1007/978-1-4939-9690-2_18, © Springer Science+Business Media, LLC, part of Springer Nature 2019

monitoring of food processing [9–11] and storage [12, 13]. Metabolite profiles generated by NMR spectroscopy can essentially be treated in two ways: an untargeted or a targeted route. Databases and software offer the opportunity to deconvolute the NMR spectra into their chemical constituents [14, 15]. A limitation of the targeted route is that the data are limited to prior knowledge of food compounds available in databases. Untargeted NMR-based metabolomics uses pattern recognition tools to explore the NMR spectral profiles. Numerous multivariate data analysis methods starting with unsupervised, principal component analysis (PCA) is used as a tool for pattern recognition in untargeted methods [16].

2 Materials

All buffers and other solutions should be prepared using ultrapure water and analytical grade reagents and solvents.

1. Amicon Ultra 0.5 mL 10 kDa spin filters (Millipore, Billerica, MA, USA).
2. D_2O containing 0.05% 3-(trimethylsilyl) propionic-2,2,3,3-d_4 acid (TSP), sodium salt.
3. 100 mM potassium phosphate pH 7.0 buffer stock solution in D_2O.
4. Bruker HRMAS disposable inserts (Bruker BioSpin, Rheinstetten, Germany).
5. High-throughput NMR tubes (multiple vendors).
6. Chemical standard compounds used to spike sample to confirm identification as needed.

3 Methods

The methodologies for NMR-based metabolomics can be split into two main categories. Beverages and other foods that are liquid by nature can either be analyzed as they are or it may be necessary to remove macromolecules. Solid foods can be analyzed using high-resolution magic-angle spinning (HR-MAS) NMR spectroscopy or can be extracted using, for example, methanol and chloroform.

3.1 Milk

The following steps describe milk analysis; however, the steps are also suitable for analysis of other types of beverages containing proteins and lipids. A representative ^1H NMR spectrum obtained for milk is shown in Fig. 1.

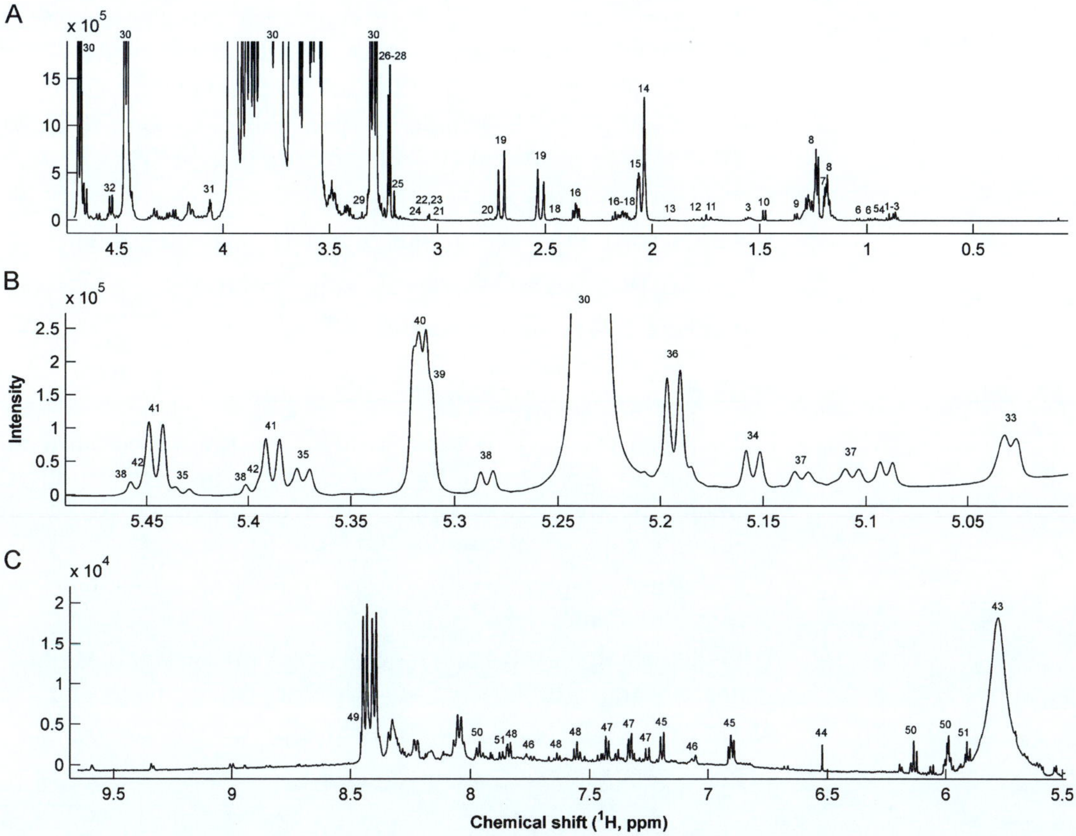

Fig. 1 Typical ^{1}H NMR spectrum of human milk. (**a**) Aliphatic region 0 to 4.6 ppm, (**b**) 5 to 5.45 ppm region showing the oligosaccharides, and (**c**) 5.5 to 9.7 ppm region highlighting the peaks in the aromatic region. Peaks: 1, caprylate; 2, caprate; 3, butyrate; 4, pantothenate; 5, leucine; 6, valine; 7, 3-BHBA; 8, fucosyl moieties; 9, lactate; 10, alanine; 11, 6′-sialyllactose; 12, 3′-sialyllactose; 13, acetate; 14, GlcNAc; 15, sialic acid; 16, glutamate; 17, methionine; 18, glutamine; 19, citrate; 20, dimethylamine; 21, 2-oxogluturate; 22, creatinine; 23, creatine; 24, malonate; 25, choline; 26, *O*-phosphocholine; 27, carnitine; 28, betaine; 29, methanol; 30, lactose; 31, gluconate; 32, galactose; 33, fuc α1,4 GlcNAc; 34, fuc α1,3 GlcNAc; 35, LNDFH II; 36, LNDFH I; 37, LNFP III; 38, LDFT; 39, LNDFP I; 40, 2′FL; 41, 3′FL; 42, LNFP V; 43, urea; 44, fumarate; 45, tyrosine; 46, methylhistidine; 47, phenylalanine; 48, hippurate; 49, formate; 50, cytidine triphosphate; 51, uridine. Reproduced with permission from [21] under a creative commons attribution 4.0 license

3.1.1 Removal of Macromolecules Interfering with NMR Analysis of Aqueous Metabolites

1. Whole or semi-fat milk samples should be skimmed prior to filtration to ensure that lipids do not clog the filtration membrane.
2. Wash 10 kDa Amicon filters at least five times with 0.5 mL water by centrifugation at 10,000 × *g* for 10 min at a temperature of 277 K (*see* **Note 1**).
3. Remove residual water bound to filter by reversing filters and centrifugation at 1000 × *g* for 30 s (*see* **Note 2**).
4. Move washed Amicon filters to a new Eppendorf tube and add 0.5 mL of the milk sample.

5. Filter proteins and residual lipids from the milk samples by centrifugation at 10,000 × *g* for 30 min at a temperature of 277 K (*see* **Note 3**).
6. Add 0.3 mL milk sample to the filter and centrifuge at 10,000 × *g* for 30 min at a temperature of 277 K.
7. Transfer 0.5 mL filtrate to NMR tubes.
8. Add 0.1 mL D_2O with 0.05% (v/w) TSP to NMR tubes.
9. Cap NMR tubes and mix by shaking (*see* **Note 4**).
10. Perform NMR spectroscopy at 298 K.

3.2 Alcoholic Beverages

Sample preparation for nonvolatile compounds.

1. Transfer 1 mL of commercial product (beer, wine, etc.) into an Eppendorf 1.5 mL safe-lock tube. For carbonated products, weigh the transferred sample on a microbalance to obtain accurate volumes of samples (*see* **Note 5**).
2. Condense samples in a centrifugal evaporator at low temperature until completely dry.
3. Redissolve the nonvolatile residue in 0.6 mL of 100 mM phosphate buffer (pH 7.0) in D_2O (*see* **Note 6**).
4. Transfer 0.5 mL filtrate to NMR tubes.
5. Perform NMR spectroscopy at 298 K.

3.3 Fruit and Vegetable Juices

1. Fresh fruits and vegetables can be pressed into juices.
2. Optional: If the juice is cloudy, centrifuge the juice at 10,000 × *g* for 30 min at a temperature of 277 K.
3. Add 0.5 mL juice to NMR tubes.
4. Add 0.1 mL D_2O with 0.05% (v/w) TSP to NMR tubes.
5. Cap NMR tubes and mix by shaking (*see* **Note 4**).
6. Perform NMR spectroscopy.

3.4 Extraction of Solid Foods: Vegetables, Cheeses, and Muscle-Based Foods

Representative ^1H NMR spectra for a carrot (solid food), obtained following aqueous and organic extractions, are shown in Fig. 2.

1. Lyophilize samples.
2. Grind lyophilized sample material.
3. Weigh 20–100 mg material.
4. Add 400 μL ice-cold MeOH to the samples and mix thoroughly, then place the sample on ice for 10 min.
5. Add 400 μL ice-cold $CHCl_3$ to the samples and mix thoroughly, then place the sample on ice for 10 min.
6. Add 400 μL ice-cold water to the samples and mix thoroughly, keep the sample aside for 8 h at 277 K for separation.

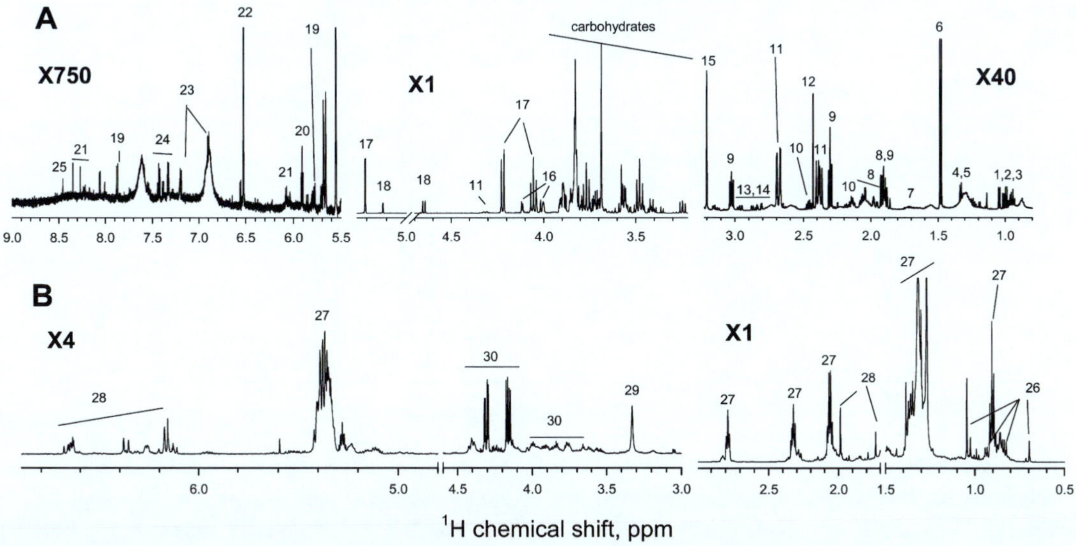

Fig. 2 Typical ^{1}H NMR spectra of carrot "Purple Haze" extracts. (**a**) Spectrum of aqueous extract and (**b**) spectrum of organic extract. Peaks: 1, isoleucine; 2, leucine; 3, valine; 4, lactic acid; 5, threonine; 6, alanine; 7, arginine; 8, quinic acid; 9, γ-aminobutyric acid; 10, glutamine; 11, malic acid; 12, succinic acid; 13, aspartic acid; 14, asparagine; 15, choline; 16, fructose; 17, sucrose; 18, glucose; 19, uracil; 20, uridine; 21, β-adenosine; 22, fumaric acid; 23, tyrosine; 24, phenylalanine; 25, formic acid; 26, sterols; 27, fatty acid chain; 28, β-carotene; 29, phospholipids; 30, glycerol moiety triglycerides. Reproduced with permission from [6]

7. Centrifuge the samples for 30 min at 1400 × *g* at 277 K.
8. Transfer the upper MeOH phase (water-soluble layer) to a new 1.5 mL Eppendorf tube, and dry by vacuum centrifugation for approximately 3 h.
9. Transfer the lower $CHCl_3$ phase (lipid extract) to a new 1.5 mL Eppendorf tube, and dry by vac. Centrifuge for approximately 1 h.
10. Optional: The dried samples can be stored at 193 K until NMR analysis (*see* **Note** 7).
11. Dissolve the dried pellet (aqueous extract) from **step 8** in 550 μL D_2O, 25 μL H_2O, and 25 μL D_2O containing 0.05% TSP and transfer to an NMR tube.
12. Dissolve the dried pellet (lipid extract) from **step 9** in 575 μL $CDCl_3$ + 25 μL $CDCl_3$ containing 0.05% TMS and transfer to an NMR tube.
13. Perform NMR spectroscopy.

3.5 HR-MAS NMR Spectroscopy of Intact Foods

A representative ^{1}H HR-MAS NMR spectrum of chicken muscle tissue obtained using the CPMG pulse sequence is shown in Fig. 3. The spectrum was obtained on a muscle tissue sample (~20 mg) cut on a dry-ice bench, packed in a HR-MAS insert and rotor.

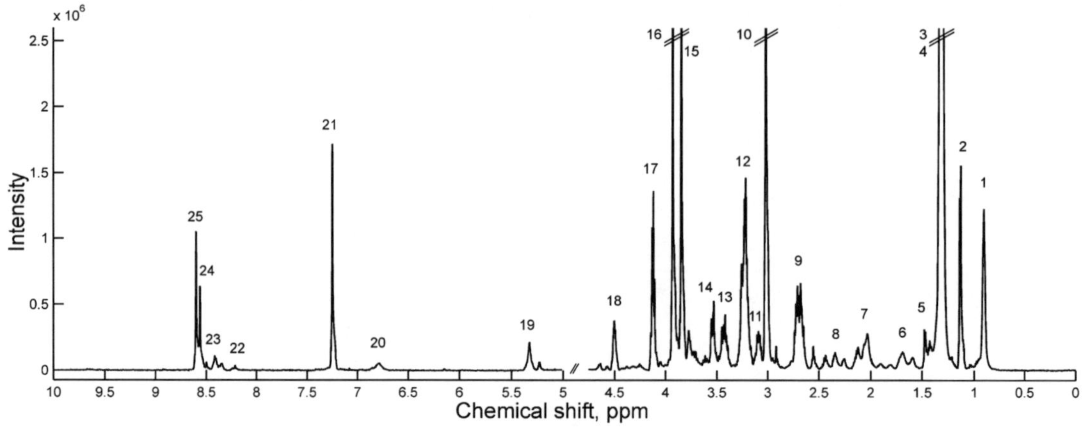

Fig. 3 ^{1}H HR-MAS CPMG NMR spectrum of chicken pectoralis muscle. Peaks: 1, CH_3 of fatty acids; 2, 2,3-butanediol; 3, CH_2 of fatty acids; 4, lactate; 5, alanine; 6, leucine/isoleucine; 7, CH_2 adjacent to CHCH in fatty acids; 8, carnitine; 9, carnosine/anserine; 10, creatine; 11, carnosine/anserine; 12, choline/phosphocholine; 13, taurine; 14, glycerol; 15, anserine; 16, creatine; 17, lactate; 18, carnosine/anserine; 19, CHCH in fatty acids; 20, tyrosine; 21, anserine/carnosine; 22, inosine; 23, nucleotides; 24, carnosine; 25, anserine. Reproduced with permission from [1]

1. Preweigh 30 μL HR-MAS inserts.
2. Optional: A dry-ice bench can be created by placing dry ice in an open Styrofoam package with a metal tray as a lid (*see* **Note 8**).
3. Optional: Cut frozen sample to a suitable size using a scalpel on a dry-ice bench (*see* **Note 8**).
4. Place sample in HR-MAS sample insert.
5. Add 10 μL cold D_2O containing 0.05% (v/w) (TSP) to insert.
6. Close insert with lid and keep it in a 1.5 mL Eppendorf tube. Centrifuge for 15 s at 1500 × *g* at 277 K to let D_2O mix with sample.
7. Weigh the insert prior to HR-MAS NMR spectroscopy.
8. Optional: Freezing step. If required, the insert with sample can be stored at 193 K until NMR spectroscopy.
9. Perform HR-MAS NMR spectroscopy.

3.6 NMR Spectroscopy

3.6.1 Liquid Phase

1. Randomize sample sequence order prior to acquisition.
2. Perform ^{1}H NMR spectroscopy at 298 K in full automation.
3. Record one-dimensional (1D) NOESY experiments with presaturation (Bruker pulse program: noesypr1d; Varian pulse program: tnnoesy) with the following settings: recycling delay of 5 s (*see* **Note 9**), mixing time of 0.1 ms, acquisition time of 4 s, spectral width of 12.02 ppm, number of scans 64, and data points 32,768 (*see* **Note 10**).

4. Spectral processing: Zero-fill the data to obtain twice the number of data points and use a line-broadening function of 0.3 Hz prior to Fourier transformation. Reference the ^{1}H NMR spectra to the TSP signal at 0 ppm.

3.6.2 Solid Phase

1. Place filled sample insert in a 4 mm zirconium rotor.
2. Set temperature to 281 K (*see* **Note 11**).
3. Inject rotor into the NMR magnet and spin rotor at 5000 Hz.
4. Manually tune and match the NMR probe.
5. Manually shim sample using line width of the added TSP as a guide (*see* **Note 12**).
6. Record one-dimensional (1D) NOESY experiments with water presaturation using the following parameters: recycling delay of 3 s (*see* **Note 9**), acquisition time of 2.25 s, and a spectral width of 12.0 ppm.
7. Record 1D CPMG experiments with the following settings: recycling delay of 3 s (*see* **Note 9**), acquisition time of 1.57 s, spectral width of 12.0 ppm, and a T_2 echo of 400 ms (*see* **Note 13**).
8. For all experiments, collect a total of 128 scans and 32,768 data points (*see* **Note 10**).
9. Spectral processing: Zero-fill the data by two times the number of data points and use a line-broadening function of 0.3 Hz prior to Fourier transformation. Reference the ^{1}H NMR spectra to the TSP signal at 0 ppm.

3.6.3 Data Analysis

Untargeted analysis of NMR spectral profiles is often employed using PCA and other exploratory multivariate data analytical approaches. These multivariate methodologies reduce the complexity of NMR data, which typically consists of thousands of data points corresponding to about 100 metabolites, to a new set of orthogonal variables [16]. The reduction in data complexity enables identification of metabolite differences between different groups of samples.

1. Import NMR data to MATLAB using RBNMR [17] or other types of script.
2. Align NMR data using Icoshift [18] or similar type of script.
3. Optional: Perform binning to reduce the number of NMR variables.
4. Perform multivariate data analysis using an appropriate software package (PLSToolbox, SIMCA, Unscrambler, Latentix, R).
5. Identify key NMR variables; the identity can be verified using literature, databases, and proprietary software.

Targeted metabolomics may involve the total deconvolution of NMR spectra by Chenomx (Chenomx Inc., Alberta, Canada), Bayesil [19], or BATMAN [20]. For targeted metabolomics, the end result is a table of known metabolites and their (ideally absolute) concentrations. Further statistical analysis can be performed using these metabolite concentrations.

4 Notes

1. Amicon 3 kDa spin filters should be washed seven times. Be careful especially if glycerol needs to be quantified. Glycerol is used in manufacturing of the filters and a thorough washing is required for its reliable quantification.
2. Do not let spin filters dry out at this stage as this will decrease filtration efficiency or cause the membrane to break.
3. Ensure spin filters are intact after the filtration process. Increased rate of breakage is noticed with milk samples compared with other biofluids.
4. Make sure that D_2O and the sample material are thoroughly mixed. Optionally, D_2O, sample, and buffer can be mixed in Eppendorf tubes and the final volume can be transferred to NMR tubes to ensure optimal mixing.
5. It can be difficult to accurately pipette volumes of carbonated liquids. Weigh the pipetted sample for increased accuracy.
6. A buffer of 100 mM concentration was chosen as higher salt concentrations elicit a sensitivity penalty (lower signal-to-noise ratio) on NMR systems using cryogenically cooled detection probes.
7. If the sample is to be stored for a longer period of time, consider substituting the headspace gas with an inert gas to reduce metabolite oxidation.
8. Make sure to keep material frozen at all times. It is difficult to place nonfrozen tissue material into HR-MAS inserts.
9. For accurate quantification, generally, a recycle delay must exceed five times the longest T_1 of the NMR signals in a sample.
10. For limited sample material, a higher signal-to-noise ratio can be achieved by increasing the number of scans. The signal-to-noise ratio is proportional to the square root of number of scans.
11. Even lower temperatures may be used depending on the spectrometer's ability to cool.
12. A line width of <3 Hz can be considered acceptable. In most cases, it should be possible to obtain a line width of <1.5 Hz.

13. The T_2 loop time depends on the extent of peak attenuation needed. It should be optimized to ensure efficient attenuation of unwanted signals.

Acknowledgements

HCB acknowledges the Independent Research Fund Denmark for financial support through her Eliteforsk 2017 grant (# 6161-00016B).

References

1. Sundekilde UK, Rasmussen MK, Young JF, Bertram HC (2017) High resolution magic angle spinning NMR spectroscopy reveals that pectoralis muscle dystrophy in chicken is associated with reduced muscle content of anserine and carnosine. Food Chem 217:151–154
2. Cappello T, Giannetto A, Parrino V, De Marco G, Mauceri A, Maisano M (2018) Food safety using NMR-based metabolomics: assessment of the Atlantic Bluefin tuna, Thunnus Thynnus, from the Mediterranean Sea. Food Chem Toxicol 115:391–397
3. Sundekilde UK, Larsen LB, Bertram HC (2013) NMR-based milk metabolomics. Meta 3(2):204–222
4. Cicero N, Corsaro C, Salvo A, Vasi S, Giofré SV, Ferrantelli V et al (2015) The metabolic profile of lemon juice by proton HR-MAS NMR: the case of the PGI Interdonato lemon of Messina. Nat Prod Res 29(20):1894–1902
5. Wishart DS (2008) Metabolomics: applications to food science and nutrition research. Trends Food Sci Technol 19:482–493
6. Clausen MR, Edelenbos M, Bertram HC (2014) Mapping the variation of the carrot metabolome using ^{1} H NMR spectroscopy and consensus PCA. J Agric Food Chem 62 (19):4392–4398
7. Straadt IK, Aaslyng MD, Bertram HC (2014) An NMR-based metabolomics study of pork from different crossbreeds and relation to sensory perception. Meat Sci 96(2):719–728
8. Kuballa T, Brunner TS, Thongpanchang T, Walch SG, Lachenmeier DW (2018) Application of NMR for authentication of honey, beer and spices. Curr Opin Food Sci 19:57–62
9. Sundekilde UK, Meier S (2016) 1H–13C NMR-based profiling of biotechnological starch utilization. Anal Chem 88 (19):9685–9690
10. Sundekilde UK, Jarno L, Eggers N, Bertram HC (2018) Real-time monitoring of enzyme-assisted animal protein hydrolysis by NMR spectroscopy – an NMR reactomics concept. LWT 95:9–16
11. García-García AB, Lamichhane S, Castejón D, Cambero MI, Bertram HC (2018) ^{1}H HR-MAS NMR-based metabolomics analysis for dry-fermented sausage characterization. Food Chem 240:514–523
12. Jansson T, Jensen HB, Sundekilde UK, Clausen MR, Eggers N, Larsen LB, Andersen HJ, Bertram HC et al (2014) Chemical and proteolysis-derived changes during long-term storage of lactose-hydrolyzed ultrahigh-temperature (UHT) milk. J Agric Food Chem 62 (46):11270–11278
13. Clausen MR, Bach V, Edelenbos M, Bertram HC (2012) Metabolomics reveals drastic compositional changes during overwintering of Jerusalem artichoke (Helianthus Tuberosus L.) tubers. J Agric Food Chem 60(37):9495–9501
14. Wishart DS, Knox C, Guo AC, Eisner R, Young N, Gautam B et al (2009) HMDB: a knowledgebase for the human metabolome. Nucleic Acids Res 37(Database):D603–D610
15. van der Hooft JJJ, Rankin N (2016) Metabolite identification in complex mixtures using nuclear magnetic resonance spectroscopy. In: Webb G (ed) Modern magnetic resonance. Springer International Publishing, Cham, pp 1–32
16. Smolinska A, Blanchet L, Buydens LMC, Wijmenga SS (2012) NMR and pattern recognition methods in metabolomics: from data acquisition to biomarker discovery: a review. Anal Chim Acta 750:82–97
17. Nyberg NT RBNMR. https://se.mathworks.com/matlabcentral/fileexchange/40332-rbnmr. Accessed 13 Jul 2018
18. Savorani F, Tomasi G, Engelsen SB (2010) Icoshift: a versatile tool for the rapid alignment of 1D NMR spectra. J Magn Reson 202 (2):190–202

19. Ravanbakhsh S, Liu P, Bjordahl TC, Mandal R, Grant JR, Wilson M et al (2015) Accurate, fully-automated NMR spectral profiling for metabolomics. PLoS One 10(5):e0124219
20. Hao J, Astle W, Iorio MD, Ebbels T (2012) BATMAN — an R package for the automated quantification of metabolites from the nuclear magnetic resonance spectra using a Bayesian model. Bioinformatics 28(15):2088–2090
21. Sundekilde U, Downey E, O'Mahony J, O'Shea CA, Ryan C, Kelly A et al (2016) The effect of gestational and lactational age on the human milk metabolome. Nutrients 8(5):304

Chapter 19

Plant Metabolomics Using NMR Spectroscopy

Denise Medeiros Selegato, Alan Cesar Pilon, and Fausto Carnevale Neto

Abstract

The major goal in plant metabolomics is to study complex extracts for the purposes of metabolic exploration and natural products discovery. To achieve this goal, plant metabolomics relies on accurate and selective acquisition of all possible chemical information, which includes maximization of the number of detected metabolites and their correct molecular assignment. Nuclear magnetic resonance (NMR) spectroscopy has been recognized as a powerful platform for obtaining the metabolite profiles of plant extracts. In this chapter, we provide a workflow for targeted and untargeted metabolite profiling of plant extracts using both 1D and 2D NMR methods. The protocol includes sample preparation, instrument operation, data processing, multivariate analysis, biomarker elucidation, and metabolite quantitation. It also addresses the annotation of plant metabolite peaks considering NMR's capabilities to cover a broad range of metabolites and elucidate structures for unknown compounds.

Key words NMR spectroscopy, Plant metabolomics, Metabolite annotation, Quantitative NMR, Natural products

1 Introduction

Plants are organisms with highly specialized cells and tissues, which produce an enormous array of chemicals to adapt their sessile lifestyle to challenging environments [1]. The chemical diversity of plants covers the primary metabolism, encompassing pathways required for survival, and "specialized" metabolism that yields chemicals to overcome biotic and abiotic stresses [1, 2].

Plants are the source of an enormous chemical diversity with metabolite concentrations ranging over nine orders of magnitude (pM to mM) [3]. Such a breadth of metabolite distribution sets plant metabolites research apart from animals and microorganisms [4]. Metabolites dynamics in plants is quite intricate; substances are continuously absorbed, recycled, and degraded as a result of cellular interactions and/or environmental stress. Plants exhibit cyclical,

Denise Medeiros Selegato and Alan Cesar Pilon are co-first authors.

G. A. Nagana Gowda and Daniel Raftery (eds.), *NMR-Based Metabolomics: Methods and Protocols*, Methods in Molecular Biology, vol. 2037, https://doi.org/10.1007/978-1-4939-9690-2_19, © Springer Science+Business Media, LLC, part of Springer Nature 2019

environmental, and ontogenetic effects; the metabolic flow generated by various stages of development and circadian rhythms hinders the recognition of metabolic patterns.

Analysis of the chemical diversity along with the wide range of metabolite concentrations in plants requires powerful analytical approaches [5]. Metabolomics is a tool capable of unraveling the complex chemical diversity of plants and improving our understanding of the metabolism and biochemistry of organisms [3]. It aims to investigate the complete set of metabolites, which are essentially nongenetically encoded substrates, intermediates, and products of biochemical pathways. Altered metabolite levels are associated with changes in the biochemical status of organisms due to stimuli such as biotic and/or abiotic stress and genetic disorders [2–4, 6–8]. To date, numerous plant metabolomics studies have focused on (1) classification and characterization of different species of plants; (2) monitoring of biochemical response to stress, wounding, herbivory, or infection of plants; (3) discrimination of different plant genotypes; and (4) exploration of chemical and bioactivity diversity in plants [9–19].

Although metabolomics has proven to be an essential strategy for plant sciences, understanding the fundamentals of plant development, physiology, resistance, and the interactions of plant metabolic networks in complex scenarios, such as in the forest environment, remains a challenge [20, 21].

NMR spectroscopy is a powerful analytical tool used in the metabolomics field for studying biological systems [3]. It offers opportunities for reliable metabolite detection with high reproducibility and exhibits flexibility for analysis of a variety of sample types [22, 23]. Its unique quantitative ability provides an unbiased view of chemical composition and allows assessment of both purity and quantity, rapidly and nondestructively [24–26]. Nevertheless, NMR has generally exhibited low sensitivity and problems with coupling to separation techniques [5]. Over the last decade, several strategies including high magnetic fields (up to 23.5T), cryoprobe technologies, pulse sequences for solvent suppression, dynamic nuclear polarization (DNP), and microcoils combined with photochemically induced dynamic nuclear polarization (photo-CIDNP) have extended the limits of NMR analysis [27]. Deconvolution approaches in NMR have also improved quality of the data, expanded the chemical coverage, and increased the number of detectable metabolites [28–30].

In this chapter, we describe NMR-based plant metabolomics protocols from sample preparation, data acquisition, data processing, multivariate analysis and metabolite quantitation to biomarker elucidation (Fig. 1).

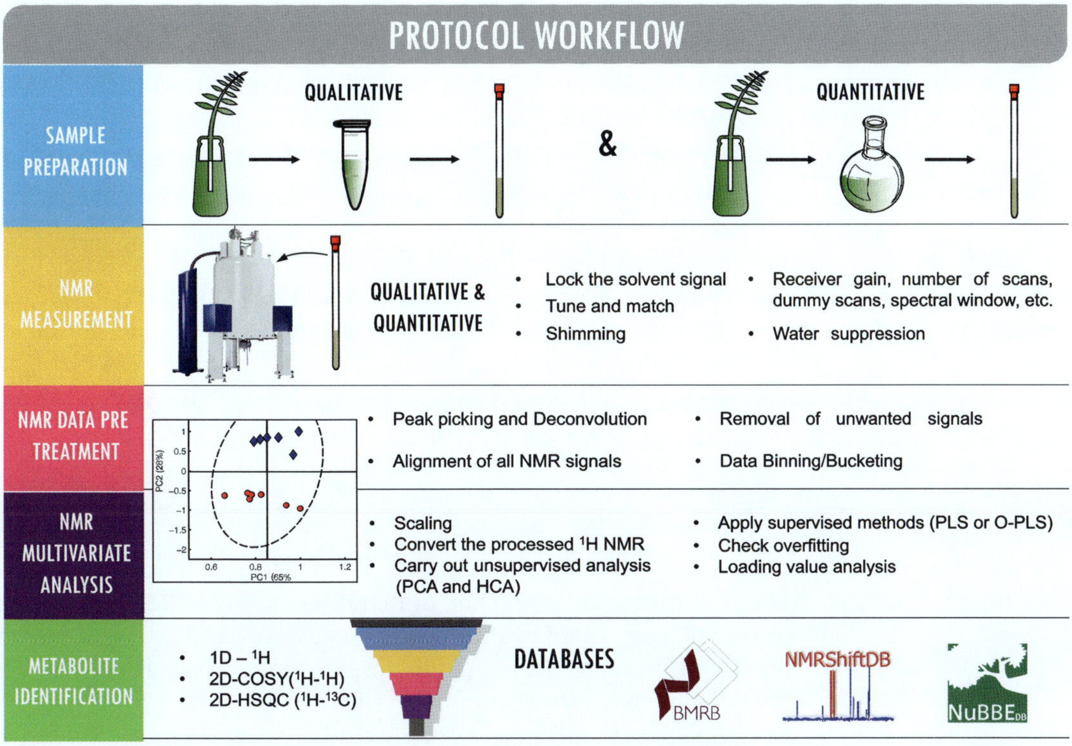

Fig. 1 Schematic workflow of the NMR-based plant metabolomics protocol

2 Materials

2.1 Chemicals and Solvents

1. KH_2PO_4, 99% (anhydrous).
2. TSP-d_4—3-(trimethylsilyl)propionic-2,2,3,3-d_4 acid sodium salt.
3. D_2O—deuterium oxide (>99.9 atom % D) + 0.1% (wt/wt) TSP.
4. CD_3OD—methanol–d_4 (99.8 atom % D) + 0.1% (wt/wt) TSP.
5. $CDCl_3$—chloroform-d (>99.8% D) + 0.1% (wt/wt) TSP.
6. NaOD—sodium deuteroxide (99.5 atom % D, 40% in D_2O).
7. N_2—Liquid nitrogen (*see* **Note 1**).

2.2 Equipment

1. General laboratory equipment: laboratory vortex mixer, automatic pipettes (50 to 1000 μL), cutting/impact analytical mill.
2. Freeze dryer for sample drying.
3. Freezer (−80 °C) for sample storage.
4. For sample preparation: ultrasonicator, microcentrifuge, 2.0 mL Eppendorf tubes, 5 mm or 3 mm NMR tubes, and pH meter with electrode.

5. For NMR data acquisition: 500 MHz or higher frequency NMR spectrometer equipped with a 5 mm or 3 mm probe, and *z*-gradient system is suitable for plant metabolomics (*see* **Note 2**).
6. Software for NMR data acquisition and processing.
7. Software for multivariate analysis (PCA, HCA, PLS-DA, etc.).

2.3 Reagents and Solutions (see Note 3)

1. Phosphate buffer: Prepare phosphate buffer (90 mM, pH 6.0) by adding 1.232 g of KH_2PO_4 and 10 mg of TSP (0.01%) to 100 mL of D_2O. Mix well to achieve complete dissolution of the components and then adjust the pH using 1.0 M NaOD.
2. NaOD solution: Prepare by adding 1 mL of NaOD (40%, 10 M) to 9 mL of D_2O and mix well.

3 Methods

Plants consist of a wide range of metabolites with different polarity, solubility, and concentration, which are significantly different from the metabolome of human, animals, and microorganisms. During chemical analysis, we recommend the use of identical protocols for all samples and replicates, which include constant sample volumes, same type of NMR tubes, identical shimming protocol, as well as spectral acquisition and processing parameters. Such protocols should avoid baseline and chemical shifts variations and minimize signal misalignment by adjusting the pH or ionic strength [31].

3.1 Sample Preparation

Sample preparation is a fundamental step in plant metabolomics. It affects the quality of acquired data and assignments as well as the accuracy of results of metabolic changes. The sample preparation procedure depends on the nature of the plant material being analyzed and the target biochemicals. The procedure typically involves harvesting, drying, extracting, and dissolving in deuterated solvent (s) [32]. If working with leaves, take into account the number of samples, leaf position, age, and conditions such as exposure to sunlight and rain, geography, and seasons and climate; also consider separating the vascular system from the mesophyll. Rapidly freeze the sample to prevent residual enzymatic activity.

A number of different sample preparation protocols have been evaluated previously for NMR-based metabolite analysis [32–35]. The protocol that we provide below achieves a broad coverage and structural information for the main constituents of plants extracts [33] and is intended for large-scale untargeted metabolomics.

1. Weigh dried samples (~50 mg) (*see* **Note 4**) into a 2 mL Eppendorf centrifuge tube (*see* Fig. 2 and **Note 5**).

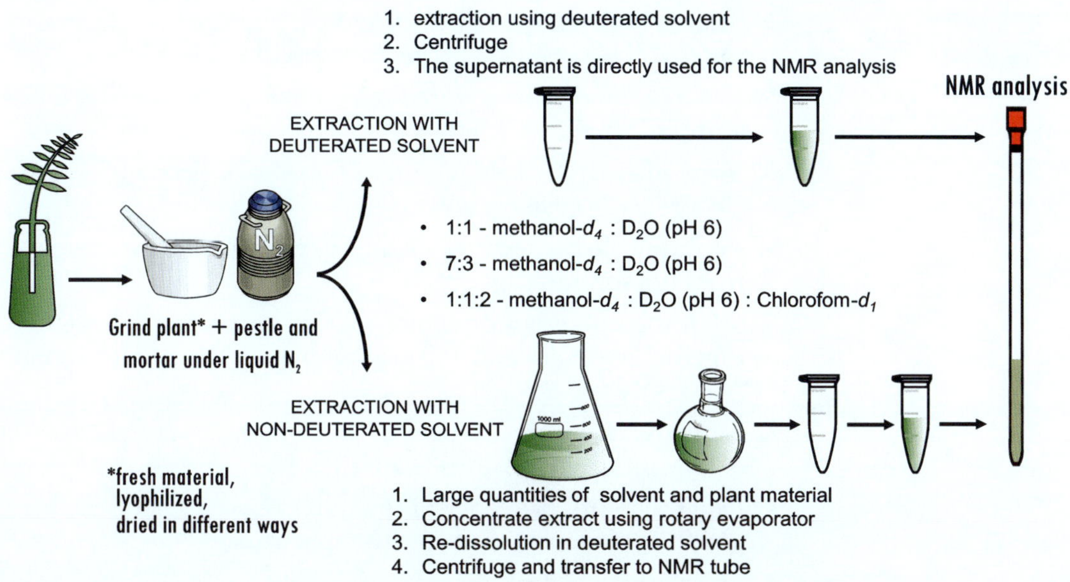

Fig. 2 General workflow for sample preparation for NMR-based plant metabolomics

2. Add 1.5 mL of deuterated solvent for metabolite extraction: we recommend 0.75 mL of CD_3OD and 0.75 mL of phosphate buffer in D_2O (pH 6.0, *see* **Notes 6–8**) containing 0.1% (wt/wt) of internal standard (IS) TSP (*see* **Notes 9** and **10**).
3. Vortex for 1 min at room temperature (20–25 °C).
4. Sonicate for 10–20 min at room temperature (20–25 °C).
5. Centrifuge at (14,000–17,000 × *g*) at room temperature for 5–10 min using a microtube to obtain a clear supernatant. For lower-speed centrifugation, more time is required to obtain a clear supernatant layer.
6. Transfer the supernatant (approx. 1 mL) to a 1.5 mL Eppendorf tube. Repeat centrifugation (17,000 × *g*, 1 min, room temperature) if the supernatant is not clear after **step 5**.
7. Transfer 600 μL of the supernatant to a 5 mm NMR tube or 180 μL to a 3-mm NMR tube.

3.2 NMR Measurements

NMR analysis must be performed as soon as possible to avoid degradation of metabolites. If NMR acquisition is focused on quantitative analysis, pulse sequence and experimental parameters need to be set accordingly [36–39]. Important steps prior to acquisition include setting the temperature, probe tuning and matching, shimming the magnet, pulse calibration, setting acquisition parameters, and receiver gain adjustment. Specific steps are given below:

1. Load the NMR tube into the spinner and introduce it in the spectrometer (*see* **Note 11**).

2. Select suitable pulse sequence and parameter for the appropriate NMR 1D or 2D experiment (*see* **Notes 12–17**).
3. Ensure sample temperature is equilibrated. Suggested temperature is 298 K (25 °C).
4. Tune and match the NMR tube. This can be performed manually or in automatic mode.
5. Perform field-frequency locking based on deuterium resonance from the solvent.
6. Shim the sample using either manual or automated method. For the first sample, manual shimming, tuning and matching may take up to 20 min.
7. Set up receiver gain, number of scans, dummy scans, relaxation delay, and spectral widths (*see* **Notes 12–17**).
8. Determine the frequency of the water resonance and set water peak on resonance (center of the spectrum).
9. Make a preliminary test on one sample using manual acquisition procedure.
10. Use these settings and start experiments in automation mode for the remaining samples.

3.3 ^1H 1D NMR Data Preprocessing

Data preprocessing is necessary to reduce the number of variables in the spectral data and to avoid peak picking issues, misalignment, and interferences. Numerous software tools are available for NMR processing (*see* **Note 18**).

For processing:

1. Apply exponential line broadening of 0.3 Hz and zero-fill to 64 k data points.
2. Apply Fourier transformation.
3. Perform phase correction and calibrate the spectrum using the internal standard or solvent residual signal (*see* **Note 19**).
4. Make baseline correction (*see* **Note 20**).
5. Perform peak picking and deconvolution (*see* **Note 21**).
6. Proceed with the alignment of the NMR signals (*see* **Note 22**).
7. Remove the solvent signals (intereferents).
8. Perform data binning/bucketing (*see* **Notes 23** and **24**).

3.4 2D NMR Data Preprocessing

1. *J*-resolved 2D data: zero-fill FID to 128 × 4096 points in F1 and F2, respectively, and apply a sine-bell-shaped window function (apodization) in both dimensions. Perform magnitude-mode two-dimensional Fourier transformation and tilt the spectrum along F2 by 45°. Correct baseline and calibrate the spectrum (TSP = 0.0 ppm).

2. ^{1}H–^{1}H COSY data: zero-fill data to 4096 × 4096 points, apply a sine-bell-shaped window function shifted by 90° in the F1 and 45° in F2 dimension. Perform Fourier transformation, phase and baseline correction and calibration (TSP = 0.0 ppm).
3. ^{1}H–^{13}C HMBC: spectral size 4096/1024; magnitude-mode Fourier transformation; use sine-bell-shaped window function shifted by 90° in F1 and 30° in F2 dimension. Calibrate spectrum (^{1}H and ^{13}C: TSP = 0.0 ppm; set ^{13}C = 49.0 ppm for CD_3OD when appropriate).
4. ^{1}H–^{13}C HSQC: spectral size 1024/1024; magnitude-mode Fourier transformation; use sine-bell-shaped window function shifted by 90° in F1 and 30° in F2 dimension. Calibrate spectrum (^{1}H and ^{13}C: TSP = 0.0 ppm; set ^{13}C = 49.0 ppm for CD_3OD when appropriate).

3.5 Multivariate Data Analysis

Multivariate data analysis (MVDA) reduces data complexity and dimensionality, prioritizes the relevant chemical features, and enables establishing the link between metabolite composition with plant phenotype. MVDA methods such as HCA, PCA, and PLS-DA are routinely used in NMR-based plant metabolomics studies (*see* **Note 25**).

1. Convert the processed ^{1}H NMR spectra to a suitable form for further multivariate analysis. AMIX software converts the raw data into ASCII format.
2. Mean centering, unit variance, autoscaling, or Pareto scaling methods are typically used for scaling (*see* **Note 26**).
3. Perform unsupervised chemometric analysis (PCA and HCA) using SIMCA-P software or equivalent (*see* **Note 27**).
4. Apply supervised methods such as PLS or O-PLS to enhance the separation among the different classes of samples after removing the outliers (*see* **Note 28**).
5. Avoid misinterpretation caused by model overfitting. Cross-validation of models such as jackknifing and permutation tests are required. Validated models should have $P > 0.98$ and $Q2 > 0.4$ (recommended).
6. Identify as many metabolites as possible based on loading values analysis. Identification can be done either by comparison with NMR signals to reference compounds or by two-dimensional NMR spectra (*see* **Note 29**).
7. Use statistical total correlation spectroscopy (STOCSY): It determines correlations and covariance across large datasets [29]. STOCSY generates a pseudo-two-dimensional NMR spectrum that exhibits correlations between various peaks in the NMR dataset.

3.6 Metabolite Annotation

1. For the annotation of previously known metabolites, compare signals from 1D and 2D spectra with reference compounds, or use NMR databases of primary and secondary metabolites (*see* **Notes 30**).
2. For the identification of novel metabolites, explore the 2D NMR measurements (*see* **Note 31**).
3. Ratio analysis of nuclear magnetic resonance spectroscopy (RANSY) can be applied to ^{1}H NMR data from plant crude extracts to select peaks from the same metabolite (*see* **Note 32**).
4. Use multiplicity, coupling constants and integral values for the peaks to assess the metabolite's identity.
5. Identified metabolites can be confirmed further using spin-selective experiments such as 1D TOCSY (*see* **Note 13**) and homonuclear decoupling (HOMODEC) to provide coupling information with high resolution and sensitivity.

3.7 Quantitative ^{1}H NMR

Experiments to determine the absolute quantitation of metabolites need to be performed using a different set of parameters [22, 40].

3.7.1 Data Acquisition

1. Prepare samples in deuterated solvent by dissolving each extract with a precise concentration of IS (*see* **Note 33**).
2. It is preferable to maintain the signal ratio of 1:1 between the IS and the peak to be quantified (*see* **Note 34**).
3. Insert sample into the magnet and perform probe tuning, matching, and shimming.
4. Perform pulse calibration using the pulsecal preform (Bruker).
5. Set up a one-pulse sequence with a 90° pulse and presaturation of the residual solvent signal (zgpr, Bruker) and optimize the spectral width.
6. Estimate the longitudinal relaxation (T_1) using the inversion-recovery pulse sequence (*see* **Note 35**).
7. Obtain ^{1}H NMR spectrum using the one-pulse sequence and optimized parameters (number of scans, receiver gain, dummy scans, spectral width), including a relaxation delay of at least $5 \times T_1$.

3.7.2 Data Processing

1. Apply an exponential function using the line broadening (LB) command of 0.3 Hz.
2. Zero-fill by a factor of 4.
3. Perform Fourier transformation, phase and baseline correction, and calibrate the spectrum.
4. Integrate the IS peak and plant signals that are used for quantitation.

5. The integral region on each side of the peak must cover 20 times the width at half-height to include 99% of the peak (10–20 Hz on each side) [54].
6. If the signal of interest is superimposed with an interfering signal, apply Global Spectral Deconvolution using pick peaking and deconvolution functions (GSD, Mestrenova, *see* **Note 36**) to all signals.

3.7.3 Quantification

1. The quantification is performed according to Eq. 1 in Subheading 4 (*see* **Note 37**) by comparing the integrals of the targeted metabolite and the IS.
2. Determine the purity (*P*) of the target analyte (*see* **Note 38**).
3. Validate the results by testing selectivity (*see* **Note 39**), metabolite stability (*see* **Note 40**), precision (*see* **Note 41**), accuracy (*see* **Note 42**), and the linearity response (*see* **Note 43**), according to previously described ^{1}H NMR protocols [41].

Alternatively, an electronic reference signal can be used for quantitation [42]. This is known as the ERETIC method (electronic reference to access in vivo concentrations). ERETIC2 is a commercial PULCON (pulse length–based concentration determination) methodology that digitally adds reference signals by an electronic device, enabling determination of absolute concentrations. Advantages of these methods are that no internal standard are needed and the reference signal frequency can be freely chosen. A calibration of the peak with an external standard solution is required for this approach.

4 Notes

1. Liquid nitrogen should be handled carefully. Gloves and eye glasses should be used for protection.
2. NMR experiments should be acquired at a constant temperature; ensure that the temperature has been calibrated using a standard sample.
3. After preparation, the deuterated buffer solutions can be stored at 4 °C.
4. For sample preparation using fresh material, make sure to inhibit the enzymatic activity by adding dry ice (−78 °C) or liquid nitrogen (−196 °C) into mortar containing the sample. Keep frozen samples at −80 °C or directly dry using a freeze dryer for 1–2 days. Grind the material with a cutting/impact mill. Use a small spatula to put the powdered sample into a centrifuge tube (2 mL) or a freezer plastic bag depending on the amount. Use a small paintbrush to clean the mill between samples.

5. NMR-based plant metabolomics have utilized two procedures for plant extraction: (a) direct extraction of vegetal material using a single or mixture deuterated solvents, followed by centrifugation and transfer of the supernatant to the NMR tube, and (b) extract using nondeuterated solvents, as shown in Fig. 2. The procedure using nondeuterated solvents is recommended to generate larger amounts of plant material, but it requires prior removal of the extraction solvents and reconstitution in deuterated solvent for NMR analysis.
6. A wide coverage of the plant metabolome is usually obtained through the combination of different solvents [43]. D_2O buffer with phosphate buffer, methanol-d_4, or their mixtures at variable proportion are used for polar extracts. Ratios of 1:1:2 or 7:3:0 methanol-d_4: D_2O buffer: $CDCl_3$ [19, 32, 44–48] are also used. Chloroform-d_1 is a widely used solvent for nonpolar extracts. Many other deuterated solvents can be used including DMSO-d_6, acetone-d_6, acetonitrile-d_3, benzene-d_6, and pyridine-d_5 depending on the needed polarity and metabolites class selectivity. Pay attention to solvent toxicity as well as potential solvent contaminants.
7. Usually the pH is controlled in aqueous-based extracts. ^1H signals in organic acids are sensitive to pH differences. Phosphate buffer (pH 6 and 7) [49] and oxalate (pH 4) [50] buffers are more often used to maintain the pH.
8. Aqueous extracts show high amount of free sugars and glycosides (~3–6 ppm) and chloroform extracts show high aliphatic hydrocarbon peaks (~1–3 ppm), which limit the dynamic range. Consider purifying using XAD (sugar) or C18 (lipids) cartridges, or fractionating methods such as two-phase extraction procedure [51]. Alternately, two-dimensional NMR, for example, *J*-resolved [12, 19, 20], HSQC, TOCSY, or HMBC [12], can be used to decrease such effects. Proteins can be depleted by removal using ultrafiltration [22].
9. Samples made using nondeuterated solvents can also be used for NMR with a small volume of deuterated solvent for locking. Such samples require suppression of the strong signals from solvents.
10. Tetramethylsilane (TMS), hexamethyldisiloxane (HMDSO), and derivatives can be used as internal standards for chemical shift reference. A known concentration of the internal standard allows to determine the absolute concentration of the metabolites.
11. NMR tubes should be handled gently while inserting into the spinner. Otherwise, the tube will break. Ensure that nitrogen gas is turned on when inserting the sample manually. Otherwise, it will cause major damage to the probe.

12. For standard 1D ^{1}H NMR spectroscopy, set the one-pulse sequence (recycle delay–90°–acquire). Generally, water signal suppression is performed by presaturation (Bruker pulse program: zgpr) or by excitation sculpting (Bruker pulse program: zgesgp). Set TD, 64 k; SW, 10 KHz (~20 ppm); temperature 298 K; 1.5 s recycle delay with presaturation turned on; transmitter frequency offset (O1) set on the water peak; and dummy scans (DS) 4. For a 50 mg dry material, typically 64–128 scans are set. ^{1}H NMR data acquisition takes about 2 to 10 min depending on the sample quantity.
13. For selective 1D TOCSY, use pulse sequence (seldigpzs, Bruker) that uses DIPSI-II isotropic mixing for coherence transfer. Set TD, 65 k; NS, 512; recycle delay (d1) 1.0 s; spectral width (SW) 20 ppm; and temperature 298 K.
14. *J*-resolved spectroscopy is a useful experiment as the chemical shifts and *J* coupling are separated into two independent dimensions [52]. It facilitates the visualization of multiplicity and coupling constant in the overlapped spectral regions and identification and quantification of metabolites. It also suppresses the signals from macromolecules that generally have short T_2 relaxation times. For this experiment, set the *J*-resolved pulse sequence (recycle delay-90°—[$t_1/2$]—180°—[$t_1/2$]-acquire) with water presaturation during a relaxation delay of 1.5 s. Use TD 16,384, 64 and SW 20, 0.15 ppm, in F2 and F1 dimension, respectively; NS 16; and temperature 298 K.
15. ^{1}H–^{1}H COSY is a useful 2D experiment for assignments. Use either the phase-sensitive or magnitude-mode experiment. Set parameters: presaturation during recycle delay of 1.5 s, 512 × 4096 points and SW 20/20 ppm in F1 × F2 dimensions, respectively. Acquisition mode: States-TPPI; NS 8; and temperature 298 K.
16. ^{1}H–^{13}C HSQC is an important 2D method for proton and carbon correlations. To perform this experiment, set acquisition parameters for ^{1}H (F2)/^{13}C (F1): TD 512/256; SW 20/230 ppm; NS 32; recycle delay 1.5 s; and temperature 298 K.
17. ^{1}H–^{13}C HMBC 2D experiment provides heteronuclear multiple bond correlations. To perform this experiment, set acquisition parameters for ^{1}H (F2)/^{13}C (F1): TD 2048/1024; SW 20/230 ppm; NS 64; recycle delay 1.5 s; and temperature 298 K.
18. Numerous software tools exist for data processing: *NMRProcFlow* is an open source package that greatly helps spectra processing and is particularly useful for metabolomics data analysis. *MestreNova* is a multivendor software suite designed

for NMR as well as LC/MS and GC/MS data. *Topspin (Bruker)* is free for academic use. It features data processing, displaying and analyzing single, and multidimensional spectra. *NMRPipe* offers data processing and analysis for multidimensional NMR data. *matNMR* is an extendable open source software which features functions such as spectral deconvolution or quadrupolar tensor fitting. It offers a viewer for 2D data and provides 2D and 3D contour, surface, and line plots in a wide range of formats. It provides an interface for processing, analyzing, and visualizing 1D and 2D NMR data. *NMRLAB in MATLAB* is a toolbox for NMR data processing in MATLAB. It features numerous functions including smoothing, zero filling, baseline correction, phase correction, linear prediction, and referencing. *AMIX* (Bruker) provides integrated MS and NMR routines for spectroscopic and statistical analyses. *Chenomx NMR Suite* enables identifying and quantifying metabolites in NMR spectra. It also features NMR data processing including phase correction, baseline correction, shim correction, water region deletion, CSI calibration, pH calibration, spectral binning, and spin simulation. *MetaboAnalyst* is a free online software for a variety of analyses including automated NMR spectral analysis, statistical analysis, biomarker analysis, and metabolic pathway analysis. The *ACD NMR Package* is commercial software, which enables NMR spectral analysis.

19. If Mestrenova is used for phase correction, set the pivot point on the reference signal. Perform zero-order phase correction for the signals near the pivot point, and first-order phase correction for other signals in the spectrum.
20. A polynomial fitting algorithm is usually used for baseline correction. It generally requires user intervention. Using adaptive iteratively reweighted Penalized Least Squares (airPLS) in R language or MATLAB [53] does not require user intervention. It works based on changing the weights of sum squares errors (SSE) between the fitted baseline and the original signals.
21. Global Spectral Deconvolution (GSD) in MestreNova utilizes generalized Lorentzian function to model and deconvolute overlapped signals [54].
22. For alignment, we recommend Interval Correlated Shifting (ICOShift) for MATLAB environment. It is an open source toolbox, provides full-resolution alignment of large datasets, and avoids downsampling steps such as binning [55].
23. "Binning" or "bucketing" reduces the number of variables in the spectrum. Use 0.04 ppm bin size [33]. Exclude buckets that have lower average integration than the signal-to-noise ratio.

24. "Binning" generally leads to loss of spectral resolution. Use "full-resolution" NMR data for statistical analysis [56] when necessary, considering that full-resolution spectra cause a significant increase in the data size (~30,000 per spectrum at ~0.15 Hz resolution).
25. Some important sources for MVDA tools are AMIX-TOOLS (Bruker Biospin GmbH), MATLAB with statistics toolbox (MathWorks, Natick, MA, USA), Minitab (Minitab, State College, PA, USA), Pirouette (Infometrix, Bothell, WA, USA), SIMCA-P (Umetrics, Umeå, Sweden), SPSS (SPSS, Chicago, IL, USA), Unscrambler (CAMO software, Woodbridge, NJ, USA), and MetaboAnalyst (Genome Canada, Canada).
26. In the mean-centering method, each scaled variable resembles the original spectra, and thus, mostly high abundance signals are considered for samples classification. In this scaling method, interpretability of loading plots improves; however, the influence of minor signals might be overlooked. In the unit variance method, each scaled variable has equal variance; thus, all signals show the same maximum intensity, independent of the original peak intensity. This scaling method is suitable for spectra with major and minor compounds; however, interpretation of the loading plots is usually more complex in this method. Autoscaling is the combination of mean centering and variance scaling. It is recommended when variables have different units. It might increase the contribution of noise variables to the analysis [57]. Pareto scaling uses the square root of the standard deviation as the scaling parameter and falls in between no-scaling and autoscaling. This method is more popular in metabolomics since it avoids overmanipulation of data [57].
27. Exploratory pattern recognition such as PCA and HCA groups samples based on the non-biased chemical features. It enables removal of potential outliers from subsequent analysis.
28. Supervised methods discriminate samples that have class information such as infected plants, healthy plants, and plants that exhibit biological activity. The results show NMR signals that contribute to maximum separation. The orthogonal filter incorporated to the PLS algorithm (O-PLS) allows removal of spectral components unrelated to sample classes. Methods such as linear discriminant analysis (LDA), independent component analysis (ICA), and multidimensional scaling (MDS) can also be applied.
29. To determine the physiological backgrounds for different classes of plants used in the study, establishment of the identity for the distinguishing NMR variables is critical.

30. Currently available NMR databases include Biological Magnetic Resonance Data Bank (BMRD: www.bmrb.wisc.edu), Human Metabolome Database (HMDB: www.hmdb.ca), NMR database of Linkoeping (MDL: http://www.liu.se/hu/mdl/main/), Magnetic Resonance Metabolomics Database (http://www.metabolomics.bioc.cam.ac.uk/metabolomics), Prime (http://prime.psc.riken.jp/), NMR Lab of Biomolecules (http://spinportal.magnet.fsu.edu), NMRShiftDB (www.ebi.ac.uk/NMRshiftdb/), DEREP-NP (https://github.com/clzani/DEREP-NP/wiki), Spectral Database for Organic Compounds (SDBS: www.riodb01.ibase.aist.go.jp), BioMagResBank (www.bmrb.wisc.edu), and NuBBE DB (http://nubbe.iq.unesp.br/portal/nubbe-search.html) [58].
31. Most primary metabolites and known secondary metabolites can be putatively identified using the following 2D NMR experiments: *COSY* shows correlations between scalar coupled hydrogen atoms; *HSQC* shows correlations between hydrogen atoms and the carbon atom to which they are directly attached; *HMBC* shows correlations between hydrogen atoms and carbon atoms that are up to three or four bonds away; *TOCSY* shows connectivity among all sequentially coupled spins; in the ***J-resolved spectrum***, chemical shift and coupling information are displayed along separate dimensions, which is especially useful for signals in crowded areas.
32. RANSY identifies all peaks of a specific metabolite in the NMR data based on the principle that the ratios between peaks from the same metabolite are constant and do not vary across samples unlike the metabolite concentrations [30].
33. Accurate weighing of both plant extract and internal standard (IS) is required. The IS should be highly pure, nonhydroscopic, nonvolatile, nonreactive, free of residual water, soluble in the deuterated solvent, and, preferably, have only a few signals. Moreover, peaks from IS should not overlap with peaks from plant extract.
34. Recommended IS compounds for chloroform solutions: hexamethyldisilane [48, 59] and pyridine [24]; for aqueous solutions: phloroglucinol [19], sodium trimethylsilyl propionate (TSP) [48, 59], and 4,4-dimethyl-4-silapentane-1-sulfonic acid [60].
35. The T_1 value can be determined using $\tau_{null} = T_1 \times \ln 2$, in which τ_{null} represents the time required for the inverted longitudinal magnetization to each of the x–y plane. To avoid the differential saturation effect, the spins should be fully relaxed between scans; set recycle delay of at least five times the T_1 of the slowest relaxing nuclei.

36. GSD is a fully automatic tool that filters relevant information, deconvolutes metabolite signals, and discards undesirable interferences in the ^{1}H NMR spectrum [54]. Recommended parameters for Mestrenova: 10 fitting cycles (refinement level 3) and high resolution.
37. The formula for determining the absolute quantity of a metabolite is given by

$$mg_{target} = \frac{PA_{target} \times ME_{target} \times mg_{internal\ std}}{ME_{internal\ std} \times PA_{internal\ std.}} \quad (1)$$

where mg = mass, PA = peak area, ME = equivalent mass (molecular mass divided by the number of nuclei associated with the quantified signal), target: target analyte/molecule and internal std.: internal standard [40].
38. Use the following equation to determine the purity (*P*) of the target analyte:

$$P\ (\%) = \frac{MW_{target} \times n_{IS} \times I_{IS} \times m_{IS}}{MW_{IS} \times n_{target} x\ I_{target} \times m_{target}} \quad (2)$$

where I = integral, MW = molecular weight, P = purity, m = mass, n = number of protons that each NMR signal corresponds to, target: target analyte/molecule and IS: internal standard.
39. The selectivity is evaluated by acquisition of a ^{1}H–^{13}C HSQC spectrum to ensure that the quantified signals are not overlapped.
40. Short-term metabolite stability is measured over a 48-h period by multiple ^{1}H NMR acquisitions, for example, every 2 h, to determine the loss, if any, due to degradation.
41. Precision is determined based on six measurements of the ^{1}H NMR signal over a short period of time by systematically changing parameters such as operator, automation method, and time.
42. Accuracy is evaluated by measurement of the IS by ^{1}H qNMR and three other validated quantification methods.
43. Linearity of the calibration curve is assessed using different concentrations of IS and extract. Calibration standard solutions of five concentrations (e.g., 1, 2, 4, 10, and 20 mM) can be prepared in triplicates.

Acknowledgements

The authors acknowledge support from the Fundação de Amparo à Pesquisa do Estado de São Paulo (FAPESP fellowship 2014/05935-0 and 2017/06466-2 to DMS, and 2016/13292-8 to ACP), and the University of Washington (to FCN).

References

1. Weng J-K (2014) The evolutionary paths towards complexity: a metabolic perspective. New Phytol 201:1141–1149
2. Hall R (2002) Plant metabolomics: the missing link in functional genomics strategies. Plant Cell 14:1437–1440
3. Kim HK, Choi YH, Verpoorte R (2011) NMR-based plant metabolomics: where do we stand, where do we go? Trends Biotechnol 29:267–275
4. Hong J, Yang L, Zhang D, Shi J (2016) Plant metabolomics: an indispensable system biology tool for plant science. Int J Mol Sci 17:767
5. Ortmayr K, Causon TJ, Hann S, Koellensperger G (2016) Increasing selectivity and coverage in LC-MS based metabolome analysis. TrAC Trends Anal Chem 82:358–366
6. Brunetti AE, Carnevale Neto F, Vera MC et al (2018) An integrative omics perspective for the analysis of chemical signals in ecological interactions. Chem Soc Rev 47:1574–1591
7. Funari CS, Castro-Gamboa I, Cavalheiro AJ, Bolzani V da S (2013) Metabolomics, an optimized approach for the rational exploitation of Brazilian biodiversity: state of the art, new scenarios, and challenges. Quim Nov 36:1605–1609
8. Fiehn O (2001) Combining genomics, metabolome analysis, and biochemical modelling to understand metabolic networks. Comp Funct Genomics 2:155–168
9. Hoffmann JF, Carvalho IR, Barbieri RL et al (2017) *Butia spp.* (Arecaceae) LC-MS-based metabolomics for species and geographical origin discrimination. J Agric Food Chem 65:523–532
10. Vargas LHG, Neto JCR, de Aquino Ribeiro JA et al (2016) Metabolomics analysis of oil palm (*Elaeis guineensis*) leaf: evaluation of sample preparation steps using UHPLC–MS/MS. Metabolomics 12: 153
11. Hantao LW, Aleme HG, Passador MM et al (2013) Determination of disease biomarkers in *Eucalyptus* by comprehensive two-dimensional gas chromatography and multivariate data analysis. J Chromatogr A 1279:86–91
12. Uarrota VG, Moresco R, Coelho B et al (2014) Metabolomics combined with chemometric tools (PCA, HCA, PLS-DA and SVM) for screening cassava (*Manihot esculenta* Crantz) roots during postharvest physiological deterioration. Food Chem 161:67–78
13. Garrett R, Schmidt EM, Pereira LFP et al (2013) Discrimination of arabica coffee cultivars by electrospray ionization Fourier transform ion cyclotron resonance mass spectrometry and chemometrics. LWT - Food Sci Technol 50:496–502
14. do Prado RM, Porto C, Nunes E et al (2018) Metabolomics and agriculture: what can be done? mSystems 3:e00156–e00117
15. Cardoso S, Maraschin M, Peruch LAM et al (2017) A chemometrics approach for nuclear magnetic resonance data to characterize the partial metabolome banana peels from southern Brazil. J Integr Bioinform 14:1–10
16. Roessner U (2001) Metabolic profiling allows comprehensive phenotyping of genetically or environmentally modified plant systems. Plant Cell 13:11–29
17. Padilla-González GF, Diazgranados M, da Costa FB (2017) Biogeography shaped the metabolome of the genus *Espeletia*: a phytochemical perspective on an andean adaptive radiation. Sci Rep 7:1–11
18. Liang YS, Choi YH, Kim HK et al (2006) Metabolomic analysis of methyl jasmonate treated *Brassica rapa* leaves by 2-dimensional NMR spectroscopy. Phytochemistry 67:2503–2511
19. Choi YH, Choi H-K, Hazekamp A et al (2003) Quantitative analysis of bilobalide and ginkgolides from *Ginkgo biloba* leaves and *Ginkgo* products using 1H-NMR. Chem Pharm Bull (Tokyo) 51:158–161
20. Sedio BE, Boya PCA, Rojas Echeverri JC (2018) A protocol for high-throughput, untargeted forest community metabolomics using mass spectrometry molecular networks. Appl Plant Sci 6:1–13
21. Pavarini DP, Da Silva DB, Carollo CA et al (2012) Application of MALDI-MS analysis of rainforest chemodiversity: a keystone for biodiversity conservation and sustainable use. J Mass Spectrom 47:1482–1485
22. Pauli GF, Gödecke T, Jaki BU, Lankin DC (2012) Quantitative1H NMR. Development and potential of an analytical method: an update. J Nat Prod 75:834–851
23. Markley JL, Brüschweiler R, Edison AS et al (2017) The future of NMR-based metabolomics. Curr Opin Biotechnol 43:34–40
24. Dagnino D, Schripsema J (2005) 1H NMR quantification in very dilute toxin solutions: application to anatoxin-a analysis. Toxicon 46:236–240
25. Larive CK, Jayawickrama D, Orfi L (1997) Quantitative analysis of peptides with NMR spectroscopy. Appl Spectrosc 51:1531–1536

26. Szasz PL (2009) Thought suppression, depressive rumination and depression: a mediation analysis. J Cogn Behav Psychother 9:199–209
27. Mompeán M, Sánchez-Donoso RM, de la Hoz A et al (2018) Pushing nuclear magnetic resonance sensitivity limits with microfluidics and photo-chemically induced dynamic nuclear polarization. Nat Commun 9:1–8
28. Simmler C, Napolitano JG, McAlpine JB et al (2014) Universal quantitative NMR analysis of complex natural samples. Curr Opin Biotechnol 25:51–59
29. Phansalkar RS, Simmler C, Bisson J et al (2017) Evolution of quantitative measures in NMR: quantum mechanical qHNMR advances chemical standardization of a red clover (*Trifolium pratense*) extract. J Nat Prod 80:634–647
30. Wei S, Zhang J, Liu L et al (2011) Ratio analysis nuclear magnetic resonance spectroscopy for selective metabolite identification in complex samples. Anal Chem 83:7616–7623
31. Defernez M, Colquhoun IJ (2003) Factors affecting the robustness of metabolite fingerprinting using 1H NMR spectra. Phytochemistry 62:1009–1017
32. Kim HK, Verpoorte R (2010) Sample preparation for plant metabolomics. Phytochem Anal 21:4–13
33. Kim HK, Choi YH, Verpoorte R (2010) NMR-based metabolomic analysis of plants. Nat Protoc 5:536–549
34. Schripsema J (2010) Application of NMR in plant metabolomics: techniques, problems and prospects. Phytochem Anal 21:14–21
35. Kruger NJ, Troncoso-Ponce MA, Ratcliffe RG (2008) 1H NMR metabolite fingerprinting and metabolomic analysis of perchloric acid extracts from plant tissues. Nat Protoc 3:1001–1012
36. Bernstein MA, Sýkora S, Peng C et al (2013) Optimization and automation of quantitative NMR data extraction. Anal Chem 85:5778–5786
37. Saude EJ, Slupsky CM, Sykes BD (2006) Optimization of NMR analysis of biological fluids for quantitative accuracy. Metabolomics 2:113–123
38. Goudarzi H, Mirsamadi ES, Ghalavand Z et al (2015) Molecular detection of metallo-beta-lactamase genes in clinical isolates of *Acinetobacter baumannii*. J Pure Appl Microbiol 9:145–151
39. Bharti SK, Sinha N, Joshi BS et al (2008) Improved quantification from1H-NMR spectra using reduced repetition times. Metabolomics 4:367–376
40. Pauli GF, Chen SN, Simmler C et al (2014) Importance of purity evaluation and the potential of quantitative1H NMR as a purity assay. J Med Chem 57:9220–9231
41. Malz F, Jancke H (2005) Validation of quantitative NMR. J Pharm Biomed Anal 38:813–823
42. Akoka S, Barantin L, Trierweiler M (1999) Concentration measurement by proton NMR using the ERETIC method. Anal Chem 71:2554–2557
43. Pilon AC, Carnevale Neto F, Freire RT et al (2016) Partial least squares model and design of experiments towards the analysis of the metabolome of *Jatropha gossypifolia* leaves: extraction and chromatographic fingerprint optimization. J Sep Sci 39:1023–1030
44. Hendrawati O, Yao Q, Kim HK et al (2006) Metabolic differentiation of *Arabidopsis* treated with methyl jasmonate using nuclear magnetic resonance spectroscopy. Plant Sci 170:1118–1124
45. Widarto HT, van der Meijden E, Lefeber AWM et al (2006) Metabolomic differentiation of *Brassica rapa* following herbivory by different insect instars using two-dimensional nuclear magnetic resonance spectroscopy. J Chem Ecol 32:2417–2428
46. Sánchez-Sampedro A, Kim HK, Choi YH et al (2007) Metabolomic alterations in elicitor treated *Silybum marianum* suspension cultures monitored by nuclear magnetic resonance spectroscopy. J Biotechnol 130:133–142
47. Simoh S, Quintana N, Kim HK et al (2009) Metabolic changes in *Agrobacterium tumefaciens*-infected *Brassica rapa*. J Plant Physiol 166:1005–1014
48. Leiss KA, Choi YH, Abdel-Farid IB et al (2009) NMR metabolomics of thrips (*Frankliniella occidentalis*) resistance in senecio hybrids. J Chem Ecol 35:219–229
49. Aranı'bar N, Ott K-H, Roongta V, Mueller L (2006) Metabolomic analysis using optimized NMR and statistical methods. Anal Biochem 355:62–70
50. Pereira GE, Gaudillere JP, Leeuwen CV et al (2006) 1H NMR metabolite fingerprints of grape berry: comparison of vintage and soil effects in Bordeaux grapevine growing areas. Anal Chim Acta 563:346–352
51. Schripsema J, Dagnino D (2018) Two-phase extraction for comprehensive analysis of the plant metabolome by NMR. In: Theodoridis GA, Gika HG, Wilson ID (eds) Metabolic profiling: methods and protocols. Springer, New York, NY, pp 195–202

52. Huang Y, Zhang Z, Chen H et al (2015) A high-resolution 2D J-resolved NMR detection technique for metabolite analyses of biological samples. Sci Rep 5:1–9
53. Zhang ZM, Chen S, Liang YZ (2010) Baseline correction using adaptive iteratively reweighted penalized least squares. Analyst 135:1138–1146
54. Cobas C, Seoane F, Sýkora S (2008) Global spectral deconvolution (GSD) of D-NMR spectra. Poster SMASH Conf 15706
55. Savorani F, Tomasi G, Engelsen SB (2010) Icoshift: a versatile tool for the rapid alignment of 1D NMR spectra. J Magn Reson 202:190–202
56. Selegato DM, Freire RT, Tannús A, Castro-Gamboa I (2016) New dereplication method applied to NMR-based metabolomics on different *Fusarium* species isolated from rhizosphere of *Senna spectabilis*. J Braz Chem Soc 27:1421–1431
57. Mortazavi-Tabatabaei SA, Fathi F, Ektefa F et al (2013) Investigation of metabonomics technique by analyze of NMR data, which method is better? Mean center or auto scale? J Paramed Sci 4:2–9
58. Pilon AC, Valli M, Dametto AC et al (2017) NuBBEDB: an updated database to uncover chemical and biological information from Brazilian biodiversity. Sci Rep 7:7215
59. Choi HK, Choi YH, Verberne M et al (2004) Metabolic fingerprinting of wild type and transgenic tobacco plants by1H NMR and multivariate analysis technique. Phytochemistry 65:857–864
60. Weljie AM, Newton J, Mercier P et al (2006) Targeted profiling: quantitative analysis of 1H NMR metabolomics data. Anal Chem 78:4430–4442

Part IX

Ultrafast and Ultrahigh-Resolution Methods

Chapter 20

Fast Quantitative 2D NMR for Untargeted and Targeted Metabolomics

Estelle Martineau and Patrick Giraudeau

Abstract

Nuclear magnetic resonance (NMR) spectroscopy exhibits a great potential for the quantitative analysis of complex biological samples such as those encountered in metabolomics. To overcome the ubiquitous problem of overlapping peaks in 1D NMR spectra of complex mixtures, acquisition of 2D NMR spectra allows a better separation between overlapped resonances while yielding accurate quantitative data when appropriate analytical protocols are implemented. The experiment duration can be made compatible with high-throughput studies on large sample collections by relying on fast acquisition methods. Here, we describe the general metabolomics workflow to acquire fast quantitative 2D NMR data with a focus on targeted or untargeted analyses.

Key words 2D NMR, Quantitative analysis, Metabolomics, Targeted, Untargeted, Fast methods

1 Introduction

Nuclear magnetic resonance (NMR) spectroscopy is a major analytical technique in the metabolomics workflow, thanks to its ability to provide both structural and quantitative information in a highly repeatable and nondestructive fashion. In numerous metabolomics studies, NMR is used as an untargeted method to record a fingerprint of complex biological samples (biofluids, extracts, food matrices, etc.), but NMR also performs as a highly accurate tool to determine the concentrations of multiple analytes in such complex samples. Unfortunately, 1D ^{1}H NMR—the far most widely used approach—suffers from the overlap of numerous peaks that limit the efficiency of NMR as a quantitative tool in metabolomics. Such overlaps are particularly exacerbated by the great diversity of metabolite concentrations within biological samples and also by issues related to the solvent peak. As a consequence, the quantification of targeted compounds can be difficult or even impossible, when no isolated peak can be found for the analyte of interest. Even in untargeted approaches that benefit from the discrimination

G. A. Nagana Gowda and Daniel Raftery (eds.), *NMR-Based Metabolomics: Methods and Protocols*, Methods in Molecular Biology, vol. 2037, https://doi.org/10.1007/978-1-4939-9690-2_20, © Springer Science+Business Media, LLC, part of Springer Nature 2019

capabilities of chemometrics, the separation between sample groups is not always optimum when the spectra are too crowded.

In this context, numerous approaches based on 2D NMR have recently been introduced in the field of metabolomics [1]. Beyond their well-known potential for the elucidation of molecular structures, 2D NMR experiments can significantly increase the separation between overlapping peaks, thus providing a powerful tool for untargeted metabolomics as well as for the targeted quantification of metabolites in biological samples. A major advantage of 2D NMR experiments is that a broad variety of homonuclear or heteronuclear pulse sequences are available, offering a diversity of tools characterized by a different level of performance in terms of resolution and sensitivity. However, 2D NMR experiments suffer from two main drawbacks that make its use for quantitative analysis not straightforward.

The first peculiarity of 2D NMR experiments is that they rely on multi-pulse sequences that create a peak-dependent analytical response of the nuclei [2]. In other words, 2D NMR peak volumes are still proportional to metabolite concentrations—as it is the case in 1D NMR—but the coefficient of proportionality between the peak volume and the concentration is different for each peak in the spectrum. Indeed, this coefficient of proportionality depends on position-specific factors such as transverse relaxation times and homo/hetero-nuclear *J*-coupling constants. For untargeted metabolomics, where relative peak intensities are compared from one sample to another, 2D peak volumes can still be used for the intercomparison of samples, provided that all the samples are prepared in similar conditions [3]. This strategy, named as "untargeted approach," will be detailed in this chapter. For targeted quantification, several strategies have been developed to yield accurate metabolite concentrations from 2D spectra. These include the development of specific pulse sequences leading to a peak-independent coefficient of proportionality between the 2D NMR signal and the concentration [4, 5]. These methods, while promising, are still limited in terms of accuracy, yielding an average error of 10%. An alternative consists in calibrating the 2D peak response as a function of concentration for all the targeted analytes, relying on an analytical chemistry approach such as an external calibration [6] or a standard addition procedure [7]. All the targeted analytes can be quantified simultaneously provided that commercial standards of known purity are available, which is often the case for major metabolites in biological samples (small amino or organic acids, sugars, lipids, etc.). This approach yields a very high accuracy and precision (errors of a few percent). The external calibration approach is potentially less accurate than the standard addition method since it does not account for matrix effects, but standard additions are very time-consuming (each sample must be spiked with a mixture of standards) while the external calibration is compatible with the high-throughput analysis of a large series of samples [8]. Both methods are described in this chapter, gathered under the name "targeted approaches."

The second limitation of 2D NMR experiments is the associated experiment duration, inherent to the need to repeat numerous 1D experiments in order to obtain 2D spectra with a sufficient resolution. The experiment duration can reach several hours per spectrum, which is incompatible with most metabolomics studies involving large sample collections. In addition, long experiments are more sensitive to spectrometer instabilities, which affects the precision of quantitative measurements. Fortunately, numerous approaches have been suggested to reduce the experiment time [9], and two of them (nonuniform sampling and ultrafast NMR) have already been used in a quantitative metabolomics context [6, 10–12].

This chapter describes a general protocol to perform fast quantitative 2D NMR experiments in a metabolomics context. It includes the sample preparation, data acquisition, and preprocessing steps, as well as the data processing. Figure 1 shows the corresponding workflow; the initial part of the workflow is common to both targeted and untargeted approaches, whereas the later part shows the workflow for targeted and untargeted approaches, separately. The two approaches are described separately in the Subheading 3.

For both targeted and untargeted methods, the workflow can be applied with any 2D NMR pulse sequence. However, we recommend a set of three fast 2D NMR pulse sequences which proved to be efficient in the context of quantitative metabolomics. Figure 2 illustrates the potential of these pulse sequences to separate overlapping resonances in complex metabolic mixtures. The first one is the ultrafast (UF) COSY pulse sequence [13], which can yield accurate results in a very short time (typically a few min). The second one is the zero-quantum filtered TOCSY (ZQF-TOCSY) pulse sequence recorded with nonuniform sampling (NUS) [10, 14]. It is longer than the UF COSY, but also more sensitive and with a higher resolution. Finally, the third approach is the ^{1}H–^{13}C HSQC experiment recorded with NUS conditions [10]. At natural ^{13}C abundance, ^{1}H–^{13}C HSQC is far less sensitive than the other two 2D methods but has a much higher ability to separate overlapping peaks. The choice of the pulse sequence will therefore depend on the best compromise between resolution and sensitivity, depending on the sample concentration, on the hardware configuration, and on the targeted metabolites.

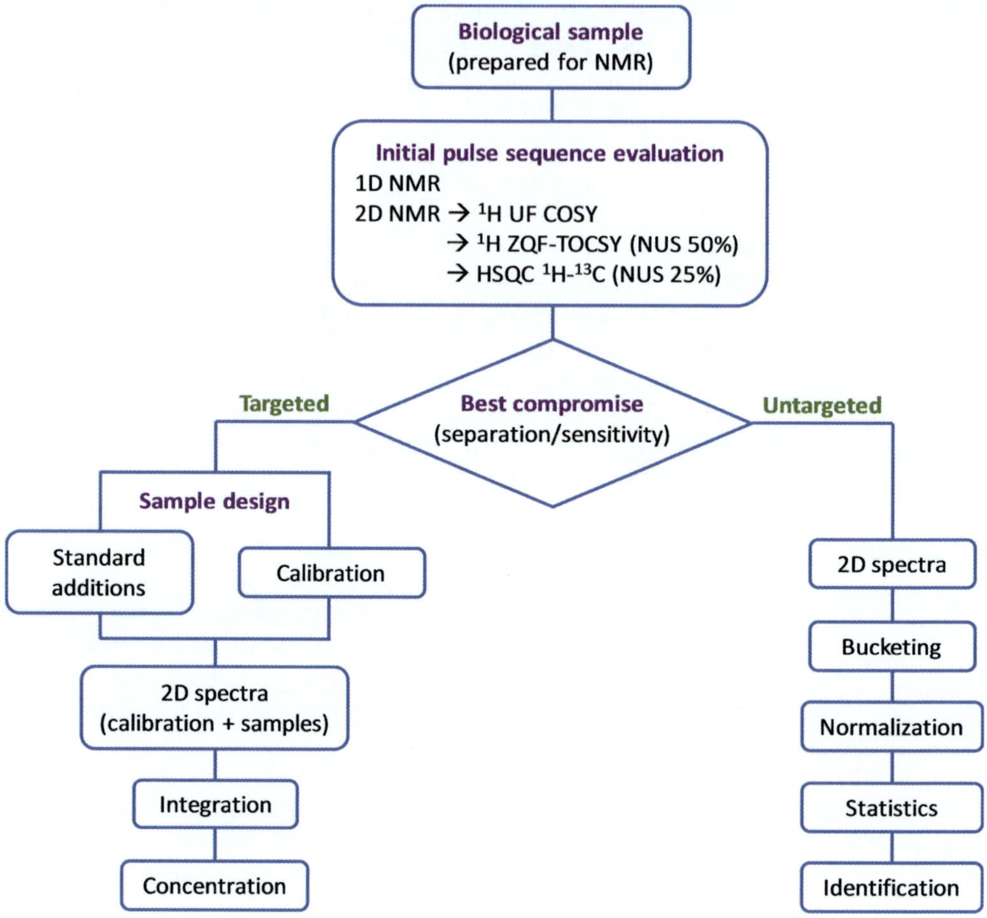

Fig. 1 Schematic diagram of the targeted and untargeted metabolomics approaches using fast 2D NMR

2 Materials

2.1 Samples

1. Suitable liquid-state biological samples under study such as urine, serum, and plasma, with volumes of approximately 500 μL.
2. Solid-state samples such as tissue or cells (approximately 20–100 mg of material) are to be extracted using organic solvents.
3. Synthetic mixtures of metabolites are also suitable for development work.

2.2 Chemical and Solvents

Solvents suitable for extraction of metabolites, such as methanol and/or chloroform, deuterated solvents to reduce solvent signals and/or provide a locking signal such as D_2O, CD_3OD, or $CDCl_3$. Internal NMR standards such as trimethylsilyl propionate (TSP), trimethylsilane (TMS), or dimethyl sulfone (DMSO2). Phosphate buffer (100 mM) is useful for pH control.

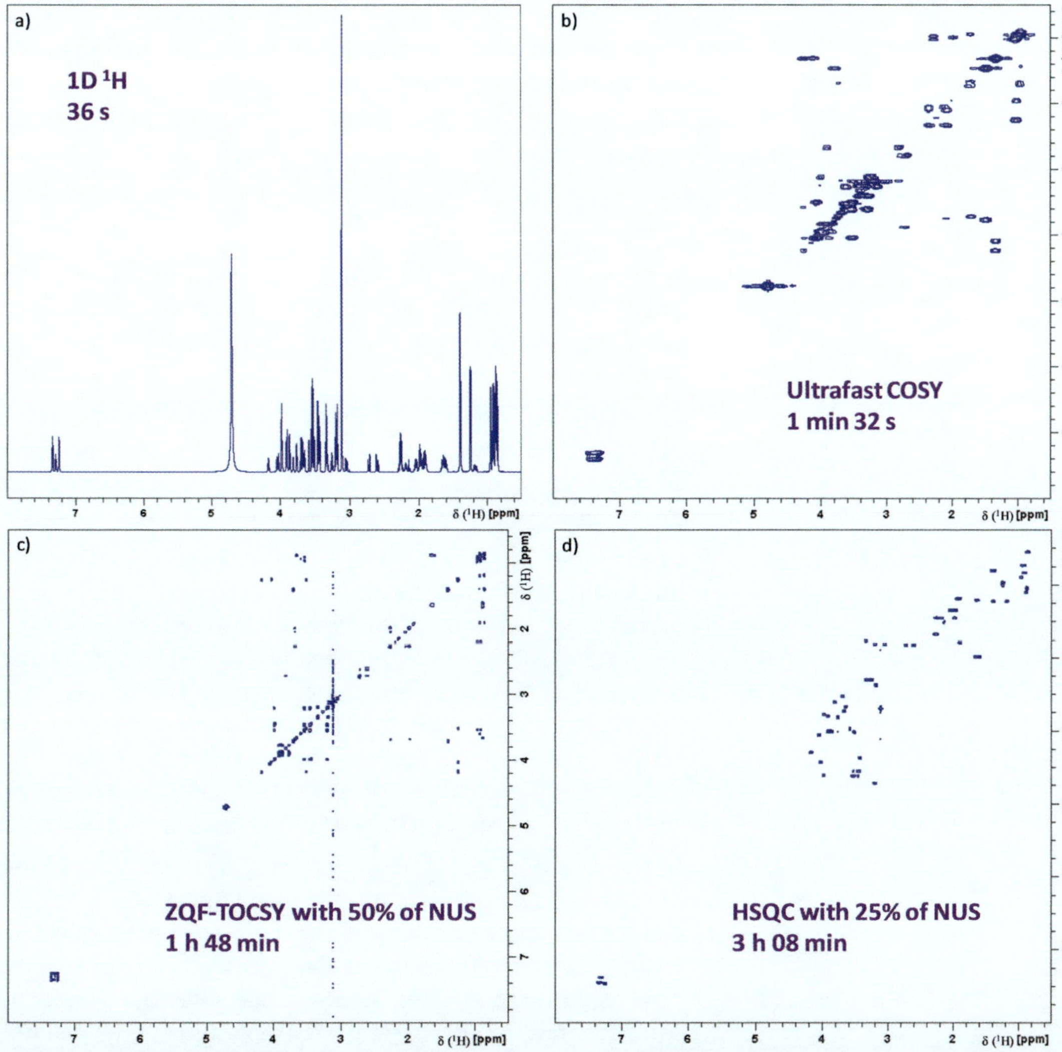

Fig. 2 Typical fast 2D NMR spectra obtained on a model mixture of metabolites: (**a**) 1D ^{1}H spectrum, (**b**) UF COSY, (**c**) ZQF-TOCSY with 50% NUS, and (**d**) ^{1}H–^{13}C HSQC with 25% NUS

2.3 Equipment

1. High-field NMR spectrometer with a probe equipped with *z*-axis gradients (*see* **Note 1**).
2. If possible, auto-sampler (*see* **Note 2**).
3. 5 mm NMR tubes (*see* **Note 3**).

3 Methods

3.1 Sample Preparation for NMR

1. Collect and prepare the biological samples for NMR analysis following protocols described in the literature, depending on the type of sample analyzed (urine, blood, extracts, tissues, etc.) (*see* **Note 4**) [15].

2. If possible, prepare some quality control samples (QCs, e.g., 3 or 4), consisting of all the collected biological samples pooled in identical quantities and mixed together.
3. Store the samples between −80 °C and −20 °C before NMR preparation.
4. Thaw the samples (15–30 min) and add NMR solvent(s) into the pill box [15].
   (a) If the samples are in the liquid state and if possible, use a ratio of 80%/20% for sample/deuterated solvent (and/or phosphate buffer) for a total volume of 700 μL (*see* **Note 5**).
   (b) If the samples are in the solid state (powder), add 700 μL of deuterated solvent (and/or phosphate buffer) or a mix of deuterated and nondeuterated solvents respecting the proportions of 80%/20% (*see* **Note 6**).
   (c) The addition of a reference is recommended to calibrate the chemical shift scale for all the spectra [8]—this is particularly important for the untargeted approach to ensure that the spectral peaks are aligned for the bucketing purpose (*see* **Note 7**). The concentration of the reference should be chosen to set its NMR signal intensity to a value smaller or equal to that of the most intense signal of interest (*see* **Note 8**). Choose a reference whose chemical shift does not overlap with a signal of interest (*see* **Note 9**).
   (d) In the case of the targeted approach, and after identifying the metabolites of interest (*see* below, Subheading 3.5.1), a stock solution containing the targeted metabolites in known concentrations has to be prepared. In the specific case of standard additions, this stock solution should be concentrated in order to add a small volume (≤10 μL) not to dilute analyzed samples.
5. Vortex and put the samples inside the NMR tubes.
6. Store samples at −20 °C if needed before analysis.
7. For the analysis of the whole cohort of samples:
   (a) If auto-sampler is not used, do not thaw all samples at the same time, but only the sample which has to be analyzed, usually between 15 and 30 min before vortexing and insertion in the magnet.
   (b) If a refrigerated auto-sampler is used, thaw all samples for 15–30 min. Then, vortex them and put them inside the auto-sampler. If possible, store the samples at 4 °C in the parking part of the auto-sampler prior to analysis (*see* **Note 10**).

3.2 Initial Evaluation (See Note 11)

1. Defrost one of the samples for 15 to 30 min. If possible, choose a sample which could be representative, for example, a control sample or a QC.
2. Vortex and put the sample inside the magnet.
3. Choose the temperature of analysis according to the nature of the sample under investigation and/or the chosen solvent (*see* **Note 12**).
4. Wait for 10 min to ensure the thermal stability of the sample.
5. Adjust the field-frequency lock and the magnetic field homogeneity (lock and shim).
6. Acquire 1D ^{1}H spectra in order to verify the quality of shimming and to determine important parameters needed before acquiring 2D homonuclear spectra:
   (a) Adapt the spectral width and the frequency offset depending on the chemical shifts of the biological samples.
   (b) Determine if a saturation of the solvent signal is necessary or not. Solvent suppression is generally required if the solvent signal is significantly intense compared to the peaks of interest and/or overlaps with targeted peaks. If a solvent saturation must be applied, different methods are available for metabolomics [16]. The parameters of the solvent suppression pulse sequence should be adapted to decrease the solvent signal as much as possible without disturbing the baseline and affecting the nearby peaks. Try to set the solvent signal intensity to a value smaller or equal to that of the most intense signal of interest to ensure that residual solvent signal intensity is not the limiting factor for receiver gain adjustment. Moreover, adjust the saturation frequency offset to the exact position of the solvent peak (*see* **Note 13**).
   (c) Calibrate the duration of the high power 90° pulse and use the same value for all the experiments (*see* **Note 14**).
   (d) Calibrate the power level corresponding to the duration of a low power 90° pulse of 30 μs. This low power pulse will be used during the spin lock of the ZQF-TOCSY pulse sequence.
7. Acquire 1D ^{13}C spectra (with ^{1}H decoupling) in order to determine important parameters needed before acquiring 2D HSQC spectra:
   (a) Adjust the spectral width and the frequency offset depending on the chemical shifts of the biological samples (*see* **Note 15**).
   (b) Calibrate the duration of the high power 90° pulse.

(c) Calibrate the power level corresponding to the duration of a decoupling pulse of 70 μs. This pulse duration will be needed for decoupling ^{13}C nuclei in the HSQC pulse sequence.

8. Acquire several types of 2D spectra and choose the most appropriate one for the study. Numerous pulse sequences are available; we recommend to record at least one homonuclear 2D ^{1}H experiment, with or without solvent saturation, an ultrafast COSY (≤30 min, *see* Fig. 2b) or a zero-quantum filtered TOCSY (ZQF-TOCSY, *see* Fig. 2c) with 50% nonuniform sampling (NUS), as well as a heteronuclear ^{1}H–^{13}C HSQC experiment with 25% of NUS (*see* Fig. 2d).

   (a) The acquisition and processing parameters for the UF COSY experiment are described in the protocol of implementation of ultrafast 2D NMR experiments, available online [17]. This protocol provides various steps to implement ultrafast 2D NMR experiments; in particular, important parameters that need to be optimized and/or used to acquire spectra are listed. It also highlights some cautions for parameter settings. More specifically, pay attention to the duration of experiments; the duration cannot exceed 30 min in order to avoid any damage to the NMR probe. Choose the number of scans accordingly. The receiver gain can be automatically calculated provided that a relaxation delay of at least 5 s is used.

   (b) Detailed acquisition and processing parameters for the ZQF-TOCSY with 50% NUS are described in recent articles [10, 14]. The classical TOCSY pulse sequence is improved by the addition of a zero-quantum filter to avoid zero-quantum coherences and to obtain better sensitivity. The use of 50% NUS allows faster acquisition without losing sensitivity appreciably (*see* **Note 16**).

   (c) Detailed acquisition and processing parameters for the HSQC ^{1}H–^{13}C pulse sequence with 25% NUS are described in the literature [10]. The classical HSQC pulse sequence is improved by the use of 25% NUS (*see* **Note 17**).

   Adjust the acquisition and processing parameters (*see* **Note 18**) to the analyzed samples. Pay particular attention to the following:

   (a) Receiver gain: the receiver gain can be automatically calculated after setting a relaxation delay of at least 5 s for UF COSY. For NUS experiments, use the value optimized for 1D acquisitions.

   (b) Number of scans: the number of scans has to be a multiple of 2 for all the pulse sequences mentioned above and must be optimized to achieve a signal-to-noise ratio (SNR) greater than or equal to 10 for the signals of interest (limit of quantification, or LOQ). For UF COSY, this value must take into account the time limitation of 30 min.

(c) Relaxation delay, which has to be optimized to take into account the time constraint for analysis of the samples and the need to acquire data with or without complete relaxation.

(d) Acquisition time, which is calculated automatically for UF COSY experiments. For ZQF-TOCSY experiments, Thrippleton et al. recommend a value of 1 s [14]. For HSQC experiments, the acquisition time must be between 0.3 and 0.5 s.

Thanks to this initial evaluation, the values of the receiver gain and the number of scans are now optimized for the untargeted approach. These values must be adjusted respectively on the most concentrated sample and the less concentrated one for the targeted approach (*see* below, Subheading 3.4.2).

3.3 Choice of the NMR Pulse Sequence(s)

1. The choice of the pulse sequence(s) depends on the biological question, the needed resolution, and sensitivity. The time needed should also be compatible with high-throughput studies, that is, analyzing a maximum number of samples in the minimum amount of time. Experimental design and pulse sequence selection also depends on the following:
   (a) Desired information: Untargeted profiling or analysis of a limited number of high concentration metabolites using the targeted approach.
   (b) In the case of the untargeted approach, experimental design should allow the detection of highest number of metabolites with sufficient sensitivity in the shortest time.
   (c) In the case of the targeted approach, the focus should be on high resolution and sensitivity so that at least one signal from each metabolite of interest is well separated to use for quantification.
2. Select the most appropriate pulse sequence according to the criteria mentioned above.
3. Create a file containing the acquisition and processing parameters which will be recalled for the analysis of the cohort.
4. Record also a shim file for later use.

3.4 Untargeted Approach

3.4.1 2D Spectra

1. Thaw the samples (15–30 min) and vortex. Thaw only one sample at a time if an auto-sampler cannot be used.
2. Insert the first sample inside the magnet (*see* **Note 19**).
3. Wait 10 min to ensure the thermal stability of the sample.
4. Perform field-frequency locking (automatic).
5. Recall the previously optimized shim file (saved during the initial evaluation). Run a 1D experiment to verify the quality of shimming.
6. Run a 2D experiment by calling the acquisition and processing parameter file recorded during the initial evaluation.
7. Use the same procedure to each sample of the cohort (**steps 1–6** in this section). When QCs are used, record QC spectra periodically, for instance, after every 5 or 10 study samples.
8. Spectral processing: multiply the FID by the appropriate apodization functions and then perform the Fourier transformation. If needed, adjust the phase manually in each dimension. Then, apply baseline correction in both dimensions (*see* **Note 20**). Calibrate all the spectra using the chemical shift of the internal reference.

3.4.2 Bucketing

1. Perform the bucketing of all the signals of interest with SNR $\geq$ 10 by using an appropriate software package, such as AMIX, Topspin, or MestReNova. For homonuclear experiments, the bucketing can also be done on diagonal signals which are well separated from the others. Draw boxes—for instance, rectangles with AMIX or Topspin, or circles with MestReNova—around the signals of interest (*see* **Note 21**). With AMIX software, the buckets on all spectra can be defined at the same time. For Topspin or MestReNova software, the buckets must be defined first on one of the spectra, then applied to others. Name and save the corresponding file, then recall it on each spectrum to pick up the values of integration areas (*see* **Note 22**).
2. Once all the buckets have been drawn, collect in an Excel file the integration values associated to each bucket "as they are", i.e. without applying any normalization at this stage.

3.4.3 Normalization

In order to compare data from different spectra, each integrated bucket can be normalized. If an internal reference is used, the data can be normalized to the bucket of the reference peak. Otherwise, the data for each spectrum can be normalized using the total sum of the integrated buckets. Other normalization methods exist and are described in the literature [8].

3.4.4 Statistics

Numerous statistical analysis software packages are available for analysis of the obtained data including SIMCA-P and MetaboAnalyst.

1. Perform principal component analysis (PCA) on all the samples in order to check the quality of the data. Note the percentage associated with each principal component (PC) on the score plot: the first two components generally explain the majority of the variability of the data. Observe if a separation trend or cluster appears between the samples (e.g., between control samples and treated samples). If the groups are separated, try to identify which PC can explain the discrimination between them. For studies containing QCs, the PCA score plot must show clustered QCs to validate the quality of the data (repeatability and reproducibility). The QCs can be removed after the verification.
2. Analyze the data with a supervised multivariate method such as two-component partial least squares discriminant analysis (PLS-DA). Verify the performance of the model by checking the following values: R^2 – which must be as close as possible to 1– and Q^2 – which must be the largest possible. Verify also the robustness of the model by applying a permutation test and a CV-ANOVA (p-value ≤ 0.05). Use a loading plot or the variable importance in the prediction (VIP) to determine which buckets can explain the discrimination between groups.
3. Use metabolomics databases (MMCD, HMDB, own database) or standard addition to identify the signal(s) of interest. Sometimes, signals of some metabolites can be overlapped. Applying a symmetrization on a homonuclear spectrum can help to identify the metabolites, but the symmetrized spectrum must not be used for bucketing.

3.5 Targeted Approach

3.5.1 Design of Standard Samples

Determine the mixture of standards that should be prepared to quantify the metabolites of interest. First, choose the metabolites which must be quantified. Each targeted metabolite should exhibit at least one isolated peak with SNR $\geq$ 10 in the 2D spectrum. Two main methods can then be considered: external calibration or standard addition (*see* **Note 23**):

(a) For external calibration: Prepare a mixture of the targeted metabolites. This mixture will serve as a stock solution. Therefore, metabolite concentration in this mixture must be the highest in the calibration range. Using this stock solution, prepare three diluted solutions. Try to make sure that the concentration of the metabolite to be measured falls within the external calibration range (*see* **Note 24**).

(b) For standard addition: Prepare a stock solution containing the target metabolites in known concentrations. This solution must be concentrated so that a small volume (≤10 μL) can be added without diluting the samples. Addition of 10 μL of stock solution at a concentration of 0.5 mmol/L will be appropriate in most cases. Standard addition is useful for analysis of small cohorts of samples.

3.5.2 2D Spectra (Calibration + Samples)

1. Thaw the samples (15–30 min) and vortex. Thaw only one sample at a time if an auto-sampler cannot be used.
2. Insert the first sample inside the magnet (*see* **Note 19**).
3. Wait for 10 min to reach thermal equilibrium of the sample.
4. Perform field-frequency locking (automatic).
5. Recall the previously optimized shim file (during the initial evaluation). Run a 1D experiment to verify the quality of shimming.
6. Run a 2D experiment by calling the acquisition and processing parameter file obtained during the initial evaluation.
   (a) For external calibration: Start with the analysis of the most concentrated sample in the calibration range in order to correctly calculate the receiver gain automatically. Then analyze the most diluted sample under the same conditions to determine the number of scans necessary to detect the metabolites with a sufficient signal-to-noise ratio. Using this same number of scans, perform experiments on all calibration samples. Using the same experimental conditions, analyze all the samples of the cohort (**steps 1–6** of this section). For studies containing QCs, analyze the QC samples regularly, for instance, every 5 or 10 study samples.
   (b) For standard addition: Based on the experimental conditions optimized during the initial evaluation (*see* Subheading 3.1), analyze the first sample of the cohort without any addition. Pay attention to the value of the receiver gain: as you will concentrate the sample after each addition, the receiver gain must not be too high in order to avoid saturation. Set the value of the receiver gain determined automatically or manually to one half its value. Then, add a small volume of the stock solution. Vortex and analyze the sample again. Repeat standard addition three times. Using the same experimental conditions, analyze all the samples of the cohort (**steps 1–6** of this section).
7. To process the acquired data, multiply the FID by an appropriate apodization function and then apply the Fourier transformation. When applicable, adjust the phase in each dimension, manually. Then, apply a baseline correction in both dimensions (*see* **Note 20**). Calibrate all the spectra using a known chemical shift (e.g., diagonal signal of alanine or lactate).

3.5.3 Integration

1. Assign signals to metabolites using metabolomics databases. For each metabolite, identify a signal of interest well separated from other signals for quantification (*see* **Note 25**).
2. Integrate the identified signals using a dedicated software package such as AMIX, MestReNova, or Topspin. Draw boxes around the signals of interest (*see* **step 1** in Subheading 3.4.2) (*see* **Note 21**). It is easier to define the integration areas based on the spectrum of the most concentrated sample of the calibration or spectrum of the sample obtained after the third addition of the stock solution.
3. Once all the integration areas are drawn, save the regions to an integration file.
4. Recall the integration file separately for each calibration spectrum and each sample before and after each standard addition and obtain integrations (*see* **Note 22**).
5. Compile the integration values associated with each integrated peak in an Excel file without applying any normalization at this stage.

3.5.4 Determination of the Concentration of Targeted Compounds

1. Calibration curve approach:
   (a) Create a table in which the concentration values C_i are listed in the first column and the corresponding integration value, V_i in the second column (*see* Fig. 3a).
   (b) For each integrated signal, create a linear regression of the type "$y = ax+b$" (where $y =$ peak integration value V_i, and x = concentration C_i). Determine the value of a (the slope) and (b) y-intercept (*see* **Note 26**). Verify that the coefficient of linear regression R^2 is close to 1. The associated graph $V_i = f(C_i)$ can be plotted to verify the model.
   (c) Using this equation of the linear regression, determine the concentration of the metabolite (C_i) from the sample inside the NMR tube by substituting y with the integration value, V_i.
   (d) If necessary, take into account the dilution factor to determine the actual concentration in each biological sample (i.e., before addition of deuterated solvent).
2. Standard addition approach:
   (a) Create a table in which the added concentration values C_i are listed in the first column and the corresponding integration value, V_i in the second column (*see* Fig. 3b).
   (b) For each integrated signal, create a linear regression of the type "$y = ax+b$" (where $y =$ peak integration value V_i, and x = concentration C_i). Determine the value of a (the slope) and (b) y-intercept. Verify that the coefficient of

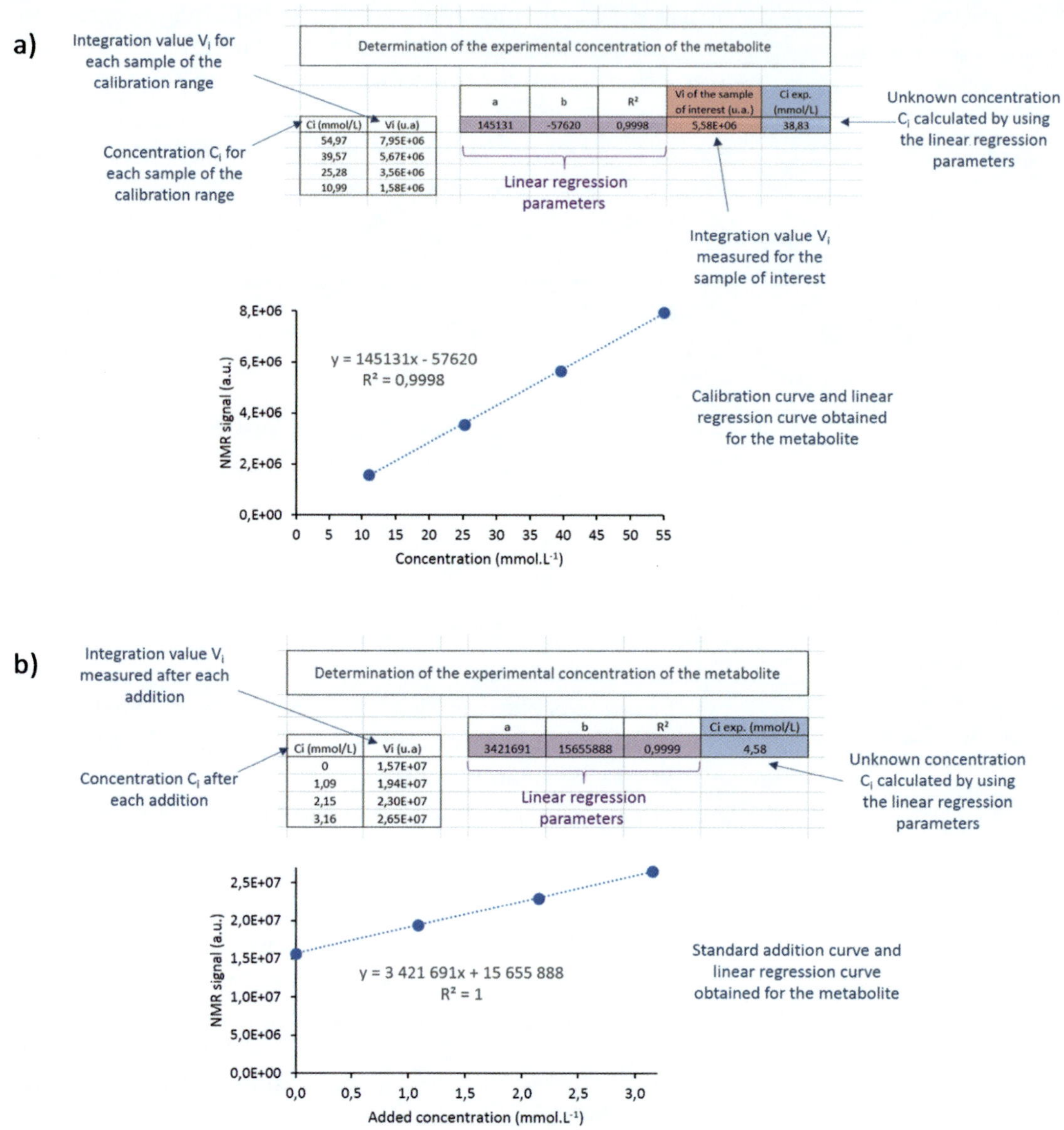

Fig. 3 Determination of the concentration of a metabolite based on (**a**) calibration curve approach and (**b**) standard addition approach

linear regression R^2 is close to 1. The linear regression curve must cross the y-axis at a value different from 0.

(c) For each integrated signal from the study sample, substitute y with 0 and use the equation of the linear regression to calculate the unknown concentration C_i. The unknown concentration can also be determined by the intersection of the linear curve regression with the x-axis (obtained after extrapolation of the regression curve).

(d) If necessary, take into account the dilution factor to determine the actual concentration in each biological sample (i.e., before addition of deuterated solvent).

4 Notes

1. Typical NMR spectrometers for metabolomics are in the 500–800 MHz range, but the approach described in this chapter is adaptable to all magnetic field strengths. Having a cryogenically cooled probe will be advantageous if the sample concentration is quite low (but it can also be detrimental for samples with high salinity).
2. A refrigerated high-capacity auto-sampler will be advantageous to analyze large cohorts of samples and/or samples which could suffer from degradation if they wait too long at ambient temperature.
3. If the quantity of sample is too small, use a 3 mm NMR tube. High-quality tubes are preferred.
4. Pay particular attention to the sample preparation:
   (a) Choose the appropriate extraction protocol depending on the targeted fraction of the metabolome—you may want to observe all detectable metabolites or just a fraction (polar metabolites, lipids, etc.).
   (b) For biological extracts or samples sensitive to pH variation, use a phosphate buffer to avoid a deviation of chemical shifts due to pH variation.
   (c) Be careful with the tubes used to collect blood or serum: avoid EDTA, citrate, and other added stabilizers since they give very large additional signals.
5. 500 μL of biological sample is optimal to acquire spectra in a reasonable time—the analyses are still possible with less sample but need additional data acquisition time.
6. At least 20 mg of biological sample (wet mass) is necessary to be able to acquire NMR spectra in a reasonable time—a mass of 100 mg of biological is the typical size for this kind of analysis. However, pay particular attention to the sample solubility in the chosen solvent—preliminary tests may be necessary.
7. This signal can also be used for the normalization of the spectrum.
8. The receiver gain of the NMR experiment is calculated based on the most intense signal. If the concentration of the reference is too large, the NMR signal from the reference determines the upper limit of the receiver gain and weak metabolite signals may not be detected under such circumstances.

9. The chemical shift of trimethylsilylpropanoic acid (TSP—soluble in D_2O) or tetramethylsilane (TMS—soluble in organic solvents such as $CDCl_3$—evaporation from 26 °C) has the advantage to be at 0 ppm, where there is no other signal of interest. According to the targeted accuracy, the use of a silicon reference may be avoided because of the silicon satellites. In the case of lipidomics, dimethyl sulfone (DMSO2—soluble in organic solvents such as $CDCl_3$) can be an appropriate reference as its signal is a singlet at a chemical shift of 3 ppm. Be careful: prior to addition of DMSO2 inside each sample, verify that there is no signal of interest close to this region.
10. If the auto-sampler cannot be refrigerated, and even if it is refrigerated, make a preliminary evaluation of the stability of the sample cohort at ambient temperature and at refrigerated temperature; make sure that NMR peaks are stable and do not vary with time.
11. Before analyzing a cohort of samples, an initial evaluation has to be done in order to determine which NMR pulse sequence(s) will be appropriate to answer the biological question. In this chapter, we assume 2D spectra can be recorded under partial saturation conditions (i.e., with a recovery delay shorter than five times the longest longitudinal relaxation time T_1) if the experiment duration is a critical factor. Indeed, unlike 1D NMR, where respecting this condition is critical to ensure that the coefficient of proportionality between peak areas and concentrations is the same for all peaks in the spectrum, 2D NMR spectra do not aim to fulfill this condition. In 2D NMR, the effect of T_1 is one factor among others (T_2, coupling constant) that affect the coefficient of proportionality between peak volumes and concentrations. Therefore, partial saturation conditions remain compatible with the approach described in this chapter, based on the condition that T_1 values do not vary significantly between samples in a study [2].
12. 303 K is temperature reported in the majority of published studies.
13. On cryogenically cooled probes, this parameter may be further optimized empirically to improve the solvent suppression.
14. This pulse duration should remain constant for a series of biological samples prepared under the same conditions.
15. Pay attention to the choice of these values: do not take into account the chemical shifts of quaternary carbons.
16. NUS can also be used to record very-high-resolution spectra without compromising the experiment time [18].
17. If the spectral width is too large, apply aliasing in the indirect dimension (^{13}C chemical shift) [19, 20]. In this case, pay particular attention to the value of the corresponding

frequency offset: a change in the pulse sequence must be done to have the frequency offset in the middle of the range (without aliasing) during the decoupling period and to have the frequency offset in the middle of the aliased range during the rest of the experiment.

18. Several apodization functions are available (exponential, Gaussian, sinusoidal, etc.). The choice of the appropriate apodization function depends on the pulse sequence, in order to observe at best the signals of interest without too much noise or unwanted signals.

19. Load all or a part of the samples into the refrigerated autosampler at a temperature of 4 °C.

20. The degree of the polynomial baseline correction must be optimized according to the baseline observed on the row and column projections. A default value of 2 is recommended in both dimensions.

21. Ensure that the rectangle boxes are large enough to include the whole signal without covering adjacent signals or noise. Modify the contour plot level to make sure that the integration box matches the signal of interest without integrating a nearby peak.

22. Normally, all the spectra should be stackable after calibration of the chemical shift axis. In practice, when the samples are not buffered (e.g., urine or plasma) or do not have exactly the same pH, the calibration cannot lead to a perfect superposition of all the spectra. In this case, the position of the buckets must be automatically or manually adjusted, without changing their size.

23. The standard addition approach is more accurate than external calibration since it is less sensitive to matrix effects like probe detuning. However, the standard addition is more time-consuming than external calibration approach as separate calibration is required for each metabolite.

24. The concentration of the targeted metabolites can be estimated from the literature. Use a low flip angle pulse so that the relaxation associated with dilution does not affect signal intensity. Otherwise, the repetition time should be chosen to be three to five times the longest T_1 in the sample.

25. It is easier to identify the signals based on the spectrum of the most concentrated calibration sample or the sample spectrum obtained after the third addition of the stock solution. Symmetrization of the spectrum can help to identify the metabolites, but the symmetrized spectrum must not be used for the quantification.

26. In principle, the linear regression curve should have the y-intercept $b = 0$. However, errors associated with sample preparation, analytical measurements, etc., can cause some deviation.

Acknowledgements

The authors acknowledge all the members of the EBSI team as well as external collaborators for fruitful discussions. Support from the CORSAIRE metabolomics platform (Biogenouest network) is also acknowledged.

References

1. Marchand J, Martineau E, Guitton Y, Dervilly-Pinel G, Giraudeau P (2017) Multidimensional NMR approaches towards highly resolved, sensitive and high-throughput quantitative metabolomics. Curr Opin Biotech 43:49–55
2. Giraudeau P (2014) Quantitative 2D liquid-state NMR. Magn Reson Chem 52:259–272
3. Van QN, Issaq HJ, Jiang Q, Li Q, Muschik GM, Waybright TJ et al (2008) Comparison of 1D and 2D NMR spectroscopy for metabolic profiling. J Proteome Res 7:630–639
4. Koskela H, Kilpeläinen I, Heikkinen S (2005) Some aspects of quantitative 2D NMR. J Magn Reson 174:237–244
5. Mauve C, Khlifi S, Gilard F, Mouille G, Farjon J (2016) Sensitive, highly resolved and quantitative ^{1}H-^{13}C NMR data in one go for tracking metabolites in vegetal extracts. Chem Commun 52:6142–6145
6. Jézéquel T, Deborde C, Maucourt M, Zhendre V, Moing A, Giraudeau P (2015) Absolute quantification of metabolites in tomato fruit extracts by fast 2D NMR. Metabolomics 11:1231–1242
7. Martineau E, Tea I, Akoka S, Giraudeau P (2012) Absolute quantification of metabolites in breast cancer cell extracts by quantitative 2D ^{1}H INADEQUATE NMR. NMR Biomed 25:985–992
8. Giraudeau P, Tea I, Remaud GS, Akoka S (2014) Reference and normalization methods: essential tools for the intercomparison of NMR spectra. J Pharm Biomed Anal 93:3–16
9. Rouger L, Gouilleux B, Giraudeau P (2016) Fast n-dimensional data acquisition methods. In: Lindon JC, Tranter GE, Koppenaal D (eds) Encyclopedia of spectroscopy and spectrometry, 3rd edn. Elsevier, pp 588–596
10. Le Guennec A, Giraudeau P, Caldarelli S (2014) Evaluation of fast 2D NMR for metabolomics. Anal Chem 86:5946–5954
11. Le Guennec A, Tea I, Antheaume I, Martineau E, Charrier B, Pathan M et al (2012) Fast determination of absolute metabolite concentrations by spatially encoded 2D NMR: application to breast cancer cell extracts. Anal Chem 84:10831–10837
12. Marchand J, Martineau E, Guitton Y, Le Bizec B, Dervilly-Pinel G, Giraudeau P (2018) A multidimensional ^{1}H NMR lipidomics workflow to address chemical food safety issues. Metabolomics 14:60–70
13. Gouilleux B, Rouger L, Giraudeau P (2016) Ultrafast multi-dimensional NMR: principles and recent applications. eMagRes 5:913–922
14. Thrippleton MJ, Keeler J (2003) Elimination of zero-quantum interference in two-dimensional NMR spectra. Angew Chem Int Ed 42:3938–3941
15. Beckonert O, Keun HC, Ebbels TMD, Bundy J, Holmes E, Lindon JC et al (2007) Metabolic profiling, metabolomic and metabonomic procedures for NMR spectroscopy of urine, plasma, serum and tissue extracts. Nat Protoc 2:2692–2703
16. Giraudeau P, Silvestre V, Akoka S (2015) Optimizing water suppression for quantitative NMR-based metabolomics: a tutorial review. Metabolomics 11:1041–1055
17. Rouger L, Charrier B, Akoka S, Giraudeau P (2017) Implementing ultrafast 2D NMR experiments on a bruker avance spectrometer. http://madoc.univ-nantes.fr/course/view.php?id=24710. Accessed 28 Aug 2017
18. Le Guennec A, Dumez J-N, Giraudeau P, Caldarelli S (2015) Resolution-enhanced 2D NMR of complex mixtures by non-uniform sampling. Magn Reson Chem 53:913–920

19. Jeannerat D (2007) Computer optimized spectral aliasing in the indirect dimension of ^{1}H-^{13}C heteronuclear 2D NMR experiments. A new algorithm and examples of applications to small molecules. J Magn Reson 186:112–122

20. Foroozandeh M, Jeannerat D (2010) Deciphered chemical shifts in aliased spectra recorded with two slightly different narrow windows or differential chemical shift evolution. Chem Phys Chem 11:2503–2505

Chapter 21

Targeted Metabolomics with Quantitative Dissolution Dynamic Nuclear Polarization

Mathilde H. Lerche, Magnus Karlsson, Jan H. Ardenkjær-Larsen, and Pernille R. Jensen

Abstract

Metabolite profiles and their isotopomer distributions can be studied noninvasively in complex mixtures with NMR. The advent of hyperpolarized ^{13}C-NMR using quantitative dissolution Dynamic Nuclear Polarization (qdDNP) and isotope enrichment add sensitivity to such metabolic studies, enabling mapping and quantification of metabolic pathways and networks. Here we describe a sample preparation method, including cell incubation, extraction, and signal enhancement, for reproducible and quantitative analysis of hyperpolarized ^{13}C-NMR metabolite spectra. We further illustrate how qdDNP can be applied to gain metabolic insights into living cells.

Key words dDNP, Hyperpolarization, ^{13}C-NMR, Glucose

1 Introduction

Targeted metabolomics or stable isotope resolved metabolomics (SIRM) measures an isotope-filtered selection of molecules aimed at understanding the dynamics and compartmentation of metabolic pathways and networks [1, 2]. The metabolic transformations of one or more isotope-labeled tracers lead to distinct labeling patterns in metabolic intermediates. Using such patterns, metabolic fluxes can be measured from temporal profiles of metabolite concentration changes. A widely used isotope-labeled tracer is [U-^{13}C] glucose, which provides valuable insight into the glycolytic pathway, the pentose phosphate pathway, labeling of intermediates in the TCA cycle, and labeling of amino acids. In conjunction with the use of stable isotope tracers, ^{13}C-NMR is a method of choice for exploring metabolic reprogramming in major metabolic diseases such as cancer and diabetes [3, 4]. ^{13}C-NMR has a number of unique advantages for metabolic studies; it is quantitative, versatile, and specific. The relative insensitivity of the method can be

G. A. Nagana Gowda and Daniel Raftery (eds.), *NMR-Based Metabolomics: Methods and Protocols*, Methods in Molecular Biology, vol. 2037, https://doi.org/10.1007/978-1-4939-9690-2_21,

compensated when combined with ^{13}C hyperpolarization in the form of dissolution dynamic nuclear polarization [dDNP] [5]. DNP make use of the transfer of electron spin polarization from a free radical to, e.g., ^{13}C nuclear spins mediated by resonant microwave irradiation [6]. This transfer is best conducted in amorphous samples that assure the homogenous distribution of electrons and nuclear spins. DNP is typically performed at low temperatures (<1.5 K) and at high magnetic fields (>3 T) where the electron spin polarization approaches 100%. Following completed hyperpolarization, hot solvent dissolves the sample, and the sample is transferred to a room temperature NMR magnet for analysis. Whereas hyperpolarization losses during this dissolution step can be kept to a minimum, a principal limitation for analyzing hyperpolarized molecules is short signal lifetime of seconds to a few minutes for non-protonated ^{13}C sites in small molecules in the liquid state, which is governed by the nuclear T_1 relaxation time. Here we demonstrate how to combine targeted metabolomics methodology with hyperpolarized ^{13}C-NMR (Fig. 1). We further show how the method can be developed to make absolute quantifications of metabolic fluxes [7] (Fig. 2). The method is illustrated in a longitudinal study on how β-cells handle increased fuel pressure (Fig. 3).

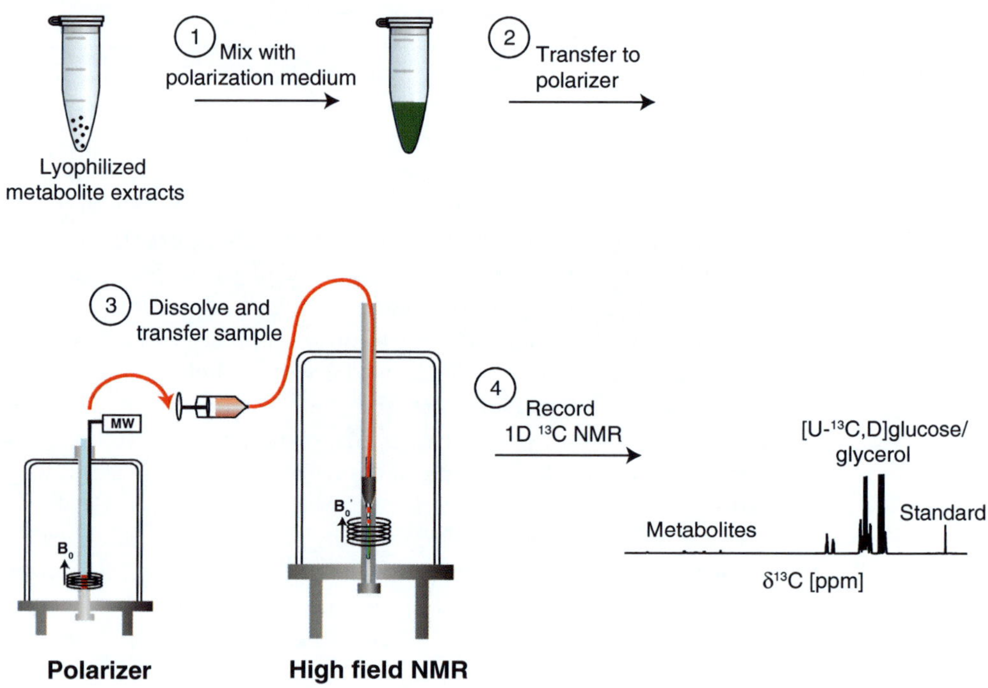

Fig. 1 Principle steps in the qdDNP-NMR analysis method. A description of the four steps in the analysis is found in Subheading 3.2

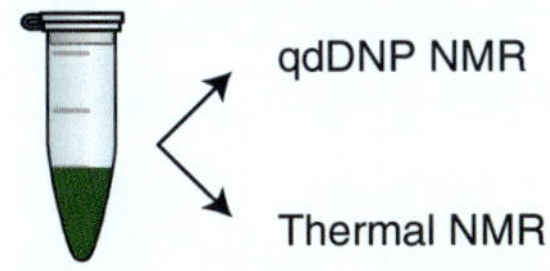

Compound (Com)	DNP ratio (Com/HP01)	NMR ratio (Com/HP01)	SLC
Lactate-C1	0.55 ± 0.03	0.67 ± 0.07	82 ± 8 %
Glutamate-C5	0.66 ± 0.03	1.34 ± 0.22	49 ± 9 %
Alanine-C1	0.60 ± 0.04	1.08 ± 0.01	55 ± 4 %
Pyruvate-C1	0.57 ± 0.09	0.53 ± 0.01	108 ± 9 %

Fig. 2 Calculation method for signal loss coefficient (SLC)

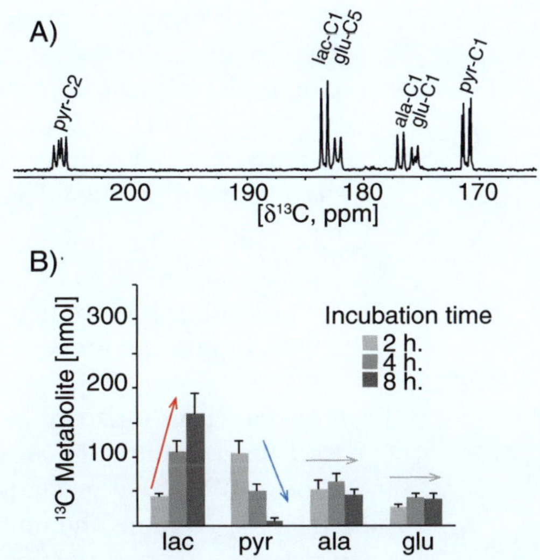

Fig. 3 Illustration of metabolite quantification in qdDNP-NMR experiments. (**a**) Typical ^{13}C NMR spectrum from the metabolite extract of β-cells incubated with 11.7 mM [U-^{13}C,D]glucose for 4 h. (**b**) Quantification of glucose-derived metabolites (lactate, pyruvate, alanine, and glutamate) in β-cells in a longitudinal study (2, 4, and 8 h, 11.7 mM). Constant production of lactate (red arrow, 0.65 nmol/min), depletion of pyruvate (blue arrow, −0.3 nmol/min), and constant pool of glutamate (TCA cycle) and alanine (grey arrows) are determined

2 Materials

2.1 Stable Isotope-Labeled Cell Culture Extract Samples and Reference Compounds

2.1.1 Chemicals

1. HP001 (1,1-bis (hydroxymethyl-D_2) cyclopropane-1-13C, D_4).
2. OX063 (trityl radical).
3. ProHance™ (MR contrast agent).
4. OmniScan™ (MR contrast agent).

2.1.2 Cell Extract Samples

1. Isotope-labeled substrate: stock solution of 120 mM [U-^{13}C, D_7]glucose (Sigma-Aldrich) dissolved in desired incubation medium.
2. Polarization medium: 70 mg OX063 (electron polarizer agent, EPA), 1227 mg glycerol (GS), 944 mg Milli-Q water, and 6 μL ProHance™ (0.5 M, electron relaxation agent, ERA).

2.1.3 Reference Samples

1. DNP sample composition.

   Stock solution is made using 0.75 μmol ^{13}C of each desired compound (corresponding to 75 μmol of natural abundance chemicals) mixed with 450 μL polarization medium, 0.75 μmol [1-^{13}C] HP001 (15 μL, 50 mM), and 5.4 mg [U-^{13}C,D] glucose. A 75 μL aliquot of DNP sample composition is hyperpolarized and dissolved as described in the methods ($n = 3$).
2. Thermal NMR sample composition.

   A 75 μL aliquot of DNP sample composition is dissolved in 400 μL phosphate buffer (40 mM, pH 7.3), 75 μL D_2O, and 15 μL OmniScan ($n = 3$).

2.2 DNP Hyperpolarization

Dynamic nuclear polarization was performed in a HyperSense (Oxford Instruments) polarizer with microwave irradiation at a frequency of 94 GHz and a power of 100 mW. All samples were polarized for 90 min. The microwave frequency and power should be optimized for the specific polarizer instrument according to established procedures.

2.3 ^{13}C-NMR

Hyperpolarized 1D ^{13}C NMR spectra were recorded on a 9.4 T NMR Varian spectrometer. Thermal 1D ^{13}C NMR spectra were acquired on a 19 T NMR Bruker spectrometer equipped with a cryoprobe.

2.4 Example: β-Cell Metabolism

Cell culture: INS-1 cells were cultured in Gibco RPMI 1640 Medium, GlutaMAX Supplement (Thermo Fisher Scientific) with 10% fetal bovine serum (FBS) (Sigma-Aldrich), 1% Penicillin-Streptomycin (Sigma-Aldrich), and 500 μL of 50 mM 2-mercaptoethanol (Thermo Fisher Scientific). This culture medium contains 11 mM glucose. The culture was maintained at

37 °C under humidified (5% CO_2, 95% air) conditions. Cell number and viability were measured by an automatic cell counter (EVE, NanoEntek). Cells were sub-cultured in 175 cm^2 flasks at a density of 3×10^6 cells for 7 days.

3 Methods

3.1 Metabolite Component Sample (see Note 1)

1. Grow human cells according to available cell-specific protocols (e.g., ATCC). Harvest cells by trypsinization. Wash and redissolve cells in desired medium (e.g., cell culture medium without glucose, PBS, or hypotonic phosphate buffer) to a concentration of 20×10^6 cells/mL (*see* **Note 2**).
2. Incubate 500 μL cells in suspension in 2 mL Eppendorf tubes with 100 μL, 120 mM uniformly ^{13}C-isotope-labeled and partially deuterated glucose in a shaking thermostat (37 °C and 500 rpm) for 30 min or longer (*see* **Note 3**).
3. Stop all enzymatic reactions immediately by addition of 400 μL ice-cold 2.2 M perchloric acid (PCA) to the 600 μL substrate incubated cell suspension and place on ice for at least 10 min. Extract the water soluble metabolites by centrifugation (10 min, 9600 × *g*, 4 °C). Neutralize the supernatant with 3 M KOH (add pH indicator for visual inspection of neutralization). In order to obtain a high recovery of metabolites, it is important not to over titrate the samples and keep them on ice to prevent heating. Leave the neutralized sample on ice for at least 10 min. Remove the precipitated salt by centrifugation (10 min, 9600 × *g*, 4 °C) (*see* **Note 4**).
4. Lyophilize the extract (*see* **Note 5**).

3.2 Dissolution Dynamic Nuclear Polarization

1. To hyperpolarize the freeze-dried extract efficiently, a glassing solvent (GS), an electron polarization agent (EPA, source of free radicals) and an electron relaxation agent (ERA, paramagnetic compound) are needed (together referred to as the polarization medium) [8]. Each lyophilized extract contained in an Eppendorf tube is dissolved in a total of 150 μL of polarization medium. To each sample, an internal standard (5 μL of a 50 mM water solution of HP001) is added. The sample is whirl-mixed and sonicated for a few minutes to ensure that the entire extract is dissolved and subsequently centrifuged (*see* **Note 6**).
2. The sample is pipetted into a sample holder and inserted into the polarizer. Microwaves are activated. The samples are fully polarized after 90 min. Since the metabolite quantification is performed relative to the added standard, it is possible to harvest the polarized samples earlier (down to 30 min polarization time) with the penalty of a lower signal-to-noise ratio.

3. After completing the hyperpolarization, the samples are dissolved in 5 mL of 40 mM phosphate buffer (pH 7.3). The sample is transferred to a 5 mm NMR tube and inserted in a NMR magnet tuned to ^{13}C for analysis (*see* **Note 7**). The presented data are based on a total sample transfer time of 10–12 s and a sample temperature starting at 50 °C immediately after dissolution and ending at 38 ± 2 °C at the time of the NMR acquisition (*see* **Note 8**).
4. A direct detection 1D ^{13}C NMR spectrum with proton decoupling is recorded with a 90° pulse, immediately after insertion of the sample into the magnet. Processing and analysis of the spectrum is performed with any available NMR analysis program, e.g., MNOVA [9].

3.3 Absolute Quantification

1. For quantification of the metabolites, an internal standard is added. The choice of standard is highly important with several requirements needing to be fulfilled: chemical inertness, a long longitudinal relaxation time (T_1), and a singlet signal not overlapping with a metabolite resonance in NMR spectrum. The ^{13}C-labeled HP001 is chosen for these reasons. With this standard, the metabolites in the cellular extract are quantified as a ratio between the integral of an identified metabolite signal relative to the integral of HP001. This ratio is stable and independent of the polarization time for individual metabolites with a coefficient of variation below 15%.
2. For absolute quantification, it is necessary to account for signal decay during the transfer between the polarizer and the NMR magnet for all carbon atoms of interest in the individual metabolites. This can be done by calculating a signal loss coefficient (SLC) for the individual carbon atoms. Samples of isotope-labeled or non-labeled compounds of interest are prepared using the polarization medium described in Subheading 3.2, **step 1**. For ^{13}C-labeled chemicals, several compounds can be analyzed in the same sample by spiking known amounts of compounds into the polarization medium. It is very important to make sure that the compounds are fully dissolved in the polarization medium; sonication and heating may aid the process. The samples are split into two parts where one part is analyzed with hyperpolarized ^{13}C NMR and the other part with thermal ^{13}C NMR. The SLCs are calculated using the respective ratios of ^{13}C NMR signals from a metabolite carbon and the internal standard. The part of the reference sample analyzed with ^{13}C hyperpolarized NMR is hyperpolarized, dissolved, and analyzed as described in Subheading 3.2, **steps 2–4**. For the other part of the reference sample, a ^{13}C thermal NMR spectrum is recorded on a high-field spectrometer with 4096 scans and a delay of 2 s between 60° pulses. Differences in

sample preparation such as varying concentrations and different labeling schemes do not influence the calculated SLC.

3. A SLC was calculated for the carbonyl atoms of four often-occurring metabolites (1-^{13}C-lactate, 5-^{13}C-glutamate, 1-^{13}C-alanine, and 1-^{13}C-pyruvate). The ratios between the individual metabolite signals and the internal standard are measured from thermal ^{13}C NMR and hyperpolarized ^{13}C NMR data, respectively. From these ratios, the individual SLC is calculated (Fig. 2) (*see* **Note 9**). These SLCs were used to calculate the absolute concentrations of these four metabolites produced by β-cells exposed to different isotope-labeled glucose loads, Fig. 3 (*see* **Note 10**).

4 Notes

1. The qdDNP methodology requires a water-soluble dry extract but does not impose other limitations to the formulated metabolite sample. The metabolite sample preparation could be done in several other ways than described here, e.g., incubation could be done in flasks without trypsinization; separation of intra- and extracellular metabolite fractions could be done; methanol or ethanol extractions could be performed instead of PCA-KOH extraction [10].
2. The stated cell number is a current minimum number of cells used to obtain the final claimed assay sensitivity. The incubation medium is chosen dependent on the study goal. It is important that the medium does not contain the incubation substrate, which is glucose in the described protocol.
3. The cell type (how glycolytic the cells are), cell number, and incubation time all influence the desired substrate concentration. The goal is to apply enough substrate to secure a linear, and small, consumption during the experiment. This requirement can easily be evaluated by changing the substrate concentrations and measuring consumption at different time points and/or cell concentrations using either ^{13}C NMR or readily available biochemical assays.
4. Not all metabolites are stable in strong acid. If in doubt, then a spiking test is recommended: Add a known concentration of the metabolite in question to the cell sample, extract the metabolite using the PCA-KOH procedure, freeze-dry the sample and redissolve it in an "NMR buffer," run conventional ^{13}C-NMR, and quantify the metabolite against an added standard.
5. At this point the sample is stable and can be kept frozen until prepared for qdDNP. For this reason, there is no requirement

for targeted metabolomics with qdDNP to have dDNP equipment in house since the metabolite component samples can be transported to a collaborator laboratory with dDNP equipment available.

6. Trityl radical (OX063) can be purchased from Albeda Research Aps, Denmark. HP001 can be purchased from Sigma-Aldrich on a quote basis (it is not available in their catalogue).
7. Any high-field NMR magnet can be used as long as the signal dispersion is good enough to separate metabolites of interest. A 10 mm NMR probe can be used advantageously to increase the available signal.
8. This part of the assay is critical. The transfer time, the sample temperature, the sample concentration, and the magnetic environment during the sample transfer all influence the resulting available metabolite signals. These parameters will vary in different hyperpolarization laboratories. The important point is to have the parameters under control such that different samples will undergo the same conditions and be measured relative to a well-characterized standard compound. *See* Subheading 3.3 on absolute quantitation. If these steps are followed, then only the signal-to-noise ratio will vary between different laboratories.
9. The sensitivity and variation in SLC between different laboratories can be improved significantly. A recent report shows that it may be highly beneficial to perform the DNP at higher field where the polarization can be expected to double [11]. Potentially the largest gain in sensitivity will come from a significant reduction in sample dissolution volume and better utilization of the available sample. A reduction from the current 5 mL to 1 mL and the use of a 10 mm probe instead of a 5 mm probe can increase the sensitivity 20 times. Implementation of these improvements is important for optimization of assay sensitivity and reproducibility. For absolute quantification, it will always be necessary to have a fully characterized metabolite signal as described above with the calculation of individual carbon SLC. As different hyperpolarization laboratories standardize their protocols, it can be envisioned that the community can produce a common relaxation database similar to the available chemical shift databases—a tool that will significantly improve the described method.
10. The qdDNP-NMR method is valuable in hypothesis-driven research, as shown in the example using β-cells that can produce easily interpreted quantitative information about cell metabolism. An important perspective of this type of derived metabolic information is its translation to clinical applications where the dDNP MRI methodology is well underway to be implemented as a metabolic contrast agent method useful for diagnosis and therapy monitoring.

Acknowledgments

The authors gratefully acknowledge funding by Danish National Research Foundation (grant DNFR124).

References

1. Nagana Gowda GA, Shanaiah N, Raftery D (2012) Isotope enhanced approaches in metabolomics. Adv Exp Med Biol 992:147–164
2. Bruntz RC, Lane AN, Higashi RM, Fan TWM (2017) Exploring cancer metabolism using stable isotope resolved metabolomics (SIRM). J Biol Chem 292(28):11601–11609
3. Fan TWM, Lorkiewicz PK, Sellers K, Moseley HN, Higashi RM, Lane AN (2012) Stable isotope-resolved metabolomics and applications for drug development. Pharmacol Ther 133(3):366–391
4. Sas KM, Karnovsky A, Michailidis G, Pennathur S (2015) Metabolomics and diabetes: analytical and computational approaches. Diabetes 64(3):718–732
5. Ardenkjær-Larsen JH, Frindlund B, Gram A, Hansson G, Hansson L, Lerche MH et al (2003) Increase in signal-to-noise ratio of > 10,000 times in liquid-state NMR. Proc Natl Acad Sci U S A 100(18):10158–10163
6. Abragam A, Goldman M (1978) Principles of dynamic nuclear polarization. Rep Prog Phys 41:395–467
7. Lerche MH, Yigit D, Frahm AB, Ardenkjær-Larsen JH, Malinowski RM et al (2018) Stable isotope-resolved analysis with quantitative dissolution dynamic nuclear polarization. Anal Chem 90(1):674–678
8. Karlsson M, Jensen PR, Duus JO, Meier S, Lerche MH (2012) Development of dissolution DNP-MR substrates for metabolic research. Appl Magn Reson 43:223–236
9. Mestrelab Research SL (2015) Mestrenova. https://www.mestrelab.com
10. Sellick CA, Knight D, Croxford AS, Maqsood AR, Stephens GM, Goodacre R et al (2010) Evaluation of extraction processes for intracellular metabolite profiling of mammalian cells: matching extraction approaches to cell type and metabolite targets. Metabolomics 6:427–438
11. Capozzi A, Patel S, Ouari O, Karlsson M, Lerche MH, Ardenkjær-larsen JH (2019) Gadolinium effect at High-Magnetic-Field DNP: 70% ^{13}C Polarization of [U-^{13}C] Glucose using Trityl. J Phys Chem Lett vol 10 (12):3420–3425

Chapter 22

Metabolic Profiling Using In Vivo High Field Flow NMR

Maryam Tabatabaei Anaraki, Daniel Lane, Monica Bastawrous, Amy Jenne, and André J. Simpson

Abstract

In vivo NMR (nuclear magnetic resonance) has the potential to monitor and record metabolic flux in close to real time, which is essential for better understanding the toxic mode of action of a contaminant and deciphering complex interconnected stress-induced pathways impacted inside an organism. Here, we describe how to construct and use a simple flow system to keep small aquatic organisms alive inside the NMR spectrometer. In living organisms, magnetic susceptibility distortions lead to severe broadening in conventional NMR. Two main approaches can be employed to overcome this issue: (1) use a pulse sequence to reduce the distortions, or (2) employ multidimensional NMR in combination with isotopic enrichment to provide the spectral dispersion required to separate peaks from overlapping resonances. Both approaches are discussed, and protocols for both approaches are provided here in the context of small aquatic organisms.

Key words Solution-state in vivo NMR, IP-ISQC, Flow system, ^{13}C enrichment, Environmental stress, Metabolomics, *Daphnia magna*

1 Introduction

NMR (nuclear magnetic resonance) based metabolomics is evolving as an important tool in environmental research. Here, in vivo NMR is defined as monitoring metabolites of an entire living organism in a noninvasive way, while the intact organism is kept alive inside the NMR magnet. This technique minimizes sample preparation and enables continuous data collection over time, providing novel information on toxic response pathways. This manuscript focuses on in vivo NMR inside a high field vertically oriented NMR spectrometer with the goal of extracting high-resolution metabolic information on small aquatic organisms. The chapter does not deal with the more conventional MRI-based imaging approaches often applied to larger organisms [1]. Furthermore, high field magic angle spinning (HR-MAS) approaches are not discussed here. MAS studies contain a wealth of metabolic

G. A. Nagana Gowda and Daniel Raftery (eds.), *NMR-Based Metabolomics: Methods and Protocols*, Methods in Molecular Biology, vol. 2037, https://doi.org/10.1007/978-1-4939-9690-2_22,

information [2, 3]; however, the considerable stress caused by the rapid spinning can lead to cell damage and organism death which limits its ability to monitor in vivo processes [2]. In this protocol, we describe a flow system that offers a low stress environment for small aquatic organisms and keeps them alive inside the NMR magnet during the entire experiment. We also discuss two approaches that can be used to afford the spectral resolution required for in vivo metabolite assignment. The first, termed "in phase- intermolecular single quantum coherence" (IP-ISQC), overcomes magnetic susceptibility distortions and produces in vivo line shape close to that of a conventional buffer extract [4]. This approach selectively targets the dissolved metabolites within the living organism and does not require isotopic enrichment. The second approach utilizes isotopic enrichments (^{13}C or $^{13}C/^{15}N$) to increase NMR sensitivity allowing for rapid acquisition of 2D NMR datasets such as heteronuclear single quantum coherence (HSQC). While the magnetic susceptibility distortions still exist, the increased spectral dispersion afforded by nD NMR allows signals from individual metabolites to be separated and assigned in vivo.

Readers should note that multidimensional flow-based in vivo NMR on aquatic living organisms is a relatively new approach, and many aspects are still evolving. With this in mind, the following protocols should be considered as a starting point, with further optimization and development likely required for many applications.

2 Materials

2.1 Model Organism

We find *Daphnia magna* to be a suitable model organism. *Daphnia* species are considered keystone organisms in the environment and are model organisms for aquatic toxicity testing [5–7]. Different *D. magna* sizes, from neonate to adult, fit inside the most common, standard 5 mm NMR tubes which are the most commonly used in water testing and exist in any NMR facility (*see* **Note 1**).

2.2 D. magna *Diet*

1. Isotopically enriched diet for *Daphnia magna:* We find *Chlamydomonas reinhardtii* to be a good source of food for ^{13}C labeling of *D. magna* [8]. Details as to the feeding protocol are given in the Subheading 3 (*see* **Notes 2–4**).
2. Unlabeled diet for *Daphnia magna:* To keep a consistent *D. magna* diet, we use unenriched *Chlamydomonas reinhardtii* as a food source while running the experiment (*see* **Note 5**).

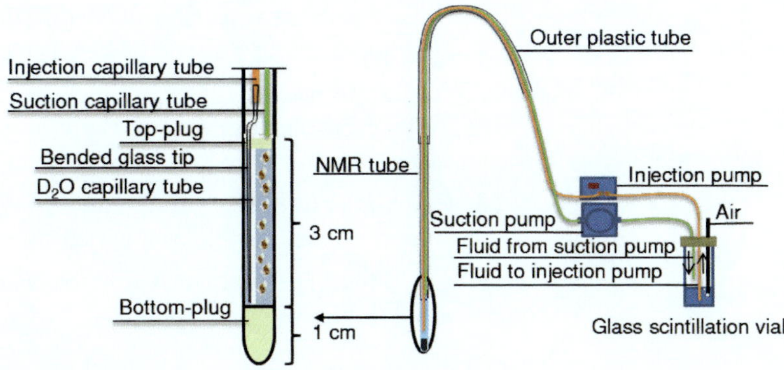

Fig 1 A schematic diagram of a flow system for in vivo NMR studies of small aquatic living organisms. The total volume of circulating fluid can be as low as 10 mL [25]. Reproduced with permission from American Chemical Society

2.3 Stressors

Stressors vary widely depending on the research study. They can be physical such as temperature, electrolytes such as salts, or actual contaminants such as contaminated environmental samples or acute/sublethal concentrations of toxicants.

2.4 Flow System

Figure 1 shows the schematic of a flow system used for in vivo NMR studies of small aquatic living organisms.

1. Two capillary tubes: use one for injection (e.g., PEEK tubing 0.76 mm ID) and one for suction (e.g., PTFE 1.5 mm ID) (*see* **Notes 6** and **7**).
2. Outer plastic tube: ~5 mm ID, elastic silicone rubber tube as an outer tube.
3. Two pumps: use one pump for injection of the fluid to the NMR tube (e.g., Knauer HPLC pump, Knauer, Berlin, Germany) and one for the suction of the fluid from the NMR tube (e.g., FMI Q pump, Cole-Parmer, Vernon Hills, IL, USA) (*see* **Notes 8** and **9**).
4. NMR tube: Thin-walled 5 mm NMR tube is suitable for *Daphnia magna* (*see* **Note 10**).
5. External lock system: A sealed 4–5 cm long Teflon glass capillary tube (with the minimum wall thickness possible and the inner diameter of $\approx$ 100 μm) filled with D_2O (*see* **Note 11**).
6. Bottom Plug: The bottom plug can be made of Teflon zirconium. The height of the plug should be ~1 cm for a 5 mm NMR tube to prevent *D. magna* from swimming outside the coil region.
7. Top plug: The top plug can be made of Teflon tape. The diameter of the plug should be big enough to help maintain the organisms in the coil region but small enough to prevent clogging and allow fluid containing food to reach the suction tip.

8. Glass tip: bent injection glass tip (use the tip of a disposable glass pipette and a torch to bend) (*see* **Note 12**).
9. Water and food reservoir: A glass container used as a water and food reservoir. The size of the container depends on the total volume of the fluid (can be as low as 10 mL).
10. Cooling system: A cooling system with adjustable temperature. This can be as simple as a wine cooler or small fridge.
11. Air/oxygen generator: An air/oxygen generator with an adjustable flow rate.

3 Methods

3.1 Daphnia magna ^{13}C Enrichment

1. Feed the *D. magna* neonates a diet of ^{13}C-enriched *Chlamydomonas reinhardtii* as their sole food source three times a week (~15 mg/100 daphnids/per day) from birth [9].
2. Change 50%–60% of the water and separate newly born neonates at the same time as feeding (*see* **Note 13**).
3. Keep the cultures in the laboratory under a 16:8 h light to dark ratio at ambient room temperature.
4. Add cobalamin and selenium as supplements to the water (1 μg L^{-1}) twice a week [10].
5. Feed *D. magna*-unenriched *C. reinhardtii* for a minimum of 4 h before starting each experiment to ensure removal of enriched food from their gut [9] (*see* **Notes 14–16**).

3.2 ^{13}C-Enriched Chlamydomonas reinhardtii Culturing

Chlamydomonas reinhardtii is cultivated in a small-scale closed loop system photobioreactor (Silantes GmbH, Germany). Fermentation is carried out autotrophically and entirely with $^{13}CO_2$ (98% ^{13}C). The environmental conditions such as media, temperature, light intensity, and pH are optimized for maximum growth. The biomass is then harvested and freeze-dried and stored at −20 °C prior to use. Comprehensive details on culturing ^{13}C-enriched algae can be found in the literature [8].

3.3 Unenriched Chlamydomonas reinhardtii Culturing

1. There are many ways to culture *Chlamydomonas reinhardtii*. We suggest to cultivate fresh, unenriched *C. reinhardtii* as described by Stein [11].
2. Centrifuge to concentrate the cultured *C. reinhardtii* (~4000 × *g*).
3. Keep the concentrated algae at 4 °C (maximum of 2 weeks) (*see* **Note 17**).

3.4 Setting Up the Flow System

1. Pass the two capillary tubes through the outer plastic tube (*see* **Note 18**).
2. Attach one side of the injection capillary tube that goes to the NMR tube to the glass tip. The glass tip should be adjusted to be a maximum 1–2 mm above the bottom plug.
3. Connect the other side of the injection capillary tube to the outlet of the injection pump.
4. Adjust the tip of the suction capillary tube that goes to the NMR tube above the top plug at 4–5 cm from the bottom of the NMR tube (*see* **Note 19**).
5. Connect the other side of the suction capillary tube to the inlet of the suction pump.
6. Put the reservoir inside the cooling system and adjust the cooling system to the desired temperature (*see* **Note 20**).
7. Connect a capillary tube (with an ID bigger than the injection capillary tube) to the inlet of the injection pump and put the other side of it inside the reservoir.
8. Connect a capillary tube (with the same ID as the suction capillary tube) to the outlet of the suction pump, and put the other side of the capillary tube inside the reservoir above the surface of the water (*see* **Note 21**).
9. Put the tip of the air/oxygen bubbler inside the water. Adjust the flow rate to the lowest possible to prevent vigorous bubbling (*see* **Note 22**).
10. Add the desired amount of unenriched food (e.g., *Chlamydomonas reinhardtii*) to water.
11. Seal the reservoir.
12. Turn on the injection pump until the injection capillary tube is filled with fluid.
13. Stop the injection pump when the fluid reaches the end of injection glass tip.
14. Insert a Teflon zirconium plug into the bottom of the NMR tube to prevent the organisms from swimming beneath the coil region.
15. Adjust the height of the top plug that holds the sealed D_2O capillary and is attached to the injection tip so that it is above the coil region (4–5 cm from the bottom of the NMR tube and below the tip of the suction tube).
16. Use a glass pipette to fill the NMR tube with water (*see* **Note 23**).
17. Insert *Daphnia* into the NMR tube using a wide mouth plastic pipette. Wait until daphnids swim to the bottom of the NMR tube (*see* **Note 24**).

18. Turn on the suction pump (*see* **Note 25**).
19. Slowly insert the injection tip, the top plug holding capillary lock, and the suction tip into the NMR tube.
20. Push 0.5 cm of the NMR tube into the outer plastic tube.
21. Secure the outer plastic tube to the NMR tube using Teflon tape.
22. Turn on the injection pump (*see* **Notes 26** and **27**).
23. Adjust the spinner using the depth gauge such that the active volume is centered in the coil.
24. Slowly insert the NMR tube and tubings into the NMR system by holding the tubing and lowering into the magnet. Ensure the sample is detected by the NMR spectrometer.

3.5 Stressors

The flow system enables environmental changes during the process of the experiment.

Add the toxicant to the fluid or apply another form of stressor such as a change in temperature (*see* **Note 28**). The simplest approach is to add a chemical stressor to the reservoir.

3.6 NMR Spectroscopy

There are a wide range of potential NMR experiments that can be performed on in vivo samples. Here, three key approaches are briefly mentioned.

1. ^{1}H SPR-W5-WATERGATE. An effective approach to water suppression that is effective with very wide water signals as is the case of in vivo studies [12, 13].
2. ^{1}H–^{13}C HSQC. A central experiment that provides ^{1}H–^{13}C correlations and relatively high spectral dispersion, reducing overlap and allowing signals from individual metabolites to be resolved and assigned (*see* **Note 29**).
3. In-Phase Intermolecular Single Quantum Coherence Correlation (IP-ISQC). A recently introduced 2D NMR experiment [4], this approach removes magnetic susceptibility distortions permitting detailed metabolite assignments from 1D ^{1}H NMR. This approach does not require the use of isotopic enrichment permitting studies of non-lab cultured organisms.

3.6.1 General

1. Make sure the probe temperature is stable at a desired temperature for the analysis.
2. Ensure that at minimum the sample passes TopShim 3D (*see* **Note 30**).
3. Set the frequency offset (for Bruker instruments: o1p) right to the center of the water peak.
4. Adjust the phase and loop parameters of the lock, or run the command “loopadj” on a Bruker spectrometer.

3.6.2 Water Suppression

For cryoprobes, the level of the water (from the bottom of the NMR tube to the water surface) should be around ~4 cm. Cryoprobes detect far water effectively (water outside the coil that is partially excited) which cannot be completely cancelled using water suppression approaches. By optimizing the water height, the far water is reduced while retaining a volume required for optimal shims. Water suppression should be performed at different water heights to determine the best height for a particular cryoprobe. We find SPR-W5-Watergate the most efficient technique in suppressing the large and broad signal of water in vivo [5, 14, 15].

1. Calibrate the 90° pulse for the spectrometer.
2. Use the high power level (i.e., the same power used for the hard 90° pulse) for all pulses in the DANTE sequence in the W5 train, i.e., set the DANTE 90° pulse the same as a hard 90 °. There is no need to reduce the power level of the DANTE pulse on modern NMR probes.
3. Set the shaped water pulse to a shape that matches the water. A 2 ms square pulse is a reasonable shape for quick implementation.
4. Set the recycle delay to 1 ms. A loop of shaped pulses replaces the recycle delays (D1 on a Bruker system). For example, if 1000 loops are set using a 2 ms shaped pulse, then the recycle delay is 1000×2 ms $= 2$ s.
5. Set the binomial delay to 125 μs. On a 500 MHz instrument, this puts the sidebands at ~11 ppm and −1 ppm. For other field strengths, this delay may need optimizing to ensure the sidebands remain outside the region of interest. The shorter the delay, the further the sidebands are shifted. However, shorter delays also lead to a wider suppression window around the water itself.
6. Set up the sequence on a 2 mM sucrose sample in 90% H_2O/ 10% D_2O (i.e., Bruker water suppression standard).
7. Use at least 16 dummy scans. If set up correctly, the water should be close to or below the noise after eight scans.

3.6.3 ^1H–^{13}C NMR

There are a wide range of heteronuclear single quantum coherence (HSQC) and heteronuclear multiple quantum coherence (HMQC) ^1H–^{13}C experiments that hold potential for in vivo NMR applications. In our group HSQC has proved to be the most effective. We commonly and experimentally use a double insensitive nuclei enhanced by polarization transfer (INEPT) experiment, with sensitivity improvement States-TPPI (time proportional phase incrementation) phase cycling and globally optimized alternating phase rectangular pulse (GARP)-4 decoupling. W5 WATERGATE is concatenated to the end of the sequence to improve water suppression in very challenging samples.

1. Calibrate the 90° pulses for the sample. Adjust the number of increments, number of scans, time domain points, and recycle delay (*see* **Note 31**).
2. Set the evolution delay for the INEPT transfer (often $^{1}J_{HC}$ of 145 Hz is a good starting point).
3. Process the data with a sine-squared function, phase shifted by 90° in both dimensions, and a zero-filling factor of 2 (*see* **Note 32**).
4. For assignment of individual compounds, our group uses the Bio Reference databases with the Bruker Biospin using analysis mixture (AMIX) package as the software interface. However, free alternatives such as the Human Metabolome Database http://www.hmdb.ca/ are extremely useful. A list of metabolite libraries can be found here https://omictools.com/nmr-spectra-category

An example of a ^{1}H–^{13}C HSQC spectra with key metabolites assigned is shown below (*see* Fig. 2).

3.6.4 IP-ISQC Instructions for High-Resolution ^{1}H NMR

Proton (^{1}H) is the most common nucleus used in solution-state NMR spectroscopy because of its high sensitivity and high natural abundance (99.98%) and because it is present in the vast majority of organic molecules [16]. Unfortunately, in vivo ^{1}H NMR of organisms is considerably broadened due to susceptibility mismatches across the sample. In simple terms, components of the organisms, cell walls, membranes, and soluble components interact with the external field differently leading to field distortions that give rise to broad signals and the loss of metabolic information. The IP-ISQC experiment overcomes these distortions by correlating the chemical shift of free water to that of the solutes over distances larger than those from the distortions. The result is a 2D NMR experiment that can separate the susceptibility distortions from the chemical shift. When applied to a living system, high-resolution NMR spectra, near identical to a buffer extract, are obtained. Figure 3 shows the approach applied to a living worm submerged under oxygenated water. As this protocol is based on ^{1}H NMR detection, it can be applied to intact aquatic living organisms extracted directly from the environment without the need for isotopic enrichment.

The main limitation is the relaxation of the free water. The longer the water can evolve during t1, the greater the resolution of the resulting experiments. The relaxation of the solutes is also a consideration, but as these are spin-locked for the entire experiment, the relaxation of the free water remains the biggest limiting factor in terms of the resolution that can be obtained. The sample must have truly dissolved components/metabolites in it for the experiment to work. Example data sets, pulse sequence, and AU programs can be obtained from the corresponding author of the IP-ISQC experiment in this communication.

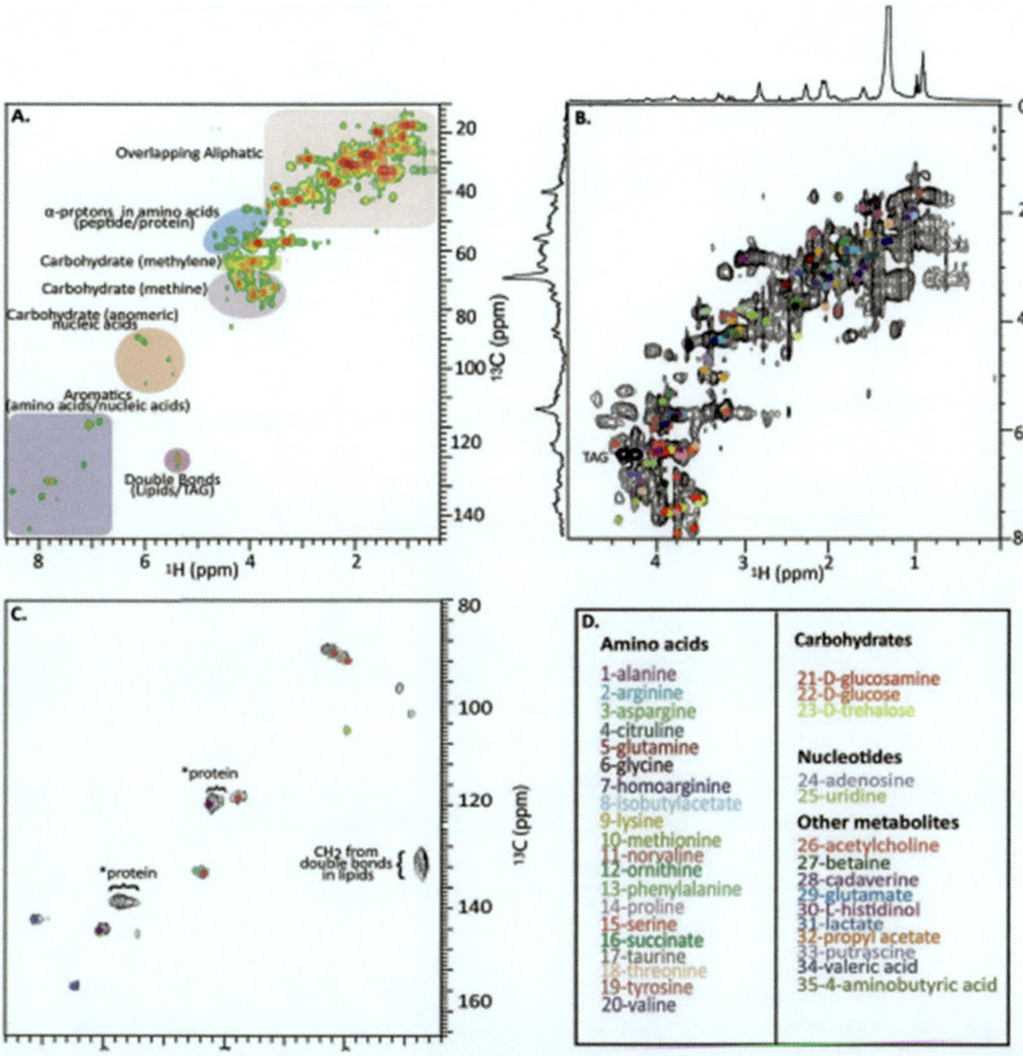

Fig. 2 In vivo ^{1}H–^{13}C HSQC of ^{13}C-enriched *Hyalella azteca* [17]. (**a**) HSQC with 40 metabolites assigned using Bruker's Bio-reference databases. (**b**) Aliphatic region, (**c**) aromatic region, (**d**) color-coded assignments corresponding to the dots in (**b**) and (**c**). Reproduced with permission from Royal Society of Chemistry

1. Optimize gpz1 for the greatest signal (*see* **Note 33**).
2. Optimize d21 for the greatest signal (*see* **Note 34**).
3. Choose the number of slices (td_f1) and the increment between slices (inf1). Both these parameters need to be entered in "eda" (*see* **Notes 35–37**).
4. Before the data processing, run the AU program proc_jres.be, which will set the key parameters in case any acquisition parameters are changed. This program can be requested from the corresponding author.

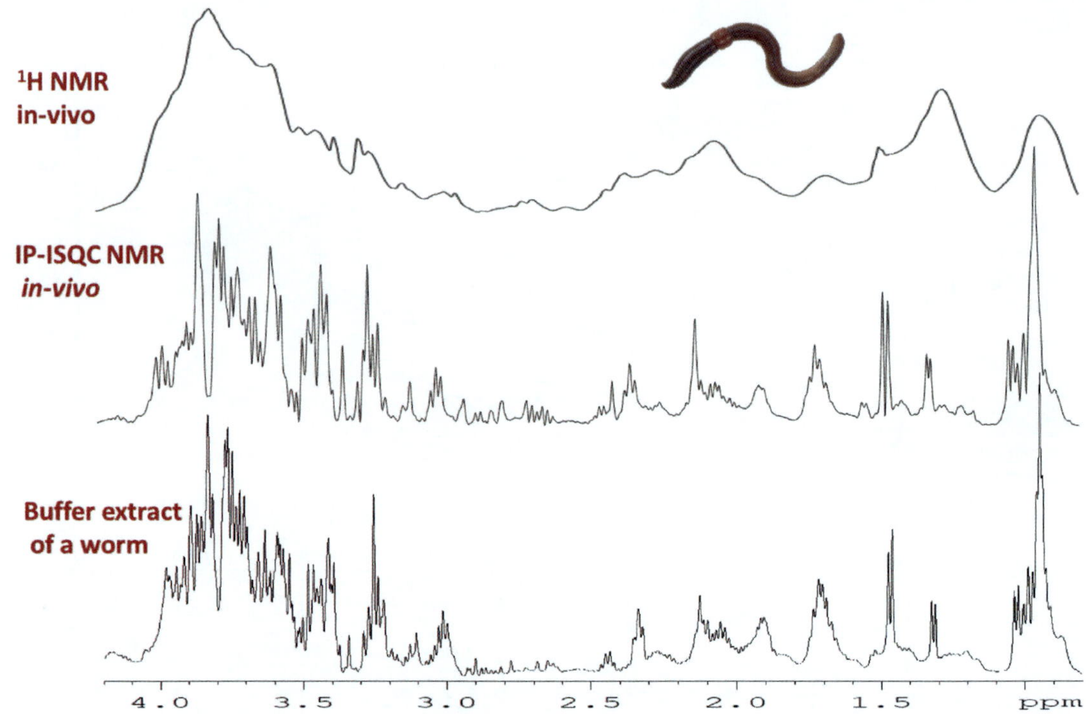

Fig. 3 Earthworm ^{1}H NMR; top = ^{1}H NMR in vivo, middle = IP-iSQC NMR in vivo, bottom = ^{1}H NMR of buffer extract of a worm [4]. Modified with permission from John Wiley and Sons

5. After running the program to further increase spectral resolution, reduce the values of NCOEF (F1) and TDoff (F1). Both are under "edp." Reduce them in pairs, for example, if the program sets them at NCOEF (F1) = 96 and TDoff (F1) −96, try 48/−48, 24/−24 12/−12 (*see* **Note 38**).
6. Type "symt" in topspin and choose "tilt along row." The value will need to be manipulated such that the spectrum tilts so that all the peaks are vertical (*see* **Note 39**).
7. There are now two choices: (a) sum the signal across the rows that have signal; (b) simply extract the central slice.

We recommend option b for in vivo and natural samples. It is quicker, easier, and less susceptible to problems with the exact tilt (*see* **Notes 40** and **41**).

4 Notes

1. In vivo NMR spectroscopy is applicable to a broad range of prominent model aquatic vertebrates and invertebrates such as *Daphnia* species [5, 9], *Hyalella azteca* shrimp [17], Japanese

medaka fish [18, 19], molluscs [20, 21], and marine worms [22, 23]. Theoretically, any organism that can fit and be maintained inside the NMR spectrometer can be studied.

2. ^{1}H–^{13}C 2D NMR helps reduce spectral overlap. It has been reported that the peak capacity of 1D NMR is 3000 whereas that of HSQC is 2,000,000 [24]. Due to the lower sensitivity of ^{1}H–^{13}C HSQC vs. 1D ^{1}H NMR, ^{13}C enrichment is required to permit the acquisition in a timely fashion.
3. It may be useful to enrich the model organism with a variety of single isotopes (^{2}H, ^{13}C, ^{15}N) or multiple enrichment (i.e., $^{13}C/^{15}N$) to increase the S/N and permit additional multidimensional NMR experiments.
4. For more information on isotopic enrichment, including duration vs. NMR signal and gut clearance, *see* reference [9].
5. The main reason for using unenriched *C. reinhardtii* is to prevent the background from the ^{13}C-enriched *D. magna* diet during the NMR experiment. At 1% natural abundance, ^{13}C in unenriched food is below the detection limit [9].
6. Capillary tube ID can vary depending on the NMR tube size and the total volume of the fluid.
7. The suction tube should have a wider ID compared to the injection tubes.
8. Make sure the injection tubing and the injection pump do not create air bubbles that can be transferred to the NMR tube.
9. Use a strong suction pump to ensure fast and complete pumping of water and food back to the reservoir. This eliminates the risk of clogging. FMI Q-pumps are strongly recommended.
10. Depending on the size of the model organism, different NMR tubes might be used.
11. The external lock system prevents D_2O from coming into contact with the organisms.
12. To place the injection tip adjacent to the NMR tube wall and maximize the space for the living organisms, we attach the injection tip to a bent glass tip.
13. Use dechlorinated municipal tap water [10] that is aged for at least 7 days and constantly aerated prior to use for *Daphnia* culturing.
14. For more information on *D. magna* culturing, *see* ref. 10.
15. It is recommended to use second or further generations of *D. magna* born from isotopically labelled parents for the test [9].

16. Depending on the model organism being used for the in vivo study, different culturing and treatment methodologies may be required before and during the in vivo NMR experiments.
17. Unenriched food is used to help purge the guts of ^{13}C-enriched ***Daphnia*** [9], for feeding the living organism during the experiment, and for studies where ^{13}C enrichment is not required.
18. The NMR tube is not sealed and instead an outer plastic tube is attached to the outside of the NMR tube with the capillary tubes passing through it.
19. Keeping the level of water just above the coil height, especially in cryoprobes (~4 cm from the bottom of NMR tube), enables efficient water suppression [25].
20. The temperature should be the same as the NMR probe temperature.
21. If the suction pump has a high flow rate or makes air bubbles, secure the tip of the tube above the surface of the water to prevent vigorous bubbling inside the water reservoir.
22. If air bubbles enter the NMR system spectral, quality can be lost; if a large air pocket forms, the organisms can die. Careful placement of the inlet tube avoids this complication.
23. To keep the total volume of fluid constant, fill the NMR tube using the water inside the reservoir.
24. The number of organisms that fit inside the NMR tube, for each test, depends on the size of the NMR tube and the size of the organism. For instance, taking into account the sensitivity of NMR, about 100 neonates or 20 ten-day-old daphnids provide optimal S/N without overcrowding the 5 mm NMR tube [9].
25. The flow rate of the suction pump should be at least twofold higher than the flow rate of the injection pump. Therefore, it can siphon off the fluid immediately keeping the level of water inside the NMR tube constant and at a desired level.
26. We find the injection flow rate of 0.5 mL/min sufficient for *Daphnia*.
27. Considering the model organism, the injection flow rate can be optimized.
28. Changes in the environment of the living organisms can be applied during the experiment and at different time points as desired.
29. In in vivo ^{1}H 1D NMR magnetic susceptibility effects (field distortions across a heterogeneous sample) broaden the ^{1}H NMR spectrum to such an extent that the vast majority of metabolite information is lost.

30. The majority of the sample is water, and shimming is critical for in vivo NMR to narrow the water line shape to the point where water suppression is effective.
31. In most ^{13}C-enriched samples, relaxation is quite rapid and can limit resolution in F_1 (indirect dimension). To assess how many indirect increments to collect, first run an experiment with a large number of increments. Then, selectively transform increments until the signal disappears in the transformed spectrum. Collecting data beyond this increment will only add noise and no additional resolution to the 2D spectrum.
32. In samples with weak signals, try replacing the sine-squared function in F2 with an exponential function corresponding to a line broadening of 10 Hz or 15 Hz.
33. This gradient breaks the symmetry in the sample allowing long range water/solute interactions to evolve. The strength of the gradient controls the distance over which the water and solutes interact. A recommended starting point is −10% using a 10A gradient amplifier. However, this should be optimized carefully and is sample specific. Poor optimization will lead to lower signal in the experiment.
34. This delay time permits the interactions between water and solute to build and equates to half the spin-lock period. The value will be sample dependent, can range from 5 ms to 120 ms, and needs careful optimization to obtain the maximum signal.
35. For samples with very broad lines, try infl at 1 ms and 24–128 slices (td_f1). For most samples, try infl at 2 ms and 24–128 slices (td_f1). For samples with sharp lines, try infl at 4 ms and 24–128 slices (td_f1).
36. The more slices and the larger the increment between slices, the better the resolution. Ultimately, however, the resolution is limited by the decay of the free water during t1. Try to balance the slices and increments so the sample's signal throughout the 2D and the signal has nearly fully decayed by the last slice.
37. Try running a quick experiment using infl at 4 ms and 128 slices with 4 scans, and then look at the FIDs to see how long the signal persists. Then, set the slices and increment based on the settings that give the best result. However, the quality of the line narrowing will depend on the relaxation of the free water in the sample.
38. Decreasing these values will increase the resolution of the central slice but at some point will result in phase twisting. Stop before this occurs. It may be necessary to do considerable phase correction in F1 each time the values are changed.

39. The spectrum will tilt depending on the degree of magnetic susceptibility distortions in the sample (i.e., more distortions, more tilt).
40. In in vivo samples, the overwhelming majority of signal occurs in the central slice. The exception may be for pure compounds in an inhomogeneous field. In this later case, the longer relaxation times lead to more elongation in F1, and summing may become more desirable.
41. When trying to sum with phase twists in the rows, you can try the Bruker AU program diff_apk2d.

Acknowledgments

Andre Simpson would like to thank the Natural Sciences and Engineering Research Council (NSERC) (Strategic (STPGP 494273-16) and Discovery Programs (RGPIN-2019-04165)), the Canada Foundation for Innovation (CFI), the Ontario Ministry of Research and Innovation (MRI), the Krembil Foundation for providing funding, and the Government of Ontario for an Early Researcher Award.

References

1. Soares DP, Law M (2009) Magnetic resonance spectroscopy of the brain: review of metabolites and clinical applications. Clin Radiol 64:12–21
2. Bunescu A, Garric J, Vollat B et al (2010) In vivo proton HR-MAS NMR metabolic profile of the freshwater cladoceran *Daphnia magna*. Mol BioSyst 6:121–125. https://doi.org/10.1039/b915417e
3. Bondu S, Kervarec N, Deslandes E, Pichon R (2008) The use of HRMAS NMR spectroscopy to study the in vivo intra-cellular carbon/nitrogen ratio of Solieria chordalis (Rhodophyta). J Appl Phycol 20:673–679. https://doi.org/10.1007/s10811-007-9231-2
4. Fugariu I, Bermel W, Lane D et al (2017) In-phase ultra high-resolution in vivo NMR. Angew Chemie - Int Ed 56:6324–6328. https://doi.org/10.1002/anie.201701097
5. Soong R, Nagato E, Sutrisno A et al (2015) In vivo NMR spectroscopy: toward real time monitoring of environmental stress. Magn Reson Chem 53:774–779. https://doi.org/10.1002/mrc.4154
6. Edison AS, Hall RD, Junot C et al (2016) The time is right to focus on model organism metabolomes. Meta 6. https://doi.org/10.3390/metabo6010008
7. Ebert D (2005) Ecology, epidemiology and evolution of parasitism in *Daphnia*
8. Akhter M, Dutta Majumdar R, Fortier-McGill B et al (2016) Identification of aquatically available carbon from algae through solution-state NMR of whole ^{13}C-labelled cells. Anal Bioanal Chem 408:4357–4370. https://doi.org/10.1007/s00216-016-9534-8
9. Majumdar RD, Akhter M, Fortier-McGill B et al (2017) In vivo solution-state NMR-based environmental metabolomics. eMagRes 6:133–148. https://doi.org/10.1002/9780470034590.emrstm1533
10. Environment Canada (2007) Biological test method: acute lethality test using *Daphnia* spp.
11. Stein JR (1973) Handbook of phycological methods: physiological and biochemical methods. Cambridge University Press, Cambridge
12. Giraudeau P, Silvestre V, Akoka S (2015) Optimizing water suppression for quantitative NMR-based metabolomics: a tutorial review. Metabolomics 11:1041–1055
13. Masoom H, Adamo A, Simpson AJ (2016) From the environment to NMR: water suppression for whole samples in their native state. Environ Chem 13:767–775. https://doi.org/10.1071/EN15139

14. Lam B, Simpson AJ (2008) Direct (1)H NMR spectroscopy of dissolved organic matter in natural waters. Analyst 133:263–269. https://doi.org/10.1039/b713457f
15. Pautler BG, Simpson AJ, Simpson MJ et al (2011) Detection and structural identification of dissolved organic matter in antarctic glacial ice at natural abundance by SPR-W5-WATERGATE ^{1}H NMR spectroscopy. Environ Sci Technol 45:4710–4717. https://doi.org/10.1021/es200697c
16. Zerbe O, Jurt S (2014) Applied NMR spectroscopy for chemists and life scientists. Wiley-VCH, Weinheim
17. Mobarhan YL, Fortier-McGill B, Soong R et al (2016) Comprehensive multiphase NMR applied to a living organism. Chem Sci 7:4856–4866. https://doi.org/10.1039/C6SC00329J
18. Viant MR, Pincetich CA, Hinton DE, Tjeerdema RS (2006) Toxic actions of dinoseb in medaka (*Oryzias latipes*) embryos as determined by in vivo ^{31}P NMR, HPLC-UV and ^{1}H NMR metabolomics. Aquat Toxicol 76:329–342. https://doi.org/10.1016/j.aquatox.2005.10.007
19. Pincetich CA, Viant MR, Hinton DE, Tjeerdema RS (2005) Metabolic changes in Japanese medaka (*Oryzias latipes*) during embryogenesis and hypoxia as determined by in vivo ^{31}P NMR. Comp Biochem Physiol Part C Toxicol Pharmacol 140:103–113. https://doi.org/10.1016/J.CCA.2005.01.010
20. Viant MR, Walton JH, Tjeerdema RS (2001) Comparative sublethal actions of 3-Trifluoromethyl-4-nitrophenol in marine molluscs as measured by in vivo ^{31}P NMR. Pestic Biochem Physiol 71:40–47. https://doi.org/10.1006/pest.2001.2554
21. Viant MR, Walton JH, TenBrook PL, Tjeerdema RS (2002) Sublethal actions of copper in abalone (*Haliotis rufescens*) as characterized by in vivo ^{31}P NMR. Aquat Toxicol 57:139–151. https://doi.org/10.1016/S0166-445X(01)00195-3
22. Kreutzer U, Jue T (1998) Metabolic response to oxygen limitation in Arenicola marina as determined with the ^{1}H NMR signals of myoglobin. Comp Biochem Physiol Part A Mol Integr Physiol 120:127–132. https://doi.org/10.1016/S1095-6433(98)10020-X
23. Bock C, Frederich M, Wittig R-M, Pörtner H-O (2001) Simultaneous observations of haemolymph flow and ventilation in marine spider crabs at different temperatures: a flow weighted MRI study. Magn Reson Imaging 19:1113–1124. https://doi.org/10.1016/S0730-725X(01)00414-3
24. Hertkorn N, Ruecker C, Meringer M et al (2007) High-precision frequency measurements: indispensable tools at the core of the molecular-level analysis of complex systems. Anal Bioanal Chem 389:1311–1327. https://doi.org/10.1007/s00216-007-1577-4
25. Tabatabaei Anaraki M, Dutta Majumdar R, Wagner N et al (2018) Development and application of a low-volume flow system for solution-state in vivo NMR. Anal Chem 90:7912–7921. https://doi.org/10.1021/acs.analchem.8b00370

Part X

Automation, Databases and Software Tools

Chapter 23

Tools for Enhanced NMR-Based Metabolomics Analysis

John L. Markley, Hesam Dashti, Jonathan R. Wedell, William M. Westler, and Hamid R. Eghbalnia

Abstract

Metabolomics is the study of profiles of small molecules in biological fluids, cells, or organs. These profiles can be thought of as the "fingerprints" left behind from chemical processes occurring in biological systems. Because of its potential for groundbreaking applications in disease diagnostics, biomarker discovery, and systems biology, metabolomics has emerged as a rapidly growing area of research. Metabolomics investigations often, but not always, involve the identification and quantification of endogenous and exogenous metabolites in biological samples. Software tools and databases play a crucial role in advancing the rigor, robustness, reproducibility, and validation of these studies. Specifically, the establishment of a robust library of spectral signatures with unique compound descriptors and atom identities plays a key role in profiling studies based on data from nuclear magnetic resonance (NMR) spectroscopy. Here, we discuss developments leading to a rigorous basis for unique identification of compounds, reproducible numbering of atoms, the compact representation of NMR spectra of metabolites and small molecules, tools for improved compound identification, quantification and visualization, and approaches toward the goal of rigorous analysis of metabolomics data.

Key words NMR, Metabolomics, Identification, Quantification, Numbering of atoms

1 Introduction

Owing to their close proximity to the functional endpoints that govern an organism's phenotype, metabolites are highly informative about functional states. As a result, the metabolome, which is the collection of small molecules associated with an organism, has become the target of a rapidly growing body of research over the past decade. Metabolomics data have been utilized for systems biology, disease diagnostics, biomarker discovery, and the broader characterization of small molecules in mixtures. Information acquired from nuclear magnetic resonance (NMR) spectroscopy, mass spectrometry (MS), databases, and published literature are combined to produce an interpretation of available data [1, 2].

G. A. Nagana Gowda and Daniel Raftery (eds.), *NMR-Based Metabolomics: Methods and Protocols*, Methods in Molecular Biology, vol. 2037, https://doi.org/10.1007/978-1-4939-9690-2_23, © Springer Science+Business Media, LLC, part of Springer Nature 2019

Although NMR has lower sensitivity when compared to MS, the utility of NMR as a rapid and nondestructive method is validated by its decades-long use in identifying and quantifying products and impurities in complex reaction mixtures [3]. NMR spectra are used routinely to identify and characterize molecules and molecular interactions in a wide range of applications [4–13]. The relatively short time needed to acquire a one-dimensional (1D) ^{1}H NMR spectrum makes this experimental modality ideal for routine, high-throughput profiling and screening procedures.

Study design, biological sample collection, and sample preparation are key considerations prior to NMR data collection. Protocols and procedures for sample collection and preparation are often designed based on the goals of the study and the specific tissue under investigation. Protocols for sample collection from human participants in a clinical study, or from animals in the field, are varied and often require additional validation steps when compared to samples collected in the laboratory setting. Samples for NMR analysis typically require minimal preparation. NMR data can be collected at different field strengths, using a diverse set of pulse sequences and experimental parameters, including delays and number of scans. Data analysis and data interpretation practices are also diverse, and one is faced with choosing from a variety of possible workflows with potentially different outcomes [14]. For metabolic flux analysis, and for studies involving stable isotope enrichment, detailed tracing of peaks and their correspondence to moieties is required. Identification and quantification is not the only modality in which NMR data are utilized for metabolomics studies. Approaches that focus on global changes in spectral data using methods such as principal component analysis (PCA), or similar multivariate methods, are also popular. These methods also require a range of mathematical transformations during data processing and specific statistical approaches for classification and discrimination [15]. As the field develops, suggested protocols are beginning to emerge from individual laboratories [16]. As more studies and protocols develop, more process-oriented databases are likely to emerge that will make the sharing of data about the findings of studies, the detailed workflow, and the computational processes more practical. Toward this goal, tools and methodologies that improve the interpretation and validation of NMR metabolomics studies are critically needed.

This chapter focuses on metabolite identification and quantification. We examine challenging elements of the workflow, including data processing, analysis, interpretation, and deposition. We discuss tools that we have developed to address known challenges. We conclude with future approaches to address needs of the field, including our plans for new tools.

2 A Brief Overview of NMR-Based Metabolomics

The peak positions and intensities from a reference NMR spectrum generally serve as the identifying signature for a compound. Reference spectra normally are collected under specific conditions of pH, temperature, and magnetic field strength, because changes in conditions can distort the identifying signatures of compounds. The general approach to NMR-based metabolomics profiling utilizes sets of chemical shifts and intensities from reference spectra, which constitute the "fingerprint" used to detect the presence of a particular compound in the mixture of small molecules and to estimate its concentration [17]. The process of identifying and quantifying a metabolite from an 1D-^{1}H NMR spectrum requires that NMR peaks from the experiment be matched against those of one- and two-dimensional reference spectra: for example, the experimental spectra archived at the Biological Magnetic Resonance Data Bank (BMRB) [18] or the Human Metabolome Database (HMDB) [19–21]. Full utilization of the reference NMR spectra in metabolic profiling requires the assignment of spectral transitions to the atoms of the molecule. In addition, effects of magnetic field strength on the positions and intensities of NMR peaks need to be taken into account (*see* Fig. 1).

Databases are being constructed to archive data from metabolomics experiments. MetabolomeXchange (http://www.metabolomexchange.org/site/) aggregates experimental data from four such repositories: Metabolites, Metabolomic Repository

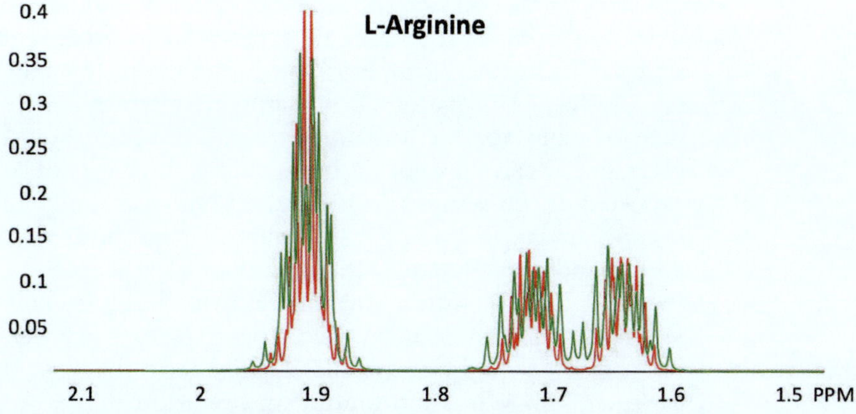

Fig. 1 Effect of magnetic field strength on a spectrum. Comparison of simulated ^{1}H NMR spectra of the 2.1–1.5 ppm region of L-arginine at 600 MHz (green) and 900 MHz (red). Aside from the improved resolution at 900 MHz, several peaks observed at 600 MHz are not present at 900 MHz, for example, those around 1.68 ppm. GISSMO software [22] was used in determining the spin system matrix for L-arginine by fitting the 500 MHz NMR spectrum of the compound from the BioMagResBank (BMRB) small molecule database [18]; the resulting spin system matrix was used subsequently to simulate the spectra at the two higher magnetic field strengths

Bordeaux, Metabolomics Workbench, and Metabolonote. The National Institutes of Health (NIH) Common Fund Metabolomics Program has supported the establishment of a Data Repository and Coordinating Center (DRCC). The DRCC initiative is a response to the burgeoning development of databases. Proper identification of metabolites is key to establishing the identity of metabolites and the ability to cross-reference these small molecules across databases. Recent progress in establishing a unique naming and labeling convention that is entirely based on the InChI standard [23] has led to some remediation in a few databases, but much work remains to be done in this arena.

As an analytical method, NMR experimental data are highly reproducible; however, their interpretation often is subject to variability and uncertainty. A robust and reliable interpretation of data must include the following: (a) a rigorous quantification of uncertainty, (b) reliable small sample performance, and (c) wide applicability to different biological samples and sample conditions. For example, rigorous quantification must enable the comparison of uncertainty for results of experiments replicated in different laboratories. Moreover, because most exploratory data are obtained for a small number of samples, estimates of metabolite concentration and related uncertainties must be reliable despite the small number of samples.

3 What Would Constitute Optimal Tools for NMR-Based Metabolomics?

Ideally, metabolomics tools should be capable of harnessing information available on small molecules of interest in biological systems federated from a wide range of databases and other sources at all stages of an investigation. For example, in the early stages of sample preparation, information about sample collection and preparation methods and applicable data from prior studies should be easily accessible. Relevant information would include lists of compounds expected to be present in different biological fluids and tissue/organ extracts, including metabolites, drugs, and environmental compounds. NMR and MS signatures of these compounds are required, as is quantum chemical information on compounds for use in studies of molecular interactions. The keys to federating such information would be a system of easily derived unique descriptors for each compound and unique and reproducible designators for each atom present in each compound.

Libraries of known metabolites must be "dynamic" in the sense that templates for identification of metabolites must account for varying conditions such as field strength, pH, relaxation, and other effects. The tools that use library templates for analyzing the metabolomics data should integrate the information about the uncertainty of identification and quantification in their output. These tools should also identify gaps in the underlying information, such as compounds not present or improperly represented in particular databases, and assist in their remediation. In addition, data and workflow should be constructed in a way that streamlines data deposition and experimental reproducibility.

4 Naming and Numbering of Compounds for NMR (ALATIS)

Studies of small molecules rely on the unique and reproducible identification of each individual compound. The problem is that compound naming is not always unique (same descriptor can be used for more than one compound or multiple descriptors are used for a single compound). What uniquely discriminates one chemical compound from another is its three-dimensional chemical structure, which can distinguish isomers, enantiomers, and isotopomers. For this reason, we have advocated the adoption of InChI strings created from three-dimensional structure files as unique and reproducible compound identifiers. We have created an algorithm called ALATIS (Atom Label Assignment Tool using InChI String) and have incorporated it into software that produces a valid InChI string from a structure file in multiple standard formats (mol, sdf, cdx, pdb) [23]. In addition to creating the InChI string, ALATIS produces unique and reproducible numbering of all atoms in the compound. The universal ALATIS atom designators can be cross-referenced to specific atom designators in use in different scientific domains. The ALATIS approach enables the construction of validated cross-references among all small molecule databases that include 3D structures. In addition, the universal atom designators provide a mechanism for consolidating atom-specific information from these databases, such as NMR chemical shifts and coupling constants. ALATIS is embodied in a publicly available webserver located at (http://alatis.nmrfam.wisc.edu/). The ALATIS naming system has been fully implemented in the BMRB small molecule database and has been adopted by the NMReData initiative [24]. Through the application of ALATIS technology, we have been developing a federated database of information on metabolites and other small molecules that now includes more than 91,600,000 entries from PubChem and more than 400,000 entries from other databases: (http://gateway.nmrfam.wisc.edu/). Although most of these databases utilize InChI strings to represent individual compounds, our analysis has revealed widespread inconsistencies in their application that suggest the need for remediation [23].

5 NMR Templates for Metabolites and Other Compounds of Interest (GISSMO)

A spin system matrix, which parameterizes relationships among transitions, provides a much richer feature set for a compound than a spectral signature based on peak positions and intensities [25, 26]. Spin system matrices expand the applicability of NMR spectral libraries beyond the specific condition under which data were collected. In addition to their capability of simulating spectra at any field strength, spin system matrix parameters can be adjusted to systematically explore alterations in chemical shift patterns due to variations in other experimental conditions, such as pH or temperature.

Several simulation software packages have been developed for the purpose of predicting NMR experimental spectra (for a list *see* http://www.east-nmr.eu/index.php/databases-and-links). Among the nonproprietary software packages, NMRdb [27], GAMMA [28], and Spinach [29] are those more commonly used. The focus of these software packages is to produce an accurate approximation of the experimental data based on empirical or quantum mechanical computations—they are not designed to build spin system matrices by matching experimental spectra. A few methods have been introduced whose goal has been to automate fitting peak shapes to experimental spectra [17, 30–32]. We have developed an approach called GISSMO (Guided Ideographic Spin System Model Optimization) [22, 33] that facilitates the derivation of NMR spin system matrices by optimizing the fit between an experimental one-dimensional ^{1}H NMR spectrum and the theoretical spin system matrix. GISSMO provides a graphical user interface (GUI) with several semi-automatic optimization modules. The GUI enables the splitting of spin system matrices so that portions can be optimized prior to merging. The GUI also allows for the consideration of couplings between ^{1}H nuclei and other types of nuclei (e.g., ^{31}P), and it supports the use of auxiliary two-dimensional spectra in refining chemical shifts and couplings. GISSMO utilizes ALATIS-derived compound identifiers and atom nomenclature [23].

The initial release of the GISSMO GUI reported spin system matrices for about 400 compounds from the BMRB archive [33]. In the current release, the number of entries from BMRB has increased to 511. In addition, the library now contains optimized spin system matrices for 666 molecular fragments from the Maybridge Ro3 fragment library (https://www.maybridge.com/), which are routinely utilized in NMR-based ligand screening for drug development. For every GISSMO entry in the library, we utilized ALATIS to identify its corresponding BMRB entry and applied an in-house text-processing module to extract the associated bio-locations of the compounds. The result, which is

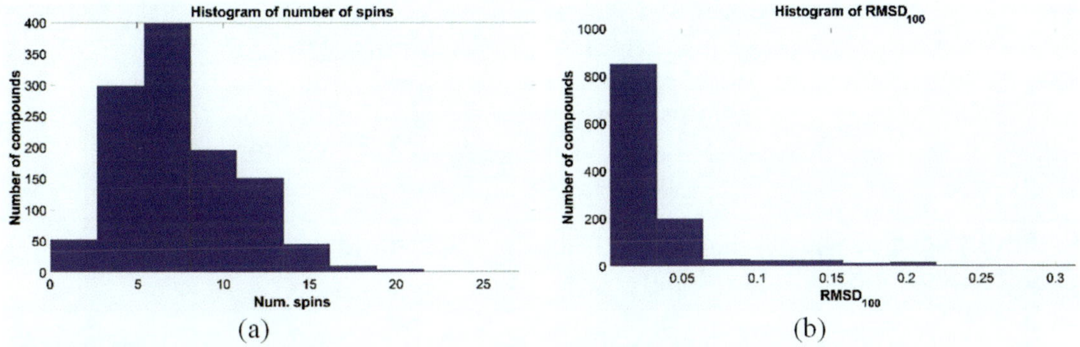

Fig. 2 Histograms showing the number of spins and normalized RMSD of the entries in the current release of GISSMO library. (**a**) This histogram shows the number of spins (*x*-axis) versus the number of compounds (*y*-axis). (**b**) Normalized RMSD between simulated and experimental spectra. While the majority of simulated spectra are accurate representations of the experimental data (RMSD <0.1), there are cases where RMSD values are higher than 0.1 owing to relaxation effects or low signal-to-noise ratio of the experimental spectra

available on the GISSMO website, indicates that 338 out of 511 (66%) of the compounds have been observed in at least one cellular or tissue location. This library of optimized spin system matrices is publicly available in XML, NMR-STAR, and NMReData formats from GISSMO website. Figure 2 displays a histogram showing compounds as a function of the number of NMR spins analyzed and a histogram with a validation of the optimized spin system matrices in terms of the normalized $RMSD_{100}$ between the experimental and fitted 1D-^{1}H NMR spectra [34].

Because the ^{1}H NMR spectra of many compounds are not strictly first order, a major challenge of databases containing reference NMR data is how to deal with their dependence on field strength. One approach is to collect reference spectra at a single field strength and require that this field strength be used for collecting experimental spectra. Alternatively, some reference databases contain data collected at more than one field strength. A more general solution to this challenge is to derive parametric representations of ^{1}H NMR spectra in terms of spin system matrices, which can be utilized to generate spectra at any desired magnetic field strength. We have taken advantage of this feature of spin system matrices to produce spectra of all compounds in the GISSMO database at ^{1}H resonance frequencies of 40, 60, 80, 90, 100, 200, 300, 400, 500, 600, 700, 750, 800, 900, 950, 1000, and 1300 MHz. These spectra, which cover the range of magnetic fields used in NMR spectroscopy and MRI, can be accessed and downloaded from the GISSMO website. To display the spectra, we utilize the open source graphing library Plotly (https://plot.ly/), which provides interactive zooming and pan features for visual investigation of the spectra. The downloadable spectra are formatted in two columns (ppm and amplitude) as a comma-separated values (CSV) file. These files can be loaded into NMR software programs such as Mestrelab Mnova (http://mestrelab.com) and NMRFx [35], which are accessible through the NMRbox project [36] (https://www.nmrbox.org/). In addition, these files can be easily loaded using any scripting and programming languages.

6 Advantages of Spectra Simulated from Spin System Matrices as References in Spectral Profiling

Spectral peak pattern matching is a common approach for profiling of small molecules. This approach involves peak picking the NMR spectra and subsequently using the resulting chemical shifts to search for matching peak patterns in a small molecule reference database. This process of identification relies strongly on the spectral peak lists, especially those archived in the reference databases. Because these databases utilize peak picking programs on their archived experimental spectra, the accuracy and reliability of the reference spectral peaks depend on a variety of factors including the correct identification of the reference compound, the presence of impurities, spectral artifacts from water signal suppression or other sources, and the signal-to-noise ratio of the experimental spectra. By contrast, the reference spectra generated from parameterized spin system matrices are noise-free, contain no impurity peaks, and are free from spectral artifacts. In addition, they serve to validate the identity of the reference compound.

6.1 Peak Lists and Searching

Generating peak lists from these highly refined spectra can be readily achieved by standard peak picking approaches. We utilized the peak picking modules of the Mnova program, under both the "Standard" and "Global Spectral Deconvolution (GSD)" options, to generate peak lists from the entire library of compounds parameterized by GISSMO at all aforementioned magnetic field strengths. We used these to generate interactive lists of chemical shifts and peak amplitudes (standard or GSD) for each compound at a selectable magnetic field strength, which are available on the GISSMO website. Clicking on a chemical shift brings up a region of the ^{1}H NMR spectrum with the corresponding peak identified.

A "Peak Search" module is now available on the GISSMO website. The search is linked to a PostgreSQL (https://www.postgresql.org/) database containing all curated spectral peak lists. The user can query the database by specifying one or more peak positions in ppm (standard or GSD) with a specified tolerance at a selected magnetic field strength. The result is a list of compounds associated with the queried peaks within the specified matching tolerance. The compounds returned from the query are sorted based on the minimum differences between the queried peaks and those archived in the database. Users can investigate the resulting compounds by browsing the corresponding GISSMO webpages, downloading the GISSMO entries, or downloading the output in CSV format. Additional details regarding the results are provided on the website.

6.2 Simulated Spectra of Compound Mixtures

^{1}H NMR is widely used to identify and quantify metabolites in biological fluids or tissue extracts. We are developing tools based on spin system matrix parameterizations for use in NMR-based metabolomics. As a first step, we have created a module for simulating ^{1}H NMR spectra of compound mixtures [33]. This module utilizes the archived spin system matrices in the GISSMO library that have been manually optimized against the reference spectra. As its input, the module accepts a list of these compounds and their corresponding concentrations. Users can upload a CSV file or provide them through an interactive webpage on GISSMO's website. The spectrum of the simulated mixture can be downloaded as a two-column CSV file. A user-controlled slider enables adjustments of the components in the simulated spectrum to achieve the best match to an experimental spectrum containing the same components. The mixture simulation module uses the Plotly library to display the spectra. As an example, Fig. 3 shows ^{1}H NMR spectra of blood plasma simulated at two magnetic resonance fields.

The mixture module [33] accepts an optional experimental spectrum in CSV format and displays an overlay of the uploaded experimental and simulated mixtures on the website. Comparisons of simulated spectra with experimental spectra of mixtures can be

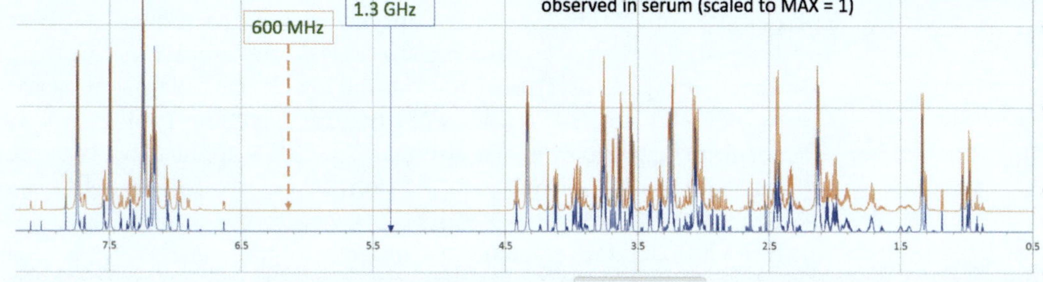

Fig. 3 Simulated spectrum of a biological mixture created from GISSMO spectral templates. Simulated ^{1}H NMR spectra at 600 MHz (red) and 1.3 GHz (blue) of the 34 most abundant metabolites of blood plasma at their relative concentrations. The difference spectrum (green) shown above documents the additional level of detail available at the higher magnetic field

used to validate prior analyses of compounds present and their relative concentrations. If experimental spectra of mixtures have been collected at multiple fields, the simulation can provide an additional layer of validation. Additional effects from pH differences and molecular interactions can, in principle, be incorporated into the ideographs representing the spin system matrix of each compound. We are currently working to expand this first step of processing mixture spectra by developing ways of optimizing spin system matrices against experimental spectra of biological samples.

7 Dynamic Metabolite Libraries: Archiving, Maintenance, and Retrieval of Data

The establishment of permanent and dynamic data records is a key requisite for robust scientific progress in the developing and expanding field of metabolomics. Owing to its flexible data model, the NMR-STAR ontology at the BMRB has evolved gracefully over time to include tags that accommodate almost all requirements for the archiving of metabolite library records [37]. The database contains more than 1400 entries that correspond to different experimental conditions (sample conditions and magnetic field strengths) of 1250 small molecules. For the majority of these entries, BMRB archives 10 different NMR spectra that can be used in metabolomics profiling procedures. Permanence and identity of records for small molecules has been achieved through the adoption of the ALATIS identifier, and entries with a corresponding GISSMO spin system matrix are cross-linked to GISSMO website. Data records for each metabolite contain cross-references to multiple relevant databases, including PubChem, KEGG, BioCyc, PDB, literature records, and other sources. In addition to references to the GISSMO record, the entries provide detailed data relevant to sample conditions. Additional tags for connecting additional experimental records, for example from MS, provide a path to full data integration. Because the NMR-STAR format is self-defining, it is a dynamic data record that can evolve over time in order to satisfy new demands on additional content relevant to each metabolite. Users can retrieve data from the NMR-STAR archive by means of query interfaces available on the BMRB website. The instant search on BMRB website provides a fast pattern matching tool to look up small molecule names. In addition, BMRB's application program interface (API) (https://github.com/uwbmrb/BMRB-API) and BMRB's FTP access (http://www.bmrb.wisc.edu/ftp/pub/bmrb/) facilitate batch download of the entries and their corresponding NMR experimental spectra. The PyNMRSTAR program (https://github.com/uwbmrb/PyNMRSTAR) has been developed for easy access and processing of NMR-STAR files. This program can be used for seamless extraction of every information archived in the corresponding NMR-STAR file of the compounds (e.g., compound name, sample condition, assigned chemical shifts, and list of experimental data associated). The reference entries of the BMRB database are constantly being expanded to cover the small molecules routinely studied in biomedical research.

8 Discussion

A number of initiatives to standardize metabolomics data formats are underway, and these efforts are expected to face the same growing pains as those in other omics fields. In the EU, the "Coordination of Standards in Metabolomics" (COSMOS) project is developing infrastructure and exchange standards for metabolomics within the European metabolomics community [38]. Along with standardization, national and international funding agencies are increasingly requesting publicly funded research data and software to become *open access*. In practice, changes in standards are the result of social, technical, economic, political, and legal activities, which include strategic decisions that can have a dramatic effect on day-to-day operations of the research enterprise. Moreover, the impact felt by changing practices is not homogeneous. For example, defining a new data exchange protocol has a significant impact on technology developers, but may have little or no impact on technology users. On the other hand, the use of standard vocabularies and protocols within defined laboratory management systems has its highest impact on technology users. Nonetheless, apt technological innovations can find balanced solutions that address the tension between standardization and the scientific endeavor.

For example, decoupling data exchange, data storage, and compute-cycle delivery would enable continued scientific innovation at the computational level, while, at the same time, meeting the goals of standardization and data preservation. One approach to decoupling data exchange and storage is to adopt a flexible data model. The Self-Defining Text Archive and Retrieval (STAR) specification [39, 40] has served as the model for the mmCIF format used by the Protein Data Bank [41] as well as the NMR-STAR format used by the BioMagResBank (BMRB) [37, 42]. NMR-STAR is a backward compatible self-defining format in which every data item has attributes that describe its features, including explicit definitions of relationships among data items. It is human-readable, and no domain knowledge is required to read the files. The format is easily converted to XML; it supports the requisite RDF (Resource Description Framework); and its flexibility provides a strong hedge against the potential of future disruptions due to format. The decoupling of compute-cycle delivery can be achieved by using a virtual machine as the technology delivery vehicle. Moreover, the use of virtual machines simplifies the long-term preservation of data

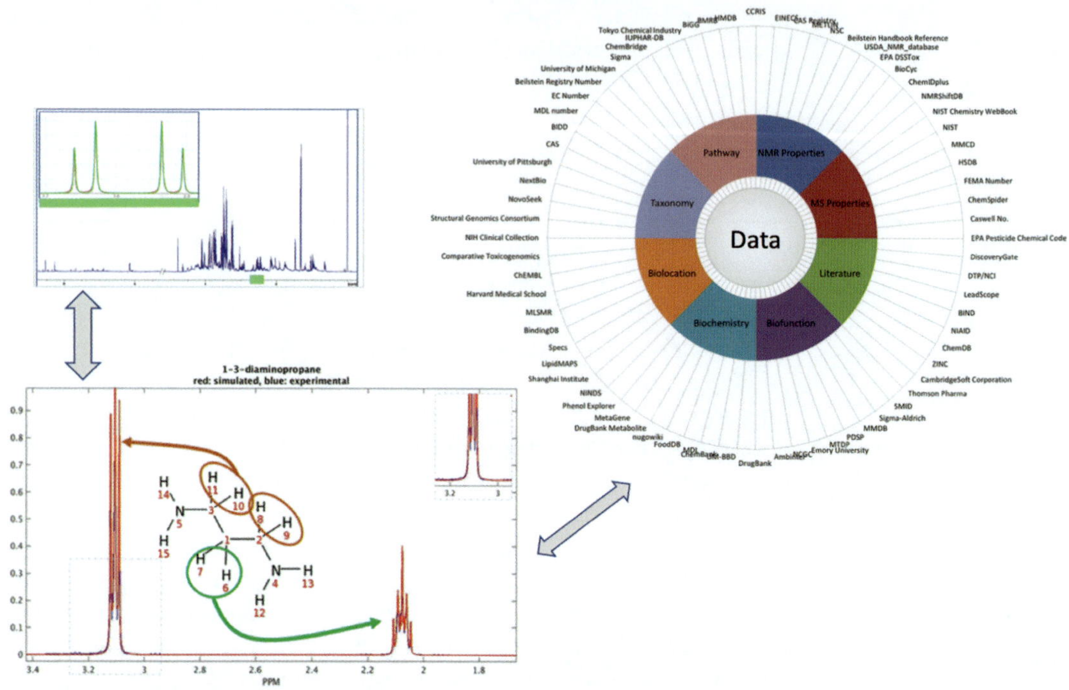

Fig. 4 Proposed federation of information from multiple databases on the basis of standard InChI strings and the ALATIS universal atom numbering system [23]. The wheel lists various databases from different scientific domains that can be integrated. The inset indicates a compound with ALATIS atom numbering and arrows indicate ^{1}H NMR signals associated with particular atoms. This information can then be used in analyzing NMR spectra of compound mixtures

and procedures. More broadly, we can improve the odds for success by adopting newer design methodologies and architectures.

We envision the use of these tools in constructing a federated database containing a vast reservoir of knowledge of relevance to metabolomics (*see* Fig. 4) [34]. Rigor within this database will be ensured through unique identification of compounds through standard InChI strings and unique and reproducible ALATIS atom numbering [23]. And we are developing an automated probabilistic approach to the analysis of NMR spectra of metabolite mixtures based on GISSMO-derived templates for individual compounds tuned to experimental conditions.

9 Conclusions

To enable the precise reproduction of results, it is important to define reporting requirements associated with experimental design, data acquisition, data processing, and downstream statistical analysis and interpretation. Ideally, this will include a complete description of the workflow and access to the versions of the software used.

The scientific enterprise is an increasingly interconnected activity in which data exchange and data preservation are both essential requirements.

Acknowledgments

This work received funding from the National Institutes of Health (NIH). Grants from the NIH National Institute of General Medical Science provided support to the National Magnetic Resonance Facility at Madison (P41 GM103399) and the Biological Magnetic Resonance Data Bank (R01 GM109046) and partial support for HRE, HD, and JRW (P41 GM111135 to the National Center for Biomolecular NMR Data Processing and Analysis). H.D. currently is supported by National Heart, Lung, and Blood Institute grant T32 HL007575.

References

1. Misra BB (2018) New tools and resources in metabolomics: 2016-2017. Electrophoresis 39 (7):909–923
2. Misra BB, Mohapatra S (2018) Tools and resources for metabolomics research community: A 2017–2018 update. Electrophoresis. doi: https://doi.org/10.1002/elps.201800428. [Epub ahead of print]
3. Halabalaki M, Vougogiannopoulou K, Mikros E, Skaltsounis AL (2014) Recent advances and new strategies in the NMR-based identification of natural products. Curr Opin Biotechnol 25:1–7
4. Ravanbakhsh S, Liu P, Bjorndahl TC, Mandal R, Grant JR, Wilson M et al (2015) Accurate, fully-automated NMR spectral profiling for metabolomics. PLoS One 10(5): e0124219
5. Clarke CJ, Haselden JN (2008) Metabolic profiling as a tool for understanding mechanisms of toxicity. Toxicol Pathol 36(1):140–147
6. Zhang S, Liu L, Steffen D, Ye T, Raftery D (2012) Metabolic profiling of gender: headspace-SPME/GC–MS and 1H NMR analysis of urine. Metabolomics 8(2):323–334
7. Beckonert O, Keun HC, Ebbels TM, Bundy J, Holmes E, Lindon JC et al (2007) Metabolic profiling, metabolomic and metabonomic procedures for NMR spectroscopy of urine, plasma, serum and tissue extracts. Nat Protoc 2(11):2692–2703
8. Jupin M, Michiels PJ, Girard FC, Spraul M, Wijmenga SS (2014) NMR metabolomics profiling of blood plasma mimics shows that medium- and long-chain fatty acids differently release metabolites from human serum albumin. J Magn Reson 239:34–43
9. Larive CK, Barding GA, Dinges MM (2015) NMR spectroscopy for metabolomics and metabolic profiling. Anal Chem 87(1):133–146
10. Elmsjo A, Rosqvist F, Engskog MK, Haglof J, Kullberg J, Iggman D et al (2015) NMR-based metabolic profiling in healthy individuals overfed different types of fat: links to changes in liver fat accumulation and lean tissue mass. Nutr Diabetes 5:e182
11. Palma M, Hernandez-Castellano LE, Castro N, Arguello A, Capote J, Matzapetakis M et al (2016) NMR-metabolomics profiling of mammary gland secretory tissue and milk serum in two goat breeds with different levels of tolerance to seasonal weight loss. Mol BioSyst 12(7):2094–2107
12. Vermathen M, Paul LEH, Diserens G, Vermathen P, Furrer J (2015) ^{1}H HR-MAS NMR based metabolic profiling of cells in response to treatment with a Hexacationic ruthenium Metallaprism as potential anticancer drug. PLoS One 10(5):e0128478
13. Whigham LD, Butz DE, Dashti H, Tonelli M, Johnson LK, Cook ME et al (2014) Metabolic evidence of diminished lipid oxidation in women with polycystic ovary syndrome. Curr Metabolomics 2(4):269–278
14. Bingol K (2018) Recent advances in targeted and untargeted metabolomics by NMR and MS/NMR methods. High Throughput 7(2): E9. https://doi.org/10.3390/ht7020009

15. Worley B, Powers R (2014) MVAPACK: a complete data handling package for NMR metabolomics. ACS Chem Biol 9 (5):1138–1144
16. Vignoli A, Ghini V, Meoni G, Licari C, Takis PG, Tenori L et al (2018) High-throughput metabolomics by 1D NMR. Angew Chem Int Ed Engl. https://doi.org/10.1002/anie.201804736e. [Epub ahead of print]
17. Napolitano JG, Lankin DC, McAlpine JB, Niemitz M, Korhonen S-P, Chen S-N et al (2013) Proton fingerprints portray molecular structures: enhanced description of the 1H NMR spectra of small molecules. J Org Chem 78(19):9963–9968
18. Ulrich EL, Akutsu H, Doreleijers JF, Harano Y, Ioannidis YE, Lin J et al (2008) BioMagResBank. Nucl Acids Res 36(suppl 1):402–408
19. Wishart DS, Jewison T, Guo AC, Wilson M, Knox C, Liu Y et al (2012) HMDB 3.0--the human metabolome database in 2013. Nucl Acids Res 41(Database issue):D801–D807
20. Wishart DS, Knox C, Guo AC, Eisner R, Young N, Gautam B et al (2008) HMDB: a knowledgebase for the human metabolome. Nucleic Acids Res 37(Database issue): D603–D610
21. Wishart DS, Tzur D, Knox C, Eisner R, Guo AC, Young N et al (2007) HMDB: the human Metabolome database. Nucleic Acids Res 35 (Database issue):D521–D526
22. Dashti H, Westler WM, Tonelli M, Wedell JR, Markley JL, Eghbalnia HR (2007) Spin system modeling of nuclear magnetic resonance spectra for applications in metabolomics and small molecule screening. Anal Chem 89 (22):12201–12208
23. Dashti H, Westler WM, Markley JL, Eghbalnia HR (2017) Unique identifiers for small molecules enable rigorous labeling of their atoms. Sci Data 4:170073
24. Pupier M, Nuzillard JM, Wist J, Schlörer Nils E, Kuhn S, Erdelyi M et al (2018) NMReDATA, a standard to report the NMR assignment and parameters of organic compounds. Magn Reson Chem 56(8):703–715
25. Anderson WA, McConnell HM (1957) Analysis of high-resolution NMR spectra. J Chem Phys 26:1496
26. Corio PL (1960) The analysis of nuclear magnetic resonance spectra. Chem Rev 60:363–429
27. Binev Y, Marques MMB, Aires-de-Sousa J (2007) Prediction of 1H NMR coupling constants with associative neural networks trained for chemical shifts. J Chem Inform Modeling 47(6):2089–2097
28. Smith SA, Levante TO, Meier BH, Ernst RR (1994) Computer simulations in magnetic resonance. An object-oriented programming approach. J Magn Reson Series A 106 (1):75–105
29. Goodwin DL, Kuprov I (2015) Auxiliary matrix formalism for interaction representation transformations, optimal control, and spin relaxation theories. J Chem Phys 143 (8):084113
30. Cheshkov DA, Sinitsyn DO, Sheberstov KF, Chertkov VA (2016) Total lineshape analysis of high-resolution NMR spectra powered by simulated annealing. J Magn Reson 272:10–19
31. Laatikainen R, Niemitz M, Weber U, Sundelin J, Hassinen T, Vepsäläinen J (1996) General strategies for total-Lineshape-type spectral analysis of NMR spectra using integral-transform iterator. J Mag Reson Series A 120(1):1–10
32. Stephenson DS, Binsch G (1980) Automated analysis of high-resolution NMR spectra. I. Principles and computational strategy. JMagReson (1969) 37(3):395–407
33. Dashti H, Wedell JR, Westler WM, Tonelli M, Aceti D, Amarasinghe GK et al (2018) Applications of parametrized NMR spin systems of small molecules. Anal Chem 90 (18):10646–10649
34. Dashti H, Wedell JR, Cornilescu G, Schwieters C, Westler WM, Markley JL et al (2018) Robust nomenclature and software for enhanced reproducibility in molecular modeling of small molecules. bioRxiv:429530. https://doi.org/10.1101/429530
35. Norris M, Fetler B, Marchant J, Johnson BA (2016) NMRFx processor: a cross-platform NMR data processing program. J Biomol NMR 65(3–4):205–216
36. Maciejewski MW, Schuyler AD, Gryk MR, Moraru I, Romero PR, Ulrich ER et al (2017) NMRbox: a resource for biomolecular NMR computation. Biophy J 112 (8):1529–1534
37. Ulrich EL, Baskaran K, Dashti H, Ioannidis YE, Livny M, Romero PR et al (2018) NMR-STAR: comprehensive ontology for representing, archiving and exchanging data from nuclear magnetic resonance spectroscopic experiments. J Biomol NMR. https://doi.org/10.1007/s10858-018-0220-3. [Epub ahead of print]
38. Salek RM, Neumann S, Schober D, Hummel J, Billiau K, Kopka J et al (2015) Coordination of standards in metabolomics (COSMOS): facilitating integrated metabolomics data access. Metabolomics 11(6):1587–1597

39. Hall SR (1991) The STAR file: a new format for electronic data transfer and archiving. J Chem Inf Comput Sci 31:326–333
40. Hall SR, Cook APF (1995) STAR dictionary definition language: initial specification. J Chem Inf Comput Sci 35:819–825
41. Fitzgerald PMD, Westbrook JD, Bourne PE, McMahon B, Watenpaugh KD, Berman HM (2005) 4.5 macromolecular dictionary (mmCIF). In: International tables for crystallography G definition and exchange of crystallographic dataedited by Hall SR, McMahon B. Springer, Dordrecht, pp 295–443
42. Ulrich EL, Argentar D, Klimowicz A, Markley JL (1996) STAR/CIF macromolecular NMR data dictionaries and data file formats. Acta Crystallogr A 52(a1):C577–C577

Chapter 24

Automated Tools for the Analysis of 1D-NMR and 2D-NMR Spectra

Matthias Lipfert, Manoj Kumar Rout, Mark Berjanskii, and David S. Wishart

Abstract

Nuclear magnetic resonance (NMR) spectroscopy is becoming increasingly automated. Most modern NMR spectrometers are now equipped with auto-tune/auto-match probes along with automated locking and shimming systems. Likewise, more and more instruments, especially for NMR-based metabolomics applications, are equipped with automated sample changers. All this instrumental automation allows NMR data to be collected at a rate of >100 samples/day. However, a continuing bottleneck in NMR-based metabolomics has been the time required to manually analyze and annotate the collected NMR spectra. In many cases, manual spectral annotation and analysis can take one or more hours per spectrum. Fortunately, over the past few years, several software tools have been developed that largely automate the spectral deconvolution or spectral annotation process. Using these tools requires that the samples must be prepared and the NMR spectra must be acquired in a very specific manner. In this chapter, we will describe the step-by-step preparation of biofluid samples along with the required protocols for acquiring optimal spectra for automated NMR metabolomics analysis. We will also discuss the use of three common tools (Chenomx NMR Suite, Bayesil, and COLMARm) for (semi-) automated profiling, and annotation of 1D- and 2D-NMR spectra of biofluids.

Key words Nuclear magnetic resonance, Metabolomics, Sample preparation, Software, Automated profiling

1 Introduction

The richness of information contained in a single high-resolution NMR spectrum combined with the ability to simultaneously identify multiple metabolites with little or no sample preparation makes NMR one of the most commonly used platforms for metabolomics studies. While NMR is relatively insensitive compared to mass spectrometry (MS), NMR spectroscopy has a number of unique advantages. Specifically, NMR spectroscopy is non-destructive, quantitative (and therefore unbiased), and it requires little or no chromatographic separation nor chemical derivatization. For novel

G. A. Nagana Gowda and Daniel Raftery (eds.), *NMR-Based Metabolomics: Methods and Protocols*, Methods in Molecular Biology, vol. 2037, https://doi.org/10.1007/978-1-4939-9690-2_24,

biofluids, NMR permits the identification of novel compounds. Furthermore, NMR is highly automatable and exceptionally reproducible, making large, high throughput metabolomics studies much more feasible with NMR rather than MS-based techniques.

NMR spectra from "metabolically rich" biofluids such as urine can contain up to 5000 detectable resonances [1]. This spectral complexity has led to the development of two distinct approaches for collecting, processing, and interpreting NMR data. One approach, called targeted metabolomics, uses spectral deconvolution software to identify and quantify fluid-specific or targeted compounds in individual NMR spectra [2]. The resulting tables of compounds and concentrations are used to identify interesting trends or features via multivariate statistics. The other approach, called untargeted metabolomics or statistical spectroscopy [3], uses spectral alignment, spectral binning, and multivariate statistics to identify interesting spectral features. Both targeted and untargeted metabolomics have their advantages and disadvantages [2]. Targeted metabolomics is more precise, more quantitative, and generally much more reproducible. However, it is generally more time-consuming (if the spectral deconvolution step is done manually), more limited in scope, and it does not normally allow one to identify novel metabolites. Untargeted metabolomics is often faster, relatively open-ended, and more useful for novel compound discovery. However, untargeted metabolomics is not especially quantitative nor is it as reproducible.

In this chapter, we will focus on describing the protocols associated with conducting a targeted NMR metabolomics experiment and performing automated and semi-automated spectral deconvolution. Every NMR spectrum collected in a targeted metabolomics study must be analyzed using specialized peak fitting software. By matching peak positions and peak intensities to reference spectra of pure compounds (i.e., spectral deconvolution), it is possible to accurately determine compound identities and concentrations in biofluids. A variety of software tools for small molecule NMR spectral deconvolution have been developed over the past 10 years that make this spectral fitting process a relatively painless process. These include commercial tools such as the Chenomx NMR Suite [4], Bruker's JuiceScreener [5], and WineScreener [6], as well as non-commercial tools such as Batman [7] and Bayesil [8]. Several other spectral deconvolution programs have also recently appeared, including the automated quantification algorithm (AQuA) [9], the automatic method for identification and quantification of metabolites (ASICS) [10], and rDolphin [11]. Most of these tools focus on analyzing 1D ^{1}H NMR spectra, although programs have also been developed to analyze 2D (two-dimensional) ^{1}H NMR spectra and 2D heteronuclear spectra [12–14].

Extensive reference spectral libraries now exist for essentially all of the NMR-detectable compounds found in biofluids which is allowing a number of the deconvolution software tools to become almost fully automated. Indeed, some software packages can be extremely fast and robust with >95% of the known compounds in a given biofluid being identified and accurately quantified within seconds to minutes [4, 7–9, 15]. However, to use these kinds of automated or semi-automated spectral deconvolution tools, one first has to prepare the samples and collect the NMR spectra in a very specific and consistent way. In this chapter, we will describe the sample preparation and spectral acquisition steps for conducting (semi-) automated metabolomics NMR studies. We will also describe the protocols to analyze and quantify 1D NMR spectra using two specific software tools: the Chenomx NMR Suite (semi-automated) and Bayesil (fully automated). Finally, we will describe the protocols needed to collect and analyze 2D NMR spectra, such as homonuclear total correlation spectroscopy (TOCSY) and heteronuclear single-quantum coherence spectra (HSQC), for metabolomics using the semi-automated COLMARm web server [16].

2 Materials

To conduct an NMR-based metabolomics experiment, it is essential to have access to a modern, high-field NMR spectrometer (500–1000 MHz), equipped with a 5 mm (or smaller) probe, and a set of suitable 5 mm or 3 mm (outside diameter) clean, high-quality NMR tubes. A microcentrifuge (max. speed ~20,000 rpm or max. g-force of 13,000 × *g*) with a 10–15 cm rotor is required for ultrafiltration of samples. NMR sample preparation can be done in almost any moderately equipped laboratory with access to a cold room or walk-in refrigerator. Most centrifugation steps should be done at 4 °C to maintain sample integrity. Therefore, a refrigerated microcentrifuge or a regular microcentrifuge placed in a cold room should be sufficient. In metabolomics, it is important to prepare all solutions using HPLC grade water and analytical grade reagents as well as to prepare and store all reagents at 4 °C (unless indicated otherwise) to reduce sample contamination and bacterial growth.

2.1 Specialized Instrumentation, Containers, and Software

1. High-field, Fourier Transform NMR spectrometer with a magnetic field for protons of, at least, 500 MHz, equipped with a 5 mm room temperature or cryogenically cooled probe.
2. NMR sample spinners that can accommodate 5 mm or 3 mm NMR tubes.
3. High-quality borosilicate NMR sample tubes with an outside diameter of 5 mm or 3 mm (*see* **Note 1**).

4. A tabletop microcentrifuge or microfuge with a maximum speed of ~20,000 rpm (or a max. g-force of 13,000 × *g*), equipped with a 10–15 cm rotor that accepts standard microfuge tubes. The microcentrifuge may be refrigerated or it may be operated in a cold room.
5. A tabletop mini-centrifuge or minifuge (with a maximum speed of ~2000 rpm or a max. g-force of 1000 × *g*) equipped with a suitable rotor that can accept standard (5 mm or 3 mm) 10–20 cm long NMR tubes.
6. Amicon 0.5 mL centrifugal ultrafilters with a 3 kDa molecular weight cutoff (MWCO) (*see* **Note 2**).
7. A modern, electronic pH meter equipped with a 5 mm pH probe.
8. An ultrasonic bath sonicator with a frequency of 40 kHz (e.g., a Branson 2210 Ultrasonic Cleaning Bath Sonicator, Fisher Scientific, USA).
9. A variable speed vortexer that accommodates standard microfuge tubes.
10. Access to the Chenomx NMR Suite software package (version 7 or higher, which is available at www.chenomx.com) [4] or to the Bayesil [8] (available at www.bayesil.ca) spectral deconvolution software web server for analysis of 1D NMR spectra or to the COLMARm web server (available at http://spin.ccic.ohio-state.edu/index.php/colmarm/index) for analysis of 2D NMR spectra [16].

2.2 NMR Buffers (See Note 3)

1. NMR buffer with a final phosphate buffer concentration of 50 mM (for serum, cerebrospinal fluid, or similar biofluids): Prepare a buffer containing 250 mM potassium phosphate (pH 7.0), 5.0 mM 2,2-dimethyl-2-silapentane-5 sulfonate (DSS-d_6), 5.84 mM, 2-chloropyrimidine-5-carboxylic acid, and D_2O 54% v/v in H_2O. To do so, weigh 2.67 g potassium phosphate dibasic, 1.3 g potassium phosphate monobasic, 0.112 g DSS-d_6, and 0.092 g 2-chloropyrimidine-5-carboxilic acid (needed for Bayesil automated processing and profiling). Dissolve all solids in 54 mL of 99.8% D_2O. Adjust the pH to 7.00 using a calibrated electronic pH meter by adding 0.1 M potassium hydroxide or 0.1 M phosphoric acid solution, dropwise. Add deionized water or HPLC grade water to bring the final volume to 100 mL. Sub-aliquot the buffer into 1.0 mL cryovials, seal each vial with a clear cap, label, and store in a −20 °C freezer until required (*see* **Note 4**).
2. NMR buffer with a final phosphate buffer concentration of 150 mM (for urine, fecal water, tissues): Prepare a buffer containing 750 mM potassium phosphate (pH 7.0), 5.00 mM 2,2-dimethyl-2-silapentane-5 sulfonate (DSS-d_6), 5.84 mM,

2-chloropyrimidine-5-carboxylic acid, and D_2O 54% v/v in H_2O. To do so, weigh 8.01 g potassium phosphate dibasic, 3.9 g potassium phosphate monobasic, 0.112 g DSS-d_6, and 0.092 g 2-chloropyrimidine-5-carboxilic acid (needed for Bayesil automated processing and profiling). Dissolve all components in 54 mL of 99.8% D_2O. Adjust the pH to 7.00 using a calibrated electronic pH meter and by adding 0.1 M potassium hydroxide or 0.1 M phosphoric acid solution, dropwise. Add deionized water or HPLC grade water to bring the final volume to 100 mL. Sub-aliquot the buffer into 1.0 mL cryovials, seal each vial with a clear cap, label, and store in a −20 °C freezer until required (*see* **Note 4**).

3 Methods

All steps are carried out at room temperature unless otherwise specified. It is assumed that the biological samples have already been collected, sub-aliquoted into cryovials, and stored at −20 °C or at −80 °C. Regardless of the sample typc, it is important to remember that most biological samples collected for metabolomics are metabolically active. That is, unless the sample is frozen, dried, sterilized, or otherwise devoid of cells or enzymes that sample will be metabolically active for many minutes to hours after collection. If not properly accounted for, this underlying metabolic activity can produce deceptive results and large variations in metabolite composition and concentrations. As a result, most metabolomics studies also include a metabolic quenching step. Metabolic quenching is the rapid arresting of all metabolic activities via physical or chemical means [17]. The best route to metabolic quenching for tissues is rapid freezing (using liquid nitrogen). For blood, rapid removal of the red and white blood cells to produce plasma or serum must be done prior to freezing (to avoid any red blood cell lysis upon thawing). For most other biofluids, moderate to rapid freezing is usually sufficient. Once frozen, the samples may be stored for months or even years at −80 °C [18]. When biological samples are ready to be analyzed via targeted NMR-based metabolomics, they must be suitably cleaned/extracted, filtered, and buffered prior to collecting the NMR spectra. The first three protocols in this sample preparation section describe these sample preparation steps. The latter protocols describe the NMR spectral acquisition and analysis steps.

3.1 Sample Preparation for Biofluids (Serum, Cerebrospinal Fluid, Urine)

1. Thoroughly rinse a 3 kDa Amicon ultrafilter membrane by adding 500 μL HPLC grade water to the filter and centrifuging at 9400 × *g* for 10 min (*see* **Note 5**).
2. Dispose of the glycerol-containing filtrate into the sink and repeat **step 1** four more times. Dry the Amicon ultrafilter

with a tissue (Kimwipe or other brand) and place it into a new 1.5 mL microcentrifuge tube.

3. Thaw the biofluid sample on ice until it fully liquifies (about 30 min).
4. Centrifuge the sample at 9400 × *g* for 3 min at 4 °C to remove any particulate material (*see* **Note 6**).
5. Transfer 450 μL of the supernatant into the Amicon ultrafilter.
6. Centrifuge the Amicon ultrafilter at 11,500 × *g* for 20 min at 4 °C (*see* **Note 7**).
7. Pipette 160 μL (for 3 mm NMR tubes, use 400 μL if using a 5 mm NMR tube) of filtrate from **step 6** to a clean 1.5 mL microcentrifuge tube.
8. Pipette 40 μL of the appropriate NMR Buffer (*see* Subheading 2.2, **items 1** or **2**) to the 1.5 mL microcentrifuge tube from **step 7** to make up a total sample volume of 200 μL (or pipette 100 μL of the NMR buffer to make a final volume of 500 μL, if using a 5 mm NMR tube). Vortex the sample briefly (4–5 s) and proceed to Subheading 3.2 (*see* **Note 8**).

3.2 Preparation of the NMR Sample and NMR Spectral Acquisition

1. Use a Pasteur pipette to carefully transfer the biofluids prepared in Subheading 3.1 (total volume 200 μL) into a 3 mm NMR tube (or 500 μL in a 5 mm NMR tube). Ensure that there are no bubbles or surface foam in the NMR sample tube. If multiple NMR samples have been prepared, repeat the transfer step to other NMR tubes.
2. Centrifuge the NMR tube in a suitable minicentrifuge for 1 min at 800 × *g*. If multiple NMR samples have been prepared, place multiple tubes into the centrifuge (*see* **Note 9**).
3. Place the NMR tube into a spinner and insert it into the high-field NMR spectrometer. If multiple NMR tubes have been prepared, they can be placed in a refrigerated NMR autosampler (e.g., Bruker Sample Jet) for automatic sample insertion (*see* **Notes 10** and **11**).
4. Ensure that the sample and magnet are properly matched, locked, and tuned (*see* **Note 12**). Shim the magnet to ensure that the DSS signal has a peak width of <1 Hz (preferably less than 0.6 Hz). Set the number of data points for the acquisition to be 64 k or 128 k so that the digital resolution is <0.25 Hz and the spectral width is 12 ppm. Determine the optimal 90° pulse width and pulse power.
5. ^{1}H-1D-NMR (quantitative): Collect a 1D ^{1}H NMR spectrum using the metnoesy pulse sequence (Fig. 1a) with a recycle delay of at least 3 s (this may vary with the sample or the spectrometer). Typically, 64–128 transients are sufficient for most biological samples, although this may vary with the concentration of the metabolites in the sample, the sample type, or

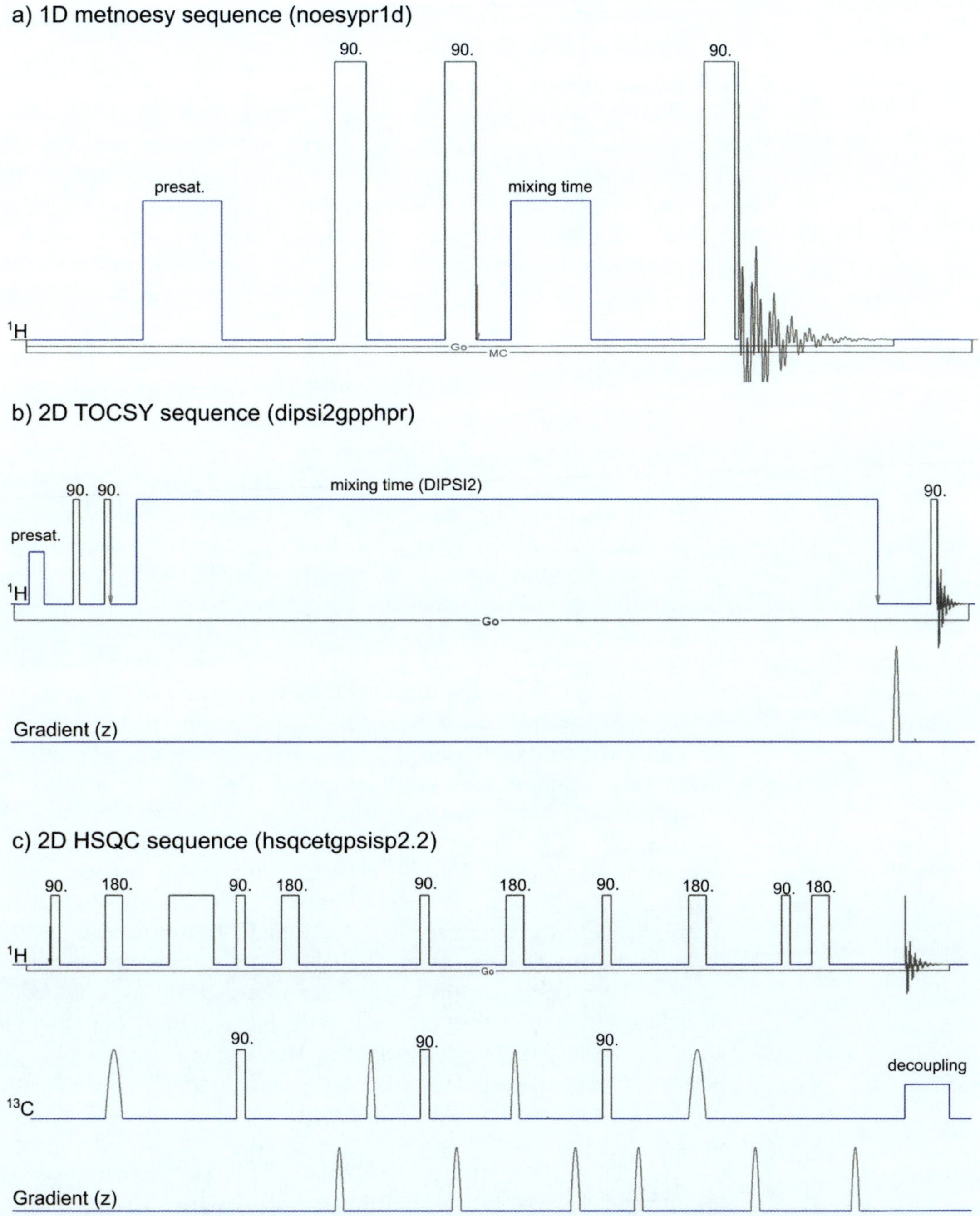

Fig. 1 Common pulse sequences used for metabolomic studies by NMR: (**a**) the metnoesy pulse sequence for 1D ^{1}H NMR studies; (**b**) a 2D ^{1}H–^{1}H TOCSY pulse sequence with presaturation; and (**c**) a 2D ^{1}H–^{13}C HSQC pulse sequence with decoupling during acquisition

instrument specification, such as type of probe installed and the magnetic field strength.

6. Homonuclear 2D-NMR (qualitative): Collect a $^{1}H–^{1}H$ 2D TOCSY NMR spectrum using a phase-sensitive TOCSY pulse sequence with presaturation for water suppression (e.g., Bruker pulse sequence dipsi2gpphpr, *see* Fig. 1b). Use a spectral width in both dimensions of 10 ppm and a mixing time of 90 ms. to ensure good quality data, collect a minimum of 1024 complex points in the direct dimension and 512 complex points in the indirect dimension. In general, a minimum of 8 transients is recommended.
7. Heteronuclear 2D-NMR (qualitative): Collect a simple $^{1}H–^{13}C$ 2D HSQC NMR spectrum with positive peaks (using adiabatic pulses, not multiplicity edited, sensitivity improved, such as Bruker pulse sequence hsqcetgpsisp2.2, *see* Fig. 1c). Use a spectral width in the proton dimension of 10–12 ppm and 160 ppm in the carbon dimension. To ensure good quality data, collect a minimum of 1024 complex points in the proton dimension and 512 complex points in the carbon dimension. In general, a minimum of 16 transients is recommended.

3.3 Spectral Analysis and Spectral Deconvolution

Spectral data obtained from NMR spectrometers need to be processed before conducting any metabolomics study or performing any kind of spectral deconvolution. The processing includes Fourier transformation after applying line broadening, phasing, referencing, and baseline correction. Most NMR spectrometers are equipped with software to perform these steps for 1D and 2D NMR spectra. Likewise, many of the 1D NMR spectral deconvolution tools (Chenomx, Bayesil, etc.) can also perform these steps (manually or semi-automatically). For processing 2D NMR spectra, freely available software such as NMRPipe [19] and nmrglue [20] can also be used. Regardless of the chosen software, phase and baseline correction are normally done interactively by observing the spectra in an interactive viewer. The objective in this interactive adjustment process is to produce pure "absorptive" peaks with near perfectly flat baselines. The final step in NMR spectral processing for spectral deconvolution involves adjusting the line broadening of the peaks in the measured NMR spectrum to match the line broadening found in the reference (or library) NMR spectra that are used for spectral deconvolution. Below, we will describe a step-by-step protocol for processing and analyzing 1D NMR spectra using a commercial software package called Chenomx NMR Suite [4] and an open-access web server called Bayesil [8].

3.3.1 Using Chenomx to Process and Analyze 1D NMR Spectra

1. Ensure that the Chenomx NMR Suite is downloaded and properly installed on a suitable desktop or laptop computer. Users may purchase the software or download a demo version (30-day free trial) from www.chenomx.com. Start the

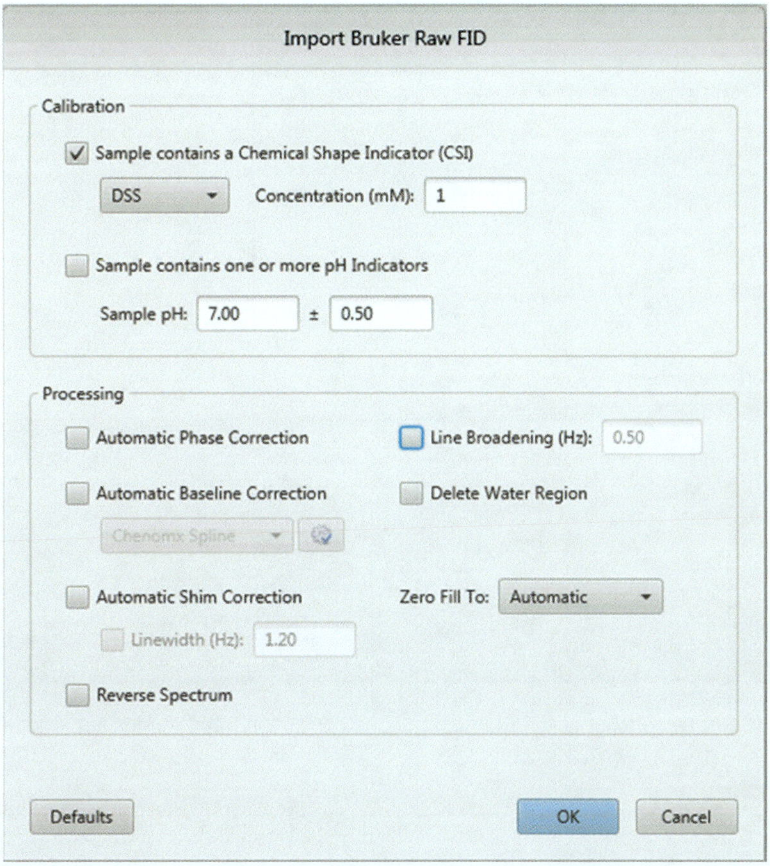

Fig. 2 A screenshot of the start-up Processor window of the Chenomx software suite

Chenomx Processor and load a raw 1D NMR data file (collected from Subheading 3.2). The start-up Processor window should appear as shown in Fig. 2. Define the Chemical Shape Indicator (CSI) and its concentration (DSS and 1 mM). Set the sample pH (pH 7.00 ± 0.50) and click "OK."

2. Click on Phase Correction (*see* Fig. 3) to adjust the spectrum's phase. Interactively adjust the zero-order and first-order phase values with the computer mouse or touchpad so that the DSS signal (0 ppm), the signal of 2-chloropyrimidine-5-carboxilic acid (~9 ppm), and other peaks in the NMR spectrum are in-phase (symmetric, Lorentzian peaks, all pointing upwards). If one is having some difficulty in phasing, the zero-order phase correction should be done to make DSS signal symmetric. Once this DSS peak is phased, use the pivot option (an option available for newer versions of the NMR Suite package) to adjust the first-order phase on DSS to make most of the peaks in the spectrum symmetric. Finally click "Accept" when the appropriate phasing criteria are reached (*see* Figs. 4 and 5a).

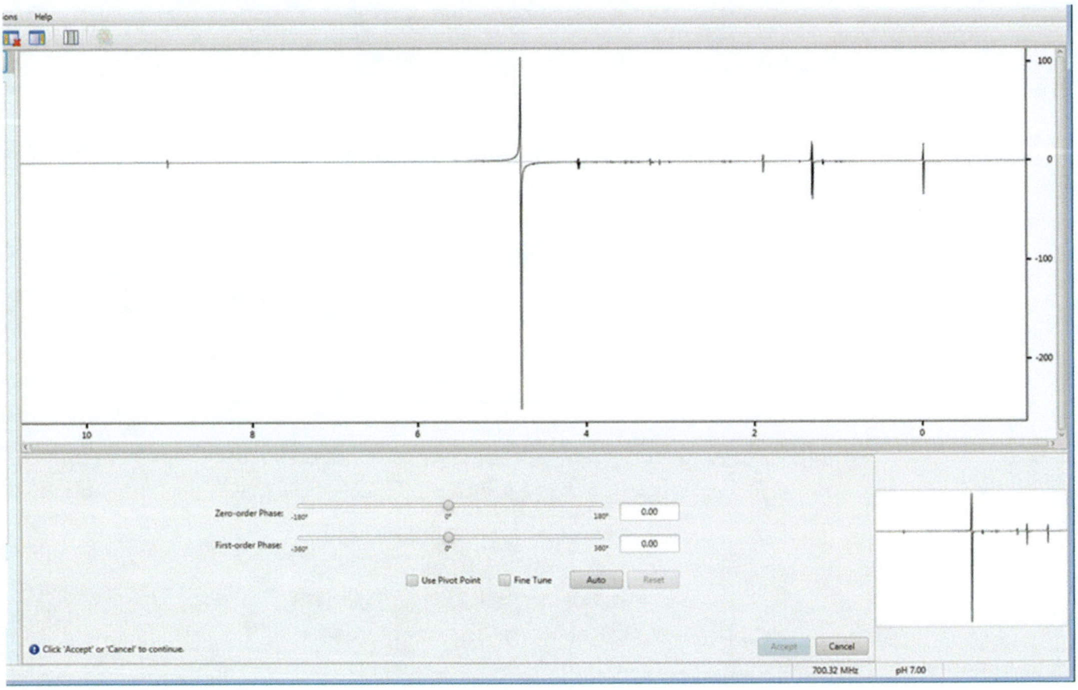

Fig. 3 A screenshot of the Chenomx phase correction interface for automated and manual phase correction

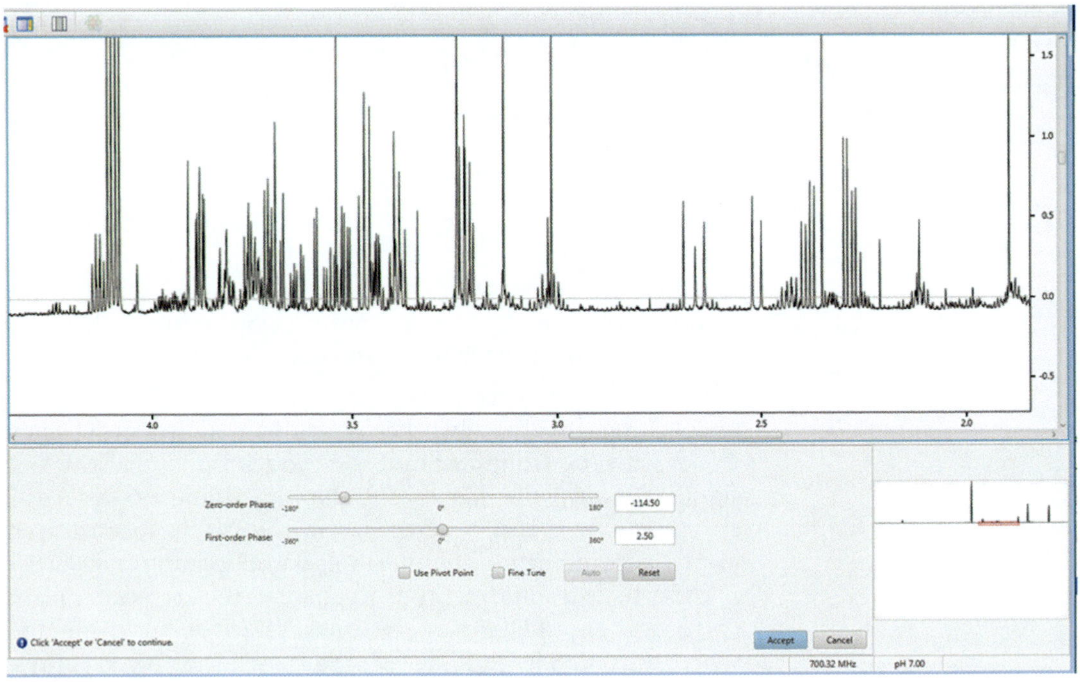

Fig. 4 A screenshot of part of a spectrum after phase correction with all peaks in the fully absorptive mode (positive) before baseline correction in Chenomx

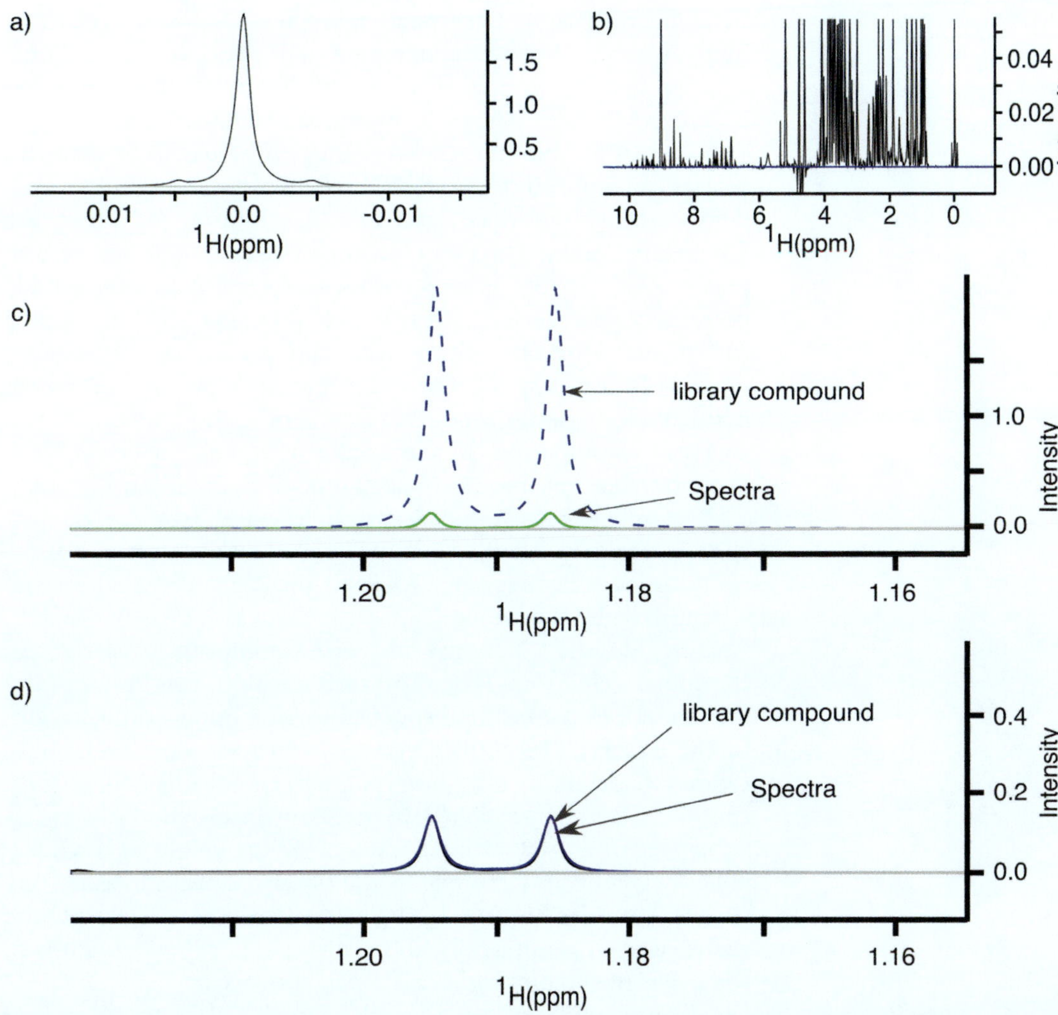

Fig. 5 The manual data processing steps used in the Chenomx NMR Suite. Phase correction to achieve symmetry of the DSS peak (**a**) and accurate baseline correction (**b**) for a serum spectra shown at the top. The initial profiling step where a cluster for a library compound is detected is shown in (**c**). The final step when the peaks corresponding to library compound have been carefully fit for accurate concentration determination as shown in (**d**)

3. Click on "Baseline Correction" to correct the baseline. Choose the Chenomx Spline option or Whittaker Spline and click "Find Baseline," Manually move or add base points (with the computer mouse or computer touchpad) to achieve a flat baseline. Once this is achieved, click "Accept" (*see* Fig. 5b).
4. Click on "Calibrate CSI" to calibrate the peak shapes automatically using the DSS peak as a peak shape reference (*see* **Note 13**).

5. Send the now-processed spectrum to the Chenomx Profiler to identify metabolites and determine their concentrations. There are two general strategies to identify metabolites in the Chenomx NMR Suite, one is biofluid based and the other is "naïve." In the biofluid-based strategy, one has to first make a reference compound spectral library of known metabolites specific to the biofluid of interest. This is done by using the Chenomx library manager. In many biofluids (such as serum, plasma, saliva, CSF), the composition of NMR-detectable metabolites is precisely known [21–23], and this knowledge can be used to limit the search and accelerate the spectral deconvolution process. Once the biofluid-specific library has been created, search for the presence of each metabolite from this set by comparing the chemical shifts of all the peaks/peak clusters from the biofluid spectrum with the library spectra. This search may be done by "right-clicking" the cursor over specific peaks in the biofluid spectrum (always start in a spectrally quiet area first) and look to see what compounds the Chenomx NMR Suite software suggests as a possible match. Ensure that the Chenomx-suggested compound matches all the appropriate peaks (chemical shift position, height multiplicity) present in the biofluid spectrum. After initial confirmation, fit the observed biofluid spectrum with individual metabolite reference spectra by dragging manually or automatically (using the keyboard "space bar") to more precisely fit the selected sub-spectrum in the biofluid. Adjusting the height of the reference spectrum peaks to match the observed peaks interactively (using a mouse or touchpad) allows one to determine the metabolite concentration. This process is repeated until all the peaks in the biofluid NMR spectrum are fit.

   The second method, "naïve" strategy, performs a search of the unassigned spectral peaks using the entire Chenomx library (~700 compounds). To perform this search, right click on an unassigned peak to make the software search for possible metabolite candidates. Visually check to see if the suggested compound (and its reference spectrum) fits the biofluid's sub-spectrum (peak positions, height, multiplicity). Once a match is found, adjust the intensity and peak positions manually as described before. Repeat these steps until all metabolites/peaks are assigned (*see* Fig. 5c, d) (*see* **Note 14**).

6. Once the spectral fitting process has been completed, the data can be exported as a .csv file as single file or as batch export with multiple files.

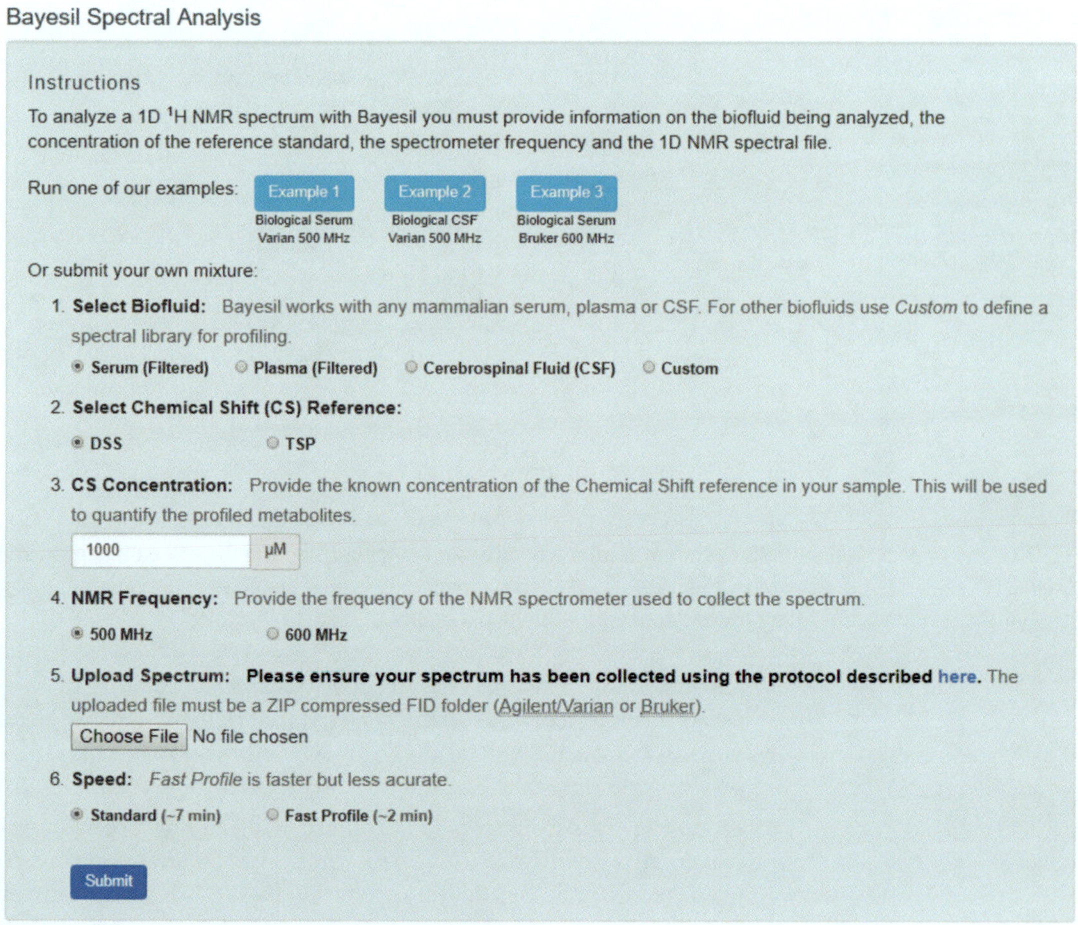

Fig. 6 A screenshot of the Bayesil submission page

3.3.2 Using the Bayesil Web Server to Automatically Analyze 1D NMR Data [12]

1. Prior to using the Bayesil web server, compress the folder containing the 1D ^{1}H NMR spectral data and the spectral acquisition parameters (collected in Subheading 3.2) into a "zip" file. The zip file must contain the FID. To start the spectral deconvolution process, go to the Bayesil website (www.bayesil.ca) and click on new submission (*see* Fig. 6). As shown in Fig. 6, the Bayesil home page provides links that describe the precise NMR spectral collection protocol and a list of available reference biofluid/spectral libraries.
2. Select the appropriate biofluid from the list (using predefined metabolites for common biofluids such as serum) or create a custom set of metabolites for a specific biospecimen. The user can also easily create a custom library by selecting "Custom" in "Select Biofluid" section and selecting individual metabolites and saving the list in the end. Select the chemical shift reference (DSS, 1 mM), select the appropriate NMR spectrometer frequency, and choose the zipped folder (prepared in **step 1**) to upload the NMR spectrum and press "Submit." Bayesil

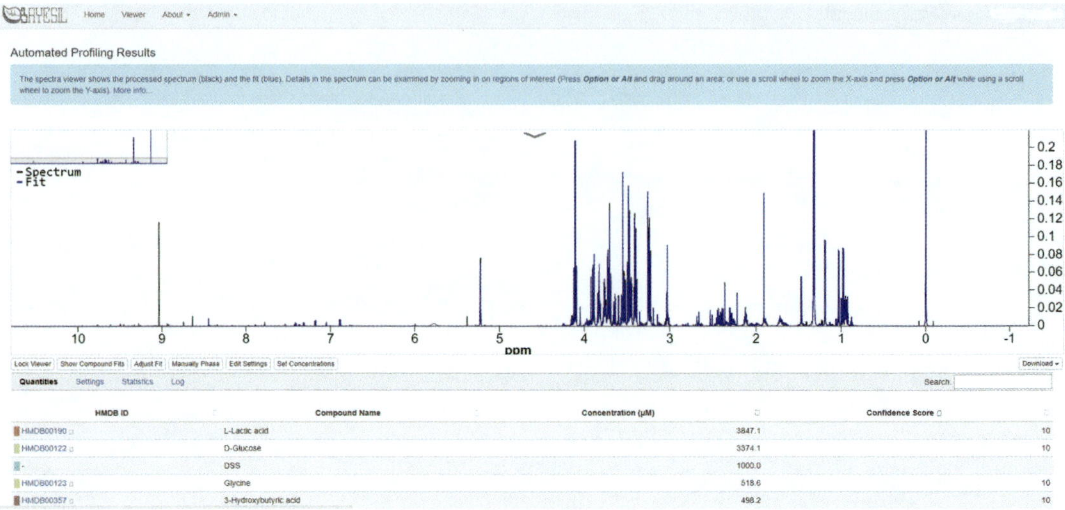

Fig. 7 Overview of a profiled NMR spectrum from a typical serum sample using the Bayesil web server. The black line represents the actual serum spectrum, while the blue line indicates the fitted spectrum. The lower panel of the window shows the concentration of the metabolites and the HMDB identifier

automatically performs the phasing, baseline correction, chemical shift reference adjustment, water signal removal, and spectral deconvolution (compound ID and quantification). The entire profiling process takes about ~7 min/spectrum (*see* Fig. 7).

3. Once the fitting process is complete, a fitted spectrum is displayed along with the identified compounds and their concentrations in a table below the spectrum (*see* Fig. 7). Users can use the computer mouse or touchpad (along with the Bayesil spectral viewer widgets) to select, expand, or zoom the fitted spectra to inspect the quality of the spectral fitting process. Generally, Bayesil is 95–100% accurate in spectral fitting. In cases where the profiling appears to be incorrect, click "Adjust fit" and select the compound, peak, or peak cluster that needs further adjustment by "Shift + left click." Fit the corresponding clusters and peaks manually by dragging or moving specific peaks (from the reference spectrum) to match the observed peaks using the mouse or touchpad in a manner similar to that done for the Chenomx NMR Suite.

4. After all the questionable metabolites or peaks have been examined or adjusted, the data can be downloaded as a .csv file. For future reference, the Bayesil file can also be saved.

3.3.3 Analyzing 2D ^{13}C–^{1}H HSQC and 2D TOCSY-Type Spectra with the COLMARm Web Server

1. Process a 2D ^{13}C–^{1}H HSQC spectrum of a given biofluid of interest and one or both 2D TOCSY-type spectra (^{1}H–^{1}H TOCSY and/or ^{13}C–^{1}H HSQC-TOCSY) using NMRPipe [19] Sparky [24], Bruker's TopSpin, or Mestrelab Mnova. Make sure that the number of points does not exceed 4 K in any of the two dimensions. The 2D ^{13}C–^{1}H HSQC spectrum is

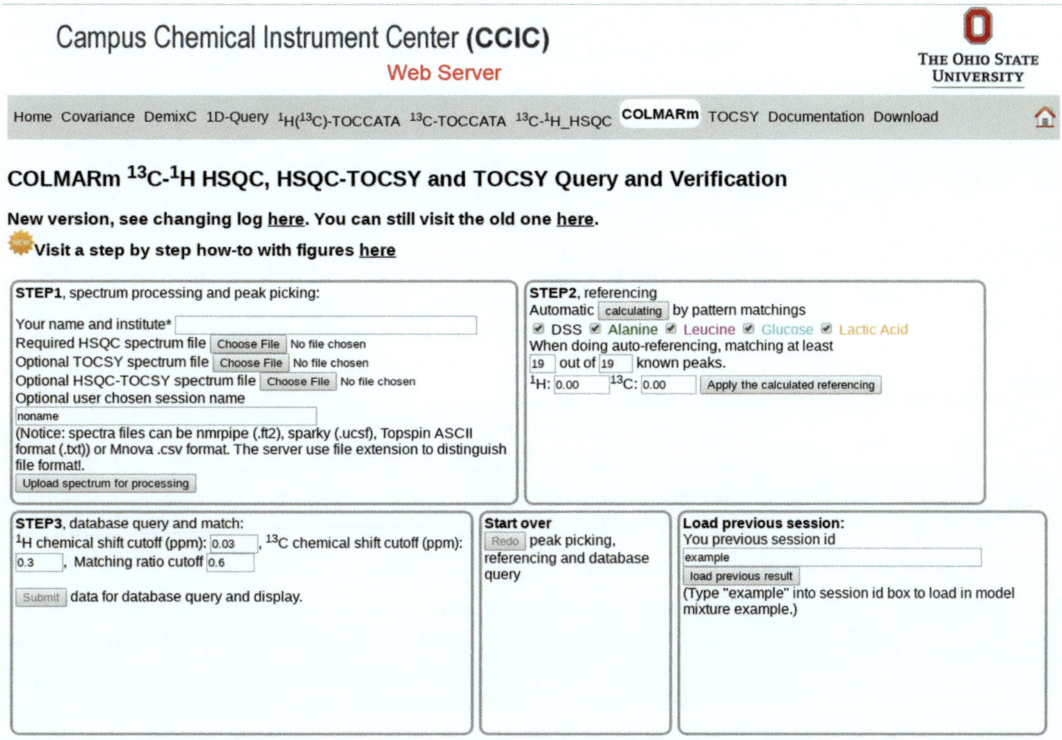

Fig. 8 A screenshot of the data entry page for the COLMARm web server. Please see the description in the text

required for COLMARm, whereas ^{1}H–^{1}H TOCSY and ^{13}C–^{1}H HSQC-TOCSY are optional spectra.

2. Save the output files for these spectra either in NMRPipe (.ft2). Sparky (.ucsf), TopSpin ASCII (.txt), or Mnova (.csv) formats. If processing with TopSpin, use the command "totxt" to save the spectra in ASCII .txt format. When saving files in MNova, use the menu option "Save as" and select NMR CSV as the format of the output file.
3. Go to the COLMARm website at http://spin.ccic.ohio-state.edu/index.php/colmarm.
4. Select the required spectral files from your computer in one of the formats shown above using the form "**step 1**" (*see* Fig. 8). Enter your name and institute. Optionally, provide a session name. Press "Upload spectrum for processing." Wait (for approximately 2 min) until the web server completes the processing and peak picking of the uploaded spectra and displays an HSQC spectrum at the bottom of the webpage.
5. Once the HSQC spectrum has been displayed, save the session ID that is located at the bottom of the displayed spectrum. This ID can be used to retrieve web server results at a later time.

6. Optionally, adjust visible peaks in the observed HSQC spectrum using the slider "HSQC Lowest contour level" on the website.
7. If necessary, remove intense noise peaks from the spectrum using the "Brush an area" feature. To remove peaks, select the option "Remove peaks within this area." To inspect peaks before removal, select the "Zoom in" option.
8. Optionally, adjust the spectrum's referencing using the web form "**step 2**, referencing" (*see* Fig. 8). Select reference molecules that are present in your sample. Currently, the web server can only use the following reference compounds: DSS, alanine, leucine, glucose, and lactic acid. If necessary, adjust the number of peaks that will be used to automatically calculate reference offsets. Press the "Calculating" button to automatically calculate reference offsets for ^{13}C and ^{1}H chemical shifts. After the reference offset values appear in the form, press the "Apply the calculating referencing" button (*see* **Note 15**).
9. To automatically identify metabolites that are present in the sample, use web form "**step 3**, database query and match" (*see* Fig. 8). If necessary, adjust parameters "^{1}H chemical shift cutoff (ppm)", "^{13}C chemical shift cutoff (ppm)", and "Matching ratio cutoff" (the ratio of the matched peaks to the total number of peaks of the metabolite). Press the "Submit" button. Wait until the results of database query are displayed (about 1–2 min).
10. After the web server has completed the matching of experimental spectra based on the database, manually inspect the results. Experimental peaks, database peaks, and expected TOCSY cross-peaks will be colored with green, red, and magenta colors, respectively (*see* Fig. 9). An overlap of green and red peaks indicates a good match. The agreement between expected and observed TOCSY cross-peaks can be used to support matching results, especially, in ambiguous cases.
11. The web server will display a table with matching results for each identified metabolite, including matching ratios as well as ^{13}C and ^{1}H chemical shift RMSDs between the database and peaks from experimental spectrum. These parameters can be used to assess the quality of matching results.
12. Save the results using options "Save peak report as," Save the compound report as," and save plot files."

3.4 Data Analysis

The protocol outlined here describes three different programs or web servers (Chenomx NMR Suite, Bayesil, and COLMARm) that can be used to (semi-) automatically identify and quantify metabolites from 1D- and 2D-NMR spectra of biofluids. It describes how samples should be prepared, how spectra should be collected and

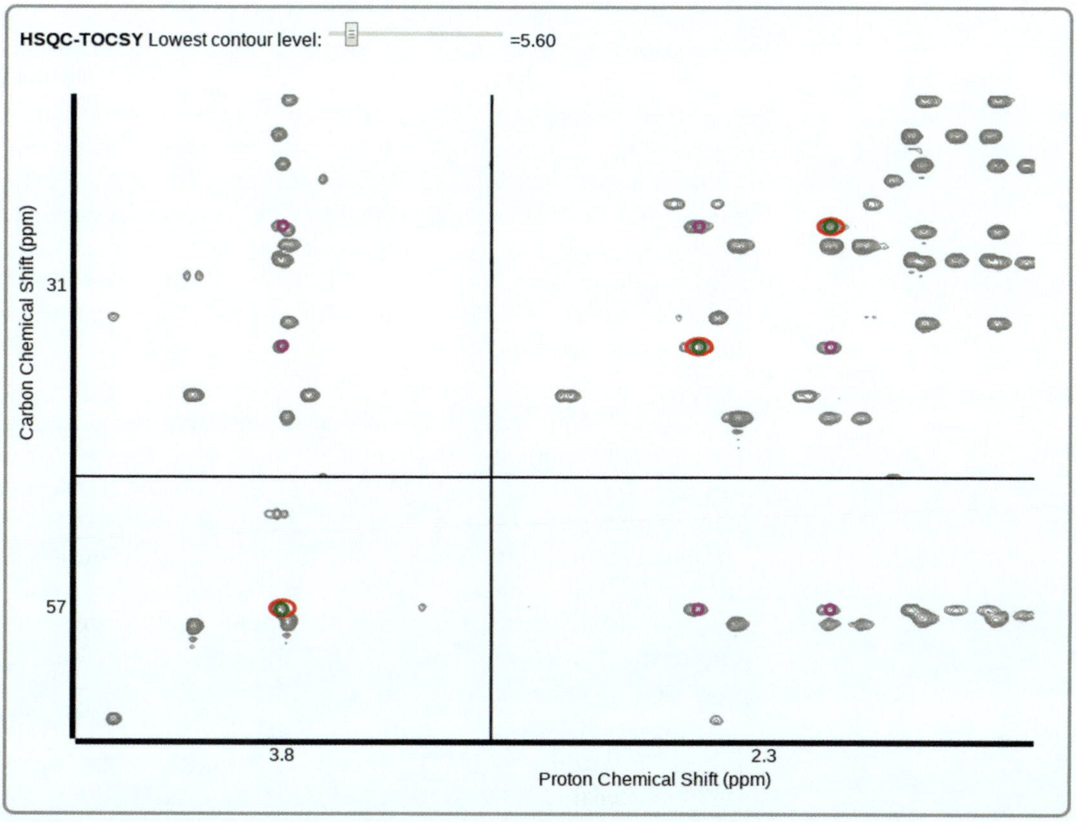

Fig. 9 An HSQC-TOCSY spectrum processed and analyzed by the COLMARm web server. Reference peaks from the COLMARm database are colored red. Matched experimental peaks are labeled in green. Expected cross-peaks are shown with magenta circles

processed, and how each of the fitting/profiling programs can be used. This protocol will allow users to generate lists of 40–60 metabolites (for 1D- and 2D-NMR data) along with their concentrations (for 1-D NMR data only) for one or more biofluid samples in an NMR-based metabolomics study. The data thus generated can be analyzed using multivariate statistical methods including DiscoverM software bundled with the Chenomx NMR Suite or online tools such as MetaboAnalyst [25].

4 Notes

1. Most samples destined for NMR metabolomics studies are placed into 5 mm borosilicate glass tubes (using a volume of 500–600 μL). However, more and more NMR laboratories are using 3 mm tubes, which require less than one-third the volume (150–200 μL) while still yielding almost the same signal to noise (S/N). Smaller tubes also reduce dielectric losses due to

high salt concentrations making them more amenable for use in cryoprobes. The use of Shigemi tubes with susceptibility matching glass plugs also allows one to use small volumes (~250 μL) while still retaining the 5 mm tube diameter that is optimal for 5 mm NMR probes and 5 mm spinners. For volume-limited samples, microprobes are now available for many high-field NMR spectrometers. These specially designed microcoil probes are able to accommodate 1.0 or 1.7 mm NMR tubes. With these tubes, it is possible to work with volumes as low as 10 μL for a 1.0 mm tube and 35 μL when using a 1.7 mm tube.

2. Analysis of biofluids using NMR spectroscopy requires the elimination of proteins and macromolecules. In order to obtain a consistent sample that has these macromolecules removed, all samples should be filtered using a 3 kDa ultrafilter. The filtrate, which contains particle size less than 3 kDa, is then mixed with the NMR buffer to be used for further analysis. Certain 3 kDa ultrafilters have a maximum *g*-force of 14,000 × *g* that cannot be exceeded without rupturing the membrane. Users must determine the appropriate rpm for their microcentrifuge to meet this criterion.
3. The purpose of using this buffer is to help adjust the pH of the sample solution to 7.0. The final concentration of the phosphate buffer in the NMR tube will be 50 mM, after adding the 250 mM buffer to the sample, in a 1:4 ratio. DSS-d_6 is used as an internal chemical shift reference as well as a concentration reference to calibrate metabolite concentrations and the chemical shifts of the analytes. 2-chloropyrimidine-5-carboxyilic acid is used to help with phasing NMR spectra. D_2O is used as a lock solvent to help properly adjust the magnetic field of the NMR spectrometer during data acquisition.
4. The NMR buffer can be stored in the dark at 4 °C for up to 2 months. For storage time longer than 2 months, keep the vials at −20 °C.
5. The 3 kDa ultrafilters contain glycerol, therefore, each filter needs to be washed prior to filtering the biological sample of interest, otherwise the glycerol signal will dominate the NMR spectrum.
6. Whole blood contains a wide range of blood cells and coagulant factors which may clot and block the filter pores. To maximize the filtrate volume, centrifuge and collect the supernatant when transferring whole blood to the filter. One may also consider using two filters for each whole blood sample and pool the filtrate.

7. The final filtrate should be clear. Cloudy filtrate indicates filter failure/damage. If the filter is damaged, repeat **steps 1** and **2** with a new filter.
8. Keep the 1.5 mL microcentrifuge tube on ice after the mixture is ready.
9. Centrifugation should remove any remaining bubbles in the NMR tube. Visually inspect the NMR tube and make sure no bubbles or scratches can be seen within the area holding the sample.
10. Make sure to use the right container to hold the NMR tube and prevent any breakage.
11. Avoid touching the lower part of NMR tube with your fingers. Wipe each NMR tube with a tissue prior to loading to the NMR spectrometer.
12. Both the NMR instrument and the sample must be appropriately locked, tuned, matched, and carefully shimmed prior to collecting NMR spectra. Most modern NMR instruments are tuned for salt-free solvents (such as deuterochloroform or pure D_2O); however, the NMR samples that are usually used in metabolomics contain relatively high salt concentrations (50–200 mM salt). If the spectrometer probe is not tuned or matched properly, these high salt levels may lead to poor performance (noticeable by low S/N, long excitation pulses, poor solvent suppression, etc.). This is particularly evident with cryogenically cooled probes. Therefore, some care and effort must be made prior to starting an NMR metabolomics experiment to ensure that the sample is well locked and optimal tuning and matching have been achieved. While tuning and matching are often done manually, newer NMR instruments now support rapid, automatic tuning and matching. These auto-tune/auto-match functions make it possible to load a wide variety of sample types (with different solvents or salt concentrations) in the same run.
13. The calibration peak should have a concentration of 1 mM and the width should be >1 Hz. If necessary, a line broadening can be applied before calibrating the CSI. The linewidth of DSS reference should be as close as possible to the reference library of Chenomx for accurate concentration determination.
14. Manual spectral fitting takes practice and it is suggested that new users practice the fitting process with some simple synthetic mixtures where the contents are precisely known. It is also recommended that prior to spectral deconvolution, a list of likely or expected NMR-detectable metabolites for the given biofluid or biospecimen should be collected to help users through the spectral fitting process. This information is readily available from the databases such as the Human Metabolome Database [26] and the Livestock Metabolome Database [27].

15. If the web server shows reference offsets of 100, it means that the automatic calculation of the reference offsets has failed. The web server can only calculate reference offsets if the referencing errors are below 2 ppm and 0.2 ppm for carbon and proton dimensions, respectively.

References

1. Nicholson JK, Wilson ID (2003) Opinion: understanding 'global' systems biology: metabonomics and the continuum of metabolism. Nat Rev Drug Discov 2(8):668–676
2. Wishart DS (2008) Quantitative metabolomics using NMR. Trac Trend Anal Chem 27 (3):228–237
3. Alonso A, Marsal S, Julia A (2015) Analytical methods in untargeted metabolomics: state of the art in 2015. Front Bioeng Biotechnol 3:23. https://doi.org/10.3389/fbioe.2015.00023
4. Mercier P, Lewis MJ, Chang D, Baker D, Wishart DS (2011) Towards automatic metabolomic profiling of high-resolution one-dimensional proton NMR spectra. J Biomol NMR 49(3–4):307–323
5. Monakhova YB, Schutz B, Schafer H, Spraul M, Kuballa T, Hahn H, Lachenmeier DW (2014) Validation studies for multicomponent quantitative NMR analysis: the example of apple fruit juice. Accred Qual Assur 19 (1):17–29
6. Spraul M, Link M, Schaefer H, Fang F, Schuetz B (2015) Wine analysis to check quality and authenticity by fully-automated 1H-NMR. Bio Web Conf 5:02022. https://doi.org/10.1051/bioconf/20150502022
7. Hao J, Liebeke M, Astle W, De Iorio M, Bundy JG, Ebbels TM (2014) Bayesian deconvolution and quantification of metabolites in complex 1D NMR spectra using BATMAN. Nat Protoc 9(6):1416–1427
8. Ravanbakhsh S, Liu P, Bjorndahl TC, Mandal R, Grant JR, Wilson M, Eisner R, Sinelnikov I, Hu X, Luchinat C, Greiner R, Wishart DS (2015) Accurate, fully-automated NMR spectral profiling for metabolomics. PLoS One 10(5):e0124219. https://doi.org/10.1371/journal.pone.0124219
9. Rohnisch HE, Eriksson J, Mullner E, Agback P, Sandstrom C, Moazzami AA (2018) AQuA: an automated quantification algorithm for high-throughput NMR-based metabolomics and its application in human plasma. Anal Chem 90(3):2095–2102
10. Tardivel PJC, Canlet C, Lefort G, Tremblay-Franco M, Debrauwer L, Concordet D et al (2017) ASICS: an automatic method for identification and quantification of metabolites in complex 1D H-1 NMR spectra. Metabolomics 13(10):ARTN 109. https://doi.org/10.1007/s11306-017-1244-5
11. Canueto D, Gomez J, Salek RM, Correig X, Canellas N (2018) rDolphin: a GUI R package for proficient automatic profiling of 1D H-1-NMR spectra of study datasets. Metabolomics 14(3):ARTN 24. https://doi.org/10.1007/s11306-018-1319-y
12. Lewis IA, Schommer SC, Markley JL (2009) rNMR: open source software for identifying and quantifying metabolites in NMR spectra. Magn Reson Chem 47:S123–S126
13. Bingol K, Li DW, Bruschweiler-Li L, Cabrera OA, Megraw T, Zhang FL et al (2015) Unified and isomer-specific NMR metabolomics database for the accurate analysis of C-13-H-1 HSQC spectra. ACS Chem Biol 10 (2):452–459
14. Bingol K, Bruschweiler-Li L, Li DW, Bruschweiler R (2014) Customized metabolomics database for the analysis of NMR H-1-H-1 TOCSY and C-13-H-1 HSQC-TOCSY spectra of complex mixtures. Anal Chem 86 (11):5494–5501
15. Zheng C, Zhang SC, Ragg S, Raftery D, Vitek O (2011) Identification and quantification of metabolites in H-1 NMR spectra by Bayesian model selection. Bioinformatics 27 (12):1637–1644
16. Bingol K, Li DW, Zhang B, Bruschweiler R (2016) Comprehensive metabolite identification strategy using multiple two-dimensional NMR spectra of a complex mixture implemented in the COLMARm web server. Anal Chem 88(24):12411–12418
17. Teng Q, Huang WL, Collette TW, Ekman DR, Tan C (2009) A direct cell quenching method for cell-culture based metabolomics. Metabolomics 5(2):199–208
18. Sellick CA, Hansen R, Maqsood AR, Dunn WB, Stephens GM, Goodacre R et al (2009) Effective quenching processes for physiologically valid metabolite profiling of suspension cultured mammalian cells. Anal Chem 81 (1):174–183

19. Delaglio F, Grzesiek S, Vuister GW, Zhu G, Pfeifer J, Bax A (1995) Nmrpipe–a multidimensional spectral processing system based on unix pipes. J Biomol NMR 6(3):277–293
20. Helmus JJ, Jaroniec CP (2013) Nmrglue: an open source python package for the analysis of multidimensional NMR data. J Biomol NMR 55(4):355–367
21. Psychogios N, Hau DD, Peng J, Guo AC, Mandal R, Bouatra S et al (2011) The human serum metabolome. PLoS One 6(2):ARTN e16957. https://doi.org/10.1371/journal.pone.0016957
22. Wishart DS, Lewis MJ, Morrissey JA, Flegel MD, Jeroncic K, Xiong YP et al (2008) The human cerebrospinal fluid metabolome. J Chromatogr B 871(2):164–173
23. Dame ZT, Aziat F, Mandal R, Krishnamurthy R, Bouatra S, Borzouie S et al (2015) The human saliva metabolome. Metabolomics 11(6):1864–1883
24. Lee W, Tonelli M, Markley JL (2015) NMRFAM-SPARKY: enhanced software for biomolecular NMR spectroscopy. Bioinformatics 31(8):1325–1327
25. Chong J, Soufan O, Li C, Caraus I, Li SZ, Bourque G et al (2018) MetaboAnalyst 4.0: towards more transparent and integrative metabolomics analysis. Nucleic Acids Res 46 (W1):W486–W494
26. Wishart DS, Feunang YD, Marcu A, Guo AC, Liang K, Vazquez-Fresno R et al (2018) HMDB 4.0: the human metabolome database for 2018. Nucleic Acids Res 46(D1): D608–D617. https://doi.org/10.1093/nar/gkx1089
27. Goldansaz SA, Guo AC, Sajed T, Steele MA, Plastow GS, Wishart DS (2017) Livestock metabolomics and the livestock metabolome: a systematic review. PLoS One 12(5):ARTN e0177675. https://doi.org/10.1371/journal.pone.0177675

Part XI

Data Processing and Statistical Methods

Chapter 25

Processing and Analysis of Untargeted Multicohort NMR Data

Timothy M. D. Ebbels, Ibrahim Karaman, and Gonçalo Graça

Abstract

NMR data from large studies combining multiple cohorts is becoming common in large-scale metabolomics. The data size and combination of cohorts with diverse properties leads to special problems for data processing and analysis. These include alignment, normalization, detection and removal of outliers, presence of strong correlations, and the identification of unknowns. Nonetheless, these challenges can be addressed with suitable algorithms and techniques, leading to enhanced data sets ripe for further data mining.

Keywords NMR, Multicohort, Data processing, Data analysis, Metabolome-wide significance level (MWSL), Subset optimization by reference matching (STORM)

1 Introduction

The metabolomics field is undergoing a transformation from a relatively niche technology to one of the standard omics tools in modern biology and medicine. As the field matures, it has become possible to generate data at much higher rates and lower cost than was previously possible. This is leading to its application in large studies, especially human epidemiological cohorts, to complement the other molecular genotype and phenotype data now routinely acquired. Today, it is quite common to study the metabolic phenotypes of thousands of individuals [1–3]. The approach has become to be known as the metabolome-wide association study (MWAS) [4] by analogy with genome-wide association studies in genomics. In these studies, global untargeted metabolic profiles are screened for associations to epidemiological outcome variables of interest, such as disease risk factors.

The underlying reasons for increased study sizes relate to the rapid evolution of analytical technologies, especially those of NMR and MS over the last 15–20 years, coupled with robotics and the data-capture capacity of modern computers. While MS is widely

G. A. Nagana Gowda and Daniel Raftery (eds.), *NMR-Based Metabolomics: Methods and Protocols*, Methods in Molecular Biology, vol. 2037, https://doi.org/10.1007/978-1-4939-9690-2_25,

used for metabolic phenotyping, NMR has a number of characteristics which are particularly advantageous when considering large-scale studies. In particular, NMR is highly reproducible [5–7], both within and between laboratories [8, 9], with only minor changes seen in high resolution spectra collected over periods of several years. NMR measurements can offer absolute quantification if required though this requires longer acquisition times and relative quantification is usually applied. A key strength of NMR is the information it yields on chemical structures. Even MS-based studies often ultimately rely on NMR to confirm structures for particularly complex unknowns. This is important because in untargeted metabolomics, there is no preselection of the metabolites to be assayed and thus annotation of spectral features to known compounds is a necessary step in the workflow. These advantages compensate for the relatively low sensitivity of NMR (detection limits of a few micromolar), as compared to MS approaches (pico to nanomolar detection limits, though heavily dependent on the metabolite and approach taken). Taken together, these aspects benefit large-scale metabolomics by allowing acquisition of data over long periods, comparison between laboratories, and reporting of accurate and well-annotated profiles.

Across epidemiology there is a trend toward larger and more complex studies, both in terms of the data collected and also the populations studied. Many studies now combine multiple cohorts, both to increase sample size and thereby improve statistical power and also for stronger biological inference, allowing for example, discoveries in one cohort to be validated in another. However, the scale and complexity of large multicohort studies brings special challenges for metabolomic analysis (detailed further in Subheading 3.1). Many of these are related to the untargeted workflow. For example, how can peaks representing the same metabolite be matched across samples if the identity of the compound is unknown? The combination of different populations also yields its own problems, since metabolites might be present in one cohort but not another or there may be systematic differences in sample collection and storage effects. All these aspects point to the need for specialized data processing and analysis techniques, which is our focus in this chapter.

2 Materials

We illustrate the challenges of large-scale multicohort NMR studies with data from the COMBI-BIO project, as described by Karaman et al. [10]. We briefly describe the source data here; for more details, please *see* [10]. Serum samples from three population-based cohorts are used: LOLIPOP [11] (The London Life Sciences Prospective Population, UK), MESA [12] (The Multi- Ethnic

Study of Atherosclerosis, USA), and ROTTERDAM, [13] (The Rotterdam Study, The Netherlands). The COMBI-BIO project combined these three cross-sectional populations with the aim of detecting combinatorial biomarkers of preclinical atherosclerosis. There were several differences in the sample collection, storage, and age ranges of these three cohorts. For example, the collection period of samples for LOLIPOP was 2002–2008, for MESA it was 2000–2002, and for ROTTERDAM it was 1997–1999. The samples of LOLIPOP and MESA were stored at −80 °C after collection, whereas the samples of ROTTERDAM were stored at −20 °C. The age range of the participants at the time of recruitment was 35–74 for LOLIPOP, 45–84 for MESA, and 55–85 for ROTTERDAM. In total, 7773 serum samples were analyzed in two phases over a period of ~1 year. Phase 1 corresponded to 3964 samples (LOLIPOP: 998, MESA: 1976, ROTTERDAM: 990) and phase 2 corresponded to 3809 (LOLIPOP: 987, MESA: 1982, ROTTERDAM: 840). The data thus consists of three cohorts and two phases considered as 6 "batches" requiring joint processing.

Two types of QC samples were used to monitor the effects of NMR data processing. QC1 samples were derived from a commercially available serum sample. In phase 2, QC2 samples were aliquoted from a pool of phase 1 LOLIPOP samples. Samples were mixed with an equal volume of phosphate buffer, centrifuged, and kept at 4 °C until analysis. In phase 1, one QC1 sample was run per 96 study samples. In phase 2, one QC1 sample and two QC2 samples were run per 192 (2 × 96) samples. All ^{1}H NMR spectra were acquired on the same Bruker DRX600 spectrometer (Bruker Biospin, Rheinstetten, Germany) operating at 600 MHz. A standard water-suppressed 1D spectrum (NOESY) and a Carr−Purcell−Meiboom−Gill (CPMG) spectrum were obtained for each sample [14]. 32 scans were collected into 131,072 frequency domain points, and a line broadening of 0.3 Hz was applied. Spectra were rejected and the sample rerun if the peak width at half-height of the alanine doublet at 1.44 ppm was >1.0 Hz.

3 Methods

3.1 Problems with Multicohort Data

Metabolic profiles are highly variable, and many different sources of variation contribute to this diversity. The total variation can be considered to be composed of analytical variation (both from sample collection, storage, preparation, and data acquisition) and from the biological variation itself. Technical variation in modern NMR spectrometers is quite low, with repeatedly measured spectra almost completely superimposable to the eye [7] and minimal error structure [15]. Large numbers of samples are usually run in smaller batches, and batch-to-batch variation can be a significant issue to deal with, both in experimental design and in data processing. The problems are compounded in multicohort studies, where all sources of variation, but especially biological, collection, and storage effects, may be different between cohorts [10]. In addition, the effects may be nonlinear, in that different batches or cohorts may exhibit not only changes in their mean profiles but also changes in variance and shape of the distribution. For example, cohorts with different age distributions will naturally show different ranges of lipoprotein levels.

Probably the largest difficulty with integrating multicohort NMR data is that of matching and alignment. This is the process of deciding which resonances derive from the same metabolite in different samples (matching) and optionally modifying the data so that chemical shifts agree (alignment). The key problem is that the assignment of each resonance is unknown a priori. That is, we do not know to which nucleus of which metabolite the resonance belongs; this is a consequence of the untargeted nature of the approach. The problem is further compounded by small but significant changes in chemical shift of the same resonance across the samples, due to small changes in physical conditions such as pH or ionic concentration [16]. This makes matching very difficult, especially in regions of the spectrum that are highly crowded with many overlapping resonances. The lack of confirmed assignments also means that matching and alignment algorithms cannot easily be benchmarked since there is no clear ground truth, and for real data, one is heavily dependent on expert judgment as to whether the correct match has been made. Many alignment algorithms require manual input and so a further issue is the difficulty of scaling them to large data sets with many thousands of spectra.

NMR metabolomic data always requires normalization, i.e., scaling each spectrum by a scalar factor to remove irrelevant variation in, e.g., sample concentration. This becomes more critical with large multibatch and multicohort data sets as the normalization procedure may be batch- or cohort-dependent. Another issue is that of outliers. The larger the study, the higher the chance of observing outliers and the greater need to remove them. It is important to understand the origin of outlying samples before considering removing them, and this can become onerous when the numbers are high. In large studies, outliers will often be observed in groups; for example, a group of patients taking a specific drug. Whether these outlying groups are to be removed or not requires careful thought and justification in the context of the aims of the overall study.

Taken together, these challenges pose many problems, which are exacerbated by the large numbers and complexity of the study design. However, many of them can be successfully addressed by suitable preprocessing techniques, which we illustrate in the next section.

3.2 Preprocessing

Standard initial processing (phasing, baseline correction, and chemical shift calibration) should be applied to each spectrum. In the example data from Karaman et al. [10], this was achieved with automated functions using Topspin version 3.2 (Bruker Biospin, Rheinstetten, Germany). When an internal standard cannot be used, chemical shift calibration can be applied using the glucose doublet at δ 5.23. This is particularly important for blood serum samples as suggested earlier [17] since internal standards such as 3-trimethylsilylpropionic acid (TSP) are affected by protein binding that results in changes in peak shape and position around δ 0.00. The chemical shift range of the spectra is also recommended to be trimmed to δ 0.50–9.00 since metabolite signals are not observed in serum outside this region. For the example data, this resulted in a data table consisting of 7869 spectra, each digitized into 30,590 intensity variables.

The conventional approach to preprocessing smaller single batch studies includes chemical shift alignment, normalization, removal of interfering spectral signals, and removal of outlying samples. For large-scale multicohort/batch studies, these steps need to be modified and cohort/batch correction should be considered in the pipeline. In the following sections, the preprocessing steps are briefly explained.

3.2.1 Removing Interfering Regions

According to NMR theory, the spectrum should consist of peaks with areas proportional to the concentration of the corresponding compounds. Some spectral regions may not follow this rule and may have negative effect on the sample normalization. A common example is the residual water signal which is usually observed between approximately δ 4.40 and δ 5.00. Although the water signal is suppressed during the data acquisition, some residual signal from the peak is still present and needs to be removed. The region δ 4.40–5.10 in the example data was removed for this purpose. In addition, spectral regions which contain interferents or contaminants may be observed in the full spectrum. Methanol, a common contaminant in clinical studies, was detected in the example data requiring removal of the region δ 3.345–3.370. Furthermore, suspected interferents in the regions δ 1.150–1.210, δ 2.214–2.231, and δ 3.630–3.680 were also removed. Selection of the interfering spectral regions is generally not straightforward and requires expert knowledge for thorough inspection. The removal of spectral regions should be applied to entire columns to retain the shape of the data table.

3.2.2 Alignment

Conventional chemical shift calibration of the spectra usually results in a data set where many resonances vary in position slightly across the set of spectra. Thus, the peaks are said to be misaligned. This is usually due to slight differences in matrix conditions (e.g., pH) across the samples. Application of an algorithm to realign each

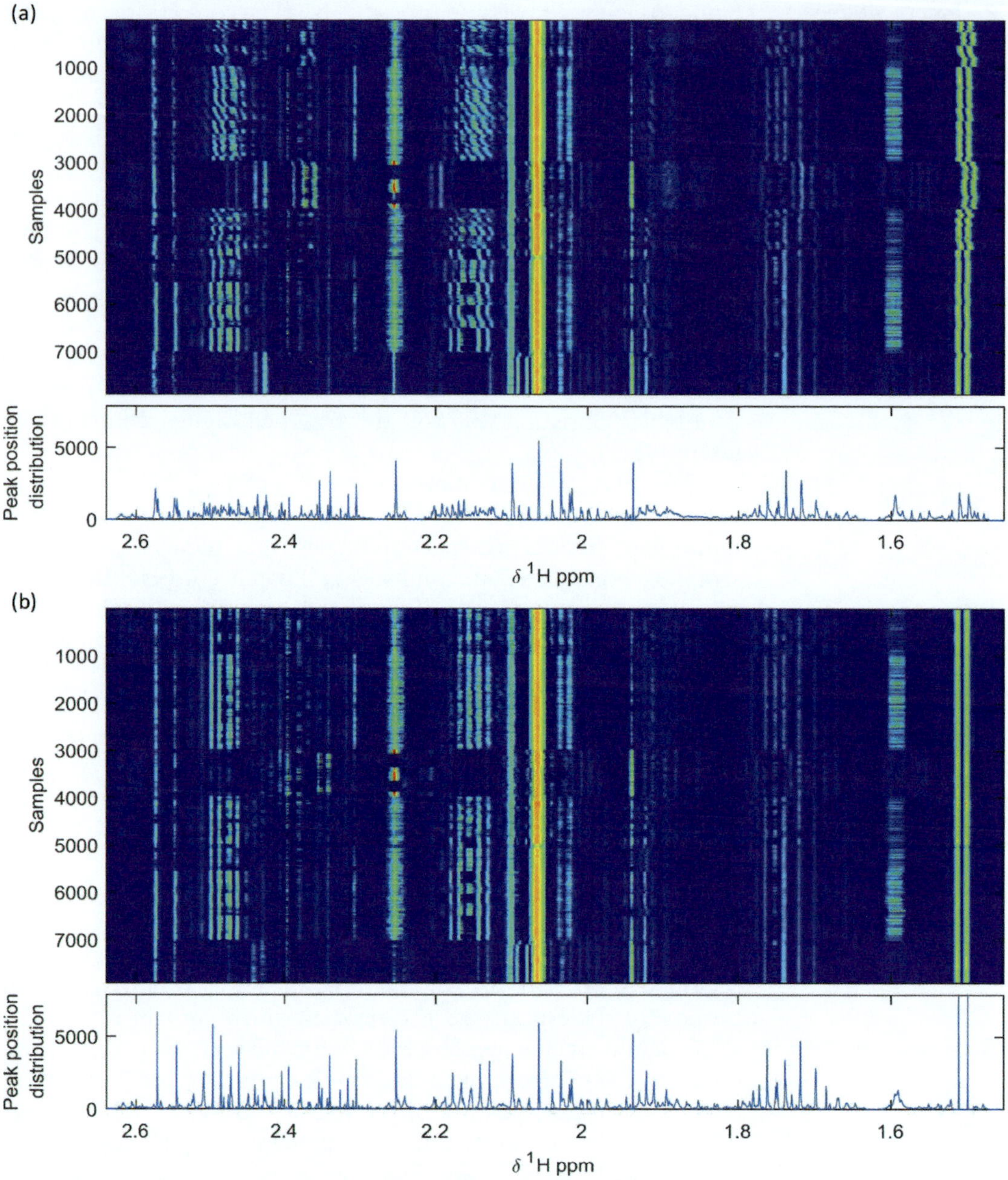

Fig. 1 Illustration of spectral peaks of the CPMG NMR data set between δ 1.45 and δ 2.64 ppm before (**a**) and after (**b**) alignment by RSPA [18]. On the peak position distribution plots (lower part of each panel), higher values indicate peaks that are better aligned. Reprinted with permission from Karaman et al. [10]. Copyright 2016 American Chemical Society

resonance can be very helpful. In Fig. 1, spectral peaks of the CPMG NMR data set between δ 1.45 and δ 2.64 ppm before and after alignment using the RSPA algorithm [18] are demonstrated in heatmaps. In panel (a), the doublet seen around δ 1.50 corresponds to alanine and it is clear that the peak intensities are not aligned across the samples. Many other peaks in this region also exhibit positional shifts like alanine, but the pattern of the shifts is non-linear. The task is complicated by true differences in composition across the multiple batches and cohorts. For example, the spectral profile of samples at approximate indexes 3000–4000 and 7000–8000 are substantially different from the remaining data, making it difficult to determine whether apparent misalignment is due to peak shift or absence of a metabolite.

Available algorithms for alignment generally locate peaks and align them to a reference spectrum. In multicohort data, selection of a reference is not straightforward as the sample set is large and complex. Depending on the spectral region of interest, a median spectrum or a sample spectrum highly correlated to the median spectrum can used as the reference. The challenge for aligning large-scale data from multicohort studies is the increased complexity and variability in the data. Running the alignment algorithms automatically on the full spectrum with the default parameters typically does not provide optimal results. Therefore, we recommend splitting the data into smaller segments, joined at regions where there is no signal. Alignment parameters can then be optimized on each segment separately until most of the peaks are aligned. Plotting only part of the spectrum will also facilitate the visualization on a regular PC. In panel (b) of Fig. 1, most peaks seem to be aligned in this spectral region. However, although not immediately clear from the figure, alignment algorithms may produce artifacts in the data, especially if peaks are crowded or overlapping. This is a major drawback of the alignment approach. Another key problem with alignment is the difficulty of objectively assessing the quality of the alignment result. While some metrics (usually based on correlation [18]) to detect alignment improvement have been proposed, they are imperfect, reflecting the fundamental problem with matching unidentified peaks discussed above. Ultimately, visual inspection by an expert is recommended to ensure a satisfactory result.

After alignment, binning methods [19, 20] can be applied to the data to reduce the resolution and to account for minor misalignments. These methods generally try to find peak locations and corresponding spectral segments. The intensities in these segments are then summed or averaged into one single value to represent one bin in a sample spectrum. Caution must be taken when binning the data because peaks may fall into wrong bins due to shifts or may be split if the assigned bin boundaries are not optimal for every sample.

Neither alignment nor binning can handle peak overlap. Overlapping peaks need to be deconvolved into pure components by more sophisticated tools such as BATMAN [21] or BAYESIL [22]. These methods use pure standard spectra of all known compounds in the sample matrix in the modeling step. Although they are computationally intensive and can be difficult to optimize, they produce quantified values for every metabolite of interest which is generally preferable for a data analyst with little or no experience in NMR spectroscopy.

3.2.3 Normalization

Normalization procedures [23–26] scale each spectrum by a spectrum-specific factor to correct for global variations, providing comparability across the samples. The variations could be due to variable dilutions or small changes in instrument calibration. Most sources of variation become more important when considering large multicohort data, especially because of the long analytical run time. We briefly mention several common normalization approaches below:

1. Addition of internal/external standard to every sample and use of the peak area/intensity of this standard as normalization factor. This method is generally not convenient because standards may not account for dilution differences and may be avoided because of complexation with other sample components as described above.
2. Total area normalization where the normalization factor is the sum of all intensities of the spectrum. This method works only if total metabolite concentration across the samples does not change. A small number of high-concentration metabolites will affect the normalization factor significantly.
3. Probabilistic quotient normalization (PQN) where a test sample is normalized by the median of fold changes of the test sample against a reference spectrum. This method assumes that the peaks affected by dilution should have the same fold changes between two samples. Use of the median means that the effect of large changes in a few metabolites is mitigated. PQN has been shown to outperform earlier methods such as total area normalization and is fast and memory efficient.

In the example data, PQN was used with the median spectrum of the full data as the reference. It is also possible to apply PQN separately to each batch/cohort using different reference spectra as batch/cohort adjustment will already be applied in the final step of preprocessing. It is important to mention that a normalization method which uses a reference must be applied after alignment, because intensity ratios prior to alignment may be influenced by uncontrolled peak shifts.

3.2.4 Outlier Detection and Removal

The presence of outlying samples can often lead to incorrect results and interpretation. Therefore, the data set needs to be thoroughly investigated for moderate and strong outliers before applying any statistical analysis. Principal component analysis (PCA) is widely used to explore the trends in the data and is an excellent tool to investigate potential outliers. Strong outliers can be found using the Hotelling's T^2 value for each sample which is calculated from the PCA scores. Samples with T^2 greater than a critical value (estimated via an approximate t distribution for a certain confidence level) are thus the potential outliers. The spectra of these outliers should be examined and discarded if they are analytical outliers, for example, due to failure of water suppression. In addition, contribution plots [27] can be used to find out wherein the spectrum the sample is outlying from the bulk of the data. With this information, a decision can be made on whether to discard the sample. Once outliers are removed, the PCA model must be recalculated. In large data sets, removal of a small number of outliers often results in a new set of weaker outliers becoming evident. Once these are removed, further outliers may be revealed in an iterative "onion skin" phenomenon. The problem is exacerbated in multicohort data, where apparently normal individuals from one cohort may appear outlying with respect to another cohort. Ultimately the decision whether to remove an outlier or outlying group must be made and justified in the light of the overall aim of the study and statistical analysis.

3.2.5 Batch and Cohort Adjustment

In large multicohort studies, the samples are generally from multiple origins and data acquisition is conducted in several batches. This causes additional variation in the integrated data that needs to be corrected. Different sample compositions, sample storage, and collection conditions between the cohorts will all contribute to this variation. In the example data, there were three cohorts, and each cohort's data were acquired in two experimental phases/batches. To adjust for batch and cohort, every variable in each batch was mean-centered so that the differences in the mean levels of the variables are removed. This operation is similar to the removal of between-subject variation in studies where each batch originates from multiple individuals [28]. This type of adjustment is justified when the objective is to find within-cohort trends which are replicated across cohorts. It should be noted, however, that it is a linear correction which does not modify the covariance structure; this is still allowed to be different from cohort to cohort.

In Fig. 2, the two panels show the PCA score plots before and after adjustment. On the left panel, QC1 and QC2 samples exhibit two small clusters showing that analytical variation within the QC samples is low compared to biological variation across the study samples. Although LOLIPOP and MESA samples overlap, ROTTERDAM samples are clustered separately. On the right panel, the scores plot show that all cohorts and batches overlap after the adjustment by mean-centering.

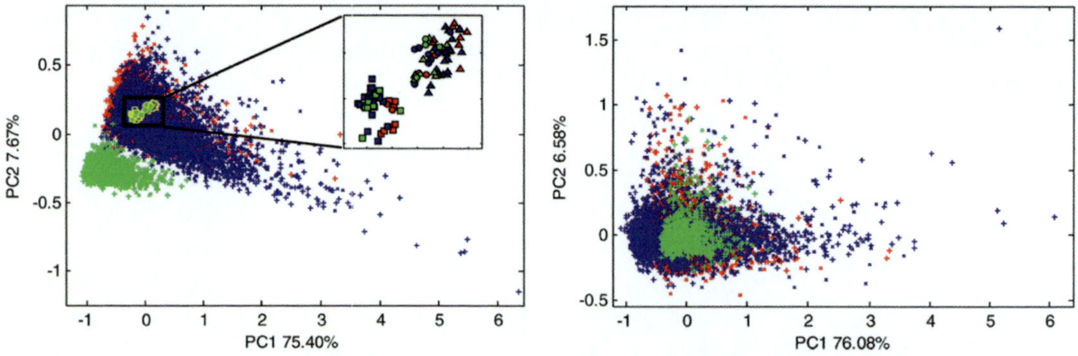

Fig. 2 Plots of the first two PCA scores for CPMG data: (left) before and (right) after adjustment for cohort and phase. Colors correspond to cohorts (red: LOLIPOP, blue: MESA, green: ROTTERDAM). Symbols for the study samples vary according to phases (×: phase 1, +: phase 2). Zoomed frame shows the QC samples (△: QC1 analyzed in phase 1, ○: QC1 analyzed in phase 2, □: QC2 analyzed in phase 2). Axis labels indicate the percentage variance explained by each principal component. Note that QC samples are not shown in the right panel because they were excluded from the data set when cohort/phase adjustment was applied. Reprinted with permission from Karaman et al. [10]. Copyright 2016 American Chemical Society

The adjustment procedure can be modified depending on the data. If the data is heteroscedastic (exhibit a non-constant variance), log transformation may help to make the data homoscedastic prior to mean-centering. In addition, the mean-centering can be followed by unit-variance scaling to remove differences in variance within each batch [29]. Furthermore, non-parametric alternatives of mean and standard deviation that are median and median absolute deviation can be considered if necessary.

3.3 Correlation Structures and Metabolome-Wide Significance

In large-scale studies, untargeted NMR analyses typically result in tens of thousands of data points (variables) for each individual. This MWAS approach, essentially screening each variable one by one, searches for correlations to an outcome variable such as disease risk. Since many thousands of statistical tests are performed, a multiple testing problem arises. While standard correction procedures, such as Bonferroni or False Discovery Rate (FDR) approaches may be applied, they are not able to deal adequately with the high degree of correlation structure existing in NMR metabolic spectra.

Correlations between NMR data points arise for a variety of reasons. The highest dependence results from nearby data points being part of the same peak (typically at least 10 data points per peak in modern spectra). This strongly local correlation is affected by the spectral resolution of the instrument and the digital resolution and smoothing of the spectrum. High dependencies also result from peaks which arise from the same compound. This correlation may be strongly non-local (e.g., between peaks from molecules with both aliphatic and aromatic resonances) but can still be very high. Further high correlations derive from biological relationships between metabolites, for example, those affected by the same

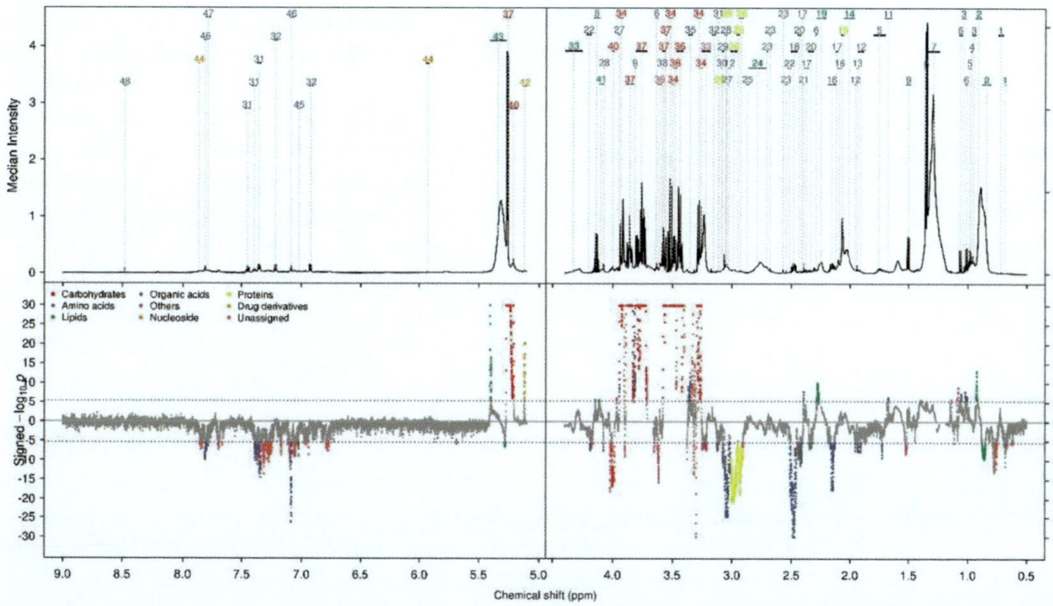

Fig. 3 A metabolic Manhattan plot visualizing associations between serum NMR metabolic variables and blood glucose levels. The top panel shows the median NMR spectrum, while the bottom panel plots the signed negative $\log_{10}$ *p*-value against NMR chemical shift. The MWSL is represented by the dashed line and significant associations are highlighted in color. *See* ref. 32 for more information and key

biological process. In contrast to the first two sources of dependence, biological correlations may be both positive and negative, reflecting the wide variety of regulatory relationships present in metabolic processes. In addition to these technical and biological correlations, NMR metabolomic data may exhibit unwanted artifactual correlations. Examples are strong negative correlations arising from sample-to-sample shifts in peak positions [30] and strong positive correlations when a single high variance metabolite dominates the spectrum and total intensity normalization is used [23]. These should be corrected as much as possible, for example, by alignment and robust normalization procedures.

One method to correct for multiple testing in the presence of strong correlations is the so-called Metabolome-Wide Significance Level (MWSL) [31]. In this approach, the outcome variable is randomly permuted many times to estimate the null distribution of the test statistic. Figure 3 illustrates the use of the MWSL to find metabolic features associated with blood glucose levels in the MESA cohort. The MWSL ($\alpha = 0.38 \times 10^{-5}$) is less conservative than the equivalent Bonferroni threshold ($\alpha = 0.16 \times 10^{-5}$), thus aiding the discovery of more associations, while maintaining robust control of the overall false positive rate. The strong advantage of the MWSL approach is that it can successfully take account of the strong correlation structures present in metabolomic data, but it does have drawbacks. It is computationally intensive, requiring a minimum of thousands of permutations; this can be time-consuming when there are tens of thousands of variables to test. Further, the MWSL threshold derived depends on the outcome variable chosen, and over correction is possible [32]. Further research is required to address these problems.

3.4 Metabolite Identification

NMR metabolite identification constitutes an essential step in untargeted metabolomics studies. Metabolite identities are necessary to be able to link statistically significant NMR features to biological or physiological interpretation. However, there are several challenges associated with metabolite identification in NMR multicohort data sets. One of these challenges, peak overlap is in fact common across all untargeted metabolomics studies making use of 1D ^{1}H NMR. This overlap results from the occurrence of several signals from different metabolites in the same spectral region. Another difficulty comes from the heterogeneity between spectra across a multicohort data set which adds another layer of complexity to feature annotation and ultimately metabolite identification. For example, in crowded spectra, a feature arising from a particular metabolite in one cohort may report mainly on a different metabolite in another cohort. An adequate strategy needs to be followed to tackle such challenges. Currently the approaches to metabolite identification are still essentially manual tasks despite recent efforts to automate such procedures [33–35]. A typical metabolite annotation/identification workflow can be divided into three stages: spectral assignment, inspection of feature variation across samples, and metabolite annotation verification which we describe in more detail in the next sections.

3.4.1 Spectral Assignment

The severe peak congestion in the 1D ^{1}H-NMR spectrum is an important challenge in NMR metabolite identification. To resolve NMR resonances, it is common to acquire additional 2D NMR spectra on representative samples. These 2D NMR spectra spread the NMR peaks into a second dimension thus resolving many overlapping peaks, and in addition, provide orthogonal information on the metabolite structure. Several types of 2D NMR experiments can be used, depending on the information required and acquisition time available. The simplest 2D experiment, ^{1}H J-resolved, separates the peak multiplicity into the second dimension, removing some of the overlap contained in 1D ^{1}H-NMR [36] (Fig. 4a). This experiment also allows the determination of a peak multiplicity (or peak splitting), a feature intrinsically related to molecular structure [36]. Other homonuclear correlation experiments such as ^{1}H–^{1}H Correlation Spectroscopy (COSY) and ^{1}H–^{1}H Total Correlation Spectroscopy (TOCSY) provide further information on signals arising from the same molecule [37]. These signals appear as off-diagonal cross peaks, thus indicating groups of closely related ^{1}H nuclei from the same molecule (Fig. 4b).

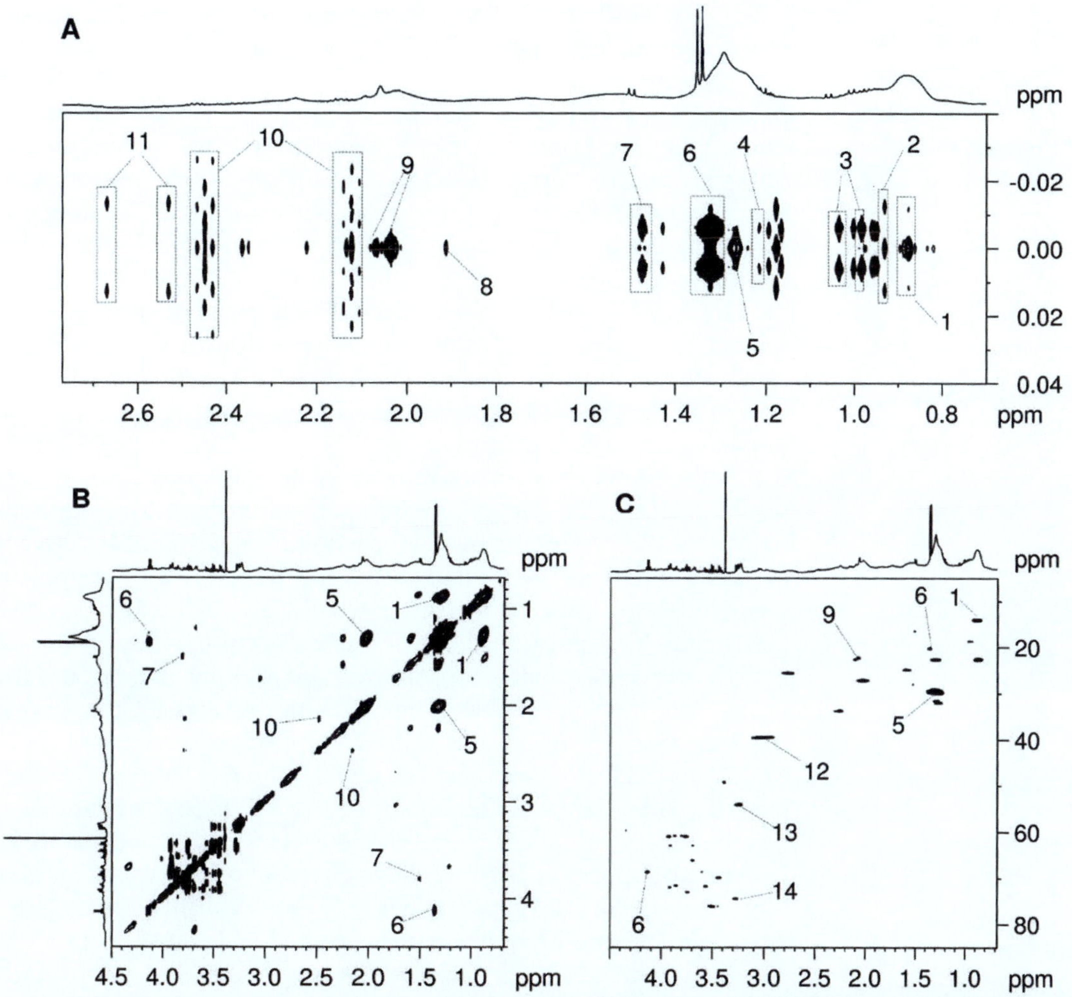

Fig. 4 2D NMR spectra of a normal human serum sample with some metabolite signals annotated. (**a**) ^{1}H J-resolved spectrum, (**b**) ^{1}H–^{1}H TOCSY spectrum, (**c**) ^{1}H–^{13}C HSQC spectrum. Legend: (1) lipid $\mathbf{CH_3}(CH_2)_n$; (2) isoleucine; (3) valine; (4) β-hydroxybutyrate; (5) lipid $CH_3\mathbf{CH_2}(CH_2)_n$; (6) lactate; (7) alanine; (8) acetate; (9) *N*-acetyl (glycoproteins); (10) glutamine; (11) citrate; (12) albumin lysyl groups; (13) choline; (14) glucose; bold symbols indicate the protons from lipid molecules giving rise to the assigned signals

Another very useful 2D NMR experiment is the ^{1}H–^{13}C heteronuclear single-quantum coherence (HSQC) which correlates ^{1}H nuclei to their directly bonded ^{13}C nuclei (Fig. 4c). The majority of 1D ^{1}H-NMR peaks can usually be assigned by using 2D NMR spectral information, thereby facilitating further metabolite identification steps.

When 2D NMR spectra are not available or cannot be easily obtained, statistical correlation methods such as Statistical Total Correlation Spectroscopy (STOCSY) and related techniques [38, 39] can be performed on the 1D ^{1}H-NMR spectral data set to look for peaks belonging to the same molecule. However, 2D

NMR data analysis is still necessary for the definitive confirmation of the spectral assignments.

3.4.2 Feature Variation Across the Data Set

Although spectral peaks can be assigned to the corresponding underlying metabolites as described earlier, several different metabolite species can contribute to the intensity of a single spectral peak, and this contribution might change across samples, making the annotation and subsequent identification of the metabolites more cumbersome.

To be able to associate a spectral data point from the NMR spectra to a given metabolite, the contribution of the underlying NMR signals must be inspected across the data set. Visual inspection of stacked spectra may be sufficient in many cases, for instance, in cases where the most intense peak is present in all samples from the multicohort data set. However, for less obvious cases, where severe peak overlap occurs or when peak intensity is low, methods such as STOCSY might be useful to highlight the most relevant metabolite signal(s) associated with a spectral point. An example is shown in Fig. 5, where the unknown feature of interest (spectral point at 1.7614 ppm, Fig. 5a) occurs in a region where the resonances from lysine, leucine, ornithine, and arginine overlap and the signal intensity is quite low compared to other signals in the same spectral area. By using the SubseT Optimization by Reference Matching (STORM) method [40], which finds a subset of spectra from the large data set exhibiting maximal STOCSY correlations, it was possible to highlight the signals correlating with the feature of interest and therefore identify the underlying metabolite (Fig. 5b). Note that the STORM approach is only applicable to large data sets where the metabolite of interest is expected to be observed only in an unknown subset of samples. Thus, STORM is particularly important when analyzing large, multicohort NMR data sets.

3.4.3 Verification and Metabolite Identification

For the annotation of an NMR feature, information from a combination of methods, such as 2D NMR and STOCSY, are necessary to achieve a high level of confidence according to the metabolomics standards initiative (MSI) [41]. However, metabolite identification requires the analysis of an authentic standard spiked into a representative sample using the same analytical conditions. The observation of the same peak shapes and position is then used to confirm the metabolite identification and grant the identification the highest confidence level of 1 according to MSI.

4 Conclusions

NMR metabolomics is becoming a widely applied assay in large and complex cohort studies. The large size of studies and complexity introduced by the presence of multiple cohorts leads to a wide

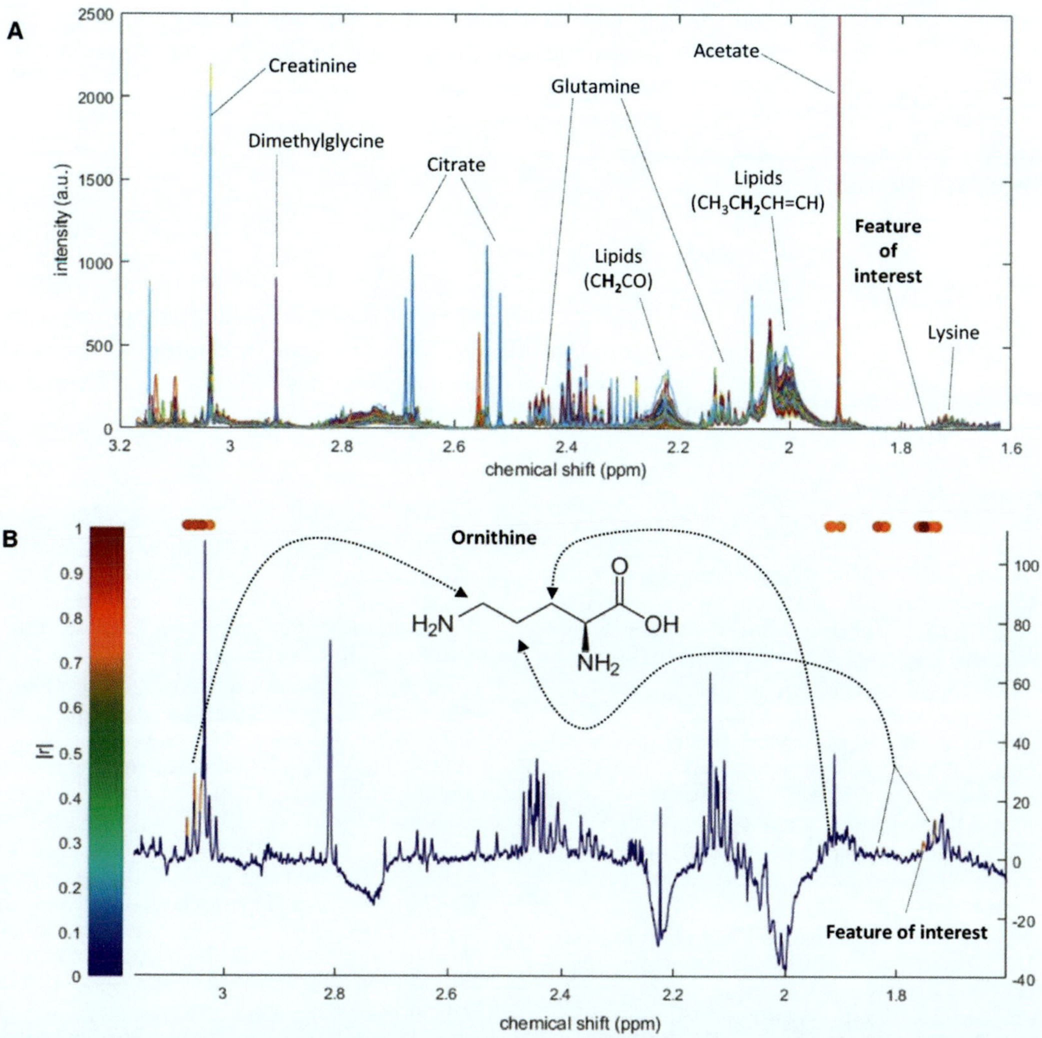

Fig. 5 Annotation of an unknown feature in the human serum NMR spectra from a large cohort. (**a**) Overlaid spectra with some peaks assigned to the corresponding metabolite and the feature of interest; (**b**) STORM analysis of the feature of interest and peak assignments to the corresponding metabolite structure of ornithine. Bold symbols indicate the protons from lipid molecules giving rise to the assigned signals

variety of challenges in analysis of the resulting NMR data. These include alignment, normalization, outlier detection, dealing with correlation, and identification of unknowns. Preprocessing and analysis procedures exist which can mitigate or overcome these difficulties, resulting in consistent data which can be interrogated for meaningful biological patterns. Nonetheless, current analysis techniques are not perfect and there is ample room for new and innovative algorithms. We expect there to be further improvements in the coming years, especially with respect to alignment and peak matching, including objective quality metrics which reduce the need for manual validation. Such advancements will enable routine analysis of large and complex data sets, leading to higher quality data and opening new avenues to biological insight.

Acknowledgments

I.K. and T.E. acknowledge support from the EU PhenoMeNal project (Horizon 2020, 654241). I.K. acknowledges support from the UK Dementia Research Institute, which is supported by the MRC, the Alzheimer's Society and Alzheimer's Research UK. T.E. and G.G. acknowledge support by National Institutes of Health (R01HL133932).

References

1. Albanes D, Moore S, Ulrich C, Stolzenberg-Solomon R, Poole E, Temprosa M et al (2017) COnsortium for METabolomics studies (COMETS): leveraging resources to accelerate scientific discovery. FASEB J 30(1):lb129
2. Soininen P, Kangas AJ, Wurtz P, Suna T, Ala-Korpela M (2015) Quantitative serum nuclear magnetic resonance metabolomics in cardiovascular epidemiology and genetics. Circ Cardiovasc Genet 8(1):192–206
3. Elliott P, Posma JM, Chan Q, Garcia-Perez I, Wijeyesekera A, Bictash M et al (2015) Urinary metabolic signatures of human adiposity. Sci Transl Med 7(285):285ra62
4. Holmes E, Loo RL, Stamler J, Bictash M, Yap IK, Chan Q et al (2008) Human metabolic phenotype diversity and its association with diet and blood pressure. Nature 453 (7193):396–400
5. Keun HC, Ebbels TM, Antti H, Bollard ME, Beckonert O, Schlotterbeck G et al (2002) Analytical reproducibility in (1)H NMR-based metabonomic urinalysis. Chem Res Toxicol 15 (11):1380–1386
6. Dumas ME, Maibaum EC, Teague C, Ueshima H, Zhou B, Lindon JC et al (2006) Assessment of analytical reproducibility of (1) H NMR spectroscopy based metabonomics for large-scale epidemiological research: the INTERMAP study. Anal Chem 78 (7):2199–2208
7. Dona AC, Jimenez B, Schafer H, Humpfer E, Spraul M, Lewis MR et al (2014) Precision high-throughput proton NMR spectroscopy of human urine, serum, and plasma for large-scale metabolic phenotyping. Anal Chem 86 (19):9887–9894
8. Viant MR, Bearden DW, Bundy JG, Burton IW, Collette TW, Ekman DR et al (2009) International NMR-based environmental metabolomics intercomparison exercise. Environ Sci Technol 43(1):219–225
9. Jimenez B, Holmes E, Heude C, Tolson RFM, Harvey N, Lodge SL et al (2018) Quantitative lipoprotein subclass and low molecular weight metabolite analysis in human serum and plasma by 1H NMR spectroscopy in a multilaboratory trial. Anal Chem 90(20):11962–11971
10. Karaman I, Ferreira DL, Boulange CL, Kaluarachchi MR, Herrington D, Dona AC et al (2016) Workflow for integrated processing of multicohort untargeted 1H NMR metabolomics data in large-scale metabolic epidemiology. J Proteome Res 15(12):4188–4194
11. Chambers JC, Obeid OA, Refsum H, Ueland P, Hackett D, Hooper J et al (2000) Plasma homocysteine concentrations and risk of coronary heart disease in UK Indian Asian and European men. Lancet 355 (9203):523–527
12. Bild DE, Bluemke DA, Burke GL, Detrano R, Diez Roux AV, Folsom AR et al (2002) Multi-ethnic study of atherosclerosis: objectives and design. Am J Epidemiol 156(9):871–881
13. Ikram MA, Brusselle GGO, Murad SD, van Duijn CM, Franco OH, Goedegebure A et al (2018) The Rotterdam study: 2018 update on objectives, design and main results. Eur J Epidemiol 32(9):807–850
14. Beckonert O, Keun HC, Ebbels TM, Bundy J, Holmes E, Lindon JC et al (2007) Metabolic profiling, metabolomic and metabonomic procedures for NMR spectroscopy of urine, plasma, serum and tissue extracts. Nat Protoc 2(11):2692–2703

15. Karakach TK, Wentzell PD, Walter JA (2009) Characterization of the measurement error structure in 1D 1H NMR data for metabolomics studies. Anal Chim Acta 636(2):163–174
16. Tredwell GD, Bundy JG, De Iorio M, Ebbels TMD (2016) Modelling the acid/base 1H NMR chemical shift limits of metabolites in human urine. Metabolomics 12(10):1–10
17. Pearce JTM, Athersuch TJ, Ebbels TMD, Lindon JC, Nicholson JK, Keun HC (2008) Robust algorithms for automated chemical shift calibration of 1D 1H NMR spectra of blood serum. Anal Chem 80(18):7158–7162
18. Veselkov K, Lindon J, Ebbels T, Volynkin V, Crockford D, Holmes E et al (2009) Recursive segment-wise peak alignment of biological 1H NMR spectra for improved metabolic biomarker recovery. Anal Chem 81(1):56–66
19. Blaise BJ, Shintu L, Elena B, Emsley L, Dumas M-E, Toulhoat P (2009) Statistical recoupling prior to significance testing in nuclear magnetic resonance based metabonomics. Anal Chem 81 (15):6242–6251
20. Sousa SAA, Magalh+úes A, Ferreira MMC (2013) Optimized bucketing for NMR spectra: three case studies. Chemom Intell Lab Syst 122 (0):93–102
21. Hao J, Astle W, De Iorio M, Ebbels TM (2012) BATMAN--an R package for the automated quantification of metabolites from nuclear magnetic resonance spectra using a Bayesian model. Bioinformatics 28(15):2088–2090
22. Ravanbakhsh S, Liu P, Bjordahl TC, Mandal R, Grant JR, Wilson M et al (2015) Accurate, fully-automated NMR spectral profiling for metabolomics. PLoS One 10(5):e0124219
23. Dieterle F, Ross A, Schlotterbeck G, Senn H (2006) Probabilistic quotient normalization as robust method to account for dilution of complex biological mixtures. Application in H-1 NMR metabonomics. Anal Chem 78 (13):4281–4290
24. Kohl S, Klein M, Hochrein J, Oefner P, Spang R, Gronwald W (2012) State-of-the art data normalization methods improve NMR-based metabolomic analysis. Metabolomics 8(1):146–160
25. Craig A, Cloarec O, Holmes E, Nicholson JK, Lindon JC (2006) Scaling and normalization effects in NMR spectroscopic metabonomic data sets. Anal Chem 78(7):2262–2267
26. Sysi-Aho M, Katajamaa M, Yetukuri L, Orešič M (2007) Normalization method for metabolomics data using optimal selection of multiple internal standards. BMC Bioinformatics 8 (1):1–17
27. Bro R, Smilde AK (2014) Principal component analysis. Anal Methods 6(9):2812–2831
28. van Velzen EJJ, Westerhuis JA, Van Duynhoven JPM, Van Dorsten FA, Hoefsloot HCJ, Jacobs DM et al (2008) Multilevel data analysis of a crossover designed human nutritional intervention study. J Proteome Res 7 (10):4483–4491
29. Karaman İ, Nørskov NP, Yde CC, Hedemann MS, Bach Knudsen KE, Kohler A (2015) Sparse multi-block PLSR for biomarker discovery when integrating data from LC–MS and NMR metabolomics. Metabolomics 11 (2):367–379
30. Couto Alves A, Rantalainen M, Holmes E, Nicholson JK, Ebbels TMD (2009) Analytic properties of statistical total correlation spectroscopy (STOCSY) based information recovery in 1H NMR metabolic data sets. Anal Chem 81(6):2075–2084
31. Chadeau-Hyam M, Ebbels TM, Brown IJ, Chan Q, Stamler J, Huang CC et al (2010) Metabolic profiling and the metabolome-wide association study: significance level for biomarker identification. J Proteome Res 9 (9):4620–4627
32. Castagne R, Boulange CL, Karaman I, Campanella G, Santos Ferreira DL, Kaluarachchi MR et al (2017) Improving visualization and interpretation of metabolome-wide association studies: an application in a population-based cohort using untargeted (1)H NMR metabolic profiling. J Proteome Res 16 (10):3623–3633
33. Filntisi A, Fotakis C, Asvestas P, Matsopoulos GK, Zoumpoulakis P, Cavouras D (2017) Automated metabolite identification from biological fluid 1H NMR spectra. Metabolomics 13(12):146
34. Takis PG, Schäfer H, Spraul M, Luchinat C (2017) Deconvoluting interrelationships between concentrations and chemical shifts in urine provides a powerful analysis tool. Nat Commun 8(1):1662
35. Tardivel PJC, Canlet C, Lefort G, Tremblay-Franco M, Debrauwer L, Concordet D et al (2017) ASICS: an automatic method for identification and quantification of metabolites in complex 1D 1H NMR spectra. Metabolomics 13(10):109
36. Ludwig C, Viant MR (2009) Two-dimensional J-resolved NMR spectroscopy: review of a key methodology in the metabolomics toolbox. Phytochem Anal 21(1):22–32
37. Dona AC, Kyriakides M, Scott F, Shephard EA, Varshavi D, Veselkov K et al (2016) A guide to the identification of metabolites in NMR-based

metabonomics/metabolomics experiments. Comput Struct Biotechnol J 14:135–153

38. Cloarec O, Dumas ME, Craig A, Barton RH, Trygg J, Hudson J et al (2005) Statistical total correlation spectroscopy: an exploratory approach for latent biomarker identification from metabolic ^{1}H NMR data sets. Anal Chem 77(5):1282

39. Robinette SL, Lindon JC, Nicholson JK (2013) Statistical spectroscopic tools for biomarker discovery and systems medicine. Anal Chem 85(11):5297–5303

40. Posma JM, Garcia-Perez I, De Iorio M, Lindon JC, Elliott P, Holmes E et al (2012) Subset optimization by reference matching (STORM): an optimized statistical approach for recovery of metabolic biomarker structural information from 1H NMR spectra of biofluids. Anal Chem 84(24):10694–10701

41. Sumner LW, Amberg A, Barrett D, Beale MH, Beger R, Daykin CA et al (2007) Proposed minimum reporting standards for chemical analysis. Metabolomics 3:211–221

Chapter 26

Advanced Statistical Methods for NMR-Based Metabolomics

Dabao Zhang and Min Zhang

Abstract

Despite the increasing popularity and applicability of metabolomics for putative biomarker identification, analysis of the data is challenged by low statistical power resulting from the small sample sizes and large numbers of metabolites and other omics information, as well as confounding demographic and clinical variables. To enhance the statistical power and improve reproducibility of the identified metabolite-based biomarkers, we advocate the use of advanced statistical methods that can simultaneously evaluate the relationship between a group of metabolites and various types of variables including other omics profiles, demographic and clinical data, as well as the complex interactions between them. Accordingly, in this chapter, we describe the method of seemingly unrelated regression that can simultaneously analyze multiple metabolites while controlling the confounding effects of demographic and clinical variables (such as gender, age, BMI, smoking status). We also introduce penalized orthogonal components regression as a screening approach that can handle millions of omics predictors in the model.

Key words Seemingly unrelated regression, Penalized orthogonal components regression, Supervised dimension reduction, Omics data

1 Introduction

Metabolomics data are now being generated rapidly and accurately [1], and increasingly are employed in biomedical research as one of the most promising sources of biomarkers for early detection, monitoring disease progression and treatment response, prognosis, and detecting disease recurrence [2, 3]. While there is consensus regarding the potential value of metabolomics data for increasing our understanding of common complex diseases [4], there are well-documented problems that challenge the statistical analysis of the data. Due to the sensitivity of metabolism to biological status, metabolite levels can potentially be affected by a number of demographic and clinical variables (age, gender, BMI, smoking status, diet, exercise, medications, etc.) that confound metabolite biomarker performance. In addition, the analysis of metabolomics

G. A. Nagana Gowda and Daniel Raftery (eds.), *NMR-Based Metabolomics: Methods and Protocols*, Methods in Molecular Biology, vol. 2037, https://doi.org/10.1007/978-1-4939-9690-2_26,

data has been largely underdeveloped due to various specific barriers including the large number of variables compared to the typical sample size, the numerous sample acquisition methods and analytical platforms that measure different metabolites, and the complicated structure of metabolomics data, among many others. Most of the current statistical methods focus on analyzing one metabolite at a time, which completely ignore the correlation between metabolites and hence lead to low power and high false positive rates. As a result, many of the reported metabolite-based biomarkers failed to replicate in comparable studies.

Recently, machine-learning approaches have been widely applied in all aspects of metabolomics data analysis such as sample classification and feature selection. For example, principal component regression (PCR) and partial least squares regression (PLS) have been utilized for visualization of data structure and selection of important features [5]; random forest (RF), support vector machine (SVM), and deep learning (DL) methods have been employed for classification [6], among the many approaches studied. While details about these machine-learning methods have been discussed previously [7], here we will focus on two advanced multivariate statistical methods, namely seemingly unrelated regression (SUR) and penalized orthogonal components regression (POCRE), which can improve the statistical power of the analysis and reduce false positive detection, especially when the sample size is relatively small. We proposed to use SUR for the simultaneous analysis of multiple metabolites with traditional demographic and clinical variables [8, 9] and developed POCRE as a supervised variable selection approach for integrative analysis to investigate the relationship between multiple metabolites and demographic information, clinical variables, and other types of omics data such as genomics, transcriptomics, and proteomics. By taking advantage of the correlation between metabolites in multivariate analysis, the statistical power can be greatly improved by borrowing information from other metabolites, as shown in both simulation studies and real-data analyses [8, 10].

2 Methods

2.1 Seemingly Unrelated Regression (See Notes 1–6)

To analyze the data with multiple response variables, Zeller proposed SUR in 1962 [11] as an extension of linear regression models. For each individual i ($i = 1, 2, \ldots, n$), the basic assumption of SUR is that there are m response variables (i.e., m metabolites) and each has its own regression model as follows:

$$Y_{ij} = \beta_{0j} + \sum_{k=1}^{p} \beta_{kj} X_{ik} + \varepsilon_{ij}, \quad \varepsilon_{ij} \sim N\left(0, \sigma_j^2\right), \qquad (1)$$

where Y_{ij} represents the level of the jth metabolite ($j = 1, 2, \ldots, m$), and X_{ik} is the value of the kth covariate ($k = 1, 2, \ldots, p$) for individual i. In the above model (1), ε_{ij} is the error term that follows a normal distribution. For metabolites j and k within individual i, we have $\text{cov}(\varepsilon_{ij}, \varepsilon_{ik}) = \sigma_{jk}$ and, on the other hand, $\text{cov}(\varepsilon_{ij}, \varepsilon_{i'k}) = 0$ for different individuals $i \neq i'$, $\sigma_{jk} = 0$. The nonzero correlation structure between error terms of different metabolites within the same individual makes the SUR model different from a simple combination of linear regression models.

Based on the SUR model, the procedure of SUR analysis is described as follows:

1. Select demographic and/or clinical variables using a backward, forward, or stepwise strategy. With all metabolites as responses, run SUR to select significant variables among all available demographic and clinical variables.
2. Classify all metabolites into different biological groups based on the metabolic pathways. For each biological group of metabolites, run SUR with the covariates selected from **step 1** and remove the variables that are not statistically significant.
3. Test the significance of each biological group of metabolites. To test whether there is any effect of a single variable (e.g., the first variable) on the levels of a group of five metabolites, we can use the likelihood ratio test that under the null hypothesis, the test statistic follows a χ_5^2 distribution.

$$H_0 : \beta_{11} = \beta_{12} = \beta_{13} = \beta_{14} = \beta_{15} = 0$$

Similarly, we can test the effect of k covariates on the levels of m metabolites using a chi-square test with $k \times m$ degrees of freedom.

2.2 Penalized Orthogonal Components Regression (See Notes 7–12)

As described previously by Chen et al. [8], SUR can improve the statistical power of the analysis by evaluating the relationship between a group of metabolites and various types of variables. However, the number of variables needs to be smaller than the sample size due to the ordinary least squares estimation method involved in SUR. With the increased trend of the availability of data from multiomics platforms, it is desirable to investigate the relationship between metabolomics and other omics profiles to facilitate better understanding of the underlying molecular mechanism.

Based on the extension of SUR to integrative analysis with multiple types of omics data, we proposed a new supervised dimension reduction method, namely penalized orthogonal components regression (POCRE), which can select a subset of variables among hundreds of thousands of candidates that are potentially associated with multiple outcomes. For example, with both genomics (i.e., single nucleotide polymorphisms, SNPs) and metabolomics data available, POCRE can be used to evaluate all SNPs across the entire genome, demographic and clinical variables, as well as their interactions, simultaneously, and select a subset of important variables that are potentially associated with a group of metabolites. After screening all variables with POCRE, SUR can be employed to evaluate the statistical significance of these selected variables. More specifically, POCRE assumes a similar but more general model to the SUR model shown in (1):

$$Y_{ij} = \beta_{0j} + \sum_{k=1}^{p} \beta_{kj} X_{ik} + \varepsilon_{ij},\ E[\varepsilon_{ij}] = 0$$

where all the covariates are uncorrelated to the error terms. Pooling all Y_{ij} into an $n \times m$ matrix Y and all X_{ij} into an $n \times p$ matrix X, we will have the following representation after centering X (i.e., $E[X] = 0$),

$$Y = \beta^T X + \varepsilon,\ E[\varepsilon] = 0.$$

Here, Y is an m-dimensional column vector, and X, centered at 0 ($E[X] = 0$), is a p-dimensional column vector that is uncorrelated to ε, and β is a $p \times m$ matrix. In the case that p is very large, for example, millions for SNP data, it is infeasible to estimate β using traditional methods. Instead, POCRE constructs orthogonal components as linear combination of all variables in X and then regresses Y on these constructed components. Such a construction is sequential where the first component has the largest correlation with the multiple outcomes, and the correlation value continues to drop until it goes to zero. Note that the components are orthogonal to each other as the residual is used after the previous construction. Upon completion, we regress Y on these orthogonal components and obtain the nonzero β corresponding to the potential important variables. These variables will be used to fit a SUR model, and p values will be obtained to evaluate statistical significance. For more details about the theoretical properties and algorithm, the readers are referred to a previous study by Zhang et al. [10].

The superior performance of the proposed method has been demonstrated through extensive simulation studies with various strengths of correlations between omics profiles and predictors with different structures. In addition, the method has been successfully utilized in the analysis of a series of high-dimensional genomics data sets [12–15].

3 A Case Study

3.1 Example Data Set

To illustrate the utility of the advanced statistical methods, we have used a data set that includes 44 patients with colorectal polyps and 58 healthy controls as an example. Briefly, serum samples were obtained from individuals after overnight fasting and bowel preparation before colonoscopy. Other information including age, gender, BMI, smoking, and alcohol status is available for both patients with polyps and healthy controls. Upon randomization of all samples, ^1H NMR experiments were performed using a Bruker DRX 500-MHz spectrometer with an HCN cryogenic probe. The detailed information about the experiments is presented in Chen et al. [8].

3.2 Data Preprocessing

The spectra were aligned using KnowItAll software (version 7.8), then binned to buckets of equal size. After getting the integrals of the buckets, the total spectral integral was employed to normalize the data. To remove potential bias, the peak integrals were rescaled to have a mean of zero and a standard derivation of one. A total of 24 metabolites were detected in all samples and subjected to statistical analysis. Note that all samples are Caucasian with ages between 45 and 65 years.

3.3 Data Analysis and Results

Based on the 24 metabolites, a subset of variables was selected with SUR using a backward elimination strategy. The result suggests that metabolite profiles are significantly different between patients with polyps and healthy controls for testing diagnosis and its interaction with several demographic covariates (*see* Table 1).

In addition, the SUR analysis of grouped metabolites offers an enhanced power for hypothesis testing. The analysis was performed by combining biologically related (*see* Table 2) metabolites in terms of the effects of the demographic covariates. As shown in Table 2, 11 groups are significantly associated with disease status, while

Table 1
List of significant variables selected using the backward elimination strategy

Variables	*P* value
Gender	9.9×10^{-8}
BMI	0.0023
Smoking	0.56
Diagnosis	0.39
Diagnosis × Gender	0.045
Diagnosis × BMI	0.012
Diagnosis × Smoking	0.0049

Note, smoking and BMI were included since their interactions with the diagnosis were significant

Table 2
Results of testing the effects of "Diagnosis" on all 15 biological groups of metabolites

Biological groups	*p*-Value	Adjusted *p*-value
Group 1: Acetate, glucose, lactate	0.014	0.023
Group 2: Isoleucine, valine	0.0046	0.012
Group 3: Alanine, glutamic acid, glutamine	0.060	0.069
Group 4: Creatinine, glutamine, urea	0.0010	0.0050
Group 5: Glutamic acid, histidine	0.058	0.072
Group 6: Acetoacetate, acetone, lactate	0.33	0.33
Group 7: Acetoacetate, citric acid, tyrosine	0.0011	0.0041
Group 8: Citric acid, formate, glutamic acid, glutamine	0.23	0.25
Group 9: Phenylalanine, tyrosine	0.0021	0.0063
Group 10: Alanine, glutamic acid, glutamine, glycine, histidine, isoleucine, lysine, phenylalanine, threonine, tyrosine, valine	1.5e–07	2.3e–06
Group 11: Alanine, citric acid, glucose, lactate	0.021	0.032
Group 12: Glycine, threonine	0.0051	0.011
Group 13: Alanine, glutamic acid, glycine, threonine	0.031	0.042
Group 14: Alanine, glutamic acid, glycine, isoleucine, threonine, valine	1.2e–05	9.1e–05
Group 15: Choline/phosphocholine, glycine, threonine	0.0057	0.011

there was no significant association between disease status and any individual metabolite. Results from this case study further confirm the superior power of simultaneous analysis of multiple metabolites using SUR. For more details of this case study, the readers are referred to Chen et al. [8, 9].

3.4 Seemingly Unrelated Regression in R

We can use the function systemfit() in R package systemfit to fit the SUR model and use the function lrtest() in R package lmtest to conduct the likelihood ratio test on each covariate.

Running the R codes in Box 1 will simulate a data set with 50 observations of five metabolites, as well as the covariates sex and BMI. The command "`summary(sfres)`" will summarize the results on fitting the SUR model for the five metabolites, with both covariates sex and BMI. The results include both the estimation of the covariate effects and hypothesis tests on both covariates in individual linear regression models for single metabolites (*see* Table 3 excluding the last row). We further fit the SUR models without the covariate sex and BMI, respectively, so as to use the

Box 1 An Example of Using SUR in R

```
require(MASS)
require(systemfit)
require(lmtest)

# Simulate Data
set.seed(3416)
sex <- rbinom(50,1,0.5)
bmi <- rexp(50,1/20)
beta <- cbind(c(1,0.1),c(-1,0.1), c(1,-.1), c(-1,-0.1))
mb <-cbind(sex,bmi)%*%beta+mvrnorm(50,rep(0,4),Sigma=10*diag(4))
mb <- cbind(mb,(mb[,1]+mb[,3])/2+rnorm(50,0,1))
colnames(mb) <- c('mb1','mb2','mb3','mb4','mb5')

# Fit SUR with Covariates sex & bmi
mmodel0 <- list(eq1=mb[,1]~sex+bmi, eq2=mb[,2]~sex+bmi, eq3=mb[,3]~
            sex+bmi,
            eq4=mb[,4]~sex+bmi, eq5=mb[,5]~sex+bmi)
sfres <- systemfit(mmodel0,method="SUR")
summary(sfres)

# Likelihood ratio test of sex
mmodel1 <- list(eq1=mb[,1]~bmi, eq2=mb[,2]~bmi, eq3=mb[,3]~bmi,
             eq4=mb[,4]~bmi, eq5=mb[,5]~bmi)
sfres1 <- systemfit(mmodel1,method="SUR")
lrtestsex <- lrtest(sfres1,sfres)
print(lrtestsex)

# Likelihood ratio test of bmi
mmodel2 <- list(eq1=mb[,1]~sex, eq2=mb[,2]~sex, eq3=mb[,3]~sex,
             eq4=mb[,4]~sex, eq5=mb[,5]~sex)
sfres2 <- systemfit(mmodel2,method="SUR")
lrtestbmi <- lrtest(sfres2,sfres)
print(lrtestbmi)
```

functions "`lrtest(sfres1,sfres)`" and "`lrtest(sfres2, sfres)`" to conduct the significance tests on the covariates sex and BMI, respectively (*see* Table 3, last row).

3.5 POCRE in R

We can use the R package POCRE to screen for covariates using the penalized orthogonal component regression. As shown in Box 2, a data set "`simdata`" of sample size 100 is included in the package, with the response values in the first column and 1000 covariates in the other columns. We can first use the function "`cvpocre(y,x, n.folds=5)`" to select the tuning parameter using the fivefold cross-validation, which suggests setting the tuning parameter

Table 3
Results of seemingly unrelated regression

	Sex		BMI	
Metabolite	**Effect**	***p*-Value**	**Effect**	***p*-Value**
mb1	−1.3836	0.1261	0.1268	0.0002
mb2	−1.5276	0.0778	0.1154	0.0003
mb3	3.5212	0.0005	−0.1659	6.86×10^{-6}
mb4	−0.6997	0.5057	−0.0974	0.0107
mb5	1.1563	0.1222	−0.0200	0.4418
All	–	0.0034	–	5.11×10^{-9}

Box 2 An Example of Using POCRE in R

```
require(POCRE)

data("simdata")
y <- simdata[,1]
x <- simdata[,2:1001]
lambda <- cvpocre(y,x,n.folds=5) # tuning parameter
pres <- pocre(y,x,lambda=lambda)
print(pres$nCmp) # number of components
plot(pres)
which(abs(pres$beta)>0.0001) # selected covariates
```

equal to 1.2. Then we run the function "`pocre(y,x,lambda=lambda)`" with the determined tuning parameter, which constructs two components as shown in Fig. 1.

Using the function "`plot(pres)`", we not only obtain the plots in Fig. 1 to show the constructed component but also the selected covariates as shown in Fig. 2. We can also use the function "`which(abs(pres$beta)>0.0001)`" to list the selected covariates, which are covariates from 1 to 12 and from 101 to 112. Note that this data set was simulated from a model with the 20 true covariates, which are from 1 to 10 and from 101 to 110.

POCRE can also be applied to select covariates for multiple metabolites at the same time. For example, the POCRE package includes the data set "`sim5ydata`" of sample size 100, with

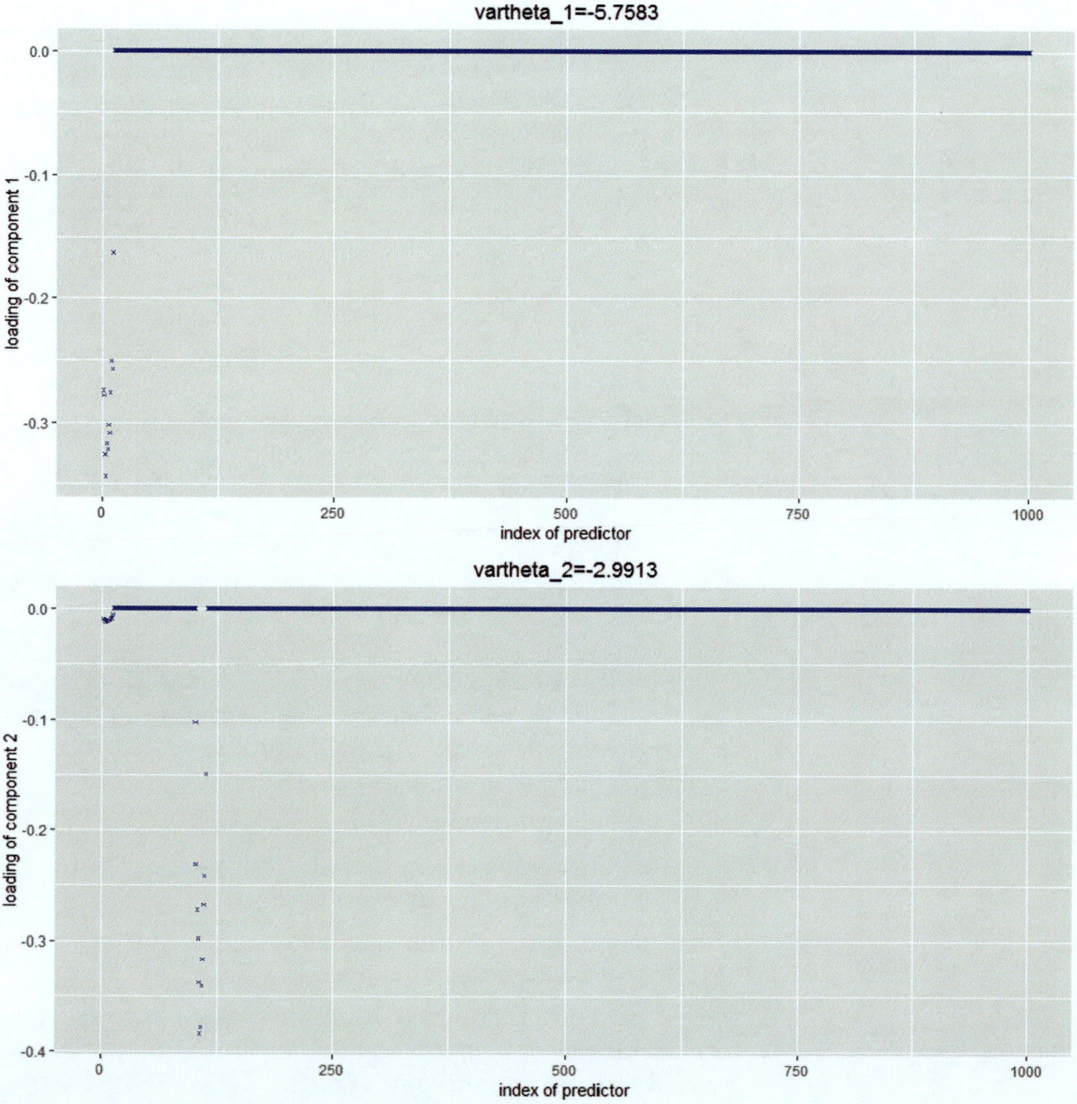

Fig. 1 Two components constructed by POCRE in R

response values in the first five columns and 1000 covariates in the other columns. Similar codes in Box 2 can be used to analyze this data set and to select covariates.

4 Notes

1. Prior to data analysis, it is important to check whether the data follow a Gaussian distribution. If not, certain transformation (such as log) needs to be performed to make sure that the data follow a Gaussian distribution.

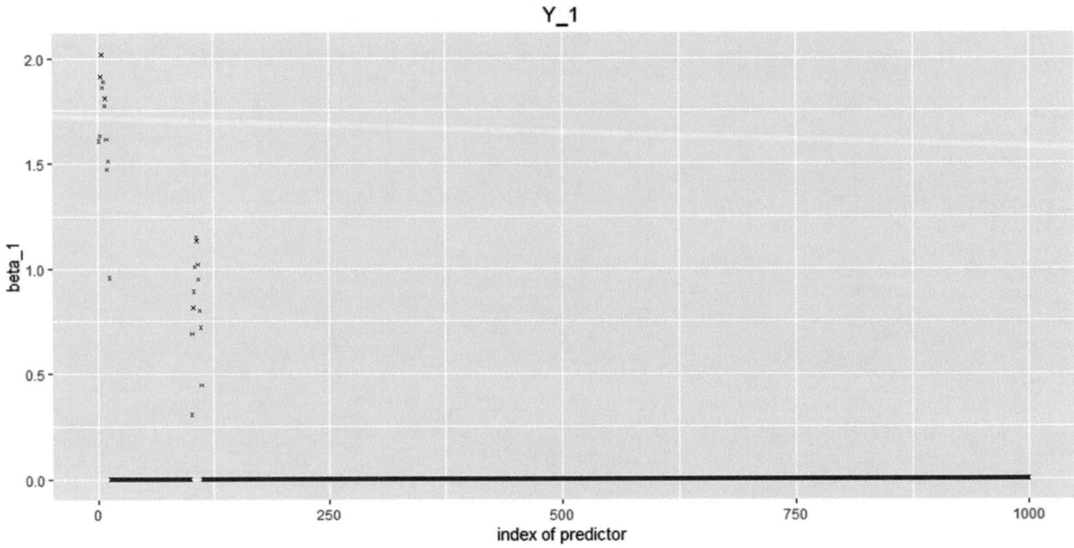

Fig. 2 Estimated coefficients of covariates from POCRE; the covariates with nonzero coefficients are selected

2. Check if there is any missing value in the data. Missing values can be imputed based on their missing mechanisms. For example, the missing genotype of a SNP can be imputed based on the linkage disequilibrium between SNPs.
3. Check if there are any outliers in the data set.
4. The biological groups of metabolites can be identified using KEGG or other existing databases.
5. For different groups of metabolites, the set of covariates included in the model can be different.
6. If there are multiple groups to be tested using SUR analysis, the family-wide error rate needs to be controlled using Bonferroni method or false discovery rate (FDR).
7. It is important to standardize both metabolite abundances and, especially, covariates in POCRE.
8. Before using POCRE to screen important covariates, try to include all available factors that may influence the level of all metabolites including demographic and clinical characteristics, disease status, treatment parameters, and the potential interactions between them.
9. When two covariates are highly correlated, POCRE tends to select or exclude both covariates at the same time. When both are included in the model, they will share the effects on the metabolite abundances. For example, when two covariates $X_1 = X_2$ after standardization, the estimated effects will be half the size of the effect when only one of them is included in the model.

10. When k-fold cross-validation is used to select the tuning parameter, a larger k will lead to better choice but usually $k = 10$ is sufficient.
11. If it is desirable to select a fixed number of covariates, the function "pocrescreen()" can be used.
12. Advanced users may explore the function "pocrepath()" and plot its results using the function "plot()" for screening covariates using different tuning parameters.

Acknowledgments

This research was partially supported by NIH (R03CA211831, P30CA015704-40) and NSF CAREER award IIS-0844945. The authors gratefully acknowledge the support of the Cancer Care Engineering (CCE) project, a joint effort between the Oncological Sciences Center (Purdue Center for Cancer Research, NCI P30CA23168) in Purdue University Discovery Park and the Indiana University Melvin and Bren Simon Cancer Center (NCI P30CA082709).

References

1. Sud M, Fahy E, Cotter D, Azam K, Vadivelu I, Burant C et al (2016) Metabolomics workbench: an international repository for metabolomics data and metadata, metabolite standards, protocols, tutorials and training, and analysis tools. Nucleic Acids Res 44(D1): D463–D470
2. Kraus WE, Muoio DM, Stevens R, Craig D, Bain JR, Grass E et al (2015) Metabolomic quantitative trait loci (mQTL) mapping implicates the ubiquitin proteasome system in cardiovascular disease pathogenesis. PLoS Genet 11(11):e1005553. https://doi.org/10.1371/journal.pgen.1005553
3. Blasco H, Nadal-Desbarats L, Pradat PF, Gordon PH, Madji Hounoum B, Patin F et al (2016) Biomarkers in amyotrophic lateral sclerosis: combining metabolomic and clinical parameters to define disease progression. Eur J Neurol 23(2):346–353
4. Nagana Gowda GA, Djukovic D (2014) Overview of mass spectrometry-based metabolomics: opportunities and challenges. Methods Mol Biol 1198:3–12
5. Gromski PS, Muhamadali H, Ellis DI, Xu Y, Correa E, Turner ML et al (2015) A tutorial review: metabolomics and partial least squares-discriminant analysis–a marriage of convenience or a shotgun wedding. Anal Chim Acta 879:10–23
6. Alakwaa FM, Chaudhary K, Garmire LX (2017) Deep learning accurately predicts estrogen receptor status in breast cancer metabolomics data. J Proteome Res 17:337–347
7. Cuperlovic-Culf M (2018) Machine learning methods for analysis of metabolic data and metabolic pathway modeling. Metabolites 8 (1): pii: E4. doi: https://doi.org/10.3390/metabo8010004
8. Chen C, Deng L, Wei S, Nagana Gowda GA, Gu H, Chiorean EG et al (2015) Exploring metabolic profile differences between colorectal polyp patients and controls using seemingly unrelated regression. J Proteome Res 14 (6):2492–2499
9. Chen C, Nagana Gowda GA, Zhu J, Deng L, Gu H, Chiorean G et al (2017) Altered metabolite levels and correlations in patients with colorectal cancer and polyps detected using seemingly unrelated regression analysis. Metabolomics 13:125. https://doi.org/10.1007/s11306-017-1265-0
10. Zhang D, Lin Y, Zhang M (2009) Penalized orthogonal-components regression for large p small n data. Electronic J Stat 3:781–796
11. Zellner A (1962) An efficient method of estimating seemingly unrelated regressions and

tests for aggregation bias. J Am Stat Assn 57 (298):348–368

12. Lin Y, Zhang M, Wang L, Pungpapong V, Fleet JC, Zhang D (2009) Simultaneous genome-wide association studies of anti-CCP in rheumatoid arthritis using penalized orthogonal-components regression. BMC Proc (Suppl 7): S20
13. Lin Y, Zhang M, Zhang D (2015) Generalized orthogonal-components regression for high-dimensional generalized linear models. Comput Stat Data Anal 88:119–127
14. Wang L, Pungpapong V, Lin Y, Zhang M, Zhang D (2011) Genome-wide case-control study in GAW17 using coalesced rare variants. BMC Proc 5(Suppl 9):S110
15. Zhang M, Lin Y, Wang L, Pungpapong V, Fleet JC, Zhang D (2009) Case-control genome-wide association studies of rheumatoid arthritis from GAW16 using POCRE-LDA. BMC Proc Suppl 7:S17

Index

G. A. Nagana Gowda and Daniel Raftery (eds.), *NMR-Based Metabolomics: Methods and Protocols*, Methods in Molecular Biology, vol. 2037, https://doi.org/10.1007/978-1-4939-9690-2, © Springer Science+Business Media, LLC, part of Springer Nature 2019

D

E

F

G

H

I

J

L

M

N

O

P

Q

R

S

T

U

V

W

X

Y

Z